Quantum Field Theory

Competitive Models

Bertfried Fauser
Jürgen Tolksdorf
Eberhard Zeidler
Editors

Birkhäuser
Basel · Boston · Berlin

Editors

Bertfried Fauser
Jürgen Tolksdorf
Eberhard Zeidler
Max-Planck-Institut für Mathematik in den Naturwissenschaften
Inselstrasse 22–26
D-04103 Leipzig
Germany
e-mail: Bertfried.Fauser@uni-konstanz.de
           Juergen.Tolksdorf@mis.mpg.de
           EZeidler@mis.mpg.de

Mathematical Subject Classification (2000). Primary: 03G30; 14J32; 28C20; 53D55; 58G26; 81Q20; 81R50; 81T05; 81T08; 81T15; 81T30; 81T99; 83C99; 83E30. Secondary: 14D20; 14J15; 14J81; 16W30; 17B37; 32G13; 35C20; 35Q40; 46F20; 55P20; 55R45; 58D27; 58D34; 60G60; 70G75; 70S10; 81P05; 81P10; 81P15; 81S40; 81T10; 81T13; 81T18; 81T40; 81T45; 81T50; 81T75; 81V17; 81V70; 83C35; 83C45; 83C47.

Library of Congress Control Number: 2008932708

Bibliographic information published by Die Deutsche Bibliothek.
Die Deutsche Bibliothek lists this publication in the Deutsche Nationalbibliografie; detailed bibliographic data is available in the Internet at http://dnb.ddb.de

ISBN 978-3-7643-8735-8 Birkhäuser Verlag AG, Basel - Boston - Berlin

© 2009 Birkhäuser Verlag AG
Basel · Boston · Berlin
P.O. Box 133, CH-4010 Basel, Switzerland
Part of Springer Science+Business Media
Printed on acid-free paper produced from chlorine-free pulp. TCF ∞
Printed in Germany

ISBN 978-3-7643-8735-8   e-ISBN 978-3-7643-8736-5
9 8 7 6 5 4 3 2 1   www.birkhauser.ch

CONTENTS

Preface

This Edited Volume is based on the workshop on "Recent Developments in Quantum Field Theory" held at the *Max Planck Institute for Mathematics in the Sciences* in Leipzig (Germany) from July 20th to 22nd, 2007. This workshop was the successor of two similar workshops held at the Heinrich-Fabri-Institute in Blaubeuren in 2003 on "Mathematical and Physical Aspects of Quantum Field Theories" and 2005 on "Mathematical and Physical Aspects of Quantum Gravity[1]".

The series of these workshops was intended to bring together mathematicians and physicists to discuss basic questions within the non-empty intersection of mathematics and physics. The general idea of this series of workshops is to cover a broad range of different approaches (both mathematical and physical) to specific subjects in mathematical physics. In particular, the series of workshops is intended to also discuss the conceptual ideas on which the different approaches of the considered issues are based.

The workshop this volume is based on was devoted to *competitive methods* in quantum field theory. Recent years have seen a certain crisis in theoretical particle physics. On the one hand there is this phenomenologically overwhelmingly successful Standard Model which is in excellent agreement with almost all of the experimental data known to date. On the other hand this model also suffers from conceptual weakness and mathematical rigorousness. In fact, almost all the experimentally confirmed statements derived from the Standard Model are based on perturbation theory. The latter, however, uses renormalization theory which actually still is not a mathematically rigorous theory. Despite recent progress, it is clear that a deeper understanding of this issue has to be achieved in order to gain a more profound understanding in elementary particle dynamics. Moreover, it seems almost embarrassing that we have no idea what more than 90% of the energy in the universe may look like. Even more demanding are the conceptual differences between the basic ideas of a given quantum field theory and general relativity. There is not yet a theory available which allows to combine the basic principles of these two cornerstones of theoretical physics and which also reproduces (at least) some of the experimentally verified predictions made by the Standard Model. A quantum theory of gravity should cover or guide an extension or re-modeling of

---

[1]See the volume "Quantum Gravity – Mathematical Models and Experimental Bounds", B. Fauser, J. Tolksdorf, and E. Zeidler (eds.), Birkhäuser Verlag, 2007.

the particle physics side. These problems are among the driving forces in recent developments in quantum field theory.

One competitive candidate for a unifying theory of quantum fields and gravity is string theory. The present volume features a particular scope on these activities. It turned out that the flow of communication between string theory and other approaches is not entirely free, despite of a great effort of the organizers to allow this to happen. The workshop showed, in very lively discussions, that there is a need for exchanging ideas and for clarifying concepts between other approaches and string theory. This might be a well suited topic for a following workshop. The first chapter, by Bert Schroer, dwells partly on some of the difficulties to achieve a better understanding between the ideas of string theory and algebraic quantum field theory. In addition to Bert Schroer's view, the editors are glad to point out, that in several chapters of this book, and especially in the long last chapter, string motivated ideas do entangle and interact with quantum field theory and provide thereby competitive approaches.

The present volume covers several approaches to generalizations of quantum field theories. A common theme of quite a number of them is the belief that the structure of space-time will change at very small distances. The basic idea is that probing space-time with quantum objects will yield a fuzzy structure of space-time and the concept of a point in a smooth manifold is difficult to maintain. Whatever the fuzzy structure of space-time may look like on a (very) small scale, any such description of fuzzy space-time would have to yield a smooth structure on a sufficiently large scale. How to resolve the discrepancy? One may start from the outset with a discrete set and expect space-time and the causal structure to be emergent phenomena. One might use q-deformation to introduce a, however rigid, discrete structure, or one might study locally deformed space-times using deformation quantization, a presently very much pursued approach. A very radical approach is proposed by the topos approach to physical theories. It allows to reestablish a (neo-)realist interpretation of quantum theories and hence goes conceptually far beyond the usual generalizations of quantum (field) theories.

Other activities in quantum field theory are tied to issues that are more mathematical in nature. While path integrals are suitable tools in particle, solid state, and statistical physics, they are notoriously ill defined. This volume contains a thorough mathematical discussion on path integrals. This discussion demonstrates under what circumstances these highly oscillatory integrals can yield rigorous results. These methods are also used in the AdS/CFT infrared problem and thus have implications for quantum holography, a major topic for discussions during the workshop.

A number of contributions to this volume discuss different aspects of perturbative quantum field theory. Approaches include causal perturbation theory, allowing to formulate quantum field theories more rigorously on curved space-time backgrounds and Hopf algebraic methods, which help to clarify the complicated process of renormalization.

The last and by far most extensive contribution to this volume presents a detailed mathematical discussion of several of the above topics. This article is motivated by string theory covering categorical issues.

The idea of the third workshop was to provide a forum to discuss different approaches to quantum field theories. The present volume provides a good cross-section of the discussions. The refereed articles are written with the intention to bring together experts working in different fields in mathematics and physics who are interested in the subject of quantum field theory. The volume provides the reader with an overview about a variety of recent approaches to quantum field theory. The articles are purposely written in a less technical style than usual to encourage an open discussion across the different approaches to the subject of the workshop.

Since this volume covers rather different perspectives, the editors thought it might be helpful to start the volume by providing a brief summary of each of the various articles. Such a summary will necessarily reflect the editors' understanding of the subject matter.

Holography, especially in the form of AdS/CFT correspondence, plays a vital role in recent developments in quantum field theory. The connection between a bulk and a boundary quantum field theory has fascinating consequences and may provide us with a pathway to a realistic interacting quantum field theory. A further important point is that it can be used to derive *area laws* much alike Bekenstein's area law for black holes.

In his discussion of holography **Bert Schroer** also highlights several critical aspects of quantum field theory. Furthermore, his contribution to this volume provides quite a bit of historical details and insights into the original motivation of the introduction of such concepts as light-cone quantization, the ancestor of holography.

Schroer's reflections on some socially driven mechanisms in the development of physics are surely subjective and controversial. His pointed contributions during the workshop made it, however, clear that his criticism should not be misunderstood as a no-go paradigm against other approaches, as also the variety of chapters in this book suggest.

A very radical way to avoid concepts like 'space-time points', which is used in general relativity but is in conflict with the uncertainty principle, is give up the assumption of a continuum. Also Bernhard Riemann, when he introduced his differential geometric concepts, was careful enough to note that the assumption of a continuum at very small scales is an untested idealization. Topos theory allows the usage of 'generalized points' in algebraic geometry. Lawvere studied elementary topoi to show that the foundation of mathematics is not necessarily tied to set theory. Moreover, Lawvere showed that the logic attached to topoi is strong enough to provide a foundation of the whole building of mathematics. The corresponding chapter by **Andreas Döring** summarizes and explains very clearly how topos theory might be useful to describe physical theories. He shows that topos theory produces an internal logic and is capable to assign to all propositions of the theory truth values. In that sense the topos approach overcomes foundational problems of quantum theory, sub-summarized by the Kochen-Specker theorem. Eventually, topos theory may also open a doorway to unify classical and quantum physics. This in turn may yield deeper insights into a quantization of gravity.

Feynman path integrals are a widely used method in quantum mechanics and quantum field theory. These integrals over a path space are relatives of Wiener integrals and provide a stochastic interpretation as also the "sum over histories" interpretation. However, path integrals in quantum field theory are known to be notoriously mathematically ill defined. In their contribution, **Sergio Albeverio** and **Sonia Mazzucchi** present an introduction to a mathematical discussion of Feynman path integrals as oscillatory integrals. Due to the oscillating integrand these integrals may converge even for functions which are not Lebesgue integrable.

Using a stochastic interpretation, constructive quantum field theory deals with path integrals of non-Gaussian, type. **Hanno Gottschalk** and **Horst Thaler** apply this stochastic interpretation of path integrals to investigate the AdS/CFT correspondence that is motivated by string theory. Especially the infra-red problem and the triviality results of ϕ^4 theory are discussed in their contribution. A comprehensive discussion of the encountered problems is presented and four possible ways to escape triviality are discussed in the conclusions of their chapter.

Originally, mirror symmetry emerged from string theory as a duality of certain 2-dimensional field theories. Mirror symmetry has very remarkable mathematical properties. In his contribution, **Karl-Georg Schlesinger** very clearly explains how mirror symmetry can be extended to the noncommutative torus. Such a generalization of mirror symmetry to a noncommutative setting is motivated, for example, by string theoretical considerations. The present work leads to decisive statements and a conjecture about the algebraic structure of cohomological field theories and deformations of the Fukaya category attached to commutative elliptic functions. Higher n-categories and fc-multi-categories appear naturally in such a development.

Using quantum field theoretical methods, Edward Witten made a number of remarkable mathematical statements. Among them he presented an expression for the volume of the moduli space of flat $SU(2)$ bundles on a compact Riemann surface of general genus. From this result follow the cohomology pairings of intersections. Many heuristically obtained results, that is using formal path-integral methods, where rigorously proved later on by mathematicians. In his contribution to the volume, **Partha Guha** presents a route to obtain similar results for flat $SU(3)$ bundles using the Verlinde formula. The results employ Euler-Zagier sums and multiple zeta values in an intriguing and surprising way.

$\theta-$ deformed space-times are another approach to quantize gravity. Such a description of space-time, however, suffers from several shortcomings. For example, it is not Lorentz invariant. Moreover, such a deformation produces 'quantum effects' on any scale, invalidating the theory on the classical level.

An interesting description of $\theta-$ deformed space-times is provided by deformation quantization. Such a description allows to introduce locally noncommutative spaces which may fit more with the physical intuition. **Stefan Waldmann** expertly reviews in his contribution the deformation quantization description of $\theta-$ deformed space-times. He also presents some motivation for the concepts used in this approach and discusses the range of validity of these concepts.

Renormalization is known to be the salt which makes quantum field theory digestible, i.e. to produce finite results. The scheme of renormalization was established by physicists in the years 1950-80. The basic ideas of renormalization have been made more mathematically concise by the work of Kreimer, Connes-Kreimer and others using Hopf algebras. However, this approach was established only on toy model QFTs. **Walter D. van Suijlekom** pushes the Hopf algebraic method into the realm of physically interesting models, like non-Abelian gauge theories. The corresponding chapter of this volume contains a clear and relatively nontechnical description how the Hopf algebra method can be applied to Ward identities and Slavnov-Taylor-identities.

Perturbative quantum field theory is well-known to be quite successful when applied to the Standard Model. However, there is some belief that perturbation theory is not fundamental. Recent developments exhibited a Hopf algebraic structure which may help to understand renormalization of Abelian and non-Abelian Yang-Mills quantum field theories. Gravity has a rather different gauge theoretical structure and is not amenable to the usual techniques used in Yang-Mills gauge theories. **Dirk Kreimer** explains similarities between perturbatively treated quantum Yang-Mills theories and Einstein's theory of gravity. These similarities might eventually allow to quantize gravity using standard perturbative methods.

Quantum field theory, celebrated presently as the fundamental approach to formulate and describe quantum systems, has weak points when applied to systems having a degenerate lowest energy sector. Such systems *do occur* in solid-state physics, for examples when studying the colors of gemstones, and cannot be treated by the usually applied standard methods of quantum field theory. **Christian Brouder** develops a method to deal with such degenerated quantum field theories. Firstly the degenerated state is described via its cumulants, then these cumulant correlations are turned into interaction terms. This extends to the edge the combinatorial complexity but reestablishes valuable tools from standard nondegenerate quantum field theory, such as the Gell-Mann Low formula. The soundness of the method exhibits itself in a short and clear proof of Hall's generalized Dyson equation.

The article by **Ferdinand Brennecke** and **Michael Dütsch** presents a summary of the present state of the art of renormalization techniques in causal perturbation theory. The main tools are the master action Ward identity and the quantum action principle, which finally allow to use local interactions in the renormalization process. A nice recipe style guide to the method is given in the conclusions.

In string theory certain dualities are known to play a crucial role in connecting strongly coupled theories with weakly coupled theories. While the strong coupled case is difficult to treat, the weakly coupled dual theory may admit a perturbative regime. One such setting is found in matrix string theory in a non-Abelian Yang-Mills setting. In his contribution to the volume, **Matthias Blau** sets up a quantum mechanical toy model to discuss the geometry behind such dualities. He shows that plane wave metrics play a certain role in the solution of the time dependent harmonic oscillator. His discussion may serve as a blue print for the much more complex noncommutative non-Abelian Yang-Mills case.

Loop quantum gravity is one of the approaches to gain insight into a theory of quantum gravity. Technical problems like the resolution of the Hamiltonian constraint make it difficult to evaluate loop quantum gravity in realistic situations. **Martin Bojowald** reviews in his article an approach that uses effective actions in canonical gravity to study quantum cosmology. He explains how solutions can be obtained for an anharmonic oscillator model using integrability. The analogous treatment of canonical quantum gravity yields a bouncing cosmological solution which allows to avoid the big bang singularity.

Many proposals have been made to establish a mathematical modeling of non-smooth structures on the Planck scale. One such model is developed by **Felix Finster** starting from a discrete set of points. All additional structures like causality, Lorentz symmetry and smoothness at large scales have to be established as emergent phenomena. Finster explains in his article how such structures may occur (in principle) in a continuum limit of a set described by a specific discrete variational principle. Several small systems of this type are analyzed and the structure of the emergent phenomena is discussed.

A recurrent theme in physics is the question: "How can space-time be mathematically modeled at very short distances?" Lattices emerging from a "q-deformation" might provide one such candidate. **Hartmut Wachter** shows in his contribution how a q-calculus approach to a non-relativistic particle can be worked out.

The method of q-deformation was originally motivated as a regularization scheme. Similarly, the idea of supersymmetry originated from the hope that supersymmetric theories may have a better ultraviolet behavior. In his contribution, **Alexander Schmidt** presents a discussion on how q-deformation can be extended to a super-symmetric setting.

The book closes with a rather long chapter by **Hisham Sati**, **Urs Schreiber** and **Jim Stasheff**. String theory replaces point-like particles by extended objects, strings and in general branes. Such objects can still be described via differential geometry on a background manifold, however, higher degree fields, like 3-form fields, emerge naturally. Higher categorical tools prove to be advantageous to investigate these higher differential geometric structures. Major ingredients are parallel n-transport, higher curvature forms etc. and therefore the algebra of invariant polynomials, which embeds into the Weil algebra, which in turn projects to the Chevalley-Eilenberg algebra. These algebras are best studied as differentially graded commutative algebras (DGCAs). On the Lie algebra level this structure is accompanied by L_∞-algebras which carry for example a (higher) Cartan-Ehresmann connection. Natural questions from differential geometry, such as classifying spaces and obstructions to lifts etc. can now be addressed. The higher category point of view generalizes, unifies and thereby explains many of the standard constructions.

The chapter is largely self contained and readable for non-experts despite being densely written. It develops the relevant structures, gives explicit proofs, and closes with an outlook how to apply these intriguing ideas to physics.

Acknowledgements

It is a great pleasure for the editors to thank all of the participants of the workshop for their contributions which have made the workshop so successful. We would like to express our gratitude also to the staff of the Max Planck Institute for Mathematics in the Sciences, especially to Mrs Regine Lübke, who managed the administrative work so excellently. The editors would like to thank the German Science Foundation (DFG) and the Max Planck Institute for Mathematics in the Sciences in Leipzig (Germany) for their generous financial support. Furthermore, they would like to thank Marc Herbstritt and Thomas Hempfling from Birkhäuser Verlag for the excellent cooperation. One of the editors (EZ) would like to thank Bertfried Fauser and Jürgen Tolksdorf for skillfully and enthusiastically organizing this stimulating workshop.

Bertfried Fauser, Jürgen Tolksdorf and Eberhard Zeidler
Leipzig, October 1, 2008

Quantum Field Theory
B. Fauser, J. Tolksdorf and E. Zeidler, Eds., 1–24

Constructive Use of Holographic Projections

Bert Schroer

Dedicated to Klaus Fredenhagen on the occasion of his 60th birthday.

Abstract. Revisiting the old problem of existence of interacting models of QFT with new conceptual ideas and mathematical tools, one arrives at a novel view about the nature of QFT. The recent success of algebraic methods in establishing the existence of factorizing models suggests new directions for a more intrinsic constructive approach beyond Lagrangian quantization. Holographic projection simplifies certain properties of the bulk theory and hence is a promising new tool for these new attempts.

Mathematics Subject Classification (2000). Primary 81T05; Secondary 81T40; 83E30; 81P05.

Keywords. AdS/CFT correspondence, holography, algebraic quantum field theory, nets of algebras, string theory, Maldacena conjecture, infrared problem, historical comments on QFT and holography, sociological aspects of science.

1. Historical background and present motivations for holography

No other theory in the history of physics has been able to cover such a wide range of phenomena with impressive precision as QFT. However its amazing predictive power stands in a worrisome contrast to its weak ontological status. In fact QFT is the only theory of immense epistemic strength which, even after more than 80 years, remained on shaky mathematical and conceptual grounds. Unlike any other area of physics, including QM, there are simply no interesting mathematically controllable interacting models, which would show that the underlying principles remain free of internal contradictions in the presence of interactions. The faith in e.g. the Standard Model is based primarily on its perturbative descriptive power; outside the perturbative domain there are more doubts than supporting arguments.

The suspicion that this state of affairs may be related to the conceptual and mathematical weakness of the method of Lagrangian quantization rather then a shortcoming indicating an inconsistency of the underlying principles in the presence of interactions can be traced back to its discoverer Pascual Jordan. It certainly was behind all later attempts of e.g. Arthur Wightman and Rudolf Haag to find a more autonomous setting away from the quantization parallelism with classical theories which culminated in Wightman's axiomatic setting in terms of vacuum correlation functions and the Haag-Kastler theory of nets of operator algebras.

The distance of such conceptual improvements to the applied world of calculations has unfortunately persisted. Nowhere is the contrast between computational triumph and conceptual misery more visible than in renormalized perturbation theory, which has remained our only means to explore the Standard Model. Most particle physicists have a working knowledge of perturbation theory and at least some of them took notice of the fact that, although the renormalized perturbative series can be shown to diverge and that in certain cases these divergent series are Borel resumable. Here I will add some more comments without going into details.

The Borel re-sumability property unfortunately does not lead to an existence proof; the correct mathematical statement in this situation is that if the existence can be established[1] by nonperturbative method then the Borel-resumed series would indeed acquire an asymptotic convergence status with respect to the solution, and one would for the first time be allowed to celebrate the numerical success as having a solid ontological basis [2]. But the whole issue of model existence attained the status of an unpleasant fact, something, which is often kept away from newcomers, so that as a result there is a certain danger to confuse the existence of a model with the ability to write down a Lagrangian or a functional integral and apply some computational recipe.

Fortunately important but unfashionable problems in particle physics never disappear completely. Even if they have been left on the wayside as "un-stringy", "unsupersymmetrizable" or too far removed from the "Holy Grail of a TOE" and therefore not really career-improving, there will be always be individuals who return to them with new ideas.

Indeed there has been some recent progress about the aforementioned existence problem from a quite unexpected direction. Within the setting of d=1+1 factorizing models the use of modular operator theory has led to a control over phase space degrees of freedom which in turn paved the way to an existence proof. Those models are distinguished by their simple generators for the wedge-localized algebra [4]; in fact these generators turned out to possess Fourier-transforms with mass-shell creation/annihilation operators, which are only slightly more complicated than free fields. An important additional idea on the way to an existence

---

[1] The existence for models with a finite wave-function renormalization constant has been established in the early 60s and this situation has not changed up to recently. The old results only include superrenormalizable models whereas the new criterion is not related to short-distance restrictions but rather requires a certain phase space behavior (modular nuclearity).

[2] This is actually the present situation for the class of d=1+1 factorizing models [5].

proof is the issue of the cardinality of degrees of freedom. In the form of the phase space in QFT as opposed to QM this issue goes back to the 60s [1] and underwent several refinements [2] (a sketch of the history can be found in [3]).

The remaining problem was to show that the simplicity of the wedge generators led to a "tame" phase space behavior, which guarantees the nontriviality as well as the additional expected properties of the double cone localized algebras obtained as intersections of wedge-localized algebras [5]. Although these models have no particle creation through on-shell scattering, they exhibit the full infinite vacuum polarization clouds upon sharpening the localization from wedges to compact spacetime regions as e.g. double cones [6]. Their simplicity is only manifest in the existence of simple wedge generators; for compact localization regions their complicated infinite vacuum polarization clouds are not simpler than in other QFT.

Similar simple-minded Ansätze for wedge algebras in higher dimensions cannot work since interactions which lead to nontrivial elastic scattering without also causing particle creation cannot exist; such a No-Go theorem for 4-dimensional QFT was established already in [7]. Nevertheless it is quite interesting to note that even if with such a simple-minded Ansatz for wedge generators in higher dimensions one does not get to compactly localized local observables, one can in some cases go to certain subwedge intersections [8, 9] before the increase in localization leads to trivial algebras.

Whereas in the Lagrangian approach one starts with local fields and their correlations and moves afterwards to less local objects such as global charges, incoming fields[3] etc., the modular localization approach goes the opposite way i.e. one starts from the wedge region (the best compromise between particles and fields) which is most close to the particle mass-shell the S-matrix and then works one's way down. The pointlike local fields only appear at the very end and play the role of *coordinatizing generators* of the double cone algebras for arbitrary small sizes.

Nonlocal models are automatically "noncommutative" in the sense that the maximal commutativity of massive theories allowed by the principles of QFT, namely spacelike commutativity, is weakened by allowing various degrees of violations of spacelike commutativity. In this context the noncommutativity associated with the deformation of the product to a star-product using the Weyl-Moyal formalism is only a very special (but very popular) case. The motivation for studying noncommutative QFT for its own sake comes from string theory, and one should not expect this motivation to be better than for string theory itself.

My motivation for having being interested in noncommutative theory during the last decade comes from the observation that noncommutative fields can

---

[3]Incoming/outgoing free fields are only local with respect to themselves. The physically relevant notion of locality is *relative locality to the interacting fields*. If incoming fields are relatively local/almost local, the theory has no interactions.

have *simpler properties than commutative ones*. More concretely: complicated two-dimensional local theories may lead to wedge-localized algebras which are generated by noncommutative fields where the latter only fulfil the much weaker wedge-locality (see above). Whereas in d=1+1 such constructions [4] may lead via algebraic intersections to nontrivial, nonperturbative local fields, it is known that in higher dimensions this simple kind of wedge generating field without vacuum polarization is not available. But interestingly enough one can improve the wedge localization somewhat [10] before the further sharpening of localization via algebraic intersections ends in trivial algebras.

These recent developments combine the useful part of the history of S-matrix theory and formfactors with very new conceptual inroads into QFT (modular localization, phase space properties of LQP). The idea to divide the difficult full problem into a collection of simpler smaller ones is also at the root of the various forms of the holography of the two subsequent sections.

The predecessor of lightfront holography was the so-called "lightcone quantization" which started in the early 70s; it was designed to focus on short-distances and forget temporarily about the rest. The idea to work with fields which are associated to the lightfront $x_- = 0$ (not the light cone which is $x^2 = 0$) as a submanifold in Minkowski spacetime looked very promising but unfortunately the connection with the original problem of analyzing the local theory in the bulk was never addressed and as the misleading name "lightcone quantization" reveals, the approach was considered as a different quantization rather then a different method for looking at the same local QFT in Minkowski spacetime. It is not really necessary to continue a separate criticism of "lightcone quantization" because its shortcomings will be become obvious after the presentation of lightfront holography (more generally *holography onto null-surfaces*).

Whereas the more elaborate and potentially more important lightfront holography has not led to heated discussions, the controversial potential of the simpler AdS/CFT holography had been enormous and to the degree that it contains interesting messages which increase our scientific understanding it will be presented in these notes.

Since all subjects have been treated in the existing literature, our presentation should be viewed as a guide through the literature with occasionally additional and (hopefully) helpful remarks.

2. Lightfront holography, holography on null-surfaces and the origin of the area law

Free fields offer a nice introduction into the bulk-holography relation which, despite its simplicity, remains conceptually non-trivial.

We seek generating fields A_{LF} for the lightfront algebra $\mathcal{A}(LF)$ by following the formal prescription $x_- = 0$ of the old "lightfront approach" [11]. Using the abbreviation $x_\pm = x^0 \pm x^3$, $p_\pm = p^0 + p^3 \simeq e^{\mp\theta}$, with θ the momentum space rapidity :

$$A_{LF}(x_+, x_\perp) := A(x)|_{x_-=0} \simeq \int \left(e^{i(p_-(\theta)x_+ + ip_\perp x_\perp} a^*(\theta, p_\perp) d\theta dp_\perp + h.c. \right) \quad (1)$$

$$\left\langle \partial_{x_+} A_{LF}(x_+, x_\perp) \partial_{x'_+} A_{LF}(x'_+, x'_\perp) \right\rangle \simeq \frac{1}{\left(x_+ - x'_+ + i\varepsilon \right)^2} \cdot \delta(x_\perp - x'_\perp)$$

$$\left[\partial_{x_+} A_{LF}(x_+, x_\perp), \partial_{x'_+} A_{LF}(x'_+, x'_\perp) \right] \simeq \delta'(x_+ - x'_+) \delta(x_\perp - x'_\perp).$$

The justification for this formal manipulation[4] follows from the fact that the equivalence class of a test function $[f]$, which has the same mass shell restriction $\tilde{f}|_{H_m}$ to the mass hyperboloid of mass m, is mapped to a unique test function f_{LF} which "lives" on the lightfront [12, 13]. It only takes the margin of a newspaper to verify the identity $A(f) = A([f]) = A_{LF}(f_{LF})$. This identity does not mean that the A_{LF} generator can be used to describe the local substructure in the bulk. The inversion involves an equivalence class and does not distinguish an individual test-function in the bulk; in fact a finitely localized test function $f(x_+, x_\perp)$ on LF corresponds to a de-localized subspace in the bulk. Using an intuitive metaphoric language one may say that a strict localization on LF corresponds to a fuzzy localization in the bulk and vice versa. Hence the pointwise use of the LF generators enforces the LF localization and the only wedge-localized operators which can be directly obtained as smeared A_{LF} fields have a noncompact extension within a wedge whose causal horizon is on LF. Nevertheless there is equality between the two operator algebras associated to the bulk W and its (upper) horizon ∂W

$$\mathcal{A}(W) = \mathcal{A}(H(W)) \subset \mathcal{A}(LF) = B(H). \quad (2)$$

These operator algebras are the von Neumann closures of the Weyl algebras generated by the smeared fields A and A_{LF} and it is only in the sense of this closure (or by forming the double commutant) that the equality holds. Quantum field theorists are used to deal with single operators. Therefore the knowledge about the equality of algebras without being able to say which operators are localized in subregion is somewhat unaccustomed. As will be explained later on, the finer localization properties in the algebraic setting can be recovered by taking suitable intersections of wedge algebras i.e. the structure of the family of all wedge algebras determines whether the local algebras are nontrivial and in case they are permits to compute the local net which contains all informations about the particular model.

This idea of taking the holographic projection of individual bulk fields can be generalized to composites of free fields (as e.g. the stress-energy tensor). In order to avoid lengthy discussions about how to interpret logarithmic chiral two-point functions in terms of restricted test functions[5] we restrict our attention to

---

[4]We took the derivatives for technical reasons (in order to write the formulas without test functions).

[5]This is a well-understood problem of chiral fields of zero scale dimension which is not directly related to holography.

Wick-composites of $\partial_{x_+} A_{LF}(x_+, x_\perp)$

$$[B_{LF}(x_+, x_\perp), C_{LF}(x'_+, x'_\perp)] =$$

$$\sum_{l=0}^{m} \delta^l(x_\perp - x'_\perp) \sum_{k(l)=0}^{n(l)} \delta^{k(l)}(x_+ - x'_+) D_{LF}^{(k(l))}(x_+, x_\perp), \quad (3)$$

where the dimensions of the composites $D_{LF}^{(k(l))}$ together with the degrees of the derivatives of the delta functions obey the standard rule of scale dimensional conservation. In the commutator the transverse and the longitudinal part both appear with delta functions and their derivatives yet there is a very important structural difference which shows up in the correlation functions. To understand this point we look at the second line in (1). The longitudinal (=lightlike) delta-functions carries the chiral vacuum polarization the transverse part consists only of products of delta functions as if it would come from a product of correlation functions of nonrelativistic Schrödinger creation/annihilation operators $\psi^*(x_\perp)$, $\psi(x_\perp)$. In other words the LF-fields which feature in this extended chiral theory are *chimera between QFT and QM*; they have one leg in QFT and n-2 legs in QM with the "chimeric vacuum" being partially a (transverse) factorizing quantum mechanical state of "nothingness" (the Buddhist nirvana) and partially the longitudinally particle-antiparticle polarized LQP vacuum state of "virtually everything" (the Abrahamic heaven).

Upon lightlike localization of LF to (in the present case) ∂W (or to a longitudinal interval) the vacuum on $\mathcal{A}(\partial W)$ becomes a radiating KMS thermal state with nonvanishing localization-entropy [13, 14]. In case of interacting fields there is no change with respect to the absence of transverse vacuum polarization, but unlike the free case the global algebra $\mathcal{A}(LF)$ or the semi-global algebra $\mathcal{A}(\partial W)$ is generally bigger than the algebra one obtains from the globalization using compactly localized subalgebras, i.e. $\overline{\cup_{\mathcal{O} \subset LF} \mathcal{A}_{LF}(\mathcal{O})} \subset \mathcal{A}(LF)$, $\mathcal{O} \subset LF$. We will return to this point at a more opportune moment.

The aforementioned "chimeric" behavior of the vacuum is related in a profound way to the conceptual distinctions between QM and QFT [16]. Whereas transversely the vacuum is tensor-factorizing with respect to the Born localization and therefore leads to the standard quantum mechanical concepts of entanglement and the related information theoretical (cold) entropy, the entanglement from restricting the vacuum to an algebra associated with an interval in lightray direction is a thermal KMS state with a genuine thermodynamic entropy. Instead of the standard quantum mechanical dichotomy between pure and entangled restricted states there are simply no pure states at all. All states on sharply localized operator algebras are highly mixed and the restriction of global particle states (including the vacuum) to the W-horizon $\mathcal{A}(\partial W)$ results in KMS thermal states. This is the result of the different nature of localized algebras in QFT from localized algebras in QM [16].

Therefore if one wants to use the terminology "entanglement" in QFT one should be aware that one is dealing with a totally intrinsic very strong form of entanglement: *all physically distinguished global pure states* (in particular finite energy states in particular the vacuum) *upon restriction to a localized algebra become intrinsically entangled and unlike in QM there is no local operation which disentangles.*

Whereas the cold (information theoretic) entanglement is often linked to the uncertainty relation of QM, the raison d'etre behind the "hot" entanglement is the phenomenon of vacuum polarization resulting from localization in quantum theories with a maximal velocity. The transverse tensor factorization restricts the Reeh-Schlieder theorem (also known as the "state-operator relation"). For a longitudinal strip (st) on LF of a finite transverse extension the LF algebra tensor-factorizes together with the Hilbert space $H = H_{st} \otimes H_{st\perp}$ and the H_{st} projected form of the Reeh-Schlieder theorem for a subalgebra localized within the strip continues to be valid.

This concept of *transverse extended chiral fields* can also be axiomatically formulated for interacting fields independently of whether those objects result from a bulk theory via holographic projection or whether one wants to study QFT on (non-hyperbolic) null-surfaces. These "lightfront fields" share some important properties with chiral fields. In both cases subalgebras localized on subregions lead to a *geometric modular theory,* whereas in the bulk this property is restricted to wedge algebras. Furthermore in both cases the symmetry groups are infinite dimensional; in chiral theories the largest possible group is (after compactification) $\text{Diff}(\dot{R})$, whereas the transverse extended version admits besides these pure lighlike symmetries also $x_\perp$-x_+ mixing ($x_\perp$-dependent) symmetry transformations which leave the commutation structure invariant.

There is one note of caution, unlike those conformal QFTs which arise as chiral projections from 2-dimensional conformal QFT, the extended chiral models of QFT on the lightfront which result from holography do not come with a stress-energy tensor and hence the diffeomorphism invariance beyond the Möbius invariance (which one gets from modular invariance, no energy momentum tensor needed) is not automatic. This leads to the interesting question if there are concepts which permit to incorporate also the diffeomorphisms beyond the Möbius transformations into a modular setting, a problem which will not be pursuit here.

We have formulated the algebraic structure of holographic projected fields for bosonic fields, but it should be obvious to the reader that a generalization to Fermi fields is straightforward. Lightfront holography is consistent with the fact that except for d=1+1 there are no operators which "live" on a lightray since the presence of the quantum mechanical transverse delta function prevents such a possibility i.e. only after transverse averaging with test functions does one get to (unbounded) operators.

It is an interesting question whether a direct "holographic projection" of *interacting* pointlike bulk fields into lightfront fields analog to (1) can be formulated, thus avoiding the algebraic steps starting with wedge algebra. The important formula which led to the lightfront generators is the *mass shell representation* of the free field; if we would have performed the $x_- = 0$ limit in the two point function the result would diverge. This suggests that we should start from the so-called Glaser-Lehmann-Zimmermann (GLZ) representation [17] which is an on-shell representation in terms of an infinite series of integrals involving the incoming particle creation/annihilation operators

$$A(x) = \sum \frac{1}{n!} \int dx_1 ... \int dx_n \; a(x; x_1, ...x_n) : A_{in}(x_1)....A(x_n) : \qquad (4)$$

$$A(x) = \sum \frac{1}{n!} \int_{H_m} dp_1 ... \int_{H_m} dp_n \; e^{ix(\sum p_i)} \tilde{a}(p_1, ...p_n) : \tilde{A}(p_1)....\tilde{A}(p_n) :$$

$$A(x)_{LF} = A(x)_{x_-=0}$$

in which the coefficient functions $a(x; x_1, ...x_n)$ are *retarded functions*. The second line shows that only the mass-shell restriction of these functions matter; the momentum space integration goes over the entire mass-shell and the two components of the mass hyperboloid H_m are associated with the annihilation/creation part of the Fourier transform of the incoming field. These mass-shell restrictions of the retarded coefficient functions are related to multi-particle formfactors of the field A. Clearly we can take $x_- = 0$ in this on-shell representation without apparently creating any problems in addition to the possibly bad convergence properties of such series (with or without the lightfront restriction) which they had from the start. The use of the on-shell representation (4) is essential, doing this directly in the Wightman functions would lead to meaningless divergences, as we already noticed in the free field case.

Such GLZ formulas amount to a representation of a local field in terms of other local fields in which the *relation between the two sets* of fields is very *nonlocal*. Hence this procedure is less intuitive than the algebraic method based on relative commutants and intersections of algebras. The use of a GLZ series also goes in some sense against the spirit of holography which is to simplify certain aspects[6] in order to facilitate the solution of certain properties of the theory (i.e. to preserve the original aim of the ill-defined lightcone quantization), whereas to arrive at GLZ representations one must already have solved the on-shell aspects of the model (i.e. know all its formfactors) before applying holography.

Nevertheless, in those cases where one has explicit knowledge of formfactors, as in the case of 2-dim. factorizing models mentioned in the previous section, this knowledge can be used to calculate the scaling dimensions of their associated holographic fields A_{LF}. These fields lead to more general plektonic (braid group) commutation relations which replace the bosonic relations of transverse extended chiral observables (3). We refer to [15] in which the holographic scaling dimensions for several fields in factorizing models will be calculated, including the Ising model for which an exact determination of the scaling dimension of the order field is

---

[6]Those aspects for which holography does not simplify include particle and scattering aspects.

possible. Although the holographic dimensions agree with those from the short distance analysis (which have been previously calculated in [18]), the conceptual status of holography is quite different from that of critical universality classes. The former is an exact relation between a 2-dimensional factorizing model (change of the spacetime ordering of a given bulk theory) whereas the latter is a passing to a different QFT in the same universality class. The mentioned exact result in the case of the Ising model strengthens the hope that GLZ representations and the closely related expansions of local fields in terms of wedge algebra generating on-shell operators [15] have a better convergence status than perturbative series.

By far the conceptually and mathematically cleanest way to pass from the bulk to the lightfront is in terms of nets of operator algebras via modular theory. This method requires to start from algebras in "standard position" i.e. a pair $(\mathcal{A}, \Omega)$ such that the operator algebra $\mathcal{A}$ acts cyclically on the state vector Ω i.e. $\overline{\mathcal{A}\Omega} = H$ and has no annihilators i.e. $A\Omega = 0 \curvearrowright A = 0$. According to the Reeh-Schlieder theorem any localized algebra $\mathcal{A}(\mathcal{O})$ forms a standard pair $(\mathcal{A}(\mathcal{O}), \Omega)$ with respect to the vacuum Ω and the best starting point for the lightfront holography is a wedge algebra since the (upper) causal horizon ∂W of the wedge W is already half the lightfront. The crux of the matter is the construction of the local substructure on ∂W. The local resolution in longitudinal (lightray) direction is done as follows.

Let W be the $x_0 - x_3$ wedge in Minkowski spacetime which is left invariant by the $x_0 - x_3$ Lorentz-boosts. Consider a family of wedges W_a which are obtained by sliding the W along the $x_+ = x_0 + x_3$ lightray by a lightlike translation $a > 0$ into itself. The set of spacetime points on LF consisting of those points on ∂W_a which are spacelike to the interior of W_b for $b > a$ is denoted by $\partial W_{a,b}$; it consists of points $x_+ \in (a, b)$ with an unlimited transverse part $x_\perp \in R^2$. These regions are two-sided transverse slabs on LF.

To get to intersections of finite size one may "tilt" these slabs by the action of certain subgroups in $\mathcal{G}$ which change the transverse directions. Using the 2-parametric subgroup $\mathcal{G}_2$ of $\mathcal{G}$ which is the restriction to LF of the two "translations" in the Wigner little group (i.e. the subgroup fixing the lightray in LF), it is easy to see that this is achieved by forming intersections with $\mathcal{G}_2$- transformed slabs $\partial W_{a,b}$

$$\partial W_{a,b} \cap g(\partial W_{a,b}), \ g \in \mathcal{G}_2. \tag{5}$$

By continuing with forming intersections and unions, one can get to finite convex regions $\mathcal{O}$ of a quite general shape.

The local net on the lightfront is the collection of all local algebras $\mathcal{A}(\mathcal{O})$, $\mathcal{O} \subset LF$ and as usual their weak closure is the global algebra $\mathcal{A}_{LF}$. For interacting systems the global lightfront algebra is generally expected to be smaller than the bulk, in particular one expects

$$\mathcal{A}_{LF}(\partial W) \subset \mathcal{A}(\partial W) = \mathcal{A}(W) \tag{6}$$
$$\mathcal{A}_{LF}(\partial W) = \cup_{\mathcal{O} \subset \partial W} \mathcal{A}_{LF}(\mathcal{O}), \ \mathcal{A}(W) = \cup_{\mathcal{C} \subset W} \mathcal{A}(\mathcal{C})$$

where the semi-global algebras are formed with the localization concept of their relative nets as indicated in the second line. The smaller left hand side accounts for the fact that the formation of relative commutants as $\mathcal{A}(\partial W_{a,b})$ may not maintain the standardness of the algebra because $\overline{\cup_{a,b}\mathcal{A}(\partial W_{a,b})\Omega} \subsetneq H$. In that case the globalization of the algebraic holography only captures a global (i.e. not localized) subalgebra of the global bulk and one could ask whether the pointlike procedure using the GLZ representation leads to generating fields which generate a bigger algebra gives more. The answer is positive since also (bosonic) fields with anomalous short distance dimensions will pass the projective holography and become anyonic fields on the lightray[7] On the other hand algebraic holography filters out bosonic fields which define the chiral obervables. These chiral observables have a DHR superselection theory. This leads to the obvious conjecture

$$Alg\{proj\ hol\} \subseteq Alg\{DHR\}. \tag{7}$$

Here the left hand side denotes the algebra generated by applying projective holography to the pointlike bulk fields and the right hand side is the smallest algebra which contains all DHR superselection sectors of the LF observable (extended chiral) algebra which resulted from algebraic holography.

It is worthwhile to emphasize that the connection between the operator algebraic and the pointlike prescription is much easier on LF than in the bulk. In the presence of conformal symmetries one has the results of Joerss [19]; looking at his theorems in the chiral setting, an adaptation to the transverse extended chiral theories on LF, should be straightforward. For consistency reasons such fields must fulfil (3) I hope to come back to this issue in a different context.

One motivation for being interested in lightfront holography is that it is expected to be helpful in dividing the complicated problem of classifying and constructing QFTs according to intrinsic principles into several less complicated steps. In the case of d=1+1 factorizing models one does not need this holographic projection onto a chiral theory on the lightray for the mere existence proof. But e.g. for the determination of the spectrum of the short distance scale dimension, it is only holography and not the critical limit which permits to maintain the original Hilbert space setting. It is precisely this property which makes it potentially interesting for structural investigations and actual constructions of higher dimensional QFT.

Now we are well-prepared to address the main point of this section: the area law for localization entropy which follows from the absence of transverse vacuum polarization. Since this point does not depend on most of the above technicalities, it may be helpful to the reader to present the conceptual mathematical origin of this unique[8] tensor-factorization property. The relevant theorem goes back to

---

[7]The standard Boson-Fermion statistics refers to spacelike distances and the lightlike statistics resulting from projective holography is determined by the anomalous short distance dimensions of the bulk field and not by their statistics.

[8]Holography on null-surfaces is the only context in which a quantum mechanical structure enters a field theoretic setting.

Borchers [20] and can be stated as follows. Let $\mathcal{A}_i \subset B(H)$, $i = 1, 2$ be two operator algebras with $[\mathcal{A}_1, U(a)\mathcal{A}_2 U(a)^*] = 0 \; \forall a$ and $U(a)$ a translation with *nonnegative* generator which fulfils the cluster factorization property (i.e. asymptotic factorization in correlation functions for infinitely large cluster separations) with respect to a unique $U(a)$-invariant state vector Ω^9. It then follows that the two algebras tensor factorize in the sense $\mathcal{A}_1 \vee \mathcal{A}_2 = \mathcal{A}_1 \otimes \mathcal{A}_2$ where the left hand side denotes the joint operator algebra.

In the case at hand the tensor factorization follows as soon as the open regions $\mathcal{O}_i \subset LF$ in $\mathcal{A}(\mathcal{O}_i)$ $i = 1, 2$ have no transverse overlap. The lightlike cluster factorization is weaker (only a power law) than its better known spacelike counterpart, but as a result of the analytic properties following from the *non-negative generator of lightlike translations* it enforces the asymptotic factorization to be valid at all distances. The resulting transverse factorization implies the transverse additivity of extensive quantities as energy and entropy and their behavior in lightray direction can then be calculated in terms of the associated auxiliary chiral theory. A well-known property for spacelike separations.

This result [13, 14] of the transverse factorization may be summarized as follows:

1. The system of LF subalgebras $\{\mathcal{A}(\mathcal{O})\}_{\mathcal{O} \subset LF}$ tensor-factorizes transversely with the vacuum being free of transverse entanglement

$$\mathcal{A}(\mathcal{O}_1 \cup \mathcal{O}_2) = \mathcal{A}(\mathcal{O}_1) \otimes \mathcal{A}(\mathcal{O}_2), \; (\mathcal{O}_1)_\perp \cap (\mathcal{O}_2)_\perp = \emptyset \qquad (8)$$
$$\langle \Omega | \mathcal{A}(\mathcal{O}_1) \otimes \mathcal{A}(\mathcal{O}_2) | \Omega \rangle = \langle \Omega | \mathcal{A}(\mathcal{O}_1) | \Omega \rangle \langle \Omega | \mathcal{A}(\mathcal{O}_2) | \Omega \rangle .$$

2. Extensive properties as entropy and energy on LF are proportional to the extension of the transverse area.

3. The area density of localization-entropy in the vacuum state for a system with sharp localization on LF diverges logarithmically

$$s_{loc} = \lim_{\varepsilon \to 0} \frac{c}{6} |\ln \varepsilon| + ... \qquad (9)$$

where ε is the size of the interval of "fuzziness" of the boundary in the lightray direction which one has to allow in order for the vacuum polarization cloud to attenuate and the proportionality constant c is (at least in typical examples) the central extension parameter of the Witt-Virasoro algebra.

The following comments about these results are helpful in order to appreciate some of the physical consequences as well as extensions to more general null-surfaces.

As the volume divergence of the energy/entropy in a heat bath thermal system results from the thermodynamic limit of a sequence of boxed systems in a Gibbs states, the logarithmic divergence in the vacuum polarization attenuation

---

[9]Locality in both directions shows that the lightlike translates $\langle \Omega | AU(a)B | \Omega \rangle$ are boundary values of entire functions and the cluster property together with Liouville's theorem gives the factorization.

distance ε plays an analogous role in the approximation of the semiinfinitely extended ∂W by sequences of algebras whose localization regions approach ∂W from the inside. In both cases the limiting algebras are monads whereas the approximands are type I analogs of the "box quantization" algebras. In fact in the present conformal context the relation between the standard heat bath thermodynamic limit and the limit of vanishing attenuation length for the localization-caused vacuum polarization cloud really goes beyond an analogy and becomes an isomorphism.

This surprising result is based on two facts [13, 14]. On the one hand conformal theories come with a natural covariant "box" approximation of the thermodynamic limit since the continuous spectrum translational Hamiltonian can be obtained as a scaled limit of a sequence of discrete spectrum conformal rotational Hamiltonians associated to global type I systems. On the other hand it has been known for some time that a heat bath chiral KMS state can always be reinterpreted as the Unruh restriction applied to a vacuum system in an larger world (a kind of inverse Unruh effect). Both facts together lead to the above formula for the area density of entropy. In fact using the conformal invariance one can write the area density formula in the more suggestive manner by identifying ε with the conformal invariant cross-ratio of 4 points

$$\varepsilon^2 = \frac{(a_2 - a_1)\,(b_1 - b_2)}{(b_1 - a_1)\,(b_2 - a_2)}$$

where $a_1 < a_2 < b_2 < b_1$ so that (a_1, b_1) corresponds to the larger localization interval and (a_2, b_2) is the approximand which goes with the interpolating type I algebras. At this point one makes contact with some interesting work on what condensed matter physicist call the "entanglement entropy"[10].

One expects that the arguments for the absence of transverse vacuum fluctuations carry over to other null-surfaces as e.g. the upper horizon ∂D of the double cone D. In the interacting case it is not possible to obtain ∂D generators through test function restrictions. For zero mass free fields there is, however, the possibility to conformally transform the wedge into the double cone and in this way obtain the holographic generators as the conformally transformed generators of $A(\partial W)$. In order to show that the resulting $A(\partial D)$ continue to play their role even when the bulk generators cease to be conformal one would have to prove that certain double-cone affiliated inclusions are modular inclusions. We hope to return to this interesting problem.

---

[10] In [21] the formula for the logarithmically increasing entropy is associated with a field theoretic cutoff and the role of the vacuum polarization cloud in conjunction with the KMS thermal properties (which is not compatible with a quantum mechanical entanglement interpretation [16]) are not noticed. Since there is no implementation of the split property, the idea of an attenuation of the vacuum polarization cloud has no conceptual place in a path integral formulation. QM and QFT are not distinguished in the functional integral setting and even on a metaphorical level there seems to be no possibility to implement the split property.

We have presented the pointlike approach and the algebraic approach next to each other, but apart from the free field we have not really connected them. Although one must leave a detailed discussion of their relation to the future, there are some obvious observations one can make. Since for chiral fields the notion of short-distance dimension and rotational spin (the action of the L_0 generator) are closely connected and since the algebraic process of taking relative commutators is bosonic, the lightfront algebras are necessarily bosonic. A field, as the chiral order variable of the Ising model with dimension $\frac{1}{16}$, does not appear in the algebraic holography, but, as mentioned above, it is the pointlike projection of the massive order variable in the factorizing Ising model in the bulk. On the other hand an integer dimensional field as the stress-energy tensor, is common to both formulations. This suggests that the anomalous dimensional fields which are missing in the algebraic construction may be recovered via representation theory of the transverse extended chiral observable algebra which arises as the image of the algebraic holography.

Since the original purpose of holography similar to that of its ill-fated light-cone quantization predecessor, is to achieve a simplified but still rigorous description (for the lightcone quantization the main motivation was a better description of certain "short distance aspects" of QFT), the question arises if one can use holography as a tool in a more ambitious program of classification and construction of QFTs. In this case one must be able to make sense of *inverse holography* i.e. confront the question whether, knowing the local net on the lightfront one can only obtain at least part of the local substructure of the bulk. It is immediately clear that one can construct that part in the bulk, which arises from intersecting the LF-affiliated wedge algebras. The full net is only reconstructible if the action of those remaining Poincaré transformations outside the 7-parametric LF covariance group is known.

The presence of the Möbius group acting on the lightlike direction on null-surfaces in curved spacetime resulting from bifurcate Killing horizons [22] has been established in [23], thus paving the way for the transfer of the thermal results to QFT in CFT. This is an illustration of symmetry enhancement, which is one of holographies "magics".

The above interaction-free case with its chiral Abelian current algebra structure (1) admits a much larger unitarily implemented symmetry group, namely the diffeomorphism group of the circle. However the unitary implementers (beyond the Möbius group) do not leave the vacuum invariant (and hence are not Wigner symmetries). As a result of the commutation relations (3) these $\mathrm{Diff}(S^1)$ symmetries are expected to appear in the holographic projection of interacting theories. These unitary symmetries act only geometrically on the holographic objects; their action on the bulk (on which they are also well-defined) is fuzzy i.e. not describable in geometric terms. This looks like an interesting extension of the new setting of local covariance [24].

The area proportionality for localization entropy is a structural property of LQP which creates an interesting and hopefully fruitful contrast with Bekenstein's area law [25] for black hole horizons. Bekenstein's thermal reading of the area behavior of a certain quantity in classical Einstein-Hilbert like field theories has been interpreted as being on the interface of QFT with QG. Now we see that the main support, namely the claim that QFT alone cannot explain an area behavior, is not correct. There remains the question whether Bekenstein's numerical value, which people tried to understand in terms of quantum mechanical level occupation, is a credible candidate for quantum entropy. QFT gives a family of area laws with different vacuum polarization *attenuation parameters* ε and it is easy to fix this parameter in terms of the Planck length so that the two values coalesce. The problem which I have with such an argument is that I have never seen a situation where a *classical* value remained intact after passing to the quantum theory. This does only happen for certain *quasiclassical* values in case the system is integrable.

3. From holography to correspondence: the AdS/CFT correspondence and a controversy

The holography onto null-surfaces addresses the very subtle relation between bulk quantum matter and the projection onto its causal/event horizon as explained in the previous section. A simpler case of holography arises if the bulk and a lower dimensional brane[11] (timelike) boundary share the same maximally possible space-time (vacuum) symmetry. The only case where this situation arises between two global Lorentz manifolds of different spacetime dimension is the famous AdS/CFT correspondence. In that case the causality leakage off a brane does not occur. In the following we will use the same terminology for the universal coverings of AdS/CFT as for the spacetimes themselves.

Already in the 60s the observation that the 15-parametric conformal symmetry which is shared between the conformal $3 + 1$-dimensional compactified Minkowski spacetime and the 5-dimensional Anti-de-Sitter space (the negative constant curvature brother of the cosmologically important de Sitter spacetime) brought a possible field theoretic relation between these theories into the foreground; in fact Fronsdal [26] suspected that QFTs on both spacetimes share more than the spacetime symmetry groups. But the modular localization theory which could convert the shared group symmetry into a relation between two *different spacetime ordering devices* (in the sense of Leibniz) for the *same abstract quantum matter substrate* was not yet in place at that time. Over several decades the main use of the AdS solution has been (similar to Gödel's cosmological model) to show that the Einstein-Hilbert field equations, besides the many desired solution (as the Robertson-Walker cosmological models and the closely related de Sitter space-time), also admit unphysical solutions (leading to timelike selfclosing worldlines, time machines, wormholes etc.) and therefore should be further restricted.

---

[11]In general the brane has a lower dimensional symmetry than its associated bulk and usually denotes a $d - 1$ dimensional subspace which contains a time-like direction. Different from null-surfaces branes have a causal leakage.

The AdS spacetime lost this role of only providing counterexamples and began to play an important role in particle physics when the string theorist placed it into the center of a conjecture about a correspondence between a particular maximally supersymmetric massless conformally covariant Yang-Mills model in d=1+3 and a supersymmetric gravitational model. The first paper was by J. Maldacena [27] who started from a particular compactification of 10-dimensional superstring theory, with 5 uncompactified coordinates forming the AdS spacetime. Since the mathematics as well as the conceptual structure of string theory is poorly understood, the string side was identified with one of the supersymmetric gravity models which in spite of its being non-renormalizable admitted a more manageable Lagrangian formulation and was expected to have a similar particle content. On the side of CFT he placed a maximally supersymmetric gauge theory of which calculations which verify the vanishing of the low order beta function already existed[12] (certainly a *necessary* prerequisite for conformal invariance). The arguments involved perturbation theory and additional less controllable approximations. The more than 4.700 follow up papers on this subject did essentially not change the status of the conjecture. But at least some aspects of the general AdS/CFT correspondence became clearer after Witten [28] exemplified the ideas in the field theoretic context of a Φ^4 coupling on AdS using a Euclidean functional integral setting.

The structural properties of the AdS/CFT correspondence came out clearly in Rehren's [30] *algebraic holography*. The setting of local quantum physics (LQP) is particularly suited for questions in which one theory is assumed as given and one wants to construct its holographic projection or its corresponding model on another spacetime. LQP can solve such problems of isomorphisms between models without being forced to actually construct a model on either side (which functional integration proposes to do but only in a metaphoric way). At first sight Rehren's setting rewritten in terms of functional integrals (with all the metaphoric caveats, but done in the best tradition of the functional trade) looked quite different from Witten's functional representation. But thanks to a functional identity (explained in the Dütsch-Rehren paper), which shows that fixing functional sources on a boundary and forcing the field values to take on a boundary value via delta function in the functional field space leads to the same result. In this way the apparent disparity disappeared [31] and there is only one AdS/CFT correspondence within QFT.

There are limits to the rigor and validity of functional integral tools in QFT. Even in QM where they are rigorous an attempt to teach a course on QM based on functional integrals would end without having been able to cover the standard material. As an interesting mental exercise just image a scenario with Feynman before Heisenberg. Since path integral representations are much closer to the old

---

[12]An historically interesting case in which the beta function vanishes in every order is the massive Thirring model. In that case the zero mass limit is indeed conformally invariant, but there is no interacting conformal theory for which a perturbation can be formulated directly, it would generate unmanageable infrared divergencies.

quasiclassical Bohr Sommerfeld formulation the transition would have been much smoother, but it would have taken a longer time to get to the operational core of quantum theory; on the other hand quasiclassical formulas and perturbative corrections thereof would emerge with elegance and efficiency.

Using the measure theoretical functional setting it is well-known that superrenormalizable polynomial couplings can be controlled this way [35]. Realistic models with infinite wave function renormalization constants (all realistic Lagrangian models in more than two spacetime dimensions have a trans-canonical short distance behavior) do not fall into this amenable category. But even in low dimensions, where there exist models with finite wave function renormalization constants and hence the short distance prerequisites are met, the functional setting of the AdS/CFT correspondence has an infrared problem[13][14] of a nasty unresolved kind [37]. As the result of lack of an analog to the operator formulation in QM the suggestive power, their close relation to classical geometric concepts and their formal elegance functional integrals have maintained their dominant role in particle physics although renormalized perturbation theory is better taken care of in the setting of "causal perturbation".[15] An operator approach which is not only capable to establish the mathematical existence of models but also permits their explicit construction exists presently only in $d = 1 + 1$; it is the previously mentioned bootstrap-formfactor or wedge-localization approach for factorizing models. Lagrangian factorizing models only constitute a small fraction.

For structural problems as holography, where one starts from a given theory and wants to construct its intrinsically defined holographic image, the use of metaphorical instruments as Euclidean functional integral representations is suggestive but not really convincing in any mathematical sense. As in the case of lightfront holography there are two mathematically controllable ways to AdS/CFT holography; either using (Wightman) fields (*projective holography*) or using operator algebras (*algebraic holography*). The result of all these different methods can be consistently related [31, 32].

The main gain in lightfront holography is a significant simplification of certain properties as compared to the bulk. Even if some of the original problems of the bulk come back in the process of holographic inversion they reappear in the more amenable form of several smaller problems rather than one big one.

The motivation for field theorists being interested in the AdS/CFT correspondence is similar, apart from the fact that the simplification obtainable through an *algebraic isomorphism* is more limited (less radical) than that of a projection.

---

[13]Infrared problems of the kind as they appear in interacting conformal theories are strictly speaking not susceptible to perturbation theoretical treatment and they also seem to pose serious (maybe insoluble) problems in functional integral representations. In those cases where on knows the exact form of the massless limit (Thirring model) this knowledge can be used to disentangle the perturbative infrared divergences.

[14]Eds. annotation: see the chapter by Gottschalk-Thaler in this volume.

[15]Eds. annotation: see the chapter by Dütsch-Brennecke in this volume.

Nevertheless it is not unreasonable to explore the possibility whether some hidden property as for example a widespread conjecture *partial integrability*[16] could become more visible after a spacetime "re-packaging" of the quantum matter substrate from CFT to AdS.

Despite many interesting analogies between chiral theories and higher dimensional QFT [36] little is known about higher-dimensional conformal QFTs. There are Lagrangian candidates as for example certain supersymmetric Yang-Mills theories which fulfil (at least in lowest order) some perturbative prerequisite of conformality which consists in a vanishing beta-function. As mentioned before perturbation theory for conformal QFT, as a result of severe infrared problems, cannot be formulated directly. The prime example for such a situation is the massive Thirring model for which there exists an elegant structural argument for $\beta(g) = 0$ and the knowledge about the non-perturbative massless version can then be used to find the correct perturbative infrared treatment.

As far as I could see (with apologies in case of having overlooked some important work) none of these two steps has been carried out for SUSY-YM, so even the conformal side of the Maldacena conjecture has remained unsafe territory.

There is one advantage which null-surface holography has over AdS/CFT type brane holography. The cardinality of degrees of freedom adjusts itself to what is *natural* for null-surfaces (as a manifold in its own right); for the lightfront holography this is the operator algebra generated from extended chiral fields (3). On the other hand this "thinning out" in holographic projections is of course the reason why inverse holography becomes more complicated and cannot be done with the QFT on one null surface only.

In the holography of the AdS/CFT correspondence the bulk degrees of freedom pass to a conformal brane; in contradistinction to the holography on null-surfaces there is *no reduction of degrees of freedom* resulting from projection. Hence the AdS/CFT isomorphism starting from a "normal" (causally complete as formally arising from Lagrangians) 5-dimensional AdS leads to a *conformal field theory with too many degrees of freedom*. Since a "thinning out" by hand does not seem to be possible, the "physically health" of such a conformal QFT is somewhat dodgy, to put it mildly.

In case one starts with a free Klein-Gordon field on AdS one finds that the generating conformal fields of the CFT are special *generalized free fields* i.e. a kind of continuous superpositions of free fields. They were introduced in the late 50s by W. Greenberg and their useful purpose was (similar to AdS in classical gravity) to *test the physical soundness of axioms of QFT* in the sense that if a system of axioms allowed such solutions, it needed to be further restricted [33] (in that case the so-called causal completion or time-slice property excluded generalized free fields). It seems that meanwhile the word "physical" has changes its meaning, it is used for anything which originated from a physicist.

---

[16]Global integrability is only possible in $d = 1 + 1$, but I am not aware of any theorem which rules out the possibility of integrable substructures.

In the opposite direction the degrees of freedom of a "normal" CFT become "diluted" on AdS in the inverse correspondence. There are not sufficient degrees of freedom for arriving at nontrivial compactly localized operators, the cardinality of degrees of freedom is only sufficient to furnish noncompact regions as AdS wedges with nontrivial operators, the compactly localized double cone algebras remain trivial (multiples of the identity). In the setting based on fields this means that the restriction on testfunction spaces is so severe that pointlike fields $A_{AdS}(x)$ at interior points $x \in \mathrm{int}(AdS)$ do not exist in the standard sense as operator-valued distributions on Schwartz spaces. They exist on much smaller test function spaces, which contain no functions with compact localizations.

Both sides of the correspondence have been treated in a mathematically rigorous fashion for free AdS (Klein-Gordon equation) theories and free (wave equation) CFT [34, 32] where the mismatch between degrees of freedom can be explicated and the structural arguments based on the principles of general QFT show that this mismatch between the transferred and the natural cardinality of the degree of freedom is really there. In terms of the better known Lagrangian formalism the statement would be that if one starts from a Lagrange theory at one side the other side cannot be Lagrangian. Of course both sides remain QFT in the more general sense of fulfilling the required symmetries, have positive energy and being consistent with spacelike commutativity. In the mentioned free field illustration an AdS Klein-Gordon field is evidently Lagrangian whereas the corresponding *conformal generalized free field* has no Lagrangian and cannot even be characterized in terms of a local hyperbolic field equation. According to the best educated guess, 4-dimensional maximally supersymmetric Yang-Mills theories (if they exist and are conformal) would be natural conformal QFTs "as we know it" and therefore cannot come from a natural QFT on AdS. Needless to say again that there are severe technical problems to set up a perturbation theory for conformally invariant interactions, the known perturbative systematics breaks down in the presence of infrared problems[17].

I belong to a generation for which not everything which is mathematically possible must have a physical realization; in particular I do not adhere to the new credo that every mathematically consistent idea is realized in some parallel world (anthropic principle): no parallel universe for the physical realization of every mathematical belch.

Generalized free fields[18] and their interacting counterparts, which arise from natural AdS free- or interacting- fields, remain in my view unphysical, but are

---

[17]A well-known problem is the massive Thirring model which leads to $\beta = 0$ in all orders. In this case one already knew the conformal limit in closed form and was able to check the correctness of the relation by consistency considerations.

[18]It is interesting to note that the Nambu-Goto Lagrangian (which describes a classical relativistic string) yields upon quantization a pointlike localized generalized free field with the well-known infinite tower mass spectrum and the appearance of a Hagedorn limit temperature. As such it is pointlike localized and there is *no intrinsic quantum concept* which permits to associate it with any stringlike localization.

of considerable mathematical interest. They do not fit into the standard causal localization setting and they do not allow thermal KMS states without a limiting Hagedorn temperature (both facts are related). Nature did not indicate that it likes to go beyond the usual localizability and thermal behavior. If string theory demands such things it is not my concern, let Max Tegmark find another universe where nature complies with string theory.

Holography is a technical tool and not a physical principle. It simplifies certain aspects of a QFT at the expense of others (i.e. it cannot achieve miracles). The use of such ideas in intermediate steps may have some technical merits, but I do not see any scientific reason to change my viewpoint about physical admissibility. The question of whether by changing the spacetime encoding one could simplify certain properties (for example detect integrable substructures) of complicated theories is of course very interesting, but in order to pursue such a line it is not necessary to physically identify the changed theory. Such attempts, where only one side needs to be physical and the role of holography would consist in exposing certain structural features which remained hidden in the original formulation, sound highly interesting to me.

There is however one deeply worrisome aspect of this whole development. Never before has there been more than 4.700 publication on such a rather narrow subject; in fact even nowadays, one decade after this gold-digger's rush about the AdS/CFT correspondence started, there is still a sizable number of papers every month by people looking for nuggets at the same place but without bringing Maldacena's gravity-gauge theory conjecture any closer to a resolution. Even with making all the allowances in comparison with earlier fashions, this phenomenon is too overwhelming in order to be overlooked. Independent of its significance for particle physics and the way it will end, the understanding of what went on and its covering by the media will be challenging to historians and philosophers of science in the years to come.

I know that it is contra bonos mores to touch on a sociological aspect in a physics paper, but my age permits me to say that at no time before was the scientific production in particle theory that strongly coupled to the Zeitgeist as during the last two decades; never before had global market forces such a decisive impact on the scientific production. Therefore it is natural to look for an explanation why thousands of articles are written on an interesting (but not clearly formulated) conjecture with hundreds of other interesting problems left aside; where does the magic attraction come from? Is it the Holy Grail of a TOE which sets into motion these big caravans? Did the critical power of past particle physics disappear in favor of acclamation? Why are the few critical but unbiased attempts only mentioned by the labels given to them and not by their scientific content?

Since commentaries about the crisis in an area of which one is part run the risk of being misunderstood, let me make perfectly clear that particle physics was a speculative subject and I uphold that it must remain this way. Therefore I have no problem whatsoever with Maldacena's paper; it is in the best tradition of particle physics, which was always a delicate blend of a highly imaginative and innovative contribution from one author with profoundly critical analysis of others. I am worried about the loss of this balance. My criticism is also not directed against the thousands of authors who enter this area in good faith believing that they are working at an epoch-forming paradigmatic problem because their peers gave them this impression. Even if they entered for the more mundane reason of carving out a career, I would not consider this as the cause of the present problem.

The real problem is with those who by their scientific qualifications and status are the intellectual leaders and the role models. If they abdicate their role as critical mediators by becoming the whips of the TOE monoculture of particle physics, then checks and balances will be lost. Would there have been almost 5000 publication, on a rather narrow theme (compared with other topics) in the presence of a more critical attitude from leading particle physicists? No way. Would particle theory, once the pride of theoretical physics with a methodological impact on many adjacent areas, have fallen into disrespect and be the object of mock within the larger physics community? The list of questions of this kind with negative answers can be continued.

It is worthwhile to look back at times when the delicate balance between the innovative and speculative on the one hand and the critical on the other was still there. Young researchers found guidance by associating themselves to "schools of thought" which where associated with geographical places and names as Schwinger, Landau, Bogoliubov, Wheeler, Wightman, Lehmann, Haag... who represented different coexisting schools of thought. Instead of scientific cross fertilization between different schools, the new globalized caravan supports the formation of a gigantic monoculture and the loss of the culture of checks and balances.

Not even string theorists can deny that this unfortunate development started with string theory. Every problem string theory addresses takes on a strange metaphoric aspect, an effect which is obviously wanted as the fondness for the use of the letter M shows. The above mentioned AdS/CFT topic gives an illustration, which, with a modest amount of mathematical physics shows, the clear structural QFT theorem as compared to the strange conjecture which even thousands of publications were not able to liberate from the metaphoric twilight.

But it is a remarkable fact that, whenever string theorist explain their ideas by QFT analogs in the setting of functional integrals, as was done by Witten in [28] for the φ^4 coupling, and on the other hand algebraic quantum field theorists present their rigorous structural method for the same model in the same setting [31], the two results agree (see also [37]).

This is good news. But now comes the bad news. Despite the agreement the Witten camp, i.e. everybody except a few individuals, claim that there exist two different types of AdS/CFT correspondences namely theirs and another one which at least some of them refer to as the "German AdS/CFT correspondence". Why is that? I think I know but I will not write it.

At this point it becomes clear that it is the abandonment of the critical role of the leaders which is fuelling this unhealthy development. Could a statement: "X-Y-Z theory is a gift of the 21st century which by chance fell into the 20 century" have come from Pauli, Schwinger, or Feynman? One would imagine that in those days people had a better awareness that mystifications like this could disturb the delicate critical counterbalance which the speculative nature of particle physics requires. The long range negative effect on particle theory of such a statement is proportional to the prominence and charisma of its author.

There have been several books which criticize string theory. Most critics emphasize that the theory has not predicted a single observable effect and that there is no reason to expect that this will change in the future. Although I sympathize with that criticism, especially if it comes from experimentalists and philosophers, I think that a theorist should focus his critique on the conceptual and mathematical structure and not rely on help from Karl Popper or dwell on the non-existent observational support. Surprisingly I could not find any scholarly article in this direction. One of the reasons may be that after 4 decades of development of string theory such a task requires rather detailed knowledge about its conceptual and mathematical basis. As a result of this unsatisfactory situation I stopped my critical article [29] from going into print and decided to re-write it in such a way that the particle physics part is strengthened at the expense of the sociological sections.

The aforementioned situation of ignoring results which shed a critical light on string theory or the string theorists version of the AdS/CFT correspondence is perhaps best understood in terms of the proverbial *executing of the messenger who brings bad news*; the unwanted message in the case at hand being the *structural* impossibility to have Lagrangian QFTs with causal propagation on both sides of the correspondence.

It seems that under the corrosive influence of more than 4 decades of string theory, Feynman's observation about its mode of arguing being based on finding excuses instead of explanations, which two decades ago was meant to be provocative, has become the norm. The quantum gravity-gauge theory conjecture is a good example of how a correct but undesired AdS/CFT correspondence is shifted to the elusive level of string theory and quantum gravity so that the degrees of freedom aspect becomes pushed underneath the rug of the elusive string theory, where it only insignificantly enlarges the already very high number of metaphors.

There have been an increasing number of papers with titles as "QCD and a Holographic Model of Hadrons", "Early Time Dynamics in Heavy Ion Collisions and AdS/CFT Correspondence", "Confinement/Deconfinement Transition in AdS/CFT", "Isospin Diffusion in Thermal AdS/CFT with Flavour", "Holographic Mesons in a Thermal Bath", "Viscous Hydrodynamics and AdS/CFT", "Heavy Quark Diffusion from AdS/CFT" ... Ads/CFT for everything? Is string theory bolstered by AdS/CFT really on the way to become a TOE for all of physics, a theory for anything which sacrifices conceptual cohesion to amok running calculations? Or are we witnessing a desperate attempt to overcome the more than 4 decade lasting physical disutility? Perhaps it is only a consequence of the "liberating" effect of following prominent leaders who have forgone their duty as critical mediators and preserver of conceptual cohesion.

4. Concluding remarks

In these notes we revisited one of the oldest and still unsolved conceptual problems in QFT, the existence of interacting models. Besides some new concrete results about the existence of factorizing models (which only exist in $d = 1 + 1$), it is the new method itself, with its promise to explore new fundamental and fully intrinsic properties of QFT, which merits attention. A particularly promising approach for the classification and construction of QFTs consists in using holographic lightfront projections (and in a later stage work one's way back into the bulk). In this situation the holographic degrees of freedom are thinned out as compared to the bulk i.e. the extended chiral fields have lesser number of degrees of freedom.

The concept of degrees of freedom used here is a dynamical one. Knowing only a global algebra[19] as the wedge algebra i.e. $\mathcal{A}(W) \subset B(H)$ as an inclusion into the full algebra one uses fewer degrees freedom than one needs in order to describe the full local substructure of $\mathcal{A}(W)$ i.e. knowing $\mathcal{A}(W)$ in the sense of a local net. The degrees of freedom emerge always from relations between algebras whereas the single algebra is a structureless monad [15]. Saying that the net $\mathcal{A}(LF)$ has less degrees of freedom than the net associated with the bulk is the same as saying that the knowledge of the LF affiliated wedges does not suffice to reconstruct the local bulk structure. In this sense the notion of degrees of freedom depends on the knowledge one has about a system; refining the net structure of localized subalgebras of a global algebra increases the degrees of freedom.

The lightfront holography is a genuine projection with a lesser cardinality of degrees of freedom i.e. without knowing how other Poincaré transformations outside the 7-parametric invariance group of the lightfront act it is not uniquely invertible. On its own, i.e. without added information, the lightfront holography cannot distinguish between massive and massless theories; a transverse extended chiral theory does not know whether the bulk was massive or massless. The knowledge of how the opposite lightray translation $U(a_-)$ acts on $\mathcal{A}(LF)$ restores uniqueness; but this action is necessarily "fuzzy" i.e. non-geometric, purely algebraic. Only upon returning to the spacetime ordering device in terms of the bulk it becomes geometric.

The hallmark of null-surface holography is an area law for localization entropy in which the proportionality constant is a product of a holographic matter dependent constant times a logarithmic dependence on the attenuation length for vacuum polarization.

By far the more popular holography has been the AdS/CFT correspondence. Here its physical utility is less clear than the mathematical structure.

Is there really a relation between a special class of conformal gauge invariant gauge theories with supersymmetric quantum gravity? Not a very probable consequence of a change of an spacetime ordering device for a given matter substrate which is what holography means. Integrable substructures within such conformal

---

[19]Knowing an operator algebra means knowing its position within the algebra $B(H)$ of all operators. Knowing its net substructure means knowing the relative position of all its subalgebras.

gauge theories which become more overt on the AdS-side? This appears a bit more realistic, but present indications are still very flimsy.

Acknowledgements

I am indebted to B. Fauser, J. Tolksdorf and E. Zeidler for the invitation to participate in a 2007 conference in Leipzig and for hospitality extended to me during my stay.

References

[1] R. Haag and J. A. Swieca, Commun. Math. Phys. **1**, (1965) 308

[2] D. Buchholz and E. H. Wichmann, Commun. Math. Phys. **106**, (1986) 321

[3] B. Schroer, *Particle physics in the 60s and 70s and the legacy of contributions by J. A. Swieca*, arXiv:0712.0371

[4] B. Schroer, Ann. Phys. **295**, (1999) 190

[5] G. Lechner, *An Existence Proof for Interacting Quantum Field Theories with a Factorizing S-Matrix*, math-ph/0601022

[6] H.-J. Borchers, D. Buchholz and B. Schroer, Commun. Math. Phys. **219**, (2001) 125, hep-th/0003243

[7] S. Åks; Journ. Math. Phys. **6** (1965) 516

[8] D. Buchholz and S. J. Summers, *String– and Brane–Localized Causal Fields in a Strongly Nonlocal Model*, arXiv:math-ph/0512060

[9] H. Grosse and G. Lechner, JETP **11**, (2007) 021

[10] Works by Buchholz and Summers as well as by Grosse and Lechner, in preparation.

[11] H. Leutwyler, Acta Phys. Austr., Suppl.V. H. Leutwyler in: Magic without Magic (J. R. Klauder ed.). Freeman, 1972. F. Jegerlehner, Helv. Phys. Acta **46**, (1974) 824

[12] W. Driessler, Acta Phys. Austr. **46**, (1977) 63

[13] B. Schroer, Class. Quant. Grav. **23** (2006) 5227, hep-th/0507038 and previous work cited therein

[14] B. Schroer, Class. Quant. Grav. **24** (2007), 1

[15] B. Schroer, *Holography on null-surfaces, entanglement and localization entropy*, arXiv:0712.4403

[16] B. Schroer, *Localization and the interface between quantum mechanics, quantum field theory and quantum gravity*, arXiv:0711.4600

[17] V. Glaser, H. Lehmann and W. Zimmermann, Nuovo Cimento **6**, (1957) 1122

[18] H. Babujian and M. Karowski, Int. J. Mod. Phys. **A1952**, (2004) 34

[19] M. Joerss, Lett. Math. Phys. **38**, (1996) 257

[20] H-J Borchers, Commun. Math. Phys. **2**, (1966) 49

[21] J.L. Cardy, O.A. Castro-Alvaredo and B. Doyon, *Form factors of branch-point twist fields in quantum integrable models and entanglement entropy*, hep-th/0706.3384 and the previous literature cited therein

[22] B. Kay and R. Wald, Phys. Rep. 207 (1991) 49

[23] D. Guido, R. Longo and H-W Wiesbrock, Commun. math. Phys. **192**, (1998) 217

[24] R. Brunetti, K. Fredenhagen and R. Verch, Commun. math. Phys. **237**, (2003) 31

[25] J.D. Bekenstein, Phys. Rev. **D 7**, (1973) 2333

[26] C. Fronsdal, Phys. Rev. **D10**, (1974) 84

[27] J. Maldacena, Adv. Theor. Math. Phys. **2** (1998) 231

[28] E. Witten, Adv. Theor. Math. Phys. 2, 253-291 (1998) 20

[29] B. Schroer, *String theory and the crisis in particle physics,* arXiv:physics/0603112

[30] K-H Rehren, Annales Henri Poincaré **1** (2000) 607, hep-th/9905179,

[31] M. Dütsch and K-H Rehren, Lett. Math. Phys. 62, 171-184 (2002)

[32] K-H Rehren, *QFT Lectures on AdS-CFT,* hep-th/0411086

[33] R. Haag and B. Schroer, J. Math. Phy. **3**, (1962) 248

[34] M. Dütsch and K-H Rehren, Ann. Henri Poincaré **4**, (2003) 613

[35] J. Glimm and A. Jaffe, Jour. Math. Phys. **13**, (1972) 1568

[36] N. M. Nikolov, K.-H. Rehren, I. Todorov, *Harmonic bilocal fields generated by globally conformal invariant scalar fields,* arXiv:0704.1960

[37] H. Gottschalk and H. Thaler, *AdS/CFT correspondence in the Euclidean context,* arXiv:math-ph/0611006

Bert Schroer
CBPF
Rua Dr. Xavier Sigaud 150
22290-180 Rio de Janeiro
Brazil
and
Institut für Theoretische Physik
der FU Berlin
Germany

Quantum Field Theory
B. Fauser, J. Tolksdorf and E. Zeidler, Eds., 25–47

Topos Theory and 'Neo-Realist' Quantum Theory

Andreas Döring

Abstract. Topos theory, a branch of category theory, has been proposed as
mathematical basis for the formulation of physical theories. In this article, we
give a brief introduction to this approach, emphasizing the logical aspects.
Each topos serves as a 'mathematical universe' with an internal logic, which
is used to assign truth-values to all propositions about a physical system. We
show in detail how this works for (algebraic) quantum theory.

Mathematics Subject Classification (2000). Primary 03G30; Secondary 81P10;
81T05; 83C45.

Keywords. topoi, internal logic, subobjects, truth values, daseinisation, contexts, 'neo-realist' quantum theory, foundation of quantum theory, quantum
gravity.

> *"The problem is all inside your head", she said to me*
> *the answer is easy if you take it logically*
>
> Paul Simon (from *'50 Ways To Leave Your Lover'*)

1. Introduction

The use of topos theory in the foundations of physics and, in particular, the foundations of quantum theory was suggested by Chris Isham more than 10 years
ago in [14]. Subsequently, these ideas were developed in an application to the
Kochen-Specker theorem (with Jeremy Butterfield, [15, 16, 18, 19], for conceptual
considerations see [17]). In these papers, the use of a multi-valued, contextual logic
for quantum theory was proposed. This logic is given by the internal logic of a certain topos of presheaves over a category of contexts. Here, contexts typically are

abelian parts of a larger, non-abelian structure. There are several possible choices for the context category. We will concentrate on algebraic quantum theory and use the category $\mathcal{V}(\mathcal{R})$ of abelian von Neumann subalgebras of the non-abelian von Neumann algebra of observables $\mathcal{R}$ of the quantum system, as first suggested in [18].

The use of presheaves over such a category of contexts is motivated by the very natural construction of the spectral presheaf $\underline{\Sigma}$, which collects all the Gel'fand spectra of the abelian subalgebras $V \in \mathcal{V}(\mathcal{R})$ into one larger structure. The Gel'fand spectra can be seen as 'local state spaces', and the spectral presheaf serves as a state space analogue for quantum theory. Interestingly, as Isham and Butterfield showed, this presheaf is not like a space: it has no points (in a category-theoretical sense), and this fact is exactly equivalent to the Kochen-Specker theorem.

The topos approach was developed considerably in the series of papers [5, 6, 7, 8] by Chris Isham and the author. In these papers, it was shown how topos theory can serve as a new mathematical framework for the formulation of physical theories. The basic idea of the topos programme is that by representing the relevant physical structures (states, physical quantities and propositions about physical quantities) in suitable topoi, one can achieve a remarkable structural similarity between classical and quantum physics. Moreover, the topos programme is general enough to allow for major generalizations. Theories beyond classical and quantum theory are conceivable. Arguably, this generality will be needed in a future theory of quantum gravity, which is expected to go well beyond our conventional theories.

In this paper, we will concentrate on algebraic quantum theory. We briefly motivate the mathematical constructions and give the main definitions.[1] Throughout, we concentrate on the logical aspects of the theory. We will show in detail how, given a state, truth-values are assigned to all propositions about a quantum system. This is independent of any measurement or observer. For that reason, we say that the topos approach gives a 'neo-realist' formulation of quantum theory.

1.1. What is a topos?

It is impossible to give even the briefest introduction to topos theory here. At the danger of being highly imprecise, we restrict ourselves to mentioning some aspects of this well-developed mathematical theory and give a number of pointers to the literature. The aim merely is to give a very rough idea of the structure and internal logic of a topos. In the next subsection, we argue that this mathematical structure may be useful in physics.

There are a number of excellent textbooks on topos theory, and the reader should consult at least one of them. We found the following books useful: [25, 9, 27, 20, 21, 1, 24].

---

[1] We suppose that the reader is familiar with the definitions of a category, functor and natural transformation.

Topoi as mathematical universes. Every (elementary) topos $\mathcal{E}$ can be seen as a *mathematical universe*. As a category, a topos $\mathcal{E}$ possesses a number of structures that generalize constructions that are possible in the category **Set** of sets and functions.[2] Namely, in **Set**, we can construct new sets from given ones in several ways: let S, T be two sets, then we can form the cartesian product $S \times T$, the disjoint union $S \amalg T$ and the exponential S^T, the set of all functions from T to S. These constructions turn out to be fundamental and can all be phrased in an abstract, categorical manner, where they are called finite limits, colimits and exponentials, respectively. By definition, a topos $\mathcal{E}$ has all of these. One consequence of the existence of finite limits is that each topos has a *terminal object*, denoted by 1. This is characterized by the property that for any object A in the topos $\mathcal{E}$, there exists exactly one arrow from A to 1. In **Set**, a one-element set $1 = \{*\}$ is terminal.[3]

Of course, **Set** is a topos, too, and it is precisely the topos which usually plays the role of our mathematical universe, since we construct our mathematical objects starting from sets and functions between them. As a slogan, we have: a topos $\mathcal{E}$ is a category similar to **Set**. A very nice and gentle introduction to these aspects of topos theory is the book [25]. Other good sources are [9, 26].

In order to 'do mathematics', one must also have a logic, including a deductive system. Each topos comes equipped with an *internal logic*, which is of *intuitionistic* type. We very briefly sketch the main characteristics of intuitionistic logic and the mathematical structures in a topos that realize this logic.

Intuitionistic logic. Intuitionistic logic is similar to Boolean logic, the main difference being that the *law of excluded middle* need not hold. In intuitionistic logic, there is *no* axiom

$$\vdash a \vee \neg a \qquad\qquad (*)$$

like in Boolean logic. Here, $\neg a$ is the negation of the formula (or proposition) a. The algebraic structures representing intuitionistic logic are *Heyting algebras*. A Heyting algebra is a pseudocomplemented, distributive lattice[4] with zero element 0 and unit element 1, representing 'totally false' resp. 'totally true'. The pseudocomplement is denoted by $\neg$, and one has, for all elements α of a Heyting algebra H,

$$\alpha \vee \neg \alpha \leq 1,$$

in contrast to $\alpha \vee \neg \alpha = 1$ in a Boolean algebra. This means that the disjunction ("Or") of a proposition α and its negation need not be (totally) true in a

---

[2] More precisely, *small* sets and functions between them. Small means that we do not have proper classes. One must take care in these foundational issues to avoid problems like Russell's paradox.

[3] Like many categorical constructions, the terminal object is fixed only up to isomorphism: any two one-element sets are isomorphic, and any of them can serve as a terminal object. Nonetheless, one speaks of *the* terminal object.

[4] Lattice is meant in the algebraic sense: a partially ordered set L such that any two elements $a, b \in L$ have a minimum (greatest lower bound) $a \wedge b$ and a maximum (least upper bound) $a \vee b$ in L. A lattice L is distributive if and only if $a \vee (b \wedge c) = (a \vee b) \wedge (a \vee c)$ as well as $a \wedge (b \vee c) = (a \wedge b) \vee (a \wedge c)$ hold for all $a, b, c \in L$.

Heyting algebra. Equivalently, one has

$$\neg\neg\alpha \geq \alpha,$$

in contrast to $\neg\neg\alpha = \alpha$ in Boolean algebras.

Obviously, Boolean logic is a special case of intuitionistic logic. It is known from Stone's theorem [29] that each Boolean algebra is isomorphic to an algebra of (clopen, i.e., closed and open) subsets of a suitable (topological) space.

Let X be a set, and let $P(X)$ be the power set of X, that is, the set of subsets of X. Given a subset $S \in P(X)$, one can ask for each point $x \in X$ whether it lies in S or not. This can be expressed by the *characteristic function* $\chi_S : X \rightarrow \{0,1\}$, which is defined as

$$\chi_S(x) := \begin{cases} 1 & \text{if } x \in S \\ 0 & \text{if } x \notin S \end{cases}$$

for all $x \in X$. The two-element set $\{0,1\}$ plays the role of a set of *truth-values* for propositions (of the form "$x \in S$"). Clearly, 1 corresponds to 'true', 0 corresponds to 'false', and there are no other possibilities. This is an argument about sets, so it takes place in and uses the logic of the topos **Set** of sets and functions. **Set** is a *Boolean topos*, in which the familiar two-valued logic and the axiom ($*$) hold. (This does not contradict the fact that the internal logic of topoi is intuitionistic, since Boolean logic is a special case of intuitionistic logic.)

In an arbitrary topos, there is a special object Ω, called the *subobject classifier*, that takes the role of the set $\{0,1\} \simeq \{\text{false,true}\}$ of truth-values. Let B be an object in the topos, and let A be a subobject of B. This means that there is a monic $A \rightarrow B$,[5] generalizing the inclusion of a subset S into a larger set X. Like in **Set**, we can also characterize A as a subobject of B by an arrow from B to the subobject classifier Ω. (In **Set**, this arrow is the characteristic function $\chi_S : X \rightarrow \{0,1\}$.) Intuitively, this characteristic arrow from B to Ω tells us how A 'lies in' B. The textbook definition is:

Definition 1. *In a category $\mathcal{C}$ with finite limits, a* **subobject classifier** *is an object Ω, together with a monic* $\text{true} : 1 \rightarrow \Omega$, *such that to every monic $m : A \rightarrow B$ in $\mathcal{C}$ there is a unique arrow χ which, with the given monic, forms a pullback square*

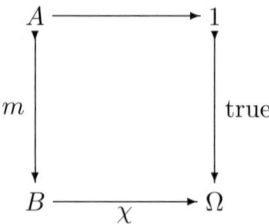

---

[5] A *monic* is the categorical version of an injective function. In the topos **Set**, monics exactly are injective functions.

In **Set**, the arrow true : $1 \rightarrow \{0,1\}$ is given by $\text{true}(*) = 1$. In general, the subobject classifier Ω need not be a set, since it is an object in the topos $\mathcal{E}$, and the objects of $\mathcal{E}$ need not be sets. Nonetheless, there is an abstract notion of *elements* (or *points*) in category theory that we can use. The elements of Ω are the truth-values available in the internal logic of our topos $\mathcal{E}$, just like 'false' and 'true', the elements of {false,true}, are the truth-values available in the topos **Set**.

To understand the abstract notion of elements, let us consider sets for a moment. Let $1 = \{*\}$ be a one-element set, the terminal object in **Set**. Let S be a set and consider an arrow e from 1 to S. Clearly, $e(*) \in S$ is one element of S. The set of all functions from 1 to S corresponds exactly to the elements of S. This idea can be generalized to other categories: if there is a terminal object 1, then we consider arrows from 1 to an object A in the category as *elements of A*. For example, in the definition of theindextrue!as element in a truth object subobject classifier the arrow true : $1 \rightarrow \Omega$ is an element of Ω. It may happen that an object A has no elements, i.e., there are no arrows $1 \rightarrow A$. It is common to consider arrows from subobjects U of A to A as *generalized elements*, but we will not need this except briefly in subsection 5.1.

As mentioned, the elements of the subobject classifier, understood as the arrows $1 \rightarrow \Omega$, are the truth-values. Moreover, the set of these arrows forms a Heyting algebra (see, for example, section 8.3 in [9]). This is how (the algebraic representation of) intuitionistic logic manifests itself in a topos. Another, closely related fact is that the subobjects of any object A in a topos form a Heyting algebra.

1.2. Topos theory and physics

A large part of the work on topos theory in physics consists in showing how states, physical quantities and propositions about physical quantities can be represented within a suitable topos attached to the system [5, 6, 7, 8]. The choice of topos will depend on the theory type (classical, quantum or, in future developments, even something completely new). Let us consider classical physics for the moment to motivate this.

Realism in classical physics. In classical physics, one has a space of states $\mathcal{S}$, and physical quantities A are represented by real-valued functions $f_A : \mathcal{S} \rightarrow \mathbb{R}$.[6] A proposition about a physical quantity A is of the form "$A \in \Delta$", which means "the physical quantity A has a value in the (Borel) set Δ". This proposition is represented by the inverse image $f_A^{-1}(\Delta) \subseteq \mathcal{S}$. In general, propositions about the physical system correspond to Borel subsets of the state space $\mathcal{S}$. If we have two propositions "$A \in \Delta_1$", "$B \in \Delta_2$" and the corresponding subsets $f_A^{-1}(\Delta_1)$, $f_B^{-1}(\Delta_2)$, then the intersection $f_A^{-1}(\Delta_1) \cap f_B^{-1}(\Delta_2)$ corresponds to the proposition

---

[6]We assume that f_A is at least measurable.

"$A \in \Delta_1$ and $B \in \Delta_2$", the union $f_A^{-1}(\Delta_1) \cup f_B^{-1}(\Delta_2)$ corresponds to "$A \in \Delta_1$ or $B \in \Delta_2$", and the complement $\mathcal{S}\backslash f_A^{-1}(\Delta_1)$ corresponds to the negation "$A \notin \Delta_1$". Moreover, given a state s, i.e., an element of the state space $\mathcal{S}$, each proposition is either true or false: if s lies in the subset of $\mathcal{S}$ representing the proposition, then the proposition is true, otherwise it is false. Every physical quantity A has a value in the state s, namely $f_A(s) \in \mathbb{R}$. Thus classical physics is a *realist* theory in which propositions have truth-values independent of measurements, observers etc. The logic is Boolean, since classical physics is based on constructions with sets and functions, i.e., it takes place in the topos **Set**. We take this as a rule: if we want to describe a physical system S as a classical system, then the topos **Set** is used. This means no departure from what is ordinarily done, but it emphasizes certain structural and logical aspects of the theory.

Instrumentalism in quantum theory. In quantum theory, the mathematical description is very different. Physical quantities A are represented by self-adjoint operators $\widehat{A}$ on a Hilbert space $\mathcal{H}$. While $\mathcal{H}$ can be called a space of states, the states $\psi \in \mathcal{H}$ play a very different role from those in classical theory. In particular, a state ψ does not assign values to all physical quantities, only to those for which ψ happens to be an eigenstate. The spectral theorem shows that propositions "$A \in \Delta$" are represented by projection operators $\widehat{E}[A \in \Delta]$ on Hilbert space. Unless ψ is an eigenstate of A, such a proposition is neither true nor false (except for the trivial cases $\widehat{E}[A \in \Delta] = \widehat{0}$, which represents trivially false propositions, and $\widehat{E}[A \in \Delta] = \widehat{1}$, which represents trivially true propositions). The mathematical formalism of quantum theory is interpreted in an *instrumentalist* manner: given a state ψ, the proposition "$A \in \Delta$" is assigned a probability of being true, given by the expectation value $p(A \in \Delta; \psi) := \langle \psi | \widehat{E}[A \in \Delta] | \psi \rangle$. This means that upon measurement of the physical quantity A, one will find the measurement result to lie in Δ with probability $p(A \in \Delta; \psi)$. This interpretation depends on measurements and an external observer. Moreover, the measurement devices (and the observer) are described in terms of classical physics, not quantum physics.

The motivation from quantum gravity. An instrumentalist interpretation cannot describe closed quantum systems, at least there is nothing much to be said about them from this perspective. A theory of quantum cosmology or quantum gravity will presumably be a quantum theory of the whole universe. Since there is no external observer who could perform measurements in such a theory, instrumentalism becomes meaningless. One of the main motivations for the topos programme is to overcome or circumvent the usual instrumentalism of quantum theory and to replace it with a more realist account of quantum systems. The idea is to use the internal logic of a topos to assign truth-values to propositions about the system.

In order to achieve this, we will sketch a new mathematical formulation of quantum theory that is structurally similar to classical physics. The details can be found in [5, 6, 7, 8] and references therein.

Plan of the paper. The starting point is the definition of a formal language $\mathcal{L}(S)$ attached to a physical system S. This is done in section 2 and emphasizes the common structure of classical and quantum physics. In section 3, we introduce the topos associated to a system S in the case of quantum theory, and in section 4 we briefly discuss the representation of $\mathcal{L}(S)$ in this topos. The representation of states and the assignment of truth-values to propositions is treated in section 5, which is the longest and most detailed section. Section 6 concludes with some remarks on related work and on possible generalizations.

2. A formal language for physics

There is a well-developed branch of topos theory that puts emphasis on the logical aspects. As already mentioned, a topos can be seen as the embodiment of (higher-order) intuitionistic logic. This point of view is expounded in detail in Bell's book [1], which is our standard reference on these matters. Other excellent sources are [24] and part D of [21]. The basic concept consists in defining a *formal language* and then finding a *representation* of it in a suitable topos. As usual in mathematical logic, the formal language encodes the syntactic aspects of the theory and the representation provides the semantics. Topoi are a natural 'home' for the representation of formal languages encoding intuitionistic logic, more precisely, *intuitionistic, higher-order, typed predicate logic with equality*. Typed means that there are several primitive species or kinds of objects (instead of just sets as primitives), from which sets are extracted as a subspecies; predicate logic means that one has quantifiers, namely an existence quantifier $\exists$ ("it exists") and a universal quantifier $\forall$ ("for all"). Higher-order refers to the fact that quantification can take place not only over variable individuals, but also over subsets and functions of individuals as well as iterates of these constructions. Bell presents a particularly elegant way to specify a formal language with these properties. He calls this type of language a *local language*, see chapter 3 of [1].

Let S denote a physical system to which we attach a higher-order, typed language $\mathcal{L}(S)$. We can only sketch the most important aspects here, details can be found in section 4 of [5]. The language $\mathcal{L}(S)$ does not depend on the theory type (classical, quantum, ...), while its representation of course does. The language contains at least the following type symbols: $1, \Omega, \Sigma$ and $\mathcal{R}$. The symbol Σ serves as a precursor of the *state!object* (see below), the symbol $\mathcal{R}$ is a precursor of the *quantity-value object*, which is where physical quantities take their values. Moreover, we require the existence of *function symbols* of the form $A : \Sigma \rightarrow \mathcal{R}$. These are the linguistic precursors of physical quantities. For each type, there exist *variables* of that type. There are a number of rules how to form terms and formulae (terms of type Ω) from variables of the various types, including the definition of logical connectives $\wedge$ ("And"), $\vee$ ("Or") and $\neg$ ("Not"). Moreover, there are axioms giving *rules of inference* that define how to get new formulae from sets of given formulae. As an example, we mention the *cut rule*: if Γ is a set of formulae and α and β are formulae, then we have

$$\frac{\Gamma : \alpha \quad \alpha, \Gamma : \beta}{\Gamma : \beta}$$

(here, any free variable in α must be free in Γ or β). This is a purely formal rule about how formulae can be manipulated in this calculus, to be read from top to bottom. In a representation, where the formulae aquire an interpretation and a 'meaning', this expresses that if Γ implies α, and α and Γ together imply β, then Γ also implies β. The axioms and rules of inference are chosen in a way such that the logical operations satisfy the laws of intuitionistic logic.

The formal language $\mathcal{L}(S)$ captures a number of abstract properties of the physical system S. For example, if S is the harmonic oscillator, then we expect to be able to speak about the physical quantity energy in all theory types, classical or quantum (or other). Thus, among the function symbols $A : \Sigma \to \mathcal{R}$, there will be one symbol $E : \Sigma \to \mathcal{R}$ which, in a representation, will become the mathematical entity describing energy. (Which mathematical object that will be depend on the theory type and thus on the representation.)

The representation of the language $\mathcal{L}(S)$ takes place in a suitable, physically motivated topos $\mathcal{E}$. The type symbol 1 is represented by the terminal object 1 in $\mathcal{E}$, the type symbol Ω is represented by the subobject classifier Ω. The choice of an appropriate object Σ in the topos that represents the symbol Σ depends on physical insight. The representing object Σ is called the *state object*, and it plays the role of a generalized state space. What actually is generalized is the space, not the states: Σ is an object in a topos $\mathcal{E}$, which need not be a topos of sets, so Σ need not be a set or space-like. However, as an object in a topos, Σ does have subobjects. These subobjects will be interpreted as (the representatives of) propositions about the physical quantities, just like in classical physics, where propositions correspond to subsets of state space. The propositions are of the form "$A \in \Delta$", where Δ now is a subobject of the object $\mathcal{R}$ that represents the symbol $\mathcal{R}$. The object $\mathcal{R}$ is called the *quantity-value object*, and this is where physical quantities take their values. Somewhat surprisingly, even for ordinary quantum theory this is *not* the real number object in the topos. Finally, the function symbols $A : \Sigma \to \mathcal{R}$ are represented by arrows between the objects Σ and $\mathcal{R}$ in the topos $\mathcal{E}$.

In classical physics, the representation is the obvious one: the topos to be used is the topos **Set** of sets and mappings, the symbol Σ is represented by a symplectic manifold $\mathcal{S}$, which is the state space, the symbol $\mathcal{R}$ is represented by the real numbers and function symbols $A : \Sigma \to \mathcal{R}$ are represented by real-valued functions $f_A : \mathcal{S} \to \mathbb{R}$. Propositions about physical quantities correspond to subsets of the state space.

3. The context category $\mathcal{V}(\mathcal{R})$ and the topos of presheaves $\mathbf{Set}^{\mathcal{V}(\mathcal{R})^{op}}$

We will now discuss the representation of $\mathcal{L}(S)$ in the case that S is to be described as a quantum system. We assume that S is a non-trivial system that −in the usual description− has a Hilbert space $\mathcal{H}$ of dimension 3 or greater, and that the physical quantities belonging to S form a von Neumann algebra $\mathcal{R}(S) \subseteq \mathcal{B}(\mathcal{H})$ that contains the identity operator $\widehat{1}$.[7]

From the Kochen-Specker theorem [23] we know that there is no state space model of quantum theory if the algebra of observables is $\mathcal{B}(\mathcal{H})$ (for the generalization to von Neumann algebras see [4]). More concretely, there is no state space $\mathcal{S}$ such that the physical quantities are real-valued functions on $\mathcal{S}$. The reason is that if there existed such a state space $\mathcal{S}$, then each point (i.e., state) $s \in \mathcal{S}$ would allow to assign values to all physical quantities at once, simply by evaluating the functions representing the physical quantities at s. One can show that under very mild and natural conditions, this leads to a mathematical contradiction.

For an *abelian* von Neumann algebra V, there is no such obstacle: the *Gel'fand spectrum* Σ_V of V can be interpreted as a state space, and the Gel'fand transforms $\overline{A}$ of self-adjoint operators $\widehat{A} \in V$, representing physical quantities, are real-valued functions on Σ_V. The Gel'fand spectrum Σ_V of an abelian von Neumann algebra V consists of the pure states λ on V (see e.g. [22]). Each $\lambda \in \Sigma_V$ also is a multiplicative state; for all $\widehat{A}, \widehat{B} \in V$, we have

$$\lambda(\widehat{A}\widehat{B}) = \lambda(\widehat{A})\lambda(\widehat{B}),$$

which, for projections $\widehat{P} \in \mathcal{P}(V)$, implies

$$\lambda(\widehat{P}) = \lambda(\widehat{P}^2) = \lambda(\widehat{P})\lambda(\widehat{P}) \in \{0,1\}.$$

Finally, each $\lambda \in \Sigma_V$ is an algebra homomorphism from V to $\mathbb{C}$. The Gel'fand spectrum Σ_V is equipped with the weak* topology and thus becomes a compact Hausdorff space.

Let $\widehat{A} \in V$ and define

$$\overline{A} : \Sigma_V \longrightarrow \mathbb{C}$$
$$\lambda \longmapsto \overline{A}(\lambda) := \lambda(\widehat{A}).$$

The function $\overline{A}$ is called the *Gel'fand transform* of $\widehat{A}$. It is a continuous function such that $\operatorname{im}\overline{A} = \operatorname{sp}\widehat{A}$. In particular, if $\widehat{A}$ is self-adjoint, then $\lambda(\widehat{A}) \in \operatorname{sp}\widehat{A} \subset \mathbb{R}$.

---

[7]There should arise no confusion between the von Neumann algebra $\mathcal{R} = \mathcal{R}(S)$ and the symbol $\mathcal{R}$ of our formal language, we hope.

The mapping

$$V \longrightarrow C(\Sigma_V)$$
$$\widehat{A} \longmapsto \overline{A}$$

is called the *Gel'fand transformation* on V. It is an isometric $*$-isomorphism between V and $C(\Sigma_V)$.[8]

This leads to the idea of considering the set $\mathcal{V}(\mathcal{R})$ of non-trivial unital abelian von Neumann subalgebras of $\mathcal{R}$.[9] These abelian subalgebras are also called *contexts*. $\mathcal{V}(\mathcal{R})$ is partially ordered by inclusion and thus becomes a category. There is an arrow $i_{V'V} : V' \to V$ if and only if $V' \subseteq V$, and then $i_{V'V}$ is just the inclusion (or the identity arrow if $V' = V$). The category $\mathcal{V}(\mathcal{R})$ is called the *context category* and serves as our index category. The process of going from one abelian algebra V to a smaller algebra $V' \subset V$ can be seen as a process of *coarse-graining*: the algebra V' contains less physical quantities (self-adjoint operators), so we can describe less physics in V' than in V. We collect all the 'local state spaces' Σ_V into one large object:

Definition 2. *The* **spectral presheaf** $\underline{\Sigma}$ *is the presheaf*[10] *over* $\mathcal{V}(\mathcal{R})$ *defined*

  a) *on objects: for all $V \in \mathcal{V}(\mathcal{R})$, $\underline{\Sigma}_V = \Sigma_V$ is the Gel'fand spectrum of V,*
  b) *on arrows: for all $i_{V'V}$, $\underline{\Sigma}(i_{V'V}) : \underline{\Sigma}_V \to \underline{\Sigma}_{V'}$ is given by restriction, $\lambda \mapsto \lambda|_{V'}$.*

The spectral presheaf was first considered by Chris Isham and Jeremy Butterfield in the series [15, 16, 18, 19] (see in particular the third of these papers). The presheaves over $\mathcal{V}(\mathcal{R})$ form a topos $\mathbf{Set}^{\mathcal{V}(\mathcal{R})^{op}}$. The arrows in this topos are natural transformations between the presheaves. Isham and Butterfield developed the idea that this is the appropriate topos for quantum theory. The object $\underline{\Sigma}$ in $\mathbf{Set}^{\mathcal{V}(\mathcal{R})^{op}}$ serves as a state space analogue. In the light of the new developments in [5]-[8], using formal languages, we identify $\underline{\Sigma}$ as the state object in $\mathbf{Set}^{\mathcal{V}(\mathcal{R})^{op}}$, i.e., the representative of the symbol Σ of our formal language $\mathcal{L}(S)$.

Isham and Butterfield showed that the Kochen-Specker theorem is exactly equivalent to the fact that the spectral presheaf $\underline{\Sigma}$ has no elements, in the sense that there are no arrows from the terminal object $\underline{1}$ in $\mathbf{Set}^{\mathcal{V}(\mathcal{R})^{op}}$ to $\underline{\Sigma}$. It is not hard to show that having an element of $\underline{\Sigma}$ would allow the assignment of real values to all physical quantities at once.

---

[8] Of course, all this holds more generally for abelian C^*-algebras. We concentrate on von Neumann algebras, since we need these in our application.
[9] The unit in each abelian subalgebra $V \in \mathcal{V}(\mathcal{R})$ is the identity operator $\widehat{1}$, which is the same unit as in $\mathcal{R}$. We exclude the trivial algebra $\mathbb{C}\widehat{1}$, which is a subalgebra of all other subalgebras.
[10] A presheaf over $\mathcal{V}(\mathcal{R})$ is a contravariant functor from $\mathcal{V}(\mathcal{R})$ to $\mathbf{Set}$, and obviously, $\underline{\Sigma}$ is of this kind. In our notation, presheaves will always be underlined.

4. Representing $\mathcal{L}(S)$ in the presheaf topos $\mathbf{Set}^{\mathcal{V}(\mathcal{R})^{op}}$

The quantity-value object for quantum theory. We already have identified the topos for the quantum-theoretical description of a system S and the state object $\underline{\Sigma}$ in this topos. Let $V \in \mathcal{V}(\mathcal{R})$ be a context, then $\downarrow V := \{V' \in \mathcal{V}(\mathcal{R}) \mid V' \subseteq V\}$ denotes the set of all subalgebras of V, equipped with the partial order inherited from $\mathcal{V}(\mathcal{R})$. It can be shown that the symbol $\mathcal{R}$ should be represented by the following presheaf [7]:

Definition 3. *The* **presheaf $\underline{\mathbb{R}^{\leftrightarrow}}$ of order-preserving and -reversing functions** *on $\mathcal{V}(\mathcal{R})$ is defined*

a) *on objects: for all $V \in \mathcal{V}(\mathcal{R})$, $\underline{\mathbb{R}^{\leftrightarrow}}_V := \{(\mu, \nu) \mid \mu : \downarrow V \to \mathbb{R}$ is order-preserving, $\nu : \downarrow V \to \mathbb{R}$ is order-reversing and $\mu \leq \nu\}$, where $\mu \leq \nu$ means that for all $V' \in \downarrow V$, one has $\mu(V') \leq \nu(V')$;*

b) *on arrows: for all $i_{V'V}$, $\underline{\mathbb{R}^{\leftrightarrow}}(i_{V'V}) : \underline{\mathbb{R}^{\leftrightarrow}}_V \to \underline{\mathbb{R}^{\leftrightarrow}}_{V'}$ is given by restriction, $(\mu, \nu) \mapsto (\mu|_{V'}, \nu|_{V'})$.*

Here, an order-preserving function $\mu : \downarrow V \to \mathbb{R}$ is a function such that $V'' \subseteq V'$ (where $V', V'' \in \downarrow V$) implies $\mu(V'') \leq \mu(V')$. Order-reversing functions are defined analogously.

The presheaf $\underline{\mathbb{R}^{\leftrightarrow}}$ is *not* the real-number object $\underline{\mathbb{R}}$ in the topos $\mathbf{Set}^{\mathcal{V}(\mathcal{R})^{op}}$, which is the constant presheaf defined by $\underline{\mathbb{R}}(V) := \mathbb{R}$ for all V and $\underline{\mathbb{R}}(i_{V'V}) : \mathbb{R} \to \mathbb{R}$ as the identity. From the Kochen-Specker theorem, we would not expect that physical quantities take their values in the real numbers. (This does not mean that the results of measurements are not real numbers. We do not discuss measurement here.) More importantly, the presheaf $\underline{\mathbb{R}^{\leftrightarrow}}$ takes into account the coarse-graining inherent in the base category $\mathcal{V}(\mathcal{R})$: at each stage V, a pair (μ, ν) consisting of an order-preserving and an order-reversing function defines a whole range or interval $[\mu(V), \nu(V)]$ of real numbers, not just a single real number. (It can happen that $\mu(V) = \nu(V)$.) If we go to a smaller subalgebra $V' \subset V$, which is a kind of coarse-graining, then we have $\mu(V') \leq \mu(V)$ and $\nu(V') \geq \nu(V)$, so the corresponding interval $[\mu(V'), \nu(V')]$ can only become larger.

The representation of function symbols $A : \Sigma \to \mathcal{R}$. In order to represent a physical quantity A belonging to the system S as an arrow from $\underline{\Sigma}$ to the presheaf $\underline{\mathbb{R}^{\leftrightarrow}}$ of 'values', we have to use a two-step process.

1. We first need the *spectral order* on self-adjoint operators in a von Neumann algebra $\mathcal{R}$ [28, 10]. This is defined for all $\widehat{A}, \widehat{B} \in \mathcal{R}_{sa}$ with spectral families $\widehat{E}^A$ resp. $\widehat{E}^B$ as

$$\widehat{A} \leq_s \widehat{B} :\Leftrightarrow (\forall \lambda \in \mathbb{R} : \widehat{E}_\lambda^A \geq \widehat{E}_\lambda^B).$$

Equipped with the spectral order, the set of self-adjoint operators in a von Neumann algebra becomes a boundedly complete lattice.

Let $\widehat{A} \in \mathcal{R}$ be the self-adjoint operator representing A. We use the spectral order on each abelian subalgebra $V \in \mathcal{V}(\mathcal{R})$ and define

$$\delta^o(\widehat{A})_V := \bigwedge \{\widehat{B} \in V_{sa} \mid \widehat{B} \geq_s \widehat{A}\},$$

$$\delta^i(\widehat{A})_V := \bigvee \{\widehat{C} \in V_{sa} \mid \widehat{C} \leq_s \widehat{A}\}.$$

We call these mappings *outer* and *inner daseinisation*, respectively.[11] The outer daseinisation $\delta^o(\widehat{A})_V$ of $\widehat{A}$ to the context V is the approximation from above by the smallest self-adjoint operator in V that is spectrally larger than $\widehat{A}$. Likewise, the inner daseinisation $\delta^i(\widehat{A})_V$ is the approximation from below by the largest self-adjoint operator in V that is spectrally smaller then $\widehat{A}$. Since the spectral order is coarser than the usual, linear order, we have, for all V,

$$\delta^i(\widehat{A})_V \leq \widehat{A} \leq \delta^o(\widehat{A})_V.$$

One can show that the spectra of $\delta^i(\widehat{A})_V$ and $\delta^o(\widehat{A})_V$ are subsets of the spectrum of $\widehat{A}$, which seems physically very sensible. If we used the approximation in the linear order, this would not hold in general. The approximation of self-adjoint operators in the spectral order was suggested by de Groote [11, 12]. If $V' \subset V$, then, by construction, $\delta^i(\widehat{A})_{V'} \leq_s \delta^i(\widehat{A})_V$ and $\delta^o(\widehat{A})_{V'} \geq_s \delta^o(\widehat{A})_V$, which implies

$$\delta^i(\widehat{A})_{V'} \leq \delta^i(\widehat{A})_V,$$

$$\delta^o(\widehat{A})_{V'} \geq \delta^o(\widehat{A})_V.$$

In this sense, the approximations to $\widehat{A}$ become coarser if the context becomes smaller.

2. Now that we have constructed a pair $(\delta^i(\widehat{A})_V, \delta^o(\widehat{A})_V)$ of operators approximating $\widehat{A}$ from below and from above for each context V, we can define a natural transformation $\breve{\delta}(\widehat{A})$ from $\underline{\Sigma}$ to $\underline{\mathbb{R}}^{\leftrightarrow}$ in the following way: let $V \in \mathcal{V}(\mathcal{R})$ be a context, and let $\lambda \in \underline{\Sigma}_V$ be a pure state of V. Then define, for all $V' \in \downarrow V$,

$$\mu_\lambda(V') := \lambda(\delta^i(\widehat{A})_{V'}) = \overline{\delta^i(\widehat{A})_{V'}}(\lambda),$$

where $\overline{\delta^i(\widehat{A})_{V'}}$ is the Gel'fand transform of the self-adjoint operator $\delta^i(\widehat{A})_{V'}$. From the theory of abelian C^*-algebras, it is known that $\lambda(\delta^i(\widehat{A})_{V'}) \in \mathrm{sp}(\delta^i(\widehat{A})_{V'})$ (see e.g. [22]). Let $V', V'' \in \downarrow V$ such that $V'' \subset V'$. We saw that $\delta^i(\widehat{A})_{V''} \leq \delta^i(\widehat{A})_{V'}$, which implies $\lambda(\delta^i(\widehat{A})_{V''}) \leq \lambda(\delta^i(\widehat{A})_{V'})$, so $\mu_\lambda :\downarrow V \to \mathbb{R}$ is an order-preserving function. Analogously, let

$$\nu_\lambda(V') := \lambda(\delta^o(\widehat{A})_{V'}) = \overline{\delta^o(\widehat{A})_{V'}}(\lambda)$$

---

[11]'Daseinisation' comes from the German word Dasein, which means existence. More specifically, we borrow from Heidegger's existential philosophy, where 'Da-sein' means 'being-there', being in the world. I hope it is needless to say that the coinage daseinisation (meaning the act of bringing into existence) is slightly tongue-in-cheek.

for all $V' \in \downarrow V$. We obtain an order-reversing function $\nu_\lambda : \downarrow V \to \mathbb{R}$. Then, for all $V \in \mathcal{V}(\mathcal{R})$, let

$$\breve{\delta}(\widehat{A})(V) : \underline{\Sigma}_V \longrightarrow \underline{\mathbb{R}^{\leftrightarrow}}_V$$
$$\lambda \longmapsto (\mu_\lambda, \nu_\lambda).$$

By construction, these mappings are the components of a natural transformation $\breve{\delta}(\widehat{A}) : \underline{\Sigma} \to \underline{\mathbb{R}^{\leftrightarrow}}$. For all $V, V' \in \mathcal{V}(\mathcal{R})$ such that $V' \subseteq V$, we have a commuting diagram

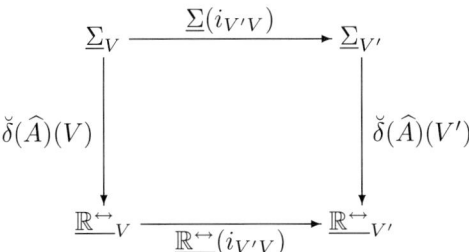

The arrow $\breve{\delta}(\widehat{A}) : \underline{\Sigma} \to \underline{\mathbb{R}^{\leftrightarrow}}$ in the presheaf topos $\mathbf{Set}^{\mathcal{V}(\mathcal{R})^{op}}$ is the representative of the physical quantity A, which is abstractly described by the function symbol $A : \Sigma \to \mathcal{R}$ in our formal language. The physical content, namely the appropriate choice of the self-adjoint operator $\widehat{A}$ from which we construct the arrow $\breve{\delta}(\widehat{A})$, is not part of the language, but part of the representation.[12]

The representation of propositions. As discussed in subsection 1.2, in classical physics the subset of state space $\mathcal{S}$ representing a proposition "$A \in \Delta$" is constructed by taking the inverse image $f_A^{-1}(\Delta)$ of Δ under the function representing A. We will use the analogous construction in the topos formulation of quantum theory: the set Δ is a subset (that is, subobject) of the quantity-value object $\mathbb{R}$ in classical physics, so we start from a subobject (in $\mathbf{Set}^{\mathcal{V}(\mathcal{R})^{op}}$) Θ of the presheaf $\underline{\mathbb{R}^{\leftrightarrow}}$. We get a subobject of the state object $\underline{\Sigma}$ by pullback along $\breve{\delta}(\widehat{A})$, which we denote by $\breve{\delta}(\widehat{A})^{-1}(\Theta)$.[13] For details see subsection 3.6 in [7] and also [13].

In both classical and quantum theory, propositions are represented by subobjects of the quantity-value object (state space $\mathcal{S}$ resp. spectral presheaf $\underline{\Sigma}$). Such subobjects are constructed by pullback from subobjects of the quantity-value object (real numbers $\mathbb{R}$ resp. presheaf of order-preserving and -reversing functions $\underline{\mathbb{R}^{\leftrightarrow}}$). The interpretation and meaning of such propositions is determined by the internal logic of the topos ($\mathbf{Set}$ resp. $\mathbf{Set}^{\mathcal{V}(\mathcal{R})^{op}}$). In the classical case, where $\mathbf{Set}$ is used, this is the ordinary Boolean logic that we are familiar with. In the quantum

---

[12]The current scheme is not completely topos-internal yet. It is an open question if every arrow from $\underline{\Sigma}$ to $\underline{\mathbb{R}^{\leftrightarrow}}$ comes from a self-adjoint operator. This is why we start from a self-adjoint operator $\widehat{A}$ to construct $\breve{\delta}(\widehat{A})$. We are working on a more internal characterization.

[13]This is a well-defined categorical construction, since the pullback of a monic is a monic, so we get a subobject of $\underline{\Sigma}$ from a subobject of $\underline{\mathbb{R}^{\leftrightarrow}}$.

case, the internal logic of the presheaf topos $\mathbf{Set}^{\mathcal{V}(\mathcal{R})^{op}}$ has to be used. This intuitionistic logic can be interpreted using Kripke-Joyal semantics, see e.g. chapter VI in [27].

The Heyting algebra structure of subobjects. In the next section, we discuss the representation of states in the topos $\mathbf{Set}^{\mathcal{V}(\mathcal{R})^{op}}$ and the assignment of truth-values to propositions. Before doing so, it is worth noting that the subobjects of $\underline{\Sigma}$ form a Heyting algebra (since the subobjects of any object in a topos do), so we have mapped propositions "$A \in \Delta$" (understood as discussed) to a *distributive* lattice with a pseudocomplement. Together with the results from the next section, we have a completely new form of quantum logic, based upon the internal logic of the presheaf topos $\mathbf{Set}^{\mathcal{V}(\mathcal{R})^{op}}$. Since this is a distributive logic and since the internal logic of a topos has powerful rules of inference, this kind of quantum logic is potentially much better interpretable than ordinary quantum logic of the Birkhoff-von Neumann kind. The latter type of quantum logic and its generalizations are based on nondistributive structures and lack a deductive system.

5. Truth objects and truth-values

In classical physics, a state is just a point of state space.[14] Since, as we saw, the spectral presheaf $\underline{\Sigma}$ has no elements (or, global elements[15]), we must represent states differently in the presheaf topos $\mathbf{Set}^{\mathcal{V}(\mathcal{R})^{op}}$.

5.1. Generalized elements as generalized states

One direct way, suggested in [13], is the following generalization: $\underline{\Sigma}$ has no global elements $\underline{1} \to \underline{\Sigma}$, but it does have subobjects $\underline{U} \hookrightarrow \underline{\Sigma}$. In algebraic geometry and more generally in category theory, such monics (and, more generally, arbitrary arrows) are called generalized elements [25]. We could postulate that these subobjects, or some of them, are 'generalized states'. Consider another subobject of $\underline{\Sigma}$ that represents a proposition "$A \in \Delta$" about the quantum system, given by its characteristic arrow $\chi_{\underline{S}} : \underline{\Sigma} \to \underline{\Omega}$. Then we can compose these arrows

$$\underline{U} \hookrightarrow \underline{\Sigma} \to \underline{\Omega}$$

to obtain an arrow $\underline{U} \to \underline{\Omega}$. This is *not* a global element $\underline{1} \to \underline{\Omega}$ of $\underline{\Omega}$, and by construction, it cannot be, since $\underline{\Sigma}$ has no global elements, but it is a generalized element of $\underline{\Omega}$. It might be possible to give a physical meaning to these arrows

---

[14] One might call this a *pure* state, though this is not customary in classical physics. Such a state actually is a point measure on state space, in contrast to more general probability measures that describe general states. We only consider pure states here and identify the point measure with the corresponding point of state space.

[15] Elements $\underline{1} \to \underline{P}$ of a presheaf $\underline{P}$ are called *global elements* or *global sections* in category theory. We follow this convention to avoid confusion with points or elements of sets.

$\underline{U} \to \underline{\Omega}$ if one can (a) give physical meaning to the subobject $\underline{U} \hookrightarrow \underline{\Sigma}$, making clear what a generalized state actually is, and (b) give a logical and physical interpretation of an arrow $\underline{U} \to \underline{\Omega}$. While a global element $\underline{1} \to \underline{\Omega}$ is interpreted as a truth-value in the internal logic of a topos, the logical interpretation of an arrow $\underline{U} \to \underline{\Omega}$ is not so clear.

We want to emphasize that mathematically, the above construction is perfectly well-defined. It remains to be worked out if a physical and logical meaning can be attached to it.

5.2. The construction of truth objects

We now turn to the construction of so-called 'truth objects' from pure quantum states ψ, see also [6]. (To be precise, a unit vector ψ in the Hilbert space $\mathcal{H}$ represents a vector state $\varphi_\psi : \mathcal{R} \to \mathbb{C}$ on a von Neumann algebra, given by $\varphi_\psi(\widehat{A}) := \langle \psi| \widehat{A} |\psi \rangle$ for all $\widehat{A} \in \mathcal{R}$. If $\mathcal{R} = \mathcal{B}(\mathcal{H})$, then every φ_ψ is a pure state.) Of course, the Hilbert space $\mathcal{H}$ is the Hilbert space on which the von Neumann algebra of observables $\mathcal{R} \subseteq \mathcal{B}(\mathcal{H})$ is represented. This is the most direct way in which Hilbert space actually enters the mathematical constructions inside the topos $\mathbf{Set}^{\mathcal{V}(\mathcal{R})^{op}}$. However, we will see how this direct appeal to Hilbert space possibly can be circumvented.

Given a subobject of $\underline{\Sigma}$ that represents some proposition, a truth object will allow us to construct a global element $\underline{1} \to \underline{\Sigma}$ of $\underline{\Sigma}$, as we will show in subsection 5.4. This means that from a proposition and a state, we *do* get an actual truth-value for that proposition in the internal logic of the topos $\mathbf{Set}^{\mathcal{V}(\mathcal{R})^{op}}$. The construction of truth objects is a direct generalization of the classical case.

For the moment, let us consider sets. Let S be a subset of some larger set X, and let $x \in X$. Then

$$(x \in S) \Leftrightarrow (S \in U(x)),$$

where $U(x)$ denotes the set of neighborhoods of x in X. The key observation is that while the l.h.s. cannot be generalized to the topos setting, since we cannot talk about points like x, the r.h.s. can. The task is to define neighborhoods in a suitable manner. We observe that $U(x)$ is a subset of the power set $PX = P(X)$, which is the same as an element of the power set of the power set $PPX = P(P(X))$.

This leads to the idea that for each context $V \in \mathcal{V}(\mathcal{R})$, we must choose an appropriate set of subsets of the Gel'fand spectrum $\underline{\Sigma}_V$ such that these sets of subsets form an element in $PP\underline{\Sigma}$. Additionally, the subsets we choose at each stage V should be clopen, since the clopen subsets $P_{cl}(\underline{\Sigma}_V)$ form a lattice that is isomorphic to the lattice $\mathcal{P}(V)$ of projections in V. If $\widehat{P} \in V$ is a projection, then the corresponding clopen subset of $\underline{\Sigma}_V$ is

$$S_{\widehat{P}} := \{\lambda \in \underline{\Sigma}_V \mid \lambda(\widehat{P}) = 1\}.$$

Conversely, given a clopen subset $S \in P_{cl}(\underline{\Sigma}_V)$, we denote the corresponding projection in $\mathcal{P}(V)$ by $\widehat{P}_S$. It is given as the inverse Gel'fand transform of the characteristic function of S.

The main difficulty lies in the fact that the spectral presheaf $\underline{\Sigma}$ has no global elements, which is equivalent to the Kochen-Specker theorem. A global element, if it existed, would pick one point λ_V from each Gel'fand spectrum $\underline{\Sigma}_V$ ($V \in \mathcal{V}(\mathcal{R})$) such that, whenever $V' \subset V$, we would have $\lambda_{V'} = \lambda_V|_{V'}$. If we had such global elements, we could define neighborhoods for them by taking, for each $V \in \mathcal{V}(\mathcal{R})$, neighborhoods of λ_V in $\underline{\Sigma}_V$.

Since no such global elements exist, we cannot expect to have neighborhoods of *points* at each stage. Rather, we will get neighborhoods of *sets* at each stage V, and only for particular V, these sets will have just one element. In any case, the sets will depend on the state ψ in a straightforward manner. We define:

Definition 4. *Let $\psi \in \mathcal{H}$ be a unit vector, let $\widehat{P}_\psi$ the projection onto the corresponding one-dimensional subspace (i.e., ray) of $\mathcal{H}$, and let $P_{cl}(\underline{\Sigma}_V)$ be the clopen subsets of the Gel'fand spectrum $\underline{\Sigma}_V$. If $S \in P_{cl}(\underline{\Sigma}_V)$, then $\widehat{P}_S \in \mathcal{P}(V)$ denotes the corresponding projection. The truth object $\mathbb{T}^\psi = (\mathbb{T}_V^\psi)_{V \in \mathcal{V}(\mathcal{R})}$ is given by*

$$\forall V \in \mathcal{V}(\mathcal{R}) : \mathbb{T}_V^\psi := \{S \in P_{cl}(\underline{\Sigma}_V) \mid \langle\psi| \widehat{P}_S |\psi\rangle = 1\}.$$

At each stage V, $\mathbb{T}_V^\psi$ collects all subsets S of $\underline{\Sigma}_V$ such that the expectation value of the projection corresponding to this subset is 1. From this definition, it is not clear at first sight that the set $\mathbb{T}_V^\psi$ can be seen as a set of neighborhoods.

Lemma 5. *We have the following equalities:*

$$\forall V \in \mathcal{V}(\mathcal{R}) : \mathbb{T}_V^\psi = \{S \in P_{cl}(\underline{\Sigma}_V) \mid \langle\psi| \widehat{P}_S |\psi\rangle = 1\}$$
$$= \{S \in P_{cl}(\underline{\Sigma}_V) \mid \widehat{P}_S \geq \widehat{P}_\psi\}$$
$$= \{S \in P_{cl}(\underline{\Sigma}_V) \mid \widehat{P}_S \geq \delta^o(\widehat{P}_\psi)_V\}$$
$$= \{S \in P_{cl}(\underline{\Sigma}_V) \mid S \supseteq S_{\delta^o(\widehat{P}_\psi)_V}\}.$$

Proof. If $\langle\psi| \widehat{P}_S |\psi\rangle = 1$, then ψ lies entirely in the subspace of Hilbert space that $\widehat{P}_S$ projects onto. This is equivalent to $\widehat{P}_S \geq \widehat{P}_\psi$. Since $\widehat{P}_S \in \mathcal{P}(V)$ and $\delta^o(\widehat{P}_\psi)_V$ is the *smallest* projection in V that is larger than $\widehat{P}_\psi$,[16] we also have $\widehat{P}_S \geq \delta^o(\widehat{P}_\psi)_V$. In the last step, we simply go from the projections in V to the corresponding clopen subsets of $\underline{\Sigma}_V$. $\qquad\square$

This reformulation shows that $\mathbb{T}_V^\psi$ actually consists of subsets of the Gel'fand spectrum $\underline{\Sigma}_V$ that can be seen as some kind of neighborhoods, not of a single point of $\underline{\Sigma}_V$, but of a certain subset of $\underline{\Sigma}_V$, namely $S_{\delta^o(\widehat{P}_\psi)_V}$. In the simplest case,

---

[16] On projections, the spectral order $\leq_s$ and the linear order $\leq$ coincide.

we have $\widehat{P}_\psi \in \mathcal{P}(V)$, so $\delta^o(\widehat{P}_\psi)_V = \widehat{P}_\psi$. Then $S_{\delta^o(\widehat{P}_\psi)_V} = S_{\widehat{P}_\psi}$, and this subset contains a single element, namely the pure state λ such that

$$\lambda(\widehat{P}_\psi) = 1$$

and $\lambda(\widehat{Q}) = 0$ for all $\widehat{Q} \in \mathcal{P}(V)$ such that $\widehat{Q}\widehat{P}_\psi = 0$. In this case, $\mathbb{T}_V^\psi$ actually consists of all the clopen neighborhoods of the point λ in $\underline{\Sigma}_V$.

In general, if $\widehat{P}_\psi$ does not lie in the projections $\mathcal{P}(V)$, then there is no subset of $\underline{\Sigma}_V$ that corresponds directly to $\widehat{P}_\psi$. We must first approximate $\widehat{P}_\psi$ by a projection in V, and $\delta^o(\widehat{P}_\psi)_V$ is the smallest projection in V larger than $\widehat{P}_\psi$. The projection $\delta^o(\widehat{P}_\psi)_V$ corresponds to a subset $S_{\delta^o(\widehat{P}_\psi)_V} \subseteq \underline{\Sigma}_V$ that may contain more than one element. However, $\mathbb{T}_V^\psi$ can still be seen as a set of neighborhoods, but now of this set $S_{\delta^o(\widehat{P}_\psi)_V}$ rather than of a single point.

It is an interesting and non-trivial point that the (outer) daseinisation $\delta^o(\widehat{P}_\psi)_V$ ($V \in \mathcal{V}(\mathcal{R})$) shows up in this construction. We did not discuss this here, but the subobjects of $\underline{\Sigma}$ constructed from the outer daseinisation of projections play a central role in the representation of a certain propositional language $\mathcal{PL}(S)$ that one can attach to a physical system S [5, 6]. Moreover, these subobjects are 'optimal' in the sense that, whenever $V' \subset V$, the restriction from $S_{\delta^o(\widehat{P})_V}$ to $S_{\delta^o(\widehat{P})_{V'}}$ is *surjective*, see Theorem 3.1 in [6]. This property can also lead the way to a more internal characterization of truth-objects, without reference to a state ψ and hence to Hilbert space.

5.3. Truth objects and Birkhoff-von Neumann quantum logic

In this short subsection, we want to consider truth objects from the point of view of ordinary quantum logic, which goes back to the famous paper [3] by Birkhoff and von Neumann. This is a small digression, since an important part of the topos approach is to replace ordinary quantum logic by the internal logic of the topos $\mathbf{Set}^{\mathcal{V}(\mathcal{R})^{op}}$. It may still be useful to understand how our constructions relate to Birkhoff-von Neumann quantum logic.

For now, let us assume that $\mathcal{R} = \mathcal{B}(\mathcal{H})$, then we write $\mathcal{P}(\mathcal{H}) := \mathcal{P}(\mathcal{B}(\mathcal{H}))$ for the lattice of projections on Hilbert space. In their paper, Birkhoff and von Neumann identify a proposition "$A \in \Delta$" about a quantum system with a projection operator $\widehat{E}[A \in \Delta] \in \mathcal{P}(\mathcal{H})$ via the spectral theorem [22] and interpret the lattice structure of $\mathcal{P}(\mathcal{H})$ as giving a quantum logic. This is very different from the topos form of quantum logic, since $\mathcal{P}(\mathcal{H})$ is a *non-distributive* lattice, leading to all the well-known interpretational difficulties.

The implication in ordinary quantum logic is given by the partial order on $\mathcal{P}(\mathcal{H})$: a proposition "$A \in \Delta_1$" implies a proposition "$B \in \Delta_2$" (where we can have $B = A$) if and only if $\widehat{E}[A \in \Delta_1] \leq \widehat{E}[B \in \Delta_2]$ holds for the corresponding projections.

The idea now is that, given a pure state ψ and the corresponding projection $\widehat{P}_\psi$ onto a ray, we can collect all the projections larger than or equal to $\widehat{P}_\psi$. We denote this by

$$T^\psi := \{\widehat{P} \in \mathcal{P}(\mathcal{H}) \mid \widehat{P} \geq \widehat{P}_\psi\}.$$

The propositions represented by these projections are exactly those propositions about the quantum system that are (totally) true if the system is in the state ψ. Totally true means 'true with probability 1' in an instrumentalist interpretation. If, for example, a projection $\widehat{E}[A \in \Delta]$ is larger than $\widehat{P}_\psi$ and hence contained in T^ψ, then, upon measurement of the physical quantity A, we will find the measurement result to lie in the set Δ with certainty (i.e., with probability 1).

T^ψ is a maximal (proper) filter in $\mathcal{P}(\mathcal{H})$. Every pure state ψ gives rise to such a maximal filter T^ψ, and clearly, the mapping $\psi \mapsto T^\psi$ is injective. We can obtain the truth object $\mathbb{T}^\psi$ from the maximal filter T^ψ simply by defining

$$\forall V \in \mathcal{V}(\mathcal{R}) : \mathbb{T}_V^\psi := T^\psi \cap V.$$

In each context V, we collect all the projections larger than $\widehat{P}_\psi$. On the level of propositions, we have all the propositions about physical quantities A *in the context V* that are totally true in the state ψ.

5.4. The assignment of truth-values to propositions

We return to the consideration of the internal logic of the topos $\mathbf{Set}^{\mathcal{V}(\mathcal{R})^{op}}$ and show how to define a global element $\underline{1} \rightarrow \underline{\Omega}$ of the subobject classifier from a clopen subobject $\underline{S}$ of $\underline{\Sigma}$ and a truth object $\mathbb{T}^\psi$. The subobject $\underline{S}$ represents a proposition about the quantum system, the truth object $\mathbb{T}^\psi$ represents a state, and the global element of $\underline{\Omega}$ will be interpreted as the truth-value of the proposition in the given state. Thus, we make use of the internal logic of the topos $\mathbf{Set}^{\mathcal{V}(\mathcal{R})^{op}}$ of presheaves over the context category $\mathcal{V}(\mathcal{R})$ to assign truth-values to all propositions about a quantum system.

It is well known that the subobject classifier $\underline{\Omega}$ in a topos of presheaves is the presheaf of *sieves* (see e.g. [27]). A sieve σ on an object A in some category $\mathcal{C}$ is a collection of arrows with codomain A with the following property: if $f : B \rightarrow A$ is in σ and $g : C \rightarrow B$ is another arrow in $\mathcal{C}$, then $f \circ g : C \rightarrow A$ is in σ, too. In other words, a sieve on A is a downward closed set of arrows with codomain A. Since the context category $\mathcal{V}(\mathcal{R})$ is a partially ordered set, things become very simple: the only arrows with codomain V are the inclusions $i_{V'V}$. Since such an arrow is specified uniquely by its domain V', we can think of the sieve σ on V as consisting of certain subalgebras V' of V. If $V' \in \sigma$ and $V'' \subset V'$, then $V'' \in \sigma$.

The restriction mappings of the presheaf $\underline{\Omega}$ are given by pullbacks of sieves. The pullback of sieves over a partially ordered set takes a particularly simple form:

Lemma 6. *If σ is a sieve on $V \in \mathcal{V}(\mathcal{R})$ and $V' \subset V$, then the pullback $\sigma \cdot i_{V'V}$ is given by $\sigma \cap \downarrow V'$. (This holds analogously for sieves on any partially ordered set, not just $\mathcal{V}(\mathcal{R})$).*

Proof. For the moment, we switch to the arrows notation. By definition, the pullback $\sigma \cdot i_{V'V}$ is given by

$$\sigma \cdot i_{V'V} := \{i_{V''V'} \mid i_{V'V} \circ i_{V''V'} \in \sigma\}.$$

We now identify arrows and subalgebras as usual and obtain (using the fact that $V'' \subseteq V'$ implies $V'' \subset V$)

$$\{i_{V''V'} \mid i_{V'V} \circ i_{V''V'} \in \sigma\} \simeq \{V'' \subseteq V' \mid V'' \in \sigma\} = \downarrow V' \cap \sigma.$$

Since $\downarrow V'$ is the maximal sieve on V', the pullback $\sigma \cdot i_{V'V}$ is given as the intersection of σ with the maximal sieve on V'. $\qquad\square$

The *name* $\ulcorner \underline{S} \urcorner$ of the subobject $\underline{S}$ is the unique arrow $1 \to P\underline{\Sigma} = \Omega^{\underline{\Sigma}}$ into the power object of $\underline{\Sigma}$ (i.e., the subobjects of $\underline{\Sigma}$) that 'picks out' $\underline{S}$ among all subobjects. $\ulcorner \underline{S} \urcorner$ is a global element of $P\underline{\Sigma}$. Here, one uses the fact that power objects behave like sets, in particular, they have global elements. Since we assume that $\underline{S}$ is a *clopen* subobject, we also get an arrow $1 \to P_{cl}\underline{\Sigma}$ into the clopen power object of $\underline{\Sigma}$, see [6]. We denote this arrow by $\ulcorner \underline{S} \urcorner$, as well.

Since $\mathbb{T}^\psi \in PP_{cl}\underline{\Sigma}$ is a collection of clopen subobjects of $\underline{\Sigma}$, it makes sense to ask if $\underline{S}$ is among them; an expression like $\ulcorner \underline{S} \urcorner \in \mathbb{T}^\psi$ is well-defined. We define, for all $V \in \mathcal{V}(\mathcal{R})$, the *valuation*

$$v(\ulcorner \underline{S} \urcorner \in \mathbb{T}^\psi)_V := \{V' \subseteq V \mid \underline{S}(V') \in \mathbb{T}^\psi_{V'}\}.$$

At each stage V, we collect all those subalgebras of V such that $\underline{S}(V')$ is contained in $\mathbb{T}^\psi_{V'}$.

In order to construct a global element of the presheaf of sieves $\underline{\Omega}$, we must first show that $v(\ulcorner \underline{S} \urcorner \in \mathbb{T}^\psi)_V$ is a sieve on V. In the proof we use the fact that the subobjects obtained from daseinisation are optimal in a certain sense.

Proposition 7. $v(\ulcorner \underline{S} \urcorner \in \mathbb{T}^\psi)_V := \{V' \subseteq V \mid \underline{S}(V') \in \mathbb{T}^\psi_{V'}\}$ *is a sieve on V.*

Proof. As usual, we identify an inclusion morphism $i_{V'V}$ with V' itself, so a sieve on V consists of certain subalgebras of V. We have to show that if $V' \in v(\ulcorner \underline{S} \urcorner \in \mathbb{T}^\psi)_V$ and $V'' \subset V'$, then $V'' \in v(\ulcorner \underline{S} \urcorner \in \mathbb{T}^\psi)_V$. Now, $V' \in v(\ulcorner \underline{S} \urcorner \in \mathbb{T}^\psi)_V$ means that $\underline{S}(V') \in \mathbb{T}^\psi_{V'}$, which is equivalent to $\underline{S}(V') \supseteq \underline{S}_{\delta^o(\widehat{P}_\psi)_{V'}}$. Here, $\underline{S}_{\delta^o(\widehat{P}_\psi)_{V'}}$ is the component at V' of the sub-object $\underline{S}_{\delta^o(\widehat{P}_\psi)} = (\underline{S}_{\delta^o(\widehat{P}_\psi)_V})_{V \in \mathcal{V}(\mathcal{R})}$ of $\underline{\Sigma}$ obtained from daseinisation of $\widehat{P}_\psi$. According to Thm. 3.1 in [6], the sub-object $\underline{S}_{\delta^o(\widehat{P}_\psi)}$ is optimal in the following sense: when restricting from V' to V'', we have

$\underline{\Sigma}(i_{V''V'})(\underline{S}_{\delta^o(\widehat{P}_\psi)_{V'}}) = \underline{S}_{\delta^o(\widehat{P}_\psi)_{V''}}$, i.e., the restriction is surjective. By assumption, $\underline{S}(V') \supseteq \underline{S}_{\delta^o(\widehat{P}_\psi)_{V'}}$, which implies

$$\underline{S}(V'') \supseteq \underline{\Sigma}(i_{V''V'})(\underline{S}(V')) \supseteq \underline{\Sigma}(i_{V''V'})(\underline{S}_{\delta^o(\widehat{P}_\psi)_{V'}}) = \underline{S}_{\delta^o(\widehat{P}_\psi)_{V''}}.$$

This shows that $V'' \in \mathbb{T}_{V''}^\psi$ and hence $V'' \in v(\ulcorner \underline{S} \urcorner \in \mathbb{T}^\psi)_V$. $\qquad\square$

Finally, we have to show that the sieves $v(\ulcorner \underline{S} \urcorner \in \mathbb{T}^\psi)_V$, $V \in \mathcal{V}(\mathcal{R})$, actually form a global element of $\underline{\Omega}$, i.e., they all fit together under the restriction mappings of the presheaf $\underline{\Omega}$:

Proposition 8. *The sieves* $v(\ulcorner \underline{S} \urcorner \in \mathbb{T}^\psi)_V$, $V \in \mathcal{V}(\mathcal{R})$, *(see Prop. 7) form a global element of* $\underline{\Omega}$.

Proof. From Lemma 6, is suffices to show that, whenever $V' \subset V$, we have $v(\ulcorner \underline{S} \urcorner \in \mathbb{T}^\psi)_{V'} = v(\ulcorner \underline{S} \urcorner \in \mathbb{T}^\psi)_V \cap \downarrow V'$. If $V'' \in v(\ulcorner \underline{S} \urcorner \in \mathbb{T}^\psi)_{V'}$, then $\underline{S}(V'') \in \mathbb{T}_{V''}^\psi$, which implies $V'' \in v(\ulcorner \underline{S} \urcorner \in \mathbb{T}^\psi)_V$. Conversely, if $V'' \in \downarrow V'$ and $V'' \in v(\ulcorner \underline{S} \urcorner \in \mathbb{T}^\psi)_V$, then, again, $\underline{S}(V'') \in \mathbb{T}_{V''}^\psi$, which implies $V'' \in v(\ulcorner \underline{S} \urcorner \in \mathbb{T}^\psi)_{V'}$. $\qquad\square$

The global element $v(\ulcorner \underline{S} \urcorner \in \mathbb{T}^\psi) = (v(\ulcorner \underline{S} \urcorner \in \mathbb{T}^\psi)_V)_{V \in \mathcal{V}(\mathcal{R})}$ of $\underline{\Omega}$ is interpreted as the truth-value of the proposition represented by $\underline{S} \in P_{cl}(\underline{\Sigma})$ if the quantum system is in the state ψ (resp. $\mathbb{T}^\psi$). This assignment of truth-values is

- contextual, since the contexts $V \in \mathcal{V}(\mathcal{R})$ play a central role in the whole construction
- global in the sense that *every* proposition is assigned a truth-value
- completely independent of any notion of measurement or observer, hence we call our scheme a 'neo-realist' formulation of quantum theory
- topos-internal, the logical structure is not chosen arbitrarily, but fixed by the topos $\mathbf{Set}^{\mathcal{V}(\mathcal{R})^{op}}$. This topos is directly motivated from the Kochen-Specker theorem
- non-Boolean, since there are (a) more truth-values than just *'true'* and *'false'* and (b) the global elements form a *Heyting* algebra, not a Boolean algebra. There is a global element 1 of $\underline{\Omega}$, consisting of the maximal sieve $\downarrow V$ at each stage V, which is interpreted as 'totally true', and there is a global element 0 consisting of the empty sieve for all V, which is interpreted as 'totally false'. Apart from that, there are many other global elements that represent truth-values between 'totally true' and 'totally false'. These truth-values are neither numbers nor probabilities, but are given by the logical structure of the presheaf topos $\mathbf{Set}^{\mathcal{V}(\mathcal{R})^{op}}$. Since the Heyting algebra of global elements of $\underline{\Omega}$, i.e., of truth-values, is a *partially* ordered set only, there are truth-values v_1, v_2 such that neither $v_1 < v_2$ nor $v_2 < v_1$, which is also different from two-valued Boolean logic where simply $0 < 1$ (i.e., 'false'<'true'). The presheaf topos $\mathbf{Set}^{\mathcal{V}(\mathcal{R})^{op}}$ has a rich logical structure.

6. Conclusion and outlook

The formulation of quantum theory within the preshcaf topos $\mathbf{Set}^{\mathcal{V}(\mathcal{R})^{op}}$ gives a theory that is remarkably similar to classical physics from a structural perspective. In particular, there is a state object (the spectral presheaf $\underline{\Sigma}$) and a quantity-value object (the presheaf $\underline{\mathbb{R}^{\leftrightarrow}}$ of order-preserving and -reversing functions). Physical quantities are represented by arrows between $\underline{\Sigma}$ and $\underline{\mathbb{R}^{\leftrightarrow}}$.

One of the future tasks will be the incorporation of dynamics. The process of daseinisation behaves well with respect to the action of unitary operators, see section 5.2 in [7], so it is conceivable that there is a 'Heisenberg picture' of dynamics. Commutators remain to be understood in the topos picture. On the other hand, it is possible to let a truth-object $\underline{\mathbb{T}}^{\psi}$ change in time by applying Schrödinger evolution to ψ. It remains to be shown how this can be understood topos-internally.

Mulvey and Banaschewski have recently shown how to define the Gel'fand spectrum of an abelian C^{*}-algebra $\mathfrak{A}$ in any Grothendieck topos, using constructive methods (see [2] and references therein). Spitters and Heunen made the following construction in [13]: one takes a *non-abelian* C^{*}-algebra $\mathfrak{A}$ and considers the topos of (covariant) functors over the category of abelian subalgebras of $\mathfrak{A}$. The algebra $\mathfrak{A}$ induces an internal *abelian* C^{*}-algebra $\underline{\mathfrak{A}}$ in this topos of functors. (Internally, algebraic operations are only allowed between commuting operators.) Spitters and Heunen observed that the Gel'fand spectrum of this internal algebra basically is the spectral presheaf.[17] It is very reassuring that the spectral presheaf not only has a physical interpretation, but also such a nice and natural mathematical one. Spitters and Heunen also discuss integration theory in the constructive context. These tools will be very useful in order to regain actual numbers and expectation values from the topos formalism.

Since the whole topos programme is based on the representation of formal languages, major generalizations are possible. One can represent the same language $\mathcal{L}(S)$ in different topoi, as we already did with $\mathbf{Set}$ for classical physics and $\mathbf{Set}^{\mathcal{V}(\mathcal{R})^{op}}$ for algebraic quantum theory. For physical theories going beyond this, other topoi will play a role. The biggest task is the incorporation of space-time concepts, which will, at the very least, necessitate a change of the base category. It is also conceivable that the 'smooth topoi' of synthetic differential geometry (SDG) will play a role.

Acknowledgements

I want to thank Chris Isham for numerous discussions and constant support. Moreover, I would like to thank Bob Coecke, Samson Abramsky, Jamie Vicary, Bas

---

[17]The change from presheaves, i.e., contravariant functors, to covariant functors is necessary in the constructive context.

Spitters and Chris Heunen for stimulating discussions. This work was supported by the DAAD and the FQXi, which is gratefully acknowledged. Finally, I want to thank the organizers of the RDQFT workshop, Bertfried Fauser, Jürgen Tolksdorf and Eberhard Zeidler, for inviting me. It was a very enjoyable experience.

References

[1] J. L. Bell, *Toposes and Local Set Theories* (Clarendon Press, Oxford 1988)

[2] B. Banaschewski, C. J. Mulvey, "A globalisation of the Gelfand duality theorem", Ann. of Pure and Applied Logic **137** (2006), 62-103

[3] G. Birkhoff, J. von Neumann, "The logic of quantum mechanics", Ann. of Math. **37** (1936), 823-843

[4] A. Döring, "Kochen-Specker Theorem for von Neumann Algebras", Int. J. Theor. Phys. **44**, no. 2 (2005), 139-160

[5] A. Döring, C. J. Isham, "A Topos Foundation for Theories of Physics: I. Formal Languages for Physics", quant-ph/0703060, to appear in J. Math. Phys. (March 2008)

[6] A. Döring, C. J. Isham, "A Topos Foundation for Theories of Physics: II. Daseinisation and the Liberation of Quantum Theory", quant-ph/0703062, to appear in J. Math. Phys. (March 2008)

[7] A. Döring, C. J. Isham, "A Topos Foundation for Theories of Physics: III. The Representation of Physical Quantities with Arrows $\delta^o(A) : \underline{\Sigma} \to \underline{\mathbb{R}^{\succeq}}$", quant-ph/0703064, to appear in J. Math. Phys. (March 2008)

[8] A. Döring, C. J. Isham, "A Topos Foundation for Theories of Physics: IV. Categories of Systems", quant-ph/0703066, to appear in J. Math. Phys. (March 2008)

[9] R. Goldblatt, *Topoi, The Categorical Analysis of Logic*, revised second edition (North-Holland, Amsterdam, New York, Oxford 1984)

[10] H. F. de Groote, "On a canonical lattice structure on the effect algebra of a von Neumann algebra", quant-ph/0410018 v2 (2005)

[11] H. F. de Groote, "Observables", math-ph/0507019

[12] H. F. de Groote, "Observables IV: The Presheaf Perspective", math-ph/0708.0677

[13] C. Heunen, B. Spitters, "A topos for algebraic quantum theory", quant-ph/0709.4364

[14] C. J. Isham "Topos Theory and Consistent Histories: The Internal Logic of the Set of all Consistent Sets", Int. J. Theor. Phys. **36** (1997), 785-814

[15] C. J. Isham, J. Butterfield, "A topos perspective on the Kochen-Specker theorem: I. Quantum states as generalised valuations", Int. J. Theor. Phys. **37** (1998), 2669-2733

[16] C. J. Isham, J. Butterfield, "A topos perspective on the Kochen-Specker theorem: II. Conceptual aspects, and classical analogues", Int. J. Theor. Phys. **38** (1999), 827-859

[17] C. J. Isham, J. Butterfield, "Some Possible Roles for Topos Theory in Quantum Theory and Quantum Gravity", Found. Phys. **30** (2000), 1707-1735

[18] C. J. Isham, J. Butterfield, J. Hamilton, "A topos perspective on the Kochen-Specker theorem: III. Von Neumann algebras as the base category", Int. J. Theor. Phys. **39** (2000), 1413-1436

[19] C. J. Isham, J. Butterfield, "A topos perspective on the Kochen-Specker theorem: IV. Interval valuations", Int. J. Theor. Phys. **41**, no. 4 (2002), 613-639

[20] P. T. Johnstone, *Sketches of an Elephant, A Topos Theory Compendium*, Vol. 1 (Clarendon Press, Oxford 2002)

[21] P. T. Johnstone, *Sketches of an Elephant, A Topos Theory Compendium*, Vol. 2 (Clarendon Press, Oxford 2002)

[22] R. V. Kadison, J. R. Ringrose, *Fundamentals of the Theory of Operator Algebras*, Vol. 1 (Academic Press, San Diego 1983; AMS reprint 2000)

[23] S. Kochen, E. P. Specker, "The problem of hidden variables in quantum mechanics", Journal of Mathematics and Mechanics **17** (1967), 59-87

[24] J. Lambek, P. J. Scott, *Introduction to higher order categorical logic* (Cambridge University Press, Cambridge, New York etc. 1986)

[25] F. W. Lawvere, R. Rosebrugh, *Sets for Mathematics* (Cambridge University Press, Cambridge 2003)

[26] S. Mac Lane, *Categories for the Working Mathematician* (Springer, New York, Berlin etc. 1971; second, revised edition 1997)

[27] S. Mac Lane, I. Moerdijk, *Sheaves in Geometry and Logic, A First Introduction to Topos Theory*, Springer (New York, Berlin etc. 1992)

[28] M. P. Olson, "The Selfadjoint Operators of a von Neumann Algebra Form a Conditionally Complete Lattice", Proc. of the AMS **28** (1971), 537-544

[29] M. H. Stone, "The theory of representations for Boolean algebras", Trans. Amer. Math. Soc. **40** (1936), 37-111

Andreas Döring
Theoretical Physics Group
Blackett Laboratory
Imperial College of Science, Technology & Medicine
Prince Consort Road
South Kensington
London SW7 2BZ

e-mail: a.doering@imperial.ac.uk

Quantum Field Theory
B. Fauser, J. Tolksdorf and E. Zeidler, Eds., 49–66
© 2009 Birkhäuser Verlag Basel/Switzerland

A Survey on Mathematical Feynman Path Integrals: Construction, Asymptotics, Applications

Sergio Albeverio and Sonia Mazzucchi

Abstract. Theory and main applications of infinite dimensional oscillatory integrals are discussed, with special attention to the mathematical realization of Feynman path integrals

Mathematics Subject Classification (2000). Primary 28C20; Secondary 35Q40; 46F20; 60G60; 81S40; 81T10; 81T45; 81P15; 35C20.

Keywords. oscillatory integrals, Feynman path integrals, stationary phase, polynomial potentials, semiclassical expansions, trace formula, Chern-Simons model.

1. Introduction

The study of oscillatory integrals of the form

$$I^{\frac{\Phi}{\epsilon}}(f) = \text{``} \int_{\mathbb{R}^N} e^{i\frac{\Phi}{\epsilon}(x)} f(x)dx \text{''} \tag{1}$$

is a classical topic, largely developed in connections with several applications in mathematics, such as the theory of Fourier integral operators [51, 66], and in physics (for instance in optics, see, e.g., [26]). In the expression on the right hand side of (1) Φ denotes a real valued "phase function", f a complex valued function and $\epsilon \in \mathbb{R}^+$ a real parameter. Well known examples of integrals of the form (1) are the Fresnel integrals

$$\int_{\mathbb{R}} e^{ix^2} f(x)dx$$

applied in the theory of wave diffraction and the Airy integrals

$$\int_{\mathbb{R}} e^{ix^3} f(x)dx$$

applied in the theory of the rainbow. The fundamental feature of the integral (1) is the oscillatory behavior of the term $e^{i\frac{\Phi}{\epsilon}}$, which allows to define the integral as an improper Riemann integral even if the function f is not summable, by exploiting the cancelations due to the alternating of the sign of the integrand. This property makes oscillatory integrals the suitable mathematical objects representing the physical concept of coherent superposition, that is of interference. It is thus not surprising that, besides optics, electromagnetism and hydrodynamics, one of the most suggestive and powerful applications of oscillatory integrals can be found in non relativistic quantum mechanics. Indeed in 1942 R. Feynman [38], inspired by a paper by Dirac [35], proposed an alternative formulation of quantum mechanics using heuristic infinite dimensional oscillatory integrals. According to Feynman the solution of the Schrödinger equation, describing the time evolution of the state of a quantum particle moving under the influence of a (real-valued) potential V

$$\begin{cases} i\hbar\frac{\partial}{\partial t}\psi = -\frac{\hbar^2}{2m}\Delta\psi + V\psi \\ \psi(0,x) = \psi_0(x) \end{cases} \tag{2}$$

(m denotes the mass of the particle and $\hbar$ the reduced Planck constant), should be given by a "sum over all possible histories". In other words the wave function of the system at time t evaluated at the point $x \in \mathbb{R}^d$ should be given by an integral on the space of continuous paths γ ending at time t at the point x:

$$\psi(t,x) = \quad `` \int_{\{\gamma|\gamma(t)=x\}} e^{\frac{i}{\hbar}S_t(\gamma)}\psi_0(\gamma(0))D\gamma \quad ''. \tag{3}$$

$S_t(\gamma)$ is the classical action of the system evaluated along the path γ

$$S_t(\gamma) = S^0(\gamma) - \int_0^t V(s,\gamma(s))ds, \qquad S^0(\gamma) = \frac{m}{2}\int_0^t |\dot{\gamma}(s)|^2 ds$$

and $D\gamma$ denotes a heuristic "flat" measure on the space of paths. The heuristic expression (3) can be regarded as an infinite dimensional analogue of the integral (1):

$$I^{\frac{\Phi}{\epsilon}} = \quad `` \int_\Gamma e^{i\frac{\Phi}{\epsilon}(\gamma)}f(\gamma)d\gamma \quad '', \tag{4}$$

that is an oscillatory integral on an (infinite dimensional) space of paths Γ with $\epsilon = \hbar$. The Feynman path integral approach of quantum mechanics is particularly suggestive as it creates a connection between the classical Lagrangian description of the physical world and the quantum one, reintroducing in quantum mechanics the concept of trajectory, which had been banned by the traditional formulation of quantum theory. Formula (3) provides a quantization method allowing, at least heuristically, to associate to each classical Lagrangian a quantum evolution. Feynman himself extended the path integral approach to the description of the dynamics of more general quantum systems, including the quantum fields, and producing a heuristic calculus that, from a physical point of view, often works even in cases where other methods fail.

Another feature of oscillatory integrals which makes representation (3) particularly interesting is the existence of a well known theory allowing one to study the asymptotic behavior of integrals (1) when ϵ is regarded as a small parameter converging to 0. Originally introduced by Stokes and Kelvin and successively developed by several mathematicians, in particular van der Corput (who, by the way, was particularly interested in applications to number theory), the "stationary phase method" provides a powerful tool to handle the asymptotics of (1) as $\epsilon \downarrow 0$. According to it, the main contribution to the asymptotic behavior of the integral should come from those points $x \in \mathbb{R}^N$ belonging to the critical manifold:

$$\{x \in \mathbb{R}^N, \mid \Phi'(x) = 0\},$$

that is the points which make stationary the phase function Φ. The asymptotic analysis of oscillatory integrals is successfully applied to different areas of mathematics such as the theory of (partial) differential equations, the singularity theory and the number theory.

The extension of these techniques to the infinite dimensional case and in particular to the Feynman formula (3) makes very intuitive the study of the semi-classical limit of quantum mechanics, that is the study of the detailed behavior of the wave function ψ in the case the Planck constant $\hbar$ is regarded as a small parameter. According to an (heuristic) application of the stationary phase method, in the limit $\hbar \downarrow 0$ the main contribution to the integral (3) should come from those paths γ which make stationary the action functional S_t. These, by Hamilton's least action principle, are exactly the classical orbits of the system. Moreover the heuristic applications of the asymptotic expansion of Feynman path integrals (and related Euclidean path integrals) in quantum field theory gives interesting results connected to the study of solitons resp. instantons, resp. in the case of certain gauge fields, of topological invariants (see, e.g., [73, 74, 59, 68] for physical discussion and e.g. , [76], [41] and [3] for Euclidean path integrals).

Despite its fascinating features, formula (3) lacks of mathematical rigor, indeed the Lebesgue "flat" measure $D\gamma$ on the space on paths does not have a mathematical meaning (it is quite simple to see that on an infinite dimensional Hilbert space it is not possible to construct a Lebesgue-type measure, that is a σ-additive regular measure which is invariant by translations and rotations and such that the measure of bounded open sets is strictly positive and finite). Feynman himself was aware of this problem as he writes "one must feel as Cavalieri must have felt calculating the volume of a pyramid before the invention of the calculus". The challenge to give meaning to Feynman's heuristic calculus and to define rigorously oscillatory integrals (4) in infinite dimension, as well as to develop an infinite dimensional version of the stationary phase method, was left to mathematicians (see also, e.g. [3, 50, 58]).

The difficulties are twofold:

- First of all one has to define an integration theory on a space of paths, that is on an infinite dimensional space. We recall that integration theory in spaces of continuous functions was present at Feynman's time thanks in particular to the work by Wiener on Brownian motion in the 20's (particularly in 1923) and much successive work by Cameron, Martin and others giving rise to the theory of stochastic processes, see e.g.[3]. However there is no mention of Wiener integrals in Feynman's papers.

- In the definition of Feynman path integrals one should exploit the oscillatory behavior of the integrand. In principle the convergence of the integral should be given by the cancelations due to this oscillatory behavior.

In the next section we shall briefly describe a solution of this problem, while in section 3 several interesting applications of Feynman path integrals to quantum mechanics and quantum fields will be mentioned. Due to limitation of space and time, we will mainly concentrate on lines of research directly connected to our own ones and with recent developments. In particular some very interesting topics have to be left out, e.g. extensions to Dirac systems see, e.g. [54, 55], hyperbolic systems [78], oscillatory complex Gaussian integrals, see also e.g., the extensive references in [3, 12, 58, 61].

2. The mathematical realization of Feynman path integrals

In this section we shall see how the definition and the main properties of classical oscillatory integrals in finite dimension can be extended to the case where the integration is performed on an infinite dimensional Hilbert space.

A systematic treatment of finite dimensional oscillatory integrals, as well as their application to the theory of Fourier integral operators, can be found in the work by Hörmander [51] (see also, e.g., [36] and [66]). According to Hörmander, the integral (1) can be computed even when the function f is not summable by exploiting the cancelations due to the oscillatory behavior of the integrand. The oscillatory integral is defined as the limit of a sequence of regularized integrals.

Definition 1. *Let* $f : \mathbb{R}^n \to \mathbb{C}$ *be a Borel function and* $\Phi : \mathbb{R}^n \to \mathbb{R}$ *a phase function. If for each test function* $\phi \in \mathcal{S}(\mathbb{R}^n)$ *such that* $\phi(0) = 1$ *the integrals*

$$I_\delta(f, \phi) := \int_{\mathbb{R}^n} (2\pi i \epsilon)^{-n/2} e^{i \frac{\Phi(x)}{\epsilon}} f(x) \phi(\delta x) dx$$

exist for all $\delta > 0$ *and* $\lim_{\delta \to 0} I_\delta(f, \phi)$ *exists and is independent of* ϕ*, then the limit is called the oscillatory integral of* f *with respect to* $\frac{\Phi}{\epsilon}$ *and denoted by*

$$\widetilde{\int_{\mathbb{R}^n}} e^{i \frac{\Phi(x)}{\epsilon}} f(x) dx \equiv I^{\frac{\Phi}{\epsilon}}(f). \tag{5}$$

In the special case where the phase function is a quadratic form, the oscillatory integral is called Fresnel integral (following the name given to certain integrals in optics, see [12]).

The existence of the integral $I^{\frac{\Phi}{\epsilon}}(f)$ can be proved for large classes of functions Φ, f [51, 52, 36, 66, 17], even though a complete direct characterization of the class for which the integral is well defined is still an open problem. However, for suitable

Φ, it is possible to find an interesting set of "integrable functions", for which the oscillatory integral $I^\Phi(f)$ can be explicitly computed in terms of an absolutely convergent integral thanks to a Parseval-type equality.

Given a (finite or infinite dimensional) real separable Hilbert space $(\mathcal{H}, \langle \, , \, \rangle)$, whose elements are

denoted by $x, y \in \mathcal{H}$, let us denote by $\mathcal{F}(\mathcal{H})$ the space of complex functions on $\mathcal{H}$ which are Fourier transforms of complex bounded variation measures on $\mathcal{H}$:

$$f \in \mathcal{F}(\mathcal{H}), \qquad f(x) = \int_{\mathcal{H}} e^{i\langle x, y\rangle} d\mu_f(y).$$

$\mathcal{F}(\mathcal{H})$ is a Banach algebra of functions, where the product is the pointwise one, the unit element is the function identically one 1, i.e. $1(x) = 1 \; \forall x \in \mathcal{H}$ and the norm of a function f is given by the total variation of the corresponding measure μ_f:

$$\|f\|_{\mathcal{F}(\mathcal{H})} = \|\mu_f\| = \sup \sum_i |\mu(E_i)|,$$

where the supremum is taken over all sequences $\{E_i\}$ of pairwise disjoint Borel subsets of $\mathcal{H}$, such that $\cup_i E_i = \mathcal{H}$. It is possible to prove [17] that if $f \in \mathcal{F}(\mathbb{R}^n)$, $f = \hat\mu_f$, and for phase functions Φ such that $F^\Phi \equiv \frac{e^{i\frac{\Phi}{\epsilon}}}{(2\pi i \epsilon)^{n/2}}$ has a Fourier transform $\hat{F}^\Phi$ which is integrable (in Lebesgue sense) with respect to μ_f, then the oscillatory integral $I^\Phi(f)$ exists and it is given by the following "Parseval formula":

$$I^\Phi(f) = \int_{\mathbb{R}^n} \hat{F}^\Phi(\alpha) d\mu_f(\alpha). \tag{6}$$

Equation (6) holds for smooth phase functions Φ of at most even polynomial growth at infinity.
Equation (6) allows, in the case Φ is an homogeneous polynomial and under regularity assumptions on the function f, to compute (see [17]) the detailed asymptotic expansion of the integral $I^{\frac{\Phi}{\hbar}}(f)$ in fractional powers of the parameters $\hbar$ around a degenerate (or non degenerate) critical point, with a strong control on the remainders. The study of the asymptotics of oscillatory integrals in the case where the critical manifold contains degenerate critical points is related with singularity theory and catastrophe theory [27, 36, 26].

The generalization of these results to the infinite dimensional case involves several technical difficulties. Indeed in the 60's Cameron [28] proved that it is not possible to realize the heuristic Feynman measure

$$d\mu_F(\gamma) \equiv \frac{e^{\frac{i}{\hbar}S^0(\gamma)} D\gamma}{\int e^{\frac{i}{\hbar}S^0(\gamma)} D\gamma}$$

as a complex measure on the space of paths, as it would have infinite total variation (even locally). This means that for Feynman integrals it is not possible to implement an integration theory in the traditional Lebesgue sense.

The main idea to overcome this problem is dualization. In other words the integral

$$\text{``}\int f(\gamma)\frac{e^{\frac{i}{\hbar}S^0(\gamma)}D\gamma}{\int e^{\frac{i}{\hbar}S^0(\gamma)}D\gamma}\text{''} = \text{``}\int f(\gamma)d\mu_F(\gamma)\text{''} := I_F(f)$$

has to be realized as a linear continuous functional on a suitable topological algebra of "integrable functions", generalizing the idea of (Radon) measure as linear functional on the space of bounded continuous functions (on locally compact spaces). Among the different approaches to the mathematical definition of the "Feynman functional", the most implemented are the theory of infinite dimensional oscillatory integrals on Hilbert (resp. Banach spaces) [12] and the white noise calculus [50]. In the following we shall extensively describe the first approach and give some elements of the latter.

Let us denote by $(\mathcal{H}, \langle \ , \ \rangle)$ an infinite dimensional real separable Hilbert space, $\Phi : \mathcal{H} \to \mathbb{R}$ a phase function and $f : \mathcal{H} \to \mathbb{C}$ a complex valued function. An infinite dimensional oscillatory integral on the Hilbert space $\mathcal{H}$ can be defined as the limit of a sequence of finite dimensional approximations, as proposed in [37, 7].

Definition 2. *A function $f : \mathcal{H} \to \mathbb{C}$ is said to be integrable with respect to the phase function $\Phi : \mathcal{H} \to \mathbb{R}$ if for any sequence P_n of projectors onto n-dimensional subspaces of $\mathcal{H}$, such that $P_n \leq P_{n+1}$ and $P_n \to 1$ strongly as $n \to \infty$ (1 being the identity operator in $\mathcal{H}$), the finite dimensional approximations*

$$\widetilde{\int_{P_n\mathcal{H}}} e^{i\frac{\Phi(P_n x)}{\epsilon}} f(P_n x)d(P_n x),$$

are well defined (in the sense of definition 1) and the limit

$$\lim_{n\to\infty} \widetilde{\int_{P_n\mathcal{H}}} e^{i\frac{\Phi(P_n x)}{\epsilon}} f(P_n x)d(P_n x) \tag{7}$$

exists and is independent of the sequence $\{P_n\}$.
In this case the limit is called oscillatory integral of f with respect to the phase function Φ and is denoted by

$$I^\Phi(f) \equiv \widetilde{\int_{\mathcal{H}}} e^{i\frac{\Phi(x)}{\epsilon}} f(x)dx.$$

As in the finite dimensional case, the basic question is the characterization of the class of functions f and Φ for which the integral $I^\Phi(f)$ is well defined. According to [12, 37, 7], in the case where f belongs to the Banach algebra $\mathcal{F}(\mathcal{H})$ of $\mathbb{C}$-valued functions on $\mathcal{H}$ which are Fourier transform of bounded complex measures on $\mathcal{H}$ and Φ is of the form $\Phi = \Phi_0 + V$, where Φ_0 is a quadratic form and $V \in \mathcal{F}(\mathcal{H})$, then $I^\Phi(f)$ can be explicitly computed in terms of a Parseval type equality. In the following the small parameter ϵ will be replaced by $\hbar$ because of its meaning in the applications to quantum mechanics.

Theorem 1. *Let $L : \mathcal{H} \to \mathcal{H}$ be a self adjoint trace class operator, such that $I - L$ is invertible. Let $f, V \in \mathcal{F}(\mathcal{H})$. Let us consider the phase function $\Phi : \mathcal{H} \to \mathbb{C}$, given by*

$$\Phi(x) = \Phi_0(x) + V(x), \qquad x \in \mathcal{H} \tag{8}$$

with $\Phi_0(x) = \frac{\langle x, (I-L)x \rangle}{2}$. Then the infinite dimensional oscillatory integral $I^\Phi(f)$ is well defined and it is given by

$$\widetilde{\int_{\mathcal{H}}} e^{\frac{i}{\hbar}\Phi(x)} f(x) dx = \widetilde{\int_{\mathcal{H}}} e^{\frac{i}{2\hbar}\langle x,(I-L)x\rangle - \frac{i}{\hbar}V(x)} f(x) dx$$

$$= (\det(I - L))^{-1/2} \int_{\mathcal{H}} e^{-\frac{i\hbar}{2}\langle x,(I-L)^{-1}x\rangle} \mu_{fe^{-\frac{i}{\hbar}V}}(dx).$$

$\det(I - L)$ *is the Fredholm determinant of the operator $(I - L)$ (that is the product of the eigenvalues of $I - L$). The right hand side is explicitly computable (e.g. by an expansion in powers of $e^{-\frac{i}{\hbar}V}$) as $fe^{-\frac{i}{\hbar}V} \in \mathcal{F}(\mathcal{H})$.*

This result has been recently generalized [18] to phase functions of the form

$$\Phi(x) = \Phi_0(x) + \lambda P(x), \qquad x \in \mathcal{H}, \tag{9}$$

where Φ_0 is of the type handled in theorem 1, $\lambda \in \mathbb{R}^+$ and P is a fourth order polynomial. In this case the integral $I^\Phi(f)$, $f \in \mathcal{F}(\mathcal{H})$, is still computable in terms of a Parseval type equality:

$$I^\Phi(f) = \widetilde{\int_{\mathcal{H}}} e^{\frac{i}{\hbar}\Phi(x)} f(x) dx = \int_{\mathcal{H}} \hat{F}^\Phi(x) d\mu_f(x) \tag{10}$$

with $\hat{F}^\Phi$ defined by

$$\hat{F}^\Phi(x) = \lim_{n\to\infty} (2\pi i\hbar)^{-n/2} \int_{P_n\mathcal{H}} e^{i\langle P_n x, P_n y \rangle} e^{\frac{i}{\hbar}\Phi(P_n y)} dP_n y$$

$$= \mathbb{E}[e^{ie^{i\frac{\pi}{4}}\sqrt{\hbar}\langle x, \cdot \rangle} e^{\frac{1}{2}\langle \cdot, L \cdot \rangle} e^{-i\lambda\hbar P(\cdot)}] \tag{11}$$

where the expectation is taken with respect to $N(0, I_\mathcal{H})$, the centered standard Gaussian measure associated with $\mathcal{H}$. Analogously one has

$$I^\Phi(f) = \mathbb{E}[f(e^{i\frac{\pi}{4}}\sqrt{\hbar} \cdot) e^{\frac{1}{2}\langle \cdot, L \cdot \rangle} e^{-i\lambda\hbar P(\cdot)}]. \tag{12}$$

It is possible to prove that the right hand side of (12) is an analytic function for $Im(\lambda) < 0$ and continuous for $Im(\lambda) = 0$. Moreover a corresponding result can be proved for $\Phi = \Phi_0 + \lambda P + V$, with $V \in \mathcal{F}(\mathcal{H})$.

The infinite dimensional oscillatory integrals have some interesting properties, which are important as they mirror some heuristic features of formal Feynman path integrals, in the case where the Hilbert space $\mathcal{H}$ is a space of paths γ. First of all they have simple transformation properties under "translations and rotations" in paths space, reflecting the fact that dx should represent a flat measure. They satisfy a Fubini-type theorem, concerning iterated integration in paths space, allowing, in the physical applications, to construct a one parameter group of unitary operators associated to the time evolution described by the Schrödinger equation.

They can be approximated by finite dimensional oscillatory integrals, allowing a sequential ("time slicing") approach very close to Feynman's original derivation. They are also related, via the Parseval type equality (12), to (Gaussian) probabilistic integrals, allowing an "analytic continuation approach" (largely developed by several authors [28, 29, 56, 57, 58, 69, 14, 8, 79]).

Moreover, the functional $I^\Phi(f)$ satisfies a "duality property", in other words the application

$$f \in \mathcal{F}(\mathcal{H}) \mapsto I^\Phi(f)$$

is continuous in the norm of the Banach algebra $\mathcal{F}(\mathcal{H})$. The duality property is central also in other approaches, e.g. [30] and [15]. It is also the main idea of the white noise calculus approach [50], where the Feynman functional $I^\Phi(f)$ is realized as the pairing between the function f and an infinite dimensional distribution T_Φ (in the framework of the Hida calculus):

$$I^\Phi(f) = \, {}_{(\mathcal{S}')}\langle T_\Phi, f \rangle_{(\mathcal{S})}$$

where the Gelfand triple is

$$(\mathcal{S}) \subset L^2(\mathcal{S}'(\mathbb{R}^s), N(0, I_{L^2(\mathbb{R}^s)})) \subset (\mathcal{S}'),$$

$N(0, I_{L^2(\mathbb{R}^s)})$ being the Gaussian white noise measure, $(\mathcal{S}) = (\mathcal{S}(\mathbb{R}^s))$ being essentially an infinite dimensional analogue of $\mathcal{S}(\mathbb{R}^s) \subset L^2(\mathbb{R}^s)$, and correspondingly $(\mathcal{S}') = (\mathcal{S}'(\mathbb{R}^s))$ an infinite dimensional analogue of $\mathcal{S}'(\mathbb{R}^s)$, the space of tempered distributions.

It is worthwhile to underline that, within all the approaches, the phase function Φ which can be handled are essentially of the form (8), except for the possibility of including functions V of the form of a Laplace transform of bounded measures, and some singular V, see, for example [12, 15, 50, 63]. The only approach which is able to handle a (quartic) polynomial term as in (9) and the corresponding method of stationary phase seems to be the infinite dimensional oscillatory integrals approach described above.

3. Applications

3.1. Quantum mechanics

The first application of the infinite dimensional oscillatory integrals is the mathematical realization of the Feynman path integral representation for the solution of the Schrödinger equation. Let us consider the Cameron-Martin space $\mathcal{H}_t$, that is the Hilbert space of absolutely continuous paths $\gamma : [0,t] \to \mathbb{R}^d$ with $\gamma(t) = 0$ and square integrable weak derivative $\int_0^t \dot\gamma(s)^2 ds < \infty$, endowed with the inner product

$$\langle \gamma_1, \gamma_2 \rangle = \int_0^t \dot\gamma_1(s) \cdot \dot\gamma_2(s) ds.$$

Let us consider Schrödinger equation (2) with a potential V which is the sum of an harmonic oscillator term and a bounded perturbation which is the Fourier transform of a complex bounded variation measure on $\mathbb{R}^d$:

$$V(x) = \frac{1}{2} x \Omega^2 x + v(x), \qquad x \in \mathbb{R}^d, v \in \mathcal{F}(\mathbb{R}^d),$$

with Ω^2 being a positive symmetric $d \times d$ matrix.

By taking as initial datum $\psi_0 \in L^2(\mathbb{R}^d)$ belonging to $\mathcal{F}(\mathbb{R}^d)$, it is possible to prove that the infinite dimensional oscillatory integral with respect to the quadratic phase function $\Phi(\gamma) = \frac{\langle \gamma, \gamma \rangle}{2\hbar}$, $\gamma \in \mathcal{H}_t$, of the functional f on the Cameron-Martin space given by

$$\gamma \mapsto f(\gamma) = e^{-\frac{i}{2\hbar} \int_0^t (\gamma(s)+x)\Omega^2(\gamma(s)+x)ds} e^{-\frac{i}{\hbar} v(\gamma(s)+x)ds)} \psi_0(\gamma(0) + x), \qquad \gamma \in \mathcal{H}_t$$

can be computed by means of a Parseval type equality and is a representation of the solution of the Schrödinger equation:

$$\psi(t, x) = \widetilde{\int_{\mathcal{H}_t}} e^{i \frac{\langle \gamma, \gamma \rangle}{2\hbar}} f(\gamma) d\gamma$$

$$= \widetilde{\int_{\mathcal{H}_t}} e^{i \frac{\langle \gamma, \gamma \rangle}{2\hbar}} e^{-\frac{i}{2\hbar} \int_0^t (\gamma(s)+x)\Omega^2(\gamma(s)+x)ds} e^{-\frac{i}{\hbar} v(\gamma(s)+x)ds)} \psi_0(\gamma(0) + x) d\gamma$$

$$= \text{``} \int_{\{\gamma | \gamma(t) = x\}} e^{\frac{i}{\hbar} S_t(\gamma)} \psi_0(\gamma(0)) D\gamma \text{''}. \tag{13}$$

The study of the asymptotics of the integral (13) in the limit $\hbar \downarrow 0$ is directly related to the study of the "semiclassical expansions" of quantum mechanics. The first rigorous results on the generalization of the method of the stationary phase to the infinite dimensional case can be found in [13]. These results were further developed in [72, 7]. By considering an infinite dimensional oscillatory integral of this form

$$I(\hbar) := \widetilde{\int_{\mathcal{H}}} e^{\frac{i}{2\hbar} \langle x, x \rangle} e^{-\frac{i}{\hbar} V(x)} f(x) dx$$

where $V, f \in \mathcal{F}(\mathcal{H})$ satisfy suitable regularity and growing conditions, one can prove that

- the phase function $\Phi(x) = \langle x, x \rangle / 2 - V(x)$ has only non degenerate stationary points
- the oscillatory integral $I(\hbar)$ is a C^∞ function of the parameter $\hbar$
- its asymptotic expansion in powers of $\hbar$ when $\hbar \to 0$ depends only on the derivatives of V and f at the critical points.

It is important to underline that, under additional assumptions on V, it is possible to prove that the asymptotic expansion is Borel summable so that it allows the unambiguous determination of the function $I(\hbar)$ itself [72]. These results can be applied to the study of the asymptotic behavior of the solution of the Schrödinger equation in the limit where the Planck constant $\hbar$ is regarded as a small parameter converging to 0. In fact, by assuming that the potential v is the Fourier transform of a complex bounded variation measure on $\mathbb{R}^d$ and that the initial datum has the following form

$$\psi_0(x) = e^{\frac{i}{\hbar}s(x)}\chi(x), \qquad x \in \mathbb{R}^d$$

with $\chi \in C_0^\infty(\mathbb{R}^d)$ and $s \in C^\infty(\mathbb{R}^d)$, it is possible to prove that the infinite dimensional oscillatory integral representation for the solution of the Schrödinger equation

$$\widetilde{\int_{\mathcal{H}_t}} e^{\frac{i}{2\hbar}\langle\gamma,\gamma\rangle} e^{-\frac{i}{2\hbar}\int_0^t (\gamma(s)+x)\Omega^2(\gamma(s)+x)ds} e^{-\frac{i}{\hbar}\int_0^t v(\gamma(s)+x)ds} e^{\frac{i}{\hbar}s(\gamma(0)+x)} \chi(\gamma(0)+x)d\gamma$$

has an asymptotic expansion in powers of $\hbar$, depending only on classical features of the system. This technique yields an independent (Feynman path integral) rigorous derivation of Maslov's results on the WKB-type asymptotics of the solution of Schrödinger equation, with, in addition, strong control on the remainders.

Another interesting application of the infinite dimensional stationary phase method is the study of the trace of the Schrödinger group

$$\mathrm{Tr}[e^{-\frac{i}{\hbar}Ht}]$$

and its asymptotic behavior when $\hbar \to 0$ [5, 6]. For potentials of the (usual) type "harmonic oscillator plus Fourier transform of measure" it is possible to prove a trace formula of Gutzwiller's type, relating the asymptotic of the trace of the Schrödinger group and the spectrum of the quantum mechanical energy operator H with the classical periodic orbits of the system. Gutwiller's trace formula, which is a basis of the theory of quantum chaotic systems, is the quantum mechanical analogue of Selberg's trace formula, relating the spectrum of the Laplace-Beltrami operator on manifolds with constant negative curvature with the periodic geodesics on those manifolds.

Infinite dimensional oscillatory integrals can be also applied to the quantum theory of open systems, in particular to the mathematical realization of the "Feynman-Vernon influence functional". Let us consider a quantum system A, with state space $L^2(\mathbb{R}^d)$, interacting with a quantum system B, with state space $L^2(\mathbb{R}^N)$, representing a reservoir. Let us assume that the total Hamiltonian of the compound system is $H_{AB} = H_A + H_B + H_I$, where H_A and H_B are both Hamiltonians describing harmonic oscillators perturbed by bounded potentials v_A, v_B belonging to $\mathcal{F}(\mathbb{R}^r)$ and $\mathcal{F}(\mathbb{R}^N)$ respectively. The interaction Hamiltonian is of the form $H_I = x_A C x_B$, where $x_A \in \mathbb{R}^d$ and $x_B \in \mathbb{R}^N$ represent the spatial coordinates of the system A and B respectively, while $C : \mathbb{R}^N \to \mathbb{R}^d$ is a matrix. Under suitable assumptions on the initial state of the compound system it is possible to prove that the reduced density operator kernel of the system A (obtained by tracing out the environmental coordinates) is heuristically given by

$$\rho_A(t,x,y) = \text{``}\int_{\substack{\gamma(t)=x \\ \gamma'(t)=y}} e^{\frac{i}{\hbar}(S_A(\gamma)-S_A(\gamma'))} F(\gamma,\gamma')\rho_A(\gamma(0),\gamma'(0))\mathrm{D}\gamma\mathrm{D}\gamma'\text{''}, \qquad (14)$$

where F is the formal Feynman-Vernon influence functional

$$F(\gamma, \gamma') \tag{15}$$
$$= \text{``} \int_{\substack{\Gamma(t)=Q \\ \Gamma'(t)=Q}} e^{\frac{i}{\hbar}(S_B(\Gamma)-S_B(\Gamma'))} e^{\frac{i}{\hbar}(S_I(\Gamma,\gamma)-S_I(\Gamma',\gamma'))} \rho_B(\Gamma(0),\Gamma'(0)) D\Gamma D\Gamma' dQ \text{''}.$$

Under suitable assumptions on the initial state of the compound system it is possible to prove that the heuristic formula (14) can be realized in terms of an infinite dimensional oscillatory integral on the Cameron-Martin space $\mathcal{H}_t$ [9]. This result has been applied to the study of the Caldeira-Leggett model for the description of the quantum Brownian motion.

An alternative description of a quantum system interacting with an external environment is the stochastic Schrödinger equation, where the influence of the reservoir is modeled by a noise term. Among the large number of stochastic Schrödinger equations proposed by several authors, we consider for instance the Belavkin equation, describing the time evolution of a quantum particle submitted to the measurement of its position:

$$\begin{cases} d\psi(t,x) = -\frac{i}{\hbar}H\psi(t,x)dt - \frac{\lambda}{2}x^2\psi(t,x)dt + \sqrt{\lambda}x\psi(t,x)dW(t) \\ \psi(0,x) = \psi_0(x) \qquad (t,x) \in [0,T] \times \mathbb{R}^d, \end{cases} \tag{16}$$

(where $\lambda > 0$ is a coupling constant and W is a d-dimensional Brownian motion). It is possible to prove that the solution of the stochastic Schrödinger equation admits a Feynman path integral representation in terms of a well defined infinite dimensional oscillatory integral [10], providing a rigorous mathematical realization of the heuristic formula [67, 16] for the state of the system in the case the observed trajectory is the path ω:

$$\psi(t,x,\omega) = \text{``} \int_{\{\gamma(t)=x\}} e^{\frac{i}{\hbar}S_t(\gamma)} e^{-\lambda \int_0^t (\gamma(s)-\omega(s))^2 ds} \psi(0,\gamma(0)) D\gamma \text{''} \tag{17}$$

One can see that, as an effect of the correction term $e^{-\lambda \int_0^t (\gamma(s)-\omega(s))^2 ds}$ due to the measurement, the paths giving the main contribution to the integral are those closer to the observed trajectory ω.

The extensions of these results to the case where the potential V in the Schrödinger equation (2) has polynomial growth, i.e. $V(x) = \frac{1}{2}x\Omega^2 x + \lambda x^{2N}$ with $\Omega : \mathbb{R}^d \to \mathbb{R}^d$, $\lambda > 0$ and $x \in \mathbb{R}^d$, has been recently obtained in the case $2N = 4$ [18], also when both Ω and λ are time-dependent [19]. The extension of the method of the stationary phase to oscillatory integrals with polynomial phase function is rather delicate and still under study. First results in this direction concerning the trace of the heat semigroup $\text{Tr}[e^{-\frac{t}{\hbar}H}]$, $t > 0$, with $H = -\frac{\hbar^2}{2}\Delta + \lambda|x|^{2N}$, can be found in [20], where the case of a degenerate critical point of the phase function is handled.

3.2. Quantum fields

Heuristic Feynman path integrals have been applied to many problems in quantum field theory. A particularly interesting application of Feynman path integrals can be found in a paper by Witten [80], who conjectured that there should be a connection between quantum gauge field theories on a 3-dimensional manifold based on the Chern-Simons action (an object originally introduced for pure differential geometric - topological considerations) and the Jones polynomial, a link invariant. In the Feynman path integral formulation of Chern-Simons theory, the integration is performed on a space of geometric objects, i.e. on a space of connections.

Let M be a smooth 3-dimensional oriented manifold without boundary, let G be a compact connected Lie group (the "gauge group") with a finite dimensional representation R (in the following the group elements will be identified with their representatives). Let us denote by g the Lie algebra of G, by Γ the space of g-valued connection 1-forms and by $A \in \Gamma$ its elements. Let $S_{CS} : \Gamma \to \mathbb{R}$ be the Chern-Simons action, defined by

$$S_{CS}(A) \equiv \frac{k}{4\pi} \int_M \mathrm{Tr}\Big(A \wedge dA + \frac{2}{3} A \wedge A \wedge A\Big), \tag{18}$$

where k is a non-zero real constant and the trace is evaluated in the given representation R. The application S_{CS} is metric independent and invariant under diffeomorphisms. The couple $(A, S(A))$ represents a classical topological gauge field.

Let us consider the functions $f : \Gamma \to \mathbb{C}$ of the form

$$f(A) \equiv \prod_{i=1}^{n} \mathrm{Tr}(Hol(A, l_i)),$$

where $(l_1, ..., l_n)$, is an n-tuple of loops in M whose arcs are pairwise disjoint and $Hol(A, l)$ denotes the holonomy of A around l. According to Witten's conjecture the heuristic Feynman integrals

$$``\int_\Gamma e^{iS_{CS}(A)} f(A) \mathrm{D}A\,\,''\tag{19}$$

should represent topological invariants.

In the case where $M = S^3$ and $G = SU(2)$, the heuristic integral (19) should give the Jones polynomials, if $G = SU(n)$ the Homfly polynomials and if $G = SO(n)$ the Kauffman polynomials (all objects of the theory of knots).

Gauge transformations can be given by differentiable functions $\chi : M \to G$ which act on a connection A by

$$A \mapsto \chi^{-1} A \chi + \chi^{-1} d\chi.$$

The Chern-Simons functional $S_{CS}(A)$ changes to

$$S_{CS}(\chi^{-1} A \chi + \chi^{-1} d\chi) = S_{CS}(A) - k W_M(\chi)$$

with

$$W_M(\chi) \equiv \frac{1}{12\pi} \int_M \mathrm{Tr}(\chi^{-1}\mathrm{d}\chi \wedge \chi^{-1}\mathrm{d}\chi \wedge \chi^{-1}\mathrm{d}\chi).$$

It can be shown that the function $e^{iS} : \Gamma \to \mathbb{C}$ is gauge invariant if and only if the quantization condition

$$k\,W_M(\chi) \subseteq 2\pi\mathbb{Z} \qquad \forall \chi \tag{20}$$

is satisfied (see, e.g., [70]). For abelian G the quantity $W_M(\chi)$ vanishes, but for a general semisimple Lie group G this quantity does not vanish and the quantization condition of the coupling constant (20) has to be required [70].

A partial mathematical realization of Witten's theory was provided in the case $M = \mathbb{R}^3$ by Fröhlich and King [39]. Rigorous "algebraic" results, without however any direct relation to Witten's heuristic path integral approach, can be found in [71].

In the framework of the theory of infinite dimensional oscillatory integrals, the rigorous mathematical realization of the integral (19)

$$`` \int_\Gamma e^{iS_{CS}(A)} f(A)\mathrm{D}A \,\, '' = I^\Phi(f)$$

can be implemented by exploiting suitable gauges, at least under some restriction on (M, G).

If M is a general manifold with $H^1(M) = 0$ and G is abelian, the integral $I^\Phi(f)$ has been constructed both as an infinite dimensional oscillatory integral [23] and as a white noise functional [64]. In this case, the integral $I^\Phi(f)$ represents the Gauss linking number of knots. The case where $H^1(M) \neq 0$ has been studied by Adams [1] by simplicial methods (related to those first pointed out in [23] on the basis of [25]).

For manifolds $M = \mathbb{R}^3$ or $M = \Sigma \times \mathbb{R}$, ($\Sigma$ being a compact manifold) and G non abelian, the construction of the Feynman functional can be implemented in the framework of the white noise calculus, by exploiting a particular gauge transformation (axial gauge) [24]. In this gauge the Chern-Simons action S_{CS} loses the cubic term and the phase function Φ in the integral becomes a quadratic form. The construction of the observables, that is the integrals (19), is rather technical and has been implemented in [47]. Analogous results have been obtained in the case where $M = S^1 \times S^2$ or $M = S^1 \times \Sigma$, where Σ is an oriented surface, and G is not abelian by exploiting the "Blau-Thompson's quasi axial gauge", also called "torus gauge"[48, 49, 34]. These computations have been performed for "general colored links". First partial results on the asymptotics in $k \to \infty$ by a rigorous infinite dimensional stationary phase method applied to a regularized Chern-Simons functional expressed in terms of Wiener integrals are in [22], where connections with Vassiliev's invariants are mentioned (these invariants appear in a heuristic asymptotic expansion of the Chern-Simons oscillatory integrals). Much remains obviously to do, but it seems that lucky combination of geometrical-algebraic ideas and rigorous infinite dimensional analysis related to oscillatory integrals (and probabilistic integrals) might lead to further exiting developments.

Acknowledgements

The first author is very grateful to the organizers of a most stimulating workshop, with much of the productive interaction between mathematics and physics which is needed for progress in its fascinating area.

References

[1] R.M. Adams, R-torsion and linking number for simplicial abelian gauge theories. hep-th/9612009 1996

[2] D.H. Adams, S. Sen, Partition function of a quadratic functional and semiclassical approximation for Witten's 3-manifold invariants hep-th/9503095 1995

[3] S. Albeverio. Wiener and Feynman Path Integrals and Their Applications. Proceedings of the Norbert Wiener Centenary Congress, 163-194, *Proceedings of Symposia in Applied Mathematics* **52**, Amer. Math. Soc., Providence, RI, 1997.

[4] S. Albeverio, Ph. Blanchard, Ph. Combe, R. Høegh-Krohn, M. Sirugue, Local relativistic invariant flows for quantum fields. Comm. Math. Phys. 90 (3), 329–351 (1983).

[5] S. Albeverio, Ph. Blanchard, R. Høegh-Krohn, Feynman path integrals and the trace formula for the Schrödinger operators, Comm. Math. Phys. 83 n.1, 49–76 (1982).

[6] S. Albeverio, A.M. Boutet de Monvel-Berthier, Z. Brzeźniak, The trace formula for Schrödinger operators from infinite dimensional oscillatory integrals, Math. Nachr. 182, 21–65 (1996).

[7] S. Albeverio and Z. Brzeźniak. Finite-dimensional approximation approach to oscillatory integrals and stationary phase in infinite dimensions. *J. Funct. Anal.*, 113(1): 177-244, 1993.

[8] S. Albeverio, Z. Brzeźniak, Z. Haba. On the Schrödinger equation with potentials which are Laplace transform of measures. *Potential Anal.* **9** n.1, 65-82, 1998.

[9] S. Albeverio, L. Cattaneo, L. Di Persio, S. Mazzucchi, An infinite dimensional oscillatory integral approach to the Feynman-Vernon influence functional I, J. Math. Phys. 48, 10, 102109 (2007)

[10] S. Albeverio, G. Guatteri, S. Mazzucchi, Representation of the Belavkin equation via Feynman path integrals, Probab. Theory Relat. Fields 125, 365–380 (2003).

[11] S. Albeverio, A. Hahn, A. Sengupta, Rigorous Feynman path integrals, with applications to quantum theory, gauge fields, and topological invariants, Stochastic analysis and mathematical physics (SAMP/ANESTOC 2002), 1–60, World Sci. Publishing, River Edge, NJ, (2004).

[12] S. Albeverio and R. Høegh-Krohn. *Mathematical theory of Feynman path integrals.* 2nd edn with S. Mazzucchi, Springer-Verlag, Berlin, 1976 and 2008. Lecture Notes in Mathematics, Vol. 523.

[13] S. Albeverio and R. Høegh-Krohn. Oscillatory integrals and the method of stationary phase in infinitely many dimensions, with applications to the classical limit of quantum mechanics. *Invent. Math.* 40(1), 59-106, 1977.

[14] S. Albeverio, J.W. Johnson, Z.M. Ma, The analytic operator-value Feynman integral via additive functionals of Brownian motion, Acta Appl. Math. 42 n.3, 267–295 (1996).

[15] S. Albeverio, A. Khrennikov, O. Smolyanov, The probabilistic Feynman-Kac formula for an infinite-dimensional Schrödinger equation with exponential and singular potentials. Potential Anal. 11, no. 2, 157–181, (1999).

[16] S. Albeverio, V. N. Kolokol'tsov, O. G. Smolyanov, Continuous quantum measurement: local and global approaches. Rev. Math. Phys. 9 (1997), no. 8, 907–920.

[17] S. Albeverio and S. Mazzucchi. Generalized Fresnel Integrals. *Bull. Sci. Math.* 129 (2005), no. 1, 1–23.

[18] S. Albeverio and S. Mazzucchi. Feynman path integrals for polynomially growing potentials. J. Funct. Anal. **221** no. 1 (2005), 83–121.

[19] S. Albeverio, S. Mazzucchi,
The time dependent quartic oscillator - a Feynman path integral approach.
J. Funct. Anal. 238 , no. 2, 471–488 (2006).

[20] S. Albeverio, S. Mazzucchi, The trace formula for the heat semigroup with polynomial potential, SFB-611-Preprint no. 332, Bonn (2007).

[21] S. Albeverio, S. Mazzucchi, Theory and applications of infinite dimensional oscillatory integrals. to appear in : "Stochastic Analysis and Applications", Proceedings of the Abel Symposium 2005 in honor of Prof. Kiyosi Ito, Springer (2007).

[22] S. Albeverio and I. Mitoma. Asymptotic Expansion of Perturbative Chern-Simons Theory via Wiener Space. SFB-611-Preprint no. 322, Bonn (2007). to appear in Bull. Sci. Math. (2007).

[23] S. Albeverio, J. Schäfer, Abelian Chern-Simons theory and linking numbers via oscillatory integrals, J. Math. Phys. 36, 2157–2169 (1995).

[24] S. Albeverio, A. Sengupta, A mathematical construction of the non-Abelian Chern-Simons functional integral, Commun. Math. Phys. 186, 563–579 (1997).

[25] S. Albeverio, B. Zegarliński, Construction of convergent simplicial approximations of quantum fields on Riemannian manifolds. Comm. Math. Phys. 132 , no. 1, 39–71 (1990).

[26] V. I. Arnold, Huygens and Barrow, Newton and Hooke. Pioneers in mathematical analysis and catastrophe theory from evolvents to quasicrystals. Birkhäuser Verlag, Basel, 1990.

[27] V. I. Arnold, S. N. Gusein-Zade, A. N. Varchenko, Singularities of differentiable maps, Vol. II Birkhäuser Verlag, Basel (1988).

[28] R.H. Cameron. A family of integrals serving to connect the Wiener and Feynman integrals. *J. Math. and Phys.* 39, 126-140, 1960.

[29] R.H. Cameron, D.A. Storvick, A simple definition of the Feynman integral with applications functionals, Memoirs of the American Mathematical Society n. 288, 1–46 (1983).

[30] P. Cartier and C. DeWitt-Morette. Functional integration. *J. Math. Phys.* 41, no. 6, 4154-4187, 2000.

[31] P. Cartier and C. DeWitt-Morette. Functional integration: action and symmetries. Cambridge Monographs on Mathematical Physics. Cambridge University Press, Cambridge, 2006.

[32] I. Daubechies, J.R. Klauder, Quantum-mechanical path integrals with Wiener measure for all polynomial Hamiltonians. Phys. Rev. Lett. 52 (1984), no. 14, 1161–1164.

[33] I. Daubechies, J.R. Klauder, Quantum-mechanical path integrals with Wiener measure for all polynomial Hamiltonians. II. J. Math. Phys. 26 (1985), no. 9, 2239–2256.

[34] S. de Haro and A. Hahn. The Chern-Simons path integral and the quantum Racah formula. arXiv:math-ph/0611084.

[35] P.A.M. Dirac, The Lagrangian in quantum mechanics, Phys. Zeitschr. d. Sowjetunion, 3, No 1, 64–72, 1933.

[36] J.J. Duistermaat. Oscillatory integrals, Lagrange immersions and unfoldings of singularities. *Comm. Pure Appl. Math.* **27**, 207-281, 1974.

[37] D. Elworthy and A. Truman. Feynman maps, Cameron-Martin formulae and anharmonic oscillators. *Ann. Inst. H. Poincaré Phys. Théor.*, 41(2), 115–142, 1984.

[38] R.P. Feynman, A.R. Hibbs. *Quantum mechanics and path integrals.* Macgraw Hill, New York, 1965.

[39] J. Fröhlich, C. King. The Chern-Simons theory and knot polynomials. Comm. Math. Phys. 126, no. 1, 167–199 (1989).

[40] D. Fujiwara, N. Kumano-go, Smooth functional derivatives in Feynman path integrals by time slicing approximation,
       Bull. Sci. math. 129, 57–79 (2005).

[41] J. Glimm, A. Jaffe, Quantum physics. A functional integral point of view. Second edition. Springer-Verlag, New York, 1987.

[42] C. Grosche, F. Steiner, Handbook of Feynman path integrals. Springer Tracts in Modern Physics, 145. Springer-Verlag, Berlin, 1998.

[43] Z. Haba, Feynman integral and random dynamics in quantum physics. A probabilistic approach to quantum dynamics. Mathematics and its Applications, 480. Kluwer Academic Publishers, Dordrecht, 1999.

[44] A. Hahn, Rigorous State Model Representations for the Wilson loop observables in Chern-Simons theory, SFB-611-Preprint, Bonn (2002).

[45] A. Hahn, Quasi-Axial Gauge Fixing for Chern-Simons models on $S^2 \times S^1$. SFB-611-Preprint, Bonn (2003).

[46] A. Hahn, Chern-Simons theory on $\mathbb{R}^3$ in axial gauge: a rigorous approach. J. Funct. Anal. 211 (2004), no. 2, 483–507.

[47] A. Hahn, The Wilson loop observables of Chern-Simons theory on $\mathbb{R}^3$ in axial gauge, Commun. Math. Phys. 248 (3), 467–499 (2004).

[48] A. Hahn, Chern-Simons models on $S^2 \times S^1$, torus gauge fixing, and link invariants I, J. Geom. Phys. 53 (3), 275–314 (2005).

[49] A. Hahn, An analytic approach to Turaevs shadow invariants, SFB 611 preprint, Bonn, submitted to Commun. Math. Phys.

[50] T. Hida, H.H. Kuo, J. Potthoff, L. Streit, *White Noise* Kluwer, Dordrecht (1995).

[51] L. Hörmander, Fourier integral operators I. *Acta Math.*, 127(1), 79-183, 1971.

[52] L. Hörmander. *The Analysis of Linear Partial Differential Operators. I. Distribution Theory and Fourier Analysis.* Springer-Verlag, Berlin/Heidelberg/New York/Tokyo, 1983.

[53] W. Ichinose, A mathematical theory of the phase space Feynman path integral of the functional. Comm. Math. Phys. 265 (2006), no. 3, 739–779

[54] T. Ichinose, Path integral for the Dirac equation in two space-time dimensions. Proc. Japan Acad. Ser. A Math. Sci. 58, no. 7, 290–293, (1982).

[55] T. Ichinose, H. Tamura, Propagation of a Dirac particle. A path integral approach. J. Math. Phys. 25, no. 6, 1810–1819 (1984).

[56] K. Itô. Wiener integral and Feynman integral. *Proc. Fourth Berkeley Symposium on Mathematical Statistics and Probability.* Vol 2, pp. 227-238, California Univ. Press, Berkeley, 1961.

[57] K. Itô. Generalized uniform complex measures in the Hilbertian metric space with their applications to the Feynman path integral. *Proc. Fifth Berkeley Symposium on Mathematical Statistics and Probability.* Vol 2, part 1, pp. 145-161, California Univ. Press, Berkeley, 1967.

[58] G.W. Johnson, M.L. Lapidus, *The Feynman integral and Feynman's operational calculus.* Oxford University Press, New York, 2000.

[59] H. Kleinert, Path integrals in quantum mechanics, statistics, polymer physics, and financial markets. Third edition. World Scientific Publishing Co., Inc., River Edge, NJ, (2004).

[60] V. N. Kolokoltsov, Complex measures on path space: an introduction to the Feynman integral applied to the Schrödinger equation. Methodol. Comput. Appl. Probab. 1 (1999), no. 3, 349–365.

[61] V. N. Kolokoltsov, Semiclassical analysis for diffusions and stochastic processes, Lecture Notes in Mathematics, 1724. Springer-Verlag, Berlin, (2000).

[62] V. N. Kolokoltsov, A new path integral representation for the solutions of the Schrödinger, heat and stochastic Schrödinger equations. Math. Proc. Cambridge Philos. Soc. 132 (2002), no. 2, 353–375.

[63] T. Kuna, L. Streit, W. Westerkamp, Feynman integrals for a class of exponentially growing potentials, J. Math. Phys. 39, no. 9, 4476–4491 (1998).

[64] S. Leukert, J. Schäfer, A Rigorous Construction of Abelian Chern-Simons Path Integral Using White Noise Analysis, Reviews in Math. Phys. 8, 445–456 (1996).

[65] P. Malliavin, S. Taniguchi, Analytic functions, Cauchy formula, and stationary phase on a real abstract Wiener space. J. Funct. Anal. 143 (1997), no. 2, 470–528.

[66] V.P. Maslov. *Méthodes Opérationelles,* Mir. Moscou, 1987.

[67] M.B. Mensky, Continuous Quantum Measurements and Path Integrals. Taylor & Francis, Bristol and Philadelphia (1993).

[68] H. J. W. Müller-Kirsten, Introduction to quantum mechanics. Schrödinger equation and path integral. World Scientific Publishing Co. Pte. Ltd., Hackensack, NJ, 2006.

[69] E. Nelson. Feynman integrals and the Schrödinger equation. *J. Math. Phys.* **5**, 332-343, 1964.

[70] M. Niemann, S. Albeverio, A. Hahn, Three remarks on the Chern-Simons quantum field model, in preparation

[71] N. Reshetkhin, V.G. Turaev, Invariants of 3-manifolds via link polynomials and quantum groups. Invent. Math. 103, no.3, 547–597 (1991).

[72] J. Rezende, The method of stationary phase for oscillatory integrals on Hilbert spaces, Comm. Math. Phys. 101, 187–206 (1985).

[73] G. Roepstorff, Path integral approach to quantum physics. An introduction. Texts and Monographs in Physics. Springer-Verlag, Berlin, (1994).

[74] L. S. Schulman, Techniques and applications of path integration. A Wiley-Interscience Publication. John Wiley & Sons, Inc., New York, 1981.

[75] A. S. Schwartz, The partition function of a degenerate functional. Comm. Math. Phys. 67 , no. 1, 1–16 (1979).

[76] B. Simon, Functional integration and quantum physics. Second edition. AMS Chelsea Publishing, Providence, RI, 2005.

[77] O.G. Smolyanov, A.Yu. Khrennikov, The central limit theorem for generalized measures on infinite-dimensional spaces. (Russian) Dokl. Akad. Nauk SSSR 281 (1985), no. 2, 279–283.

[78] F. Takeo, Generalized vector measures and path integrals for hyperbolic systems. Hokkaido Math. J. 18 (1989), no. 3, 497–511.

[79] H. Thaler. Solution of Schrödinger equations on compact Lie groups via probabilistic methods. *Potential Anal.* **18**, n.2, 119-140, 2003.

[80] E. Witten, Quantum field theory and the Jones polynomial, Commun. Math. Phys. 121, 353–389 (1989).

[81] T. J. Zastawniak, Fresnel type path integral for the stochastic Schrödinger equation. Lett. Math. Phys. 41, no. 1, 93–99 (1997).

Sergio Albeverio
Institut für Angewandte Mathematik
Wegelerstr. 6
D–53115 Bonn
and
Dip. Matematica
Università di Trento
I–38050 Povo
and BiBoS; IZKS; SFB611; CERFIM (Locarno); Acc. Arch. (Mendrisio)

Sonia Mazzucchi
Institut für Angewandte Mathematik
Wegelerstr. 6
D–53115 Bonn
and
Dip. Matematica
Università di Trento
I–38050 Povo

Quantum Field Theory
B. Fauser, J. Tolksdorf and E. Zeidler, Eds., 67–81

A Comment on the Infra-Red Problem in the AdS/CFT Correspondence

Hanno Gottschalk and Horst Thaler

Abstract. In this note we report on some recent progress in proving the AdS/CFT correspondence for quantum fields using rigorously defined Euclidean path integrals. We also comment on the infra-red problem in the AdS/CFT correspondence and argue that it is different from the usual IR problem in constructive quantum field theory. To illustrate this, a triviality proof based on hypercontractivity estimates is given for the case of an ultraviolet regularized potential of type $: \phi^4 :$. We also give a brief discussion on possible renormalization strategies and the specific problems that arise in this context.

Mathematics Subject Classification (2000). Primary 81T08; Secondary 81T30.

Keywords. AdS/CFT correspondence, infra-red problem, constructive quantum field theory, triviality results, functional integrals, generating functionals.

1. Introduction

Often, the AdS/CFT correspondence between string theory or some other theory including quantized gravity on bulk AdS and super-symmetric Yang-Mills theory on its conformal boundary [12, 17] is formulated in terms of Euclidean path integrals. In the absence of mathematically rigorous approaches to path integrals of string type (see however [1]) or even gravity, it seems to be reasonable to use the well-established theory of constructive quantum field theory (QFT) [5] as a testing lab for some aspects of the more complex original AdS/CFT conjecture. That such simplified versions of the AdS/CFT correspondence are in fact possible was already noted by Witten [17] (see also [8]) and further elaborated by [4]. In [6] we give a mathematically rigorous version of the latter work (in [9] one finds some related ideas), leaving however the infra-red (IR) problem open. In this note we come back to the IR problem and we show how the difference between the IR problem in the AdS/CFT correspondence as compared with the usual IR problem in constructive QFT leads to somewhat unexpected results.

The authors would like to underline that, in contrast to [6], the present article is rather focused on ideas and thus leaves space for the interpretation of the validity of the results. We will comment on that in several places.

The article is organized as follows: In the following section we introduce the mathematical framework of AdS/CFT correspondence and define rigorous probabilistic path integrals on AdS. In Section 3 we recall the main results from [4, 6], i.e. that the generating functional that is obtained from imposing certain boundary conditions at the conformal boundary (which is the way generating functionals are defined in string theory) can in fact be written as a usual generating functional of some other field theory. From the latter form it is then easy to extract structural properties, e.g. reflection positivity of the functional, in the usual way. Somewhat unexpectedly, it is not clear whether a functional integral can be associated to the boundary theories. These statements hold for all sorts of interactions with a IR-cut-off. In Section 4 the IR-problem in this version of the AdS/CFT correspondence is discussed on a heuristic level. We also sketch the proof of triviality of the generating functional of the conformally invariant theory on the conformal boundary of AdS for the case of an UV-regularized : ϕ^4 : interaction. We briefly survey strategies that might be candidates to overcome the triviality obstacle at a non-rigorous level and we comment on specific problems with such strategies. The final section gives some preliminary conclusions and an outlook on open research problems in understanding further the mathematical basis of AdS/CFT.

2. Functional integrals on AdS

Let us consider the $d+2$ dimensional ambient space $\mathbb{R}^{d,2} = \mathbb{R}^{d+2}$ with inner product of signature $(-, +, \ldots, +, -)$, i.e. $\zeta^2 = -\zeta_1^2 + \zeta_2^2 + \cdots + \zeta_{d+1}^2 - \zeta_{d+2}^2$ where $\zeta \in \mathbb{R}^{d,2}$. Then the submanifold defined by $\{\zeta \in \mathbb{R}^{d,2} : \zeta^2 = -1\}$ is a $d + 1$ dimensional Lorentz manifold with metric induced by the ambient metric. It is called the $d+1$ dimensional Anti de Sitter (AdS) space. Formal Wick rotation $\zeta_1 \to i\zeta_1$ converts the ambient space into the space $\mathbb{R}^{d+1,1}$ with signature $(+, \ldots, +, -)$. Under Wick rotation, the AdS space is converted to the Hyperbolic space $\mathbb{H}^{d+1} : \{\zeta \in \mathbb{R}^{d+1} : \zeta^2 = -1, \zeta^d > 0\}$, which is a Riemannian submanifold of the ambient $d + 2$ dimensional Minkowski space. We call $\mathbb{H}^{d+1}$ the Euclidean AdS space.

It has been established with full mathematical rigor that Euclidean random fields that fulfil the axioms of invariance, ergodicity and reflection positivity give rise, via an Osterwalder–Schrader reconstruction theorem, to local quantum field theories on the universal covering of the relativistic AdS, cf. [3, 10] justifying the above sketched formal Wick rotation. Hence a constructive approach with reflection positive Euclidean functional integrals is viable.

It is convenient to work in the so called half-space model of Euclidean AdS (henceforth the word Euclidean will be dropped). This coordinate system is obtained via the change of variables $\zeta_i = x_i/z, i = 1, \ldots, d, \zeta_{d+1} = -(z^2 + x^2 - 1)/2z$, $\zeta_{d+2} = (z^2 + x^2 + 1)/2z$ which maps $\mathbb{R}_+^{d+1} = \{(z, x) \in \mathbb{R}^{d+1} : z > 0\}$ to $\mathbb{H}^{d+1}$.

We will use the notation $\underline{x}$ for $(z, x_1, \ldots, x_d) \in \mathbb{R}_+^{d+1}$. The metric on $\mathbb{R}_+^{d+1}$ is given by $g = (dz^2 + dx_1^2 + \cdots + dx_d^2)/z^2$ which implies that the canonical volume form is $d_g\underline{x} = z^{-d-1}dz \wedge dx_1 \wedge \cdots \wedge dx_d$. The conformal boundary of $\mathbb{H}^{d+1}$ then is the d-dimensional Euclidean space $\mathbb{R}^d$ with metric $ds^2 = dx_1^2 + \cdots + dx_d^2$ which is obtained via the limit $z \to 0$ and a conformal transformation of the AdS metric. Of course, the upshot of the AdS/CFT correspondence is that the action of the Lorentz group on the AdS space $\mathbb{H}^{d+1}$ gives rise to an action of the conformal group transformations on the conformal boundary. One thus expects an AdS symmetric QFT (or string/quantum gravity... theory) on the bulk $\mathbb{H}^{d+1}$ to give, if properly restricted to the conformal boundary, a conformally invariant theory on $\mathbb{R}^d$.

We will now make this precise. On the hyperbolic space $\mathbb{H}^{d+1}$ one has two invariant Green's functions ("bulk-to-bulk propagators") for the operator $-\Delta_g + m^2$, with Δ_g the Laplacian and m^2 a real number suitably bounded from below, that differ by scaling properties towards the conformal boundary

$$G_\pm(z, x; z', x') = \gamma_\pm(2u)^{-\Delta_\pm}F(\Delta_\pm, \Delta_\pm + \tfrac{1-d}{2}; 2\Delta_\pm + 1 - d; -2u^{-1}) \quad (2.1)$$

Here F is the hypergeometric function, $u = \frac{(z-z')^2+(x-x')^2}{2zz'}$, $\Delta_\pm = \frac{d}{2} \pm \frac{1}{2}\sqrt{d^2+4m^2}$ $=: \frac{d}{2} \pm \nu, \nu > 0$ and $\gamma_\pm = \frac{\Gamma(\Delta_\pm)}{2\pi^{d/2}\Gamma(\Delta_\pm + 1 - \frac{d}{2})}$ [4, 6]. Taking pointwise scaling limits for $z \to 0$ in one or two of the arguments, the bulk-to-boundary and boundary-to-boundary propagators are obtained

$$H_\pm(z, x; x') = \lim_{z' \to 0} z'^{-\Delta_\pm}G_\pm(z, x; z', x') = \gamma_\pm\left(\frac{z}{z^2 + (x-x')^2}\right)^{\Delta_\pm} \quad (2.2)$$

and

$$\alpha_\pm(x, x') = \lim_{z \to 0} z^{-\Delta_\pm}H_\pm(z, x; x') = \gamma_\pm(x - x')^{-2\Delta_\pm}. \quad (2.3)$$

If (2.2) or (2.3) do not define locally integrable functions, the expressions on the right hand side are defined via analytic continuation in the weights $\Delta_\pm$. An important relation between G_+, G_-, H_+ and α_- is the covariance splitting formula for G_- given by

$$G_-(\underline{x}, \underline{x}') = G_+(\underline{x}, \underline{x}') + \int_{\mathbb{R}^d} \int_{\mathbb{R}^d} H_+(\underline{x}, y)c^2\alpha_-(y, y')H_+(\underline{x}', y')dydy', \quad (2.4)$$

with $c = 2\nu$.

We now pass on to the description of mathematically well-defined functional integrals. Let $\mathcal{D} = \mathcal{D}(\mathbb{H}^{d+1}, \mathbb{R})$ be the infinitely differentiable, compactly supported functions on $\mathbb{H}^{d+1}$ endowed with the topology of compact convergence. The propagator G_+ is the resolvent function to the Laplacian Δ_g with Dirichlet boundary conditions at conformal infinity, from which it follows that G_+ is stochastically positive, $\langle f, f\rangle_{-1} = G_+(\bar{f}, f) = \int_{\mathbb{H}^{d+1} \times \mathbb{H}^{d+1}} G_+(\underline{x}, \underline{x}')\bar{f}(\underline{x})f(\underline{x}')\,d_g\underline{x}d_g\underline{x}' \geq 0$ $\forall f \in \mathcal{D}$, and reflection positive as long as $m^2 > -\frac{d^2}{4}$. The latter value is determined by the lower bound of the spectrum of Δ_g on $\mathbb{H}^{d+1}$. In explicit, if

$\theta : (z, x_1, x_2, \ldots, x_d) \to (z, -x_1, x_2, \ldots, x_d)$ is the reflection in x_1-direction, then for any $f \in \mathcal{D}_+ = \{h \in \mathcal{D} : h(\underline{x}) = 0 \text{ if } x_1 \le 0\}$ we have

$$\int_{\mathbb{H}^{d+1} \times \mathbb{H}^{d+1}} G_+(\underline{x}, \underline{x}') \bar{f}_\theta(\underline{x}) f(\underline{x}') \, d_g\underline{x} d_g\underline{x}' \ge 0,$$

cf. [5]. Here, $f_u(\underline{x}) = f(u^{-1}\underline{x})$ for $u \in \mathrm{Iso}(\mathbb{H}^{d+1})$.

Consequently, via application of Minlos theorem, there exists a unique probability measure μ_{G_+} on the measurable space $(\mathcal{D}', \mathcal{B})$, where $\mathcal{D}'$ is the topological dual space of $\mathcal{D}$ and $\mathcal{B}$ the associated Borel sigma algebra, such that $\int_{\mathcal{D}'} e^{\langle \phi, f \rangle} d\mu_{G_+}(\phi) = e^{\frac{1}{2}\langle f, f \rangle_{-1}}$. By setting $\varphi(f)(\phi) = \phi(f)$ we define the canonical random field associated with μ_{G_+}, i.e. a random variable valued distribution. In the following we omit the distinction between φ and ϕ and write ϕ for both.

Let $\mathcal{B}_\Lambda$, $\Lambda \subseteq \mathbb{H}^{d+1}$ be the smallest sigma algebra generated by the functions $\mathcal{D}' \ni \phi \to \langle \phi, f \rangle$, $\mathrm{supp} f \in \Lambda$ and $M(\Lambda)$ be the functions that are $\mathcal{B}_\Lambda$-measurable. We use the special abbreviations $\mathcal{B}_+ = \mathcal{B}_{\{\underline{x} \in \mathbb{H}^{d+1} : x_1 > 0\}}$ and $M_+ = M(\mathcal{B}_+)$. Then μ_{G_+} is reflection positive, i.e.

$$\int_{\mathcal{D}'} \Theta\bar{F}(\phi)F(\phi) \, d\mu_{G_+}(\phi) \ge 0, \ \forall F \in M_+. \tag{2.5}$$

The reflection $\Theta F(\phi)$ is defined as $F(\phi_\theta)$ with $\langle \phi_u, f \rangle = \langle \phi, f_{u^{-1}} \rangle \ \forall \phi \in \mathcal{D}'$, $f \in \mathcal{D}$ and $u \in \mathrm{Iso}(\mathbb{H}^{d+1})$. $\langle ., . \rangle$ is the duality between $\mathcal{D}'$ and $\mathcal{D}$ induced by the $L^2(\mathbb{H}^{d+1}, d_g\underline{x})$ inner product.

Let $\{V_\Lambda\} : \mathcal{D}' \to \mathbb{R}$ be a set of interaction potentials indexed by the net of bounded, measurable subsets Λ in $\mathbb{H}^{d+1}$. In particular these sets have finite volume $|\Lambda| = \int_\Lambda d_g\underline{x}$. We require that the following conditions hold:

(i) Integrability: $e^{-V_\Lambda} \in L^1(\mathcal{D}', d\mu_{G_+}) \ \forall\Lambda$;
(ii) Locality: $V_\Lambda \in M(\mathcal{B}_\Lambda)$;
(iii) Invariance: $V_\Lambda(\phi_u) = V_{u^{-1}\Lambda}(\phi) \ \mu_{G_+}$-a.s..
(iv) Additivity: $V_\Lambda + V_{\Lambda'} = V_{\Lambda\cup\Lambda'}$ for $\Lambda \cap \Lambda' = \emptyset$.
(v) Non-degeneracy: $V_\Lambda = 0 \ \mu_{G_+}$-a.s. if $|\Lambda| = 0$.

Then, using (i), we obtain a family of interacting measures on $(\mathcal{D}', \mathcal{B})$, indexed by the net $\{\Lambda\}$, by setting $d\mu_{G_+,\Lambda} = e^{-V_\Lambda} d\mu_{G_+}/Z_\Lambda$ with $Z_\Lambda = \int_{\mathcal{D}'} e^{-V_\Lambda} d\mu_{G_+}$. Furthermore, using (ii)–(v) we get whenever $\theta\Lambda = \Lambda$

$$\int_{\mathcal{D}'} \Theta\bar{F}F \, d\mu_{G_+,\Lambda} = \frac{1}{Z_\Lambda} \int_{\mathcal{D}'} \Theta\left(\bar{F}e^{-V_{\Lambda_+}}\right)\left(Fe^{-V_{\Lambda_+}}\right) d\mu_{G_+} \ge 0, \ \forall F \in M_+, \tag{2.6}$$

where $\Lambda_+ = \Lambda \cap \{x \in \mathbb{H}^{d+1} : x_1 > 0\}$. Hence reflection positivity is preserved under the perturbation. Furthermore, from the invariance of μ_{G_+} under $\mathrm{Iso}(\mathbb{H}^{d+1})$ we get that $u_*\mu_{G_+,\Lambda} = \mu_{G_+,u\Lambda}$. Here $u \in \mathrm{Iso}(\mathbb{H}^{d+1})$ induces an action on $\mathcal{D}'$ via $\phi \to \phi_u$ and u_* is the pushforward under this action. Consequently, if the limit (in distribution) $\mu_{G_+,\mathbb{H}^{d+1}} = \lim_{\Lambda \nearrow \mathbb{H}^{d+1}} \mu_{G_+,\Lambda}$ exists and is unique, the limiting measure is invariant under $\mathrm{Iso}(\mathbb{H}^{d+1})$ and reflection positive. Invariance follows from the equivalence of the nets $\{\Lambda\}$ and $\{u\Lambda\}$ and the postulated uniqueness of the limit over the net $\{\Lambda\}$.

Let us next consider functional integrals associated with the Green's function G_-. In the case when $2\nu < d$ ($\Leftrightarrow m^2 < 0$) we get that α_- is stochastically positive since $\alpha_-(\bar{f}, f) = \int_{\mathbb{R}^d \times \mathbb{R}^d} \alpha_-(x, x') \bar{f}(x) f(x') \, dx dx' = C_{-\nu} \int_{\mathbb{R}^d} |k|^{-2\nu} |\hat{f}(k)|^2 \, dk \geq 0$. $\hat{f}$ denotes the Fourier transform of f wrt. x, $\hat{f}(k) = (2\pi)^{-d/2} \int_{\mathbb{R}^d} e^{ik \cdot x} f(x) \, dx$. Furthermore, α_- is reflection positive in x_1-direction (in the usual sense, cf. [5]) if and only if $-\nu > -1$, which is also known as the unitarity bound. It is clear from the decomposition (2.4) that G_- is stochastically positive if G_+ and α_- are both stochastically positive. The reflection positivity of G_- does not follow from the reflection positivity of G_+ and α_- due to the non-local effect of H_+. We will however not need it here. We thus conclude that for $\sup \operatorname{spec}(\Delta_g) < m^2 < 0$ a unique probability measure μ_{G_-} on $(\mathcal{D}', \mathcal{B})$ with Laplace transform $\int_{\mathcal{D}'} e^{\langle \phi, f \rangle} d\mu_{G_-}(\phi) = e^{\frac{1}{2}\langle f, f \rangle_{-1,-}}$ exists. Here $\langle f, f \rangle_{-1,-} = G_-(f, f)$. The perturbation of μ_{G_-} with an interaction can now be discussed in analogy with the above case – where however the reflection positivity for the perturbed measure remains open, as reflection positivity of the free measure does not necessarily hold.

3. Two generating functionals

On the string theory side of the AdS/CFTcorrespondence, generating functionals for the boundary theory are calculated fixing boundary conditions at the conformal boundary (so called Dirichlet boundary conditions). Little is known about the mathematical properties of such kinds of generating functionals. E.g. their stochastic and reflection positivity is far from obvious, leaving the linkage to path integrals and relativistic physics open. It was noticed by Dütsch and Rehren [4] that such kinds of generating functionals can however be re-written in terms of ordinary generating functionals, from which the structural properties can be read of in the usual way. These ideas in [6] have been made fully rigorous in the context of constructive QFT. We will now briefly review these results.

The generating functional $Z(f)/Z(0)$, $f \in \mathcal{S}(\mathbb{R}^d, \mathbb{R})$, the space of Schwartz functions, in the AdS/CFT correspondence from a string theoretic point of view can be described as follows: Let ϕ be some scalar quantum field that is included in the theory (e.g. the dilaton field) and let V_Λ be the (IR and eventually UV-regularized) effective potential for that field obtained via integrating out the remaining degrees of freedom (leaving open the question how such an "integral" can be defined). To simplify the model and for the sake of concreteness we will sometimes assume that V_Λ is of polynomial type. Formally,

$$Z(f) = \int_{\phi_0 = \phi|_{\partial\mathbb{H}^{d+1}} = f} e^{-S_0(\phi) - V_\Lambda(\phi)} \, d\phi = \int \delta(\phi_0 - f) e^{-S_0(\phi) - V_\Lambda(\phi)} \, d\phi \quad (3.1)$$

where $S_0 = |\nabla\phi|^2 + m^2\phi^2$, $\phi_0 = \phi|_{\partial\mathbb{H}^{d+1}}$ are suitably rescaled boundary values of the field ϕ and $d\phi$ is the heuristic flat measure on the space of all field configurations. The first step in making this formal expression rigorous is to replace

$e^{-S_0(\phi)}\, d\phi$ with a well-defined probabilistic path integral. It turns out that $d\mu_{G_-}(\phi)$ is the right candidate and hence for the moment restriction to $m^2 < 0$ is necessary.

In a second step we have to make sense out of the boundary condition $\phi_0 = f$ or the functional delta distribution on the boundary values of the field, respectively. Using the covariance splitting formula (2.4) we obtain the splitting $\phi_-(\underline{x}) = \phi_+(\underline{x}) + \int_{\mathbb{R}^d} H_+(\underline{x}, x')\phi_{\alpha_-}(x')\, dx'$, where $\phi_\pm$ are the canonical random fields associated with $G_\pm$ and ϕ_{α_-} is the canonical random field associated to the functional measure μ_{α_-}, i.e. the Gaussian measure with generating functional $e^{\frac{1}{2}\alpha_-(f,f)}$ living on the conformal boundary of $\mathbb{H}^{d+1}$.

The following step is to construct a finite dimensional approximation ψ_{α_-} of the boundary field ϕ_{α_-} by projecting it via a basis expansion to $\mathbb{R}^n$. Thereafter, one can implement the delta distribution as a delta distribution on $\mathbb{R}^n$. Finally one can remove the finite dimensional approximation via a limit $n \to \infty$. It turns out that this limit exists and is unique up to a diverging multiplicative constant. This constant however drops out in the quotient $Z(f)/Z(0)$. With the projection to the first n terms of the basis expansion denoted by p_n and η a linear mapping from this space to $\mathbb{R}^n$ we get

$$C_{A_-}\int_{\mathbb{R}^n}\int_{D'}\delta(\psi_{\alpha_-} - \eta p_n f)e^{-V_\Lambda(\phi_+ + cH_+(\eta^{-1}\psi_{\alpha_-}))}d\mu_{G_+}(\phi_+)e^{-\frac{1}{2}(\psi_{\alpha_-},A_-\psi_{\alpha_-})}d\psi_{\alpha_-}$$

$$= C_{A_-}e^{-\frac{1}{2}(f,(p_n\alpha - p_n)^{-1}f)}\int_{D'}e^{-V_\Lambda(\phi_+ + cH_+(p_n f))}d\mu_{G_+}(\phi_+) =: Z_n(f), \qquad (3.2)$$

where $A_- := (\eta p_n \alpha - p_n \eta^{-1})^{-1}$ and $C_{A_-} = \dfrac{|\det A_-|^{\frac{1}{2}}}{(2\pi)^{\frac{d}{2}}}$. One can then show that

$$Z(f)/Z(0) := \lim_{n\to\infty} Z_n(f)/Z_n(0) = e^{-\frac{1}{2}(f,\alpha_-^{-1}f)}\frac{\int_{D'}e^{-V_\Lambda(\phi_+ + cH_+ + f)}d\mu_{G_+}(\phi_+)}{\int_{D'}e^{-V_\Lambda(\phi_+)}d\mu_{G_+}(\phi_+)}$$

$$\qquad (3.3)$$

converges under rather weak continuity requirements on V_Λ that are fulfilled e.g. for UV-regularized potentials in arbitrary dimension and for $P(\phi)_2$ potentials without UV cut-offs in $d+1 = 2$. Obviously, the limit does not depend on the details of the finite dimensional approximation. For the details we refer to [6]. We now realize that the right hand side of (3.3) also makes sense for $m^2 \geq 0$ and we adopt (3.3) as a definition of (3.1).

At this point one would like to associate a boundary field theory to the generating functional $\mathcal{C}(f) = Z(f)/Z(0)$. In order to obtain a functional integral associated to $\mathcal{C} : \mathcal{S} = \mathcal{S}(\mathbb{R}^d, \mathbb{R}) \to \mathbb{R}$ we require that $\mathcal{C}$ is continuous wrt. the Schwartz topology, normalized, $\mathcal{C}(0) = 1$ and stochastically positive, $\sum_{j,l=1}^{n}\bar{z}_j z_l \mathcal{C}(f_j + f_l) \geq 0 \; \forall \; n \in \mathbb{N}, f_j \in \mathcal{S}, z_j \in \mathbb{C}$. Furthermore, in order to have a well defined passage from Euclidean time to real time QFT one requires reflection positivity $\sum_{j,l=1}^{n}\bar{z}_j z_l \mathcal{C}(f_{j,\theta} + f_l) \geq 0 \; \forall \; n \in \mathbb{N}, f_j \in \mathcal{S}_+, z_j \in \mathbb{C}$. Here $\mathcal{S}_+ = \{f \in \mathcal{S} : \operatorname{supp} f \subseteq \{x \in \mathbb{R}^d : x_1 > 0\}\}$. Finally, the theory obtained at the boundary should be conformally invariant, provided the IR cut-off Λ is removed from V_Λ via taking the limit of the generating functionals wrt. the net $\{\Lambda\}$.

It has been pointed out in [4, 11, 16] that an alternative representation of the functional (3.3) answers a number of the questions raised above. Let $\phi(\underline{x}) = \phi(z, x)$ be the canonical random field associated with the measure μ_{G_+}. The idea is to smear $\phi(z, x)$ in the x-variable with a test function $f \in \mathcal{D}(\mathbb{R}^d, \mathbb{R})$ and then scale $z \to 0$. In the light of (2.1), one has to multiply $\phi(z, f) = \langle \phi, \delta_z \otimes f \rangle$ with a factor $z^{-\Delta_+}$ in order to obtain a finite result in the limit. We set

$$Y_z(f) = \int_{\mathcal{D}'} e^{\langle \phi, z^{-\Delta_+} \delta_z \otimes f \rangle} e^{-V_\Lambda(\phi)} \, d\mu_{G_+}(\phi). \tag{3.4}$$

Clearly, under the conditions on V_Λ given in the preceding section and for $\Lambda = \theta\Lambda$, $Y_z(f)/Y_z(0)$ defines a continuous, normalized, stochastically positive and reflection positive generating functional for all $z > 0$. Using the fact that $G_+(\delta_z \otimes f)$ is in the Cameron-Martin space of the measure μ_{G_+}, one gets with $f_z = z^{-\Delta_+} \delta_z \otimes f$, cf. [6],

$$Y_z(f)/Y_z(0) = e^{\frac{1}{2} G_+(f_z, f_z)} \int_{\mathcal{D}'} e^{-V_\Lambda(\phi+G_+f_z)} d\mu_{G_+}(\phi)/Y_z(0). \tag{3.5}$$

We now want to take the limit $z \to 0$. Using (2.2) one can show under rather weak continuity requirements on V_Λ that the functional integral on the rhs of (3.5) converges to $\int_{\mathcal{D}'} e^{-V_\Lambda(\phi+H+f)} d\mu_{G_+}(\phi)$. The prefactor however diverges. The reason is that the limit in (2.3) is only a pointwise limit for $x \neq x'$ and not a limit in the sense of tempered distributions. One can however show that [6]

$$\int_{\mathbb{R}^d} \int_{\mathbb{R}^d} \alpha_+(x, y) f(x) f(y) dx dy =$$

$$\lim_{z \to 0} z^{-2\Delta_+} \int_{\mathbb{R}^d} \int_{\mathbb{R}^d} G_+(z, x; z, y) f(x) f(y) dx dy -$$

$$\frac{1}{(2\pi)^{\frac{d}{2}}} \left(\frac{2^{1-\nu}}{\sqrt{\pi}\Gamma(\nu + \frac{1}{2})} \right)^2 \sum_{j=0}^{[\nu]} z^{-2(\nu-j)} (-1)^j a_j \int_{\mathbb{R}^d} |\hat{f}(k)|^2 |k|^{2j} dk.$$

$$=: \lim_{z \to 0} z^{-2\Delta_+} \int_{\mathbb{R}^d} \int_{\mathbb{R}^d} G_+(z, x; z, y) f(x) f(y) dx dy - (\text{Corr}(z)f, f). \tag{3.6}$$

Here $a_j = \int_0^\infty (\int_0^1 \cos(\omega t)(1-t^2)^{\nu-\frac{1}{2}} dt)^2 \omega^{2(\nu-j)-1} d\omega$. Thus, the right hand side of (3.5) multiplied with $e^{-\frac{1}{2}(\text{Corr}(z)f, f)}$ converges and we obtain the limiting functional

$$\begin{aligned} \tilde{\mathcal{C}}(f) &= \lim_{z \to 0} e^{-\frac{1}{2}(\text{Corr}(z)f, f)} (Y_z(f)/Y_z(0)) \\ &= \lim_{z \to 0} e^{\frac{1}{2}[G_+(f_z, f_z) - (\text{Corr}(z)f, f)]} \int_{\mathcal{D}'} e^{-V_\Lambda(\phi+G_+f_z)} d\mu_{G_+}(\phi)/Y_z(0) \\ &= e^{\frac{1}{2}\alpha_+(f, f)} \frac{\int_{\mathcal{D}'} e^{-V_\Lambda(\phi+H+f)} d\mu_{G_+}(\phi)}{\int_{\mathcal{D}'} e^{-V_\Lambda(\phi)} d\mu_{G_+}(\phi)} \end{aligned} \tag{3.7}$$

This, together with $\alpha_-^{-1} = -c^2\alpha_+$, establishes the crucial identity [4, 6]

$$\mathcal{C}(f) = \tilde{\mathcal{C}}(cf), \quad \forall f \in \mathcal{S}(\mathbb{R}^d, \mathbb{R}). \tag{3.8}$$

Let us now investigate the structural properties of the generating functional $\mathcal{C}$: $\mathcal{S} \to \mathbb{R}$. If there were not the correction factor $(\mathrm{Corr}(z)f, f)$, $\mathcal{C}$ would be stochastically positive and reflection positive as the limit of functionals with that property, since we can combine (3.7) and (3.8) for a representation of $\mathcal{C}$. However, due to the signs in (3.6) $\mathcal{S} \ni f \to e^{-\frac{1}{2}(\mathrm{Corr}(z)f,f)} \in \mathbb{R}$ is not stochastically positive and consequently the stochastic positivity of $e^{-\frac{1}{2}(\mathrm{Corr}(z)f,f)}(Y_z(f)/Y_z(0))$ is at least unclear. Hence we do not have any reason to believe that the limiting functional $\mathcal{C}$ is stochastically positive and can be associated with a probabilistic functional integral. An exception is the case where $V_\Lambda \equiv 0$ where we can dwell on the fact that $\mathcal{S} \ni f \to e^{\frac{1}{2}\alpha_+(f,f)} \in \mathbb{R}$ is manifestly stochastically positive since $\hat{a}(k) = C_{-\nu}\left(\frac{|k|}{2}\right)^{2\nu} \in \mathbb{R}$ with $C_{-\nu} > 0$. It is therefore questionable if one can use the AdS/CFT correspondence to generate conformally invariant models in statistical mechanics.

We next investigate the question of reflection positivity. Since the correlation length of the distributional kernels of $\mathrm{Corr}(z)$ is zero, we get that $(\mathrm{Corr}(z)(f_{j,\theta} + f_l), (f_{j,\theta} + f_l)) = (\mathrm{Corr}(z)f_{j,\theta}, f_{j,\theta}) + (\mathrm{Corr}(z)f_l, f_l) = (\mathrm{Corr}(z)f_j, f_j) + (\mathrm{Corr}(z)f_l, f_l)$ for $f_j \in \mathcal{S}_+$. Consequently, $\forall f_j \in \mathcal{S}_+, z_1, \ldots, z_n \in \mathbb{C}$ and Λ such that $\theta\Lambda = \Lambda$ we get

$$\sum_{j,l=1}^{n} \mathcal{C}(f_{j,\theta} + f_l)\bar{z}_j z_l = \lim_{z\to 0} \sum_{j,l=1}^{n} (Y_z(cf_{j,\theta} + cf_l)/Y_z(0))\,\bar{z}'_j z'_l \geq 0 \qquad (3.9)$$

with $z'_j = z_j e^{-\frac{1}{2}(\mathrm{Corr}(z)cf_j, cf_j)}$. For a proof that the reflection positivity of generating functionals implies the reflection positivity of Schwinger functions [5] also in the absence of stochastic positivity, cf. [7]. As in [4, 6, 16], we thus come to the conclusion that the crucial property for the existence of a relativistic theory is preserved in the AdS/CFT correspondence.

Finally we address the invariance properties of the limiting generating functional $\mathcal{C}$. For being the generating functional of a CFT, we require invariance under conformal transformations, i.e. $\mathcal{C}(f) = \mathcal{C}(\lambda_u^{-1} f_u) \,\forall f \in \mathcal{S}$ where u is an element of the conformal group on $\mathbb{R}^d$ and

$$\lambda_u(x) = \left|\det\left(\frac{\partial u(x)}{\partial x}\right)\right|^{-\frac{\Delta_+}{d}}. \qquad (3.10)$$

Certainly, as long as an interaction with IR cut-off is included in the definition of $\mathcal{C} = \mathcal{C}_\Lambda$, conformal invariance can not hold. Using the identification of $\mathrm{Iso}(\mathbb{H}^{d+1})$ and the conformal group on $\mathbb{R}^d$, we get that H_+ intertwines the respective representations on function spaces, i.e. [6]

$$H_+(u(z,x); x') = \left|\det\left(\frac{\partial u^{-1}(x')}{\partial x'}\right)\right|^{\frac{\Delta_+}{d}} H_+(z, x; u^{-1}(x')). \qquad (3.11)$$

Combining this, the conformal invariance of α_+ under the given representation of the conformal group and (3.3) we obtain

$$\mathcal{C}_\Lambda(\lambda_u^{-1} f_u) = \mathcal{C}_{u\Lambda}(f) \; \forall f \in \mathcal{S}. \tag{3.12}$$

Hence, if the generating functionals $\{\mathcal{C}_\Lambda\}$ have a unique limit $\mathcal{C}$ wrt. the net $\{\Lambda\}$, then $\mathcal{C}$ is reflection positive and conformally invariant and hence is the generating functional of a boundary CFT.

4. The infra-red problem and triviality

In this section we investigate the net limit of $\{\mathcal{C}_\Lambda\}$ which is needed to establish the full AdS/CFT correspondence. This problem has been left open in [6] and we will show that this kind of IR problem behaves somewhat wired.

The reason is the following: When we identified the generating functionals $\mathcal{C}_\Lambda$ and $\tilde{\mathcal{C}}_\Lambda$, we have seen from the latter functional that it originated from a usual QFT generating functional with $z^{-\Delta_+} \delta_z \otimes f$ giving rise to a source term which needs to be considered in the limit $z \to 0$. As (3.6) shows, this source term corresponds to an interaction of an "exterior field" with the quantum field ϕ which, already for the free field, has zero expectation but infinite fluctuations in the limit $z \to 0$. Without any correction term, this would have led to a generating functional which converges to zero for any $f \neq 0$. We already then needed an ultra-local correction term to deal with the prescribed infinite energy fluctuations.

If we now switch on the interaction, a shift term $H_+ f$ in the bulk theory is generated, cf. (3.7). If we for example restrict to polynomial interactions, this shift leads to re-defined f-dependent couplings that diverge towards the conformal boundary. This again leads to an infinite energy transfer and it is probable that this infinite amount of energy plays havoc with the generating functional. Here we will show that in some situations this indeed happens.

Let us first investigate the behavior of the shift $H_+ f$ towards the conformal boundary. Let $f \in \mathcal{S}$ be such that $f(0) \neq 0$. Choosing spherical coordinates, we denote by $f_{\mathrm{rad}}(r)$ the integral of $f(x)$ over the angular coordinates. We get from (2.2) via a change of coordinates

$$H_+ f(z,0) = \gamma_+ z^{-\Delta_+ + d} \int_0^\infty \left(\frac{1}{1+r^2} \right)^{\Delta_+} f_{\mathrm{rad}}(zr) r^{d-1} dr \tag{4.1}$$

and we see that the integral on the rhs converges to $f(0) \int_0^\infty \left(\frac{1}{1+r^2} \right)^{\Delta_+} r^{d-1} dr = f(0) \times \Gamma(\Delta_+ - d/2)\Gamma(d/2)/2\Gamma(\Delta_+)$, hence $H_+ f(z,x) \sim z^{-\Delta_+ + d}$ if $f(x) \neq 0$ by translation invariance.

Let us now work with the generating functional as defined by (3.7). The prefactor on the rhs is independent of Λ, hence we have to investigate the behavior of

$$\mathcal{C}'_\Lambda(f) = \frac{\int_{\mathcal{D}'} e^{-V_\Lambda(\phi + H_+ f)} d\mu_{G_+}(\phi)}{\int_{\mathcal{D}'} e^{-V_\Lambda(\phi)} d\mu_{G_+}(\phi)}. \tag{4.2}$$

We restrict ourselves to the simplest possible case - an ultra-violet regularized ϕ^4 potential in arbitrary dimensions $d + 1$

$$V_\Lambda(\phi) = \lambda \int_\Lambda : \phi_\kappa^4 : (\underline{x}) \, d_g \underline{x} \tag{4.3}$$

where ϕ_κ denotes the random field ϕ with UV-cut off κ. Due to this cut-off, the locality axiom in Section 2 will in general be violated. This however does not matter in the following discussion. We furthermore require that $G_+^\kappa(\underline{x}, \underline{x}') = \mathbb{E}[\phi_\kappa(\underline{x})\phi_\kappa(\underline{x}')]$ is a bounded function in $\underline{x}$ and $\underline{x}'$. $\mathbb{E}$ stands for the expectation wrt. μ_{G_+}. The Wick ordering in (4.3) is taken wrt. G_+, for simplicity. The shifted potential then is given by

$$V_\Lambda(\phi + H_+ f) = \lambda \int_\Lambda \sum_{j=0}^4 \binom{4}{j} : \phi_\kappa^j : (\underline{x})(H_+ f)^{4-j}(\underline{x}) \, d_g \underline{x}. \tag{4.4}$$

Taking the expected value of the shifted potential wrt. μ_{G_+}, one obtains $\lambda \times$ $\times \int_\Lambda (H_+ f)^4 d_g \underline{x}$ which in the light of (4.1) clearly diverges as $\Lambda \nearrow \mathbb{H}^{d+1}$ whenever $f \neq 0$.

Let us now focus on a specific class of cut-offs of the form $\Lambda(z_0) = \Lambda(z_0, l) = [z_0, A] \times [-l, l]^{\times d}$ where we keep $l > 0, A > 0$ arbitrarily large but fixed. Let $V(z_0, f)(\phi) = V_{\Lambda(z_0)}(\phi + H_+ f)$. Since $d_g \underline{x} = z^{-d-1} dz dx$ we obtain the scaling of the expected shifted interaction energy

$$
\begin{aligned}
E(z_0, f) &= \mathbb{E}[V(z_0, f)] = \lambda \int_{[z_0, A]} \int_{[-l,l]^{\times d}} (H_+ f)^4(z, x) \, dx z^{-d-1} dz \\
&\sim z_0^{-d-4(\Delta_+ - d)} \text{ as } z_0 \to 0.
\end{aligned} \tag{4.5}
$$

Let us next investigate the fluctuations in the shifted energy as $z_0 \to 0$. Denoting the standard deviation of $V(z_0, f)$ with $\sigma(z_0, f)$, we obtain using (4.4) and $\mathbb{E}[: \phi_\kappa^a : (\underline{x}) : \phi_\kappa^b : (\underline{y})] = a! \, \delta_{a,b} G_+^\kappa(\underline{x}, \underline{y})^a$, $a, b \in \mathbb{N}$,

$$
\begin{aligned}
\sigma(z_0, f) &= \Big[24 \int_{\Lambda(z_0)^{\times 2}} G_+^\kappa(\underline{x}, \underline{y})^4 d_g \underline{x} d_g \underline{y} \\
&+ 96 \int_{\Lambda(z_0)^{\times 2}} H_+ f(\underline{x}) H_+ f(\underline{y}) G_+^\kappa(\underline{x}, \underline{y})^3 d_g \underline{x} d_g \underline{y} \\
&+ 72 \int_{\Lambda(z_0)^{\times 2}} (H_+ f)^2(\underline{x})(H_+ f)^2(\underline{y}) G_+^\kappa(\underline{x}, \underline{y})^2 d_g \underline{x} d_g \underline{y} \\
&+ 16 \int_{\Lambda(z_0)^{\times 2}} (H_+ f)^3(\underline{x})(H_+ f)^3(\underline{y}) G_+^\kappa(\underline{x}, \underline{y}) d_g \underline{x} d_g \underline{y}\Big]^{1/2} \\
&\sim z_0^{-d-3(\Delta_+ - d)} \text{ or slower as } z_0 \to 0,
\end{aligned} \tag{4.6}
$$

where we took the factors G_+^κ out of the integral and replaced them with a majorizing constant in order to obtain an upper bound on the scaling. Apparently, the quotient $\gamma(z_0, f) = 2\sigma(z_0, f)/E(z_0, f) \sim z_0^{\Delta_+ - d}$ scales down to zero if $m^2 > 0$. Using the Chebychev inequality $\mu_{G_+}(|V(z_0, f) - E(z_0, f)| \leq E(z_0, f)/2) \leq \gamma(z_0, f)^2$ we see from this that $V(z_0, f) \to \infty$ μ_{G_+}-a.s..

To determine the behavior of $\mathcal{C}'_{\Lambda(z_0)}(f)$ for $f \neq 0$ we however need an argument based on the hypercontractivity estimate $\|F\|_p \leq (p-1)^{n/2}\|F\|_2$ $\forall F$ that are in the $L^p(\mathcal{D}', \mathcal{B}, \mu_{G_+})$-closure of the span of Wick monomials $: \phi(f_1) \cdots \phi(f_s) :$ with $s \leq n$. Applying this to $V(z_0, f) = V_{\Lambda(z_0)}(\phi + H_+ f)$ with $n = 4$ one obtains

$$\mu_{G_+}\left(V(z_0, f) \leq \frac{E(z_0, f)}{2}\right) \leq \mu_{G_+}\left(|V(z_0, f) - E(z_0, f)| \geq \frac{E(z_0, f)}{2}\right)$$

$$\leq \frac{2^p}{E(z_0, f)^p}\|V(z_0, f) - E(z_0, f)\|_p^p$$

$$\leq \frac{2^p}{E(z_0, f)^p}(p-1)^{2p}\|V(z_0, f) - E(z_0, f)\|_2^p$$

$$= \gamma(z_0, f)^p(p-1)^{2p}. \tag{4.7}$$

The next step is to optimize this estimate wrt. p for $z_0 \to 0$. Equivalently, one can ask for the minimum of the logarithm of the rhs wrt. to p. Taking the p-derivative of this expression and setting it zero yields $0 = \log \gamma(z_0, f) + \frac{2p(z_0)}{p(z_0) - 1} + 2\log(p(z_0) - 1)$ with $p(z_0)$ the optimal p. Apparently, $p(z_0) \to \infty$ as $z_0 \to 0$ and thus $2p(z_0)/(p(z_0) - 1) \to 2$, hence $p(z_0)$ scales as

$$p(z_0) \sim e^{-1} \times \gamma(z_0, f)^{-1/2} \sim Ce^{-1} \times z_0^{-(\Delta_+ - d)/2}. \tag{4.8}$$

Combining (4.7) and (4.8) yields

$$\mu_{G_+}\left(V(z_0, f) < \frac{E(z_0, f)}{2}\right) \leq$$

$$\gamma(z_0, f)^{e^{-1} \times \gamma(z_0, f)^{-1/2}}\left(e^{-1} \times \gamma(z_0, f)^{-1/2} - 1\right)^{2e^{-1} \times \gamma(z_0, f)^{-1/2}}$$

$$\sim e^{-2e^{-1} \times \gamma(z_0, f)^{-1/2}}$$

$$\sim e^{-2Ce^{-1} \times z_0^{(d - \Delta_+)/2}} \tag{4.9}$$

We have thus seen that the portion of the probability space where $V(z_0, f)$ does not get large as $z_0 \to 0$ has a rapidly falling probability. We need an estimate that controls the negative values on this exceptional set. The ultra-violet cut-off implies $: \phi_\kappa^4 : (\underline{x}) \geq -Bc_\kappa^2$, B independent of κ, $c_\kappa = \sup_{\underline{x}, \underline{y}} |G_\kappa(\underline{x}, \underline{y})|$, μ_{G_+}-a.s., which provides us with a pointwise lower bound for $V(z_0, f)$ that is depending on z_0 as

$$V(z_0, f) \geq -\lambda Bc_\kappa^2 |\Lambda(z_0)| = -[\lambda Bc_\kappa^2 (2l)^d] \times (z_0^{-d} - A^{-d})/d \quad \mu_{G_+} - \text{a.s.} \tag{4.10}$$

Combination of (4.9) and (4.10) gives for z_0 sufficiently small

$$\mathbb{E}\left[e^{-V(z_0, f)}\right] \leq e^{-\frac{1}{2}E(z_0, f)} + e^{[\lambda Bc_\kappa^2 (2l)^d] \times (z_0^{-d} - A^{-d})/d - 2Ce^{-1} \times z_0^{(d - \Delta_+)/2}} \to 0 \tag{4.11}$$

if $\Delta_+ > 3d \Leftrightarrow m^2 > 6d^2$. Furthermore, by Jensen's inequality and $\mathbb{E}[V(z_0,0)] = 0$,

$$\mathbb{E}[e^{-V(z_0,0)}] \geq e^{-\mathbb{E}[V(z_0,0)]} = 1, \tag{4.12}$$

which implies that for m^2 sufficiently large

$$\mathcal{C}'_{\Lambda(z_0)}(f) = \frac{\mathbb{E}[e^{-V(z_0,f)}]}{\mathbb{E}[e^{-V(z_0,0)}]} \to 0 \text{ as } z_0 \to 0, \tag{4.13}$$

We have thus obtained the following result:

Theorem 4.1. *If the generating functional $\mathcal{C}(f) = \lim_\Lambda \mathcal{C}_\Lambda(f)$ exists for the UV-regularized $: \phi^4 :$-interaction and is unique (as required in order to obtain conformal invariance from AdS-invariance) it is also trivial ($\mathcal{C}(f) = 0$ if $f \neq 0$) provided $m^2 \geq 6d^2$.*

The above triviality result relies on three crucial assumptions.

   (i) The potential is quartic, cf. (4.3);
  (ii) There is a UV-cut-off;
 (iii) The mass is sufficiently large.

In order to assess the relevance of the triviality result for the general case, let us give some short comments on the role of each of these assumptions:

   (i) At the cost of a more restrictive mass bound, assumption (i) can easily be relaxed from quartic to polynomial interactions. For non-polynomial interactions, however, the hypercontractivity estimate can not be used. This might be of relevance, if we consider V as an effective potential, which in general will be non polynomial.

   (ii) The fact that there is a UV-cut-off enters our triviality argument via (4.10). When removing the UV-cut-off at least in dimension $d+1 = 2$, we therefore have to modify the triviality argument. It turns out that the bound obtained from the hypercontractivity estimate [5, 6] for the UV-problem is not good enough to reproduce the above argument. It seems to be necessary to combine UV and IR-hypercontractivity bounds in a single estimate in order to obtain triviality without cut-offs in $d + 1$ dimensions. We will come back to this point elsewhere.

   (iii) The mass bound to us rather seems to be a technical consequence of the methods used and not so much a true necessity for the onset of triviality. Different methods, e.g. based on decoupling via Dirichlet- and Neumann boundary conditions on a partition of $\mathbb{H}^{d+1}$ [5] e.g. combined with large deviation methods might very well lead to less restrictive mass bounds or eliminate them completely.

   On a heuristic level, the problem that expectation and variance of the shifted potential and the non shifted potential will have different scalings under the limit $\Lambda \nearrow \mathbb{H}^{d+1}$ prevails for a large class of polynomial and non-polynomial interactions with and without cut-offs. Thus, in our eyes, the three assumptions (i)–(iii) are not essential but rather technical. The result above therefore should be taken rather as an example of what can happen in the AdS/CFT correspondence than a definite mathematical statement. Of course, at the present and very preliminary state of the affair, everybody is free to think differently.

5. Conclusions and outlook

In this section we give an essentially non-technical discussion on repair strategies that would cure the obstacle of triviality.

(i) coupling constant renormalization: The simplest way to deal with the divergences in the potential energy $V(z_0, f)$ would be to make λ a z_0-dependent quantity. In fact, a naive guess at the scaling behavior suggests that $\lambda(z_0) \sim z_0^{d+4(\Delta_+ - d)}$ would compensate for the increase in the expected value of the interaction energy $V(z_0, f)$ such that with the modified coupling $\lim_{z_0 \to 0} \mathbb{E}[V(z_0, f)] = \lambda C \int_{\mathbb{R}^d} f^4 \, dx$ converges to a constant with $C = (\gamma_+ \Gamma(\Delta_+ - d/2)\Gamma(d/2)/2\Gamma(\Delta_+))^4$, cf. (4.1) and the paragraph thereafter. Furthermore, one can expect that the subleading terms ($j = 1 \ldots 4$ in (4.4)) converge to zero and do not affect the generating functional. It thus seems reasonable that with this renormalization the generating functional gives in the limit $z_0 \to 0$

$$\mathcal{C}(f) = e^{\frac{1}{2}\alpha_+(f,f) - \lambda C \int_{\mathbb{R}^d} f^4 \, dx} \tag{5.1}$$

which is reflection positive as a limit of reflection positive functionals (it is manifestly not stochastically positive for all $\lambda > 0$ and hence gives a nice illustration for the destruction of stochastic positivity due to the correction term in (3.6) and (3.7). The problem with this functional however is that the additional term in the interaction is an ultra local term and hence does not influence the corresponding real time CFT – which is a free theory determined by the analytic continuation of α_+. Hence this sort of renormalization only trades in another kind of triviality for the triviality observed in Section 4.

(ii) bulk counterterms: Such terms can simply be added to the (formal) Lagrangian. The problem to use this method in the AdS/CFT correspondence is twofold: Firstly, the infra-red divergences that are occurring in $V(z_0, f)$ are f-dependent. If we however want to cure them with f-dependent counterterms, the renormalization description of $\mathcal{C}_{\Lambda(z_0)}$ becomes f-dependent. Bulk counterterms however only preserve the structural properties of stochastic and reflection positivity, if the same renormalization prescription is chosen for all f. Hence, f-dependent counterterms would lead to a limiting functional, for which it is not known, whether it is reflection positive or not. The situation is worsened from the observation that, unlike in other IR problems, in the AdS/CFT correspondence the divergences in the nominator and denominator scale differently - as seen in our triviality result. This means for bulk counterterms, that, if they are working out fine for the nominator, they probably create new divergences in the denominator. Different renormalizations for the potential in the nominator and in the denominator in the limit might lead to a non normalizable vacuum for the boundary theory, which does not make sense.

(iii) boundary counterterms: The problems described above for bulk counterterms also have to be taken into account for boundary counterterms. Furthermore, while bulk counterterms, at least if they are not f-dependent, do not spoil the conformal invariance of the boundary theory, boundary counterterms theoretically might do so. Hence one needs a separate argument to show that they don't. But there is still another problem with boundary counterterms. We have seen that we can not take it for granted that a limiting functional measure exists for the boundary theory. But if the boundary theory is not described by a functional integral $\mu_{\text{bd.}}$, it is not clear how to define boundary counterterms on a mathematical basis: recall that a counterterm (at a finite value of the cut-off z_0 is defined by $d\mu_{\text{bd.,ren},z_0}(\varphi) = e^{-\mathcal{L}_{\text{ren}}(z_0,\varphi)}d\mu_{\text{bd.}}(\varphi)/\int_{\mathcal{D}(\mathbb{R}^d)} e^{-\mathcal{L}_{\text{ren}}(z_0,\varphi')}d\mu_{\text{bd.}}(\varphi')$ and it is not obvious how this can be defined if $\mu_{\text{bd.}}$ is not a measure.

(iv) giving up generating functionals: The triviality result of Section 4 relied on the scaling behavior of the expected value of $V(z_0, f)$ under the limit $z_0 \to 0$. This expected value can be associated with the Witten graph $\bigotimes$ which gives rise to the first order contribution to the four point function $\int_{\mathbb{H}^{d+1}} \prod_{l=1}^{4} H_+(\underline{x}, f_l)\, d_g\underline{x}$ which is converging as long as $\text{supp} f_j \cap \text{supp} f_l = \emptyset$ if $j \neq l$, cf (2.2) and (4.1) (see also [11] for concrete calculations). One may thus hope that the triviality result of Section 4 is an artefact of using generating functionals which makes it necessary to evaluate Schwinger functions at unphysical coinciding points. A reasonable approach to the infra-red problem in AdS/CFT would thus be to use (3.7) to define reflection positive Schwinger functions with cut off and then remove the cut-off for the Schwinger functions at physical (non coinciding) points, only. This might then work out without further renormalization along the lines of [5], as divergences might only occur on the diagonal. If this is true, triviality does only occur on the level of generating functionals – which are reminiscent of the Laplace transform of a functional measure for the boundary theory that might not exist in the present context.

Acknowledgement

The authors gratefully acknowledge interesting discussions with Sergio Albeverio, Matthias Blau, Michael Dütsch and Gordon Ritter. The first named author also would like to thank the organizers of the conference "Recent developments in QFT" for creating a very interesting meeting.

References

[1] Albeverio, S., Jost, J., Paycha, S., Scarlatti, S, A Mathematical Introduction to String Theory: Variational Problems, Geometric and Probabilistic Methods, London Mathematical Society Lecture Note Series, Cambridge University Press 1997.

[2] Bertola M., Bros J., Moschella U., Schaeffer R.: Decomposing quantum fields on branes. Nuclear Physics **B 581**, 575-603 (2000)

[3] Bros J., Epstein H., Moschella U.: Towards a general theory of quantized fields on the anti-de Sitter spacetime. Commun. Math. Phys. **231**, 481-528 (2002)

[4] Dütsch M., Rehren K.H.: A comment on the dual field in the AdS-CFT correspondence. Lett. Math. Phys. **62**, 171-184 (2002)

[5] Glimm J., Jaffe A.: *Quantum Physics. A Functional Integral Point of View*, 2nd edition. Springer, New-York 1987

[6] Gottschalk, H., Thaler, H.: *AdS/CFT correspondence in the Euclidean context*, math-ph/0611006, to appear in Commun. Math. Phys.

[7] Gottschalk H.: *Die Momente gefalteten Gauß-Poissonschen weißen Rauschens als Schwingerfunktionen*. Diploma thesis, Bochum 1995

[8] Gubser S.S., Klebanov I.R., Polyakov A.M.: Gauge theory correlators from noncritical string theory. Phys. Lett. **B 428**, 105-114 (1998)

[9] Haba Z.: Quantum field theory on manifolds with a boundary. J. Phys. **A 38**, 10393-10401 (2005)

[10] Jaffee A., Ritter G.: Quantum field theory on curved backgrounds II: Spacetime symmetries. arXiv:0704.0052v1 [hep-th]

[11] Kniemeyer O.: *Untersuchungen am erzeugenden Funktional der AdS-CFT-Korrespondenz*. Diploma thesis, Univ. Göttingen 2002

[12] Maldacena J.: The large N limit of superconformal field theories and supergravity. Adv. Theor. Math. Phys. **2**, 231-252 (1998)

[13] Osterwalder K., Schrader R.: Axioms for Euclidean Green's functions. Comm. Math. Phys. **31**, 83-112 (1973)

[14] Osterwalder K., Schrader R.: Axioms for Euclidean Green's functions. II. With an appendix by Stephen Summers. Comm. Math. Phys. **42**, 281–305 (1975)

[15] Rehren K.-H.: Algebraic holography. Ann. Henri Poincarè **1**, 607-623 (2000)

[16] Rehren, K.-H.: QFT lectures on AdS-CFT, arXiv:hep-th/0411086v1

[17] Witten E.: Anti-de Sitter space and holography. Adv. Theor. Math. Phys. **2**, 253-291 (1998)

Hanno Gottschalk
Institut für Angewandte Mathematik
Wegelerstr. 6
D–53115 Bonn
e-mail: gottscha@wiener.iam.uni-bonn.de

Horst Thaler
Dipartimento di Matematica e Informatica
Università di Camerino
I–62032 Camerino
e-mail: horst.thaler@unicam.it

Quantum Field Theory
B. Fauser, J. Tolksdorf and E. Zeidler, Eds., 83–94
© 2009 Birkhäuser Verlag Basel/Switzerland

Some Steps Towards Noncommutative Mirror Symmetry on the Torus

Karl-Georg Schlesinger

Abstract. Starting from motivating examples, we discuss some aspects of the question how to extend mirror symmetry to the case of the noncommutative torus. Following the - by now classical - approach of Dijkgraaf to mirror symmetry on elliptic curves, we will see how elliptic deformations of special functions (especially, the elliptic gamma function) arise. In the final section we indicate a possible way how these results might relate to deformations of Fukaya categories of elliptic curves.

Mathematics Subject Classification (2000). Primary 14J32; Secondary 14J15; 14J81.

Keywords. mirror symmetry, elliptic curves, noncommutative torus, string duality, Fukaya category, Gromov-Witten invariants.

1. Introduction

Mirror symmetry is a duality symmetry between certain two dimensional superconformal field theories which was discovered in the context of string theory in the early 1990s. Applied to sigma models, this leads to a completely new symmetry between Calabi-Yau spaces (of, in general, even differing topology). It is an open question if and how this duality might extend to the case of noncommutative spaces. There is far reaching work using the language of noncommutative algebraic geometry (see e.g. [11], [12], [19] and literature cited therein, to give a non-exhaustive selection), dealing with this question. We will follow a much more modest goal in these notes, restricting completely to the case of noncommutative elliptic curves (i.e. a noncommutative torus equipped with some additional structure). We still think that an approach focused on only one class of examples is justified since even in the commutative case the elliptic curve provides a highly nontrivial example for homological mirror symmetry.

The extension of mirror symmetry to noncommutative spaces is definitely not an example of generalization for the sake of generalization. We restrict, here, to two motivating examples (for additional motivation, see e.g. [12]).

In [20] the geometric Langlands correspondence is related to mirror symmetry for sigma models with a Hitchin moduli space target, for the special case of vanishing B-field. If a B-field is turned on in the sigma model, one gets a deformation to a noncommutative setting (twisted D-modules, quantum geometric Langlands correspondence). More generally, coupling of a B-field to both sides of the duality, the A- and the B-model, needs a noncommutative extension of mirror symmetry (see [24]).

Another concrete example of this type is provided by the $D5$-brane gauge theory in type IIB string theory. Here, the generalization to the noncommutative case is not just an option but is necessary for the existence of a space-filling coisotropic brane (which is decisive for the existence of a D-module structure on A-branes in [20]). We refer to [34] for the details of this example.

These notes are organized as follows: First, we review some needed results for the classical case of (commutative) elliptic curves. Next, we give our results for the noncommutative case (for the details we refer, again, to [34]). Especially, we suggest a formal extension of Gromov-Witten invariants to the case of noncommutative elliptic curves. In the final section, we give some preliminary arguments how our results might relate to certain exotic deformations of the Fukaya category of an elliptic curve. This is work in progress.

2. Elliptic curves

Mirror symmetry for elliptic curves was extensively studied in the mid 1990s (see especially [8], [9], [10], [16], [31]). It was a prime motivating source for the homological mirror symmetry conjecture of [17]. A proof of homological mirror symmetry for elliptic curves was established in [29]. The Strominger-Yau-Zaslow conjecture ([39]), proposing another mathematical formulation of mirror symmetry, was proved for elliptic curves (and, more generally, for Abelian varieties) in [18].

An elliptic curve $E_{t,\tau}$ is a smooth 2-torus equipped with a holomorphic and a symplectic structure. The holomorphic structure - parameterized by τ - is given by the representation of the elliptic curve as

$$\mathbb{C}/\left(\mathbb{Z} \oplus \mathbb{Z}\tau\right)$$

with $\tau = \tau_1 + i\tau_2 \in \mathbb{C}$ from the upper half plane $\mathbb{H}$, i.e. $\tau_2 > 0$. The symplectic structure - parameterized by $t \in \mathbb{H}$ - is given by the complexified Kähler class $[\omega] \in H^2\left(E_{t,\tau}, \mathbb{C}\right)$ with

$$\omega = -\frac{\pi t}{\tau_2} dz \wedge d\overline{z}$$

and for $t = t_1 + it_2$ the area of the elliptic curve is given by t_2. Mirror symmetry relates the elliptic curves $E_{t,\tau}$ and $E_{\tau,t}$.

For the symplectic structure, we have the Gromov-Witten invariants F_g, defined as the generating functions for counting d-fold connected covers of $E_{t,\tau}$ in genus g. One can combine the functions F_g into a two-variable partition function

$$Z(q, \lambda) = \exp \sum_{g=1}^{\infty} \lambda^{2g-2} F_g(q)$$

with $q = e^{2\pi i t}$. This is the famous partition function of Hurwitz which he used already in [14] to count connected covers of elliptic curves.

The partition function $Z(q, \lambda)$ can be calculated in three different ways (see Theorem 1 - Theorem 3 of [8]). The first case is a large N calculation in terms of $U(N)$ Yang-Mills theory on $E_{t,\tau}$. The second possibility is a calculation in terms of a Dirac fermion on the elliptic curve. Starting from Dirac spinors b, c on the elliptic curve with action

$$S = \int_{E_{t,\tau}} \left(b\bar{\partial}c + \lambda b\partial^2 c \right)$$

one shows that the operator product expansion defines a fermionic representation of the $W_{1+\infty}$ algebra. The partition function can be calculated as a generalized trace (as defined in [2]) of this algebra, leading to

$$Z(q, \lambda) = q^{-\frac{1}{24}} \oint \frac{dz}{2\pi i z} \prod_{p \in \mathbb{Z}_{\geq o} + \frac{1}{2}} \left(1 + zq^p e^{\lambda p^2} \right) \left(1 + \frac{1}{z} q^p e^{-\lambda p^2} \right)$$

where p runs over the positive half integers. For the action and the partition function above - and for the sequel of this paper - we have changed the notation to denote the parameter values of the mirror elliptic curve by t and τ. It is this representation of $Z(q, \lambda)$ which leads to the famous theorem of Dijkgraaf, Kaneko, Zagier stating that the functions $F_g(q)$ are quasi-modular forms (i.e. $F_g \in \mathbb{Q}[E_2, E_4, E_6]$ where E_2, E_4, E_6 are the classical Eisenstein series of weight 2, 4, and 6, respectively) and have weight $6g - 6$.

Finally, as in the case of Calabi-Yau 3-folds, by mirror symmetry $Z(q, \lambda)$ can be calculated as the partition function of a Kodaira-Spencer theory. In the case of elliptic curves, this is given by the action of a simple real bosonic field with $(\partial\varphi)^3$ interaction term, i.e. by the action

$$S(\varphi) = \int_{E_{\tau,t}} \left(\frac{1}{2} \partial\varphi\bar{\partial}\varphi + \frac{\lambda}{6} (-i\partial\varphi)^3 \right)$$

(see [8], [9] for the details).

3. Noncommutative elliptic curves

In this section, we will consider the question how the fermionic representation of $Z(q, \lambda)$ generalizes to the case of the noncommutative torus. The first question we have to consider is how a holomorphic structure can be introduced on the noncommutative torus since the smooth noncommutative torus – as in the commutative case – is not sufficient to consider mirror symmetry. A number of different approaches to this topic exist in the literature. Since the partition function $Z(q, \lambda)$ involves the modular parameter q, ranging over the whole family of elliptic curves, we will follow the approach of [35], [37] where noncommutative elliptic curves are introduced by extending the range of q, i.e. we get a larger family including the case of noncommutative elliptic curves. Concretely, in [35], [37] it is shown that one can view the noncommutative torus as the degenerate limit $|q| \to 1$ of classical elliptic curves (observe that since $t \in \mathbb{H}$ and $q = e^{2\pi it}$, $|q| < 1$ for classical elliptic curves). The noncommutative smooth torus (see [6]) can be introduced as a degenerate limit of a family of foliations where the foliations are defined by classical elliptic curves. In the limit, the foliation space becomes non-Hausdorff as a classical topological space but the algebras of functions still have a well-behaved limit. The limit of the algebras of functions is the noncommutative algebra defining the noncommutative smooth torus. In much the same way, one can introduce noncommutative elliptic curves as the degenerate limit of foliations defined by classical elliptic curves by studying the category of coherent sheaves instead of the algebra of functions. Again, the category of coherent sheaves has a well-behaved limit, leading to the notion of a noncommutative elliptic curve. We will make use of these results by using the fact that we can study the question of the generalization of $Z(q, \lambda)$ to noncommutative elliptic curves by considering the limit $|q| \to 1$. We would like to stress at this point that the category of coherent sheaves on noncommutative elliptic curves as introduced in [37] differs from the approach followed in [26] where holomorphic vector bundles on the smooth noncommutative torus are used.

Obviously, we can not directly perform this limit. This is very much related to the fact that there exist only very few results on q-analysis for $|q| = 1$. We will make use of the fact that there exists an elliptic deformation of the q-deformed gamma function and this elliptic gamma function (which has two deformation parameters) allows to take a limit in which a single unimodular deformation parameter arises ([32], [33]). In this sense, the elliptic gamma function includes the q-gamma function case with $|q| = 1$. We will therefore consider the problem of taking the limit $|q| \to 1$ for $Z(q, \lambda)$ in the more general form of looking for an elliptic analogue of $Z(q, \lambda)$. For simplicity, we will completely restrict to the case $\lambda = 0$. The general case can be treated in a completely analogous way (one has to replace the classical Jacobi theta function with the generalized theta functions of [16] for $\lambda \neq 0$, see [34]).

Using the substitution $q \mapsto q^2$, we have

$$
\begin{aligned}
Z(q, 0) &= q^{-\frac{1}{12}} \oint \frac{dz}{2\pi i z} \prod_{j \geq 0} \left(1 + zq^{2j+1}\right) \left(1 + \frac{1}{z} q^{2j+1}\right) \\
&= q^{-\frac{1}{12}} \oint \frac{dz}{2\pi i z} \left(-zq; q^2\right)_\infty \left(-\frac{q}{z}; q^2\right)_\infty
\end{aligned}
$$

where

$$(a;q)_n = \prod_{j=0}^{n-1} \left(1 - aq^j\right)$$

is the q-shifted factorial and

$$(a;q)_\infty = \prod_{j=0}^{\infty} \left(1 - aq^j\right)$$

the limit $n \to \infty$ which exists for $|q| < 1$. Remember that the classical Jacobi theta function

$$\vartheta(z,q) = \sum_{n=-\infty}^{n=+\infty} z^n q^{n^2}$$

can be expressed in the form of the Jacobi triple product as

$$\vartheta(z,q) = \left(-zq;q^2\right)_\infty \left(-\frac{q}{z};q^2\right)_\infty \left(q^2;q^2\right)_\infty$$

i.e. $Z(q,0)$ is basically given by an integral over the first two factors of $\vartheta(z,q)$. Rewriting $(a;q)_\infty$ in terms of the function Γ_q with

$$\Gamma_q(x) = \frac{q^{-\frac{x^2}{16}}}{\left(-q^{\frac{1}{2}(x+1)};q\right)_\infty}$$

(see [36]), we have

$$(a;q)_\infty = \frac{q^{-\left(\frac{2\frac{\log(a)}{\log(q)}-1}{4}\right)^2}}{\Gamma_q\left(2\frac{\log(a)}{\log(q)} - 1\right)}$$

and

$$Z(q,0) = q^{-\frac{1}{12}} \oint \frac{dz}{2\pi i z} \frac{q^{-\left(\frac{\log(-z)}{2\log(q)}\right)^2}}{\Gamma_{q^2}\left(\frac{\log(-z)}{\log(q)}\right) \Gamma_{q^2}\left(-\frac{\log(-z)}{\log(q)}\right)}.$$

For $q,p \in \mathbb{C}$ with $|q|,|p| < 1$ let

$$\Gamma(z;q,p) = \prod_{j,k=0}^{\infty} \frac{1 - z^{-1}q^{j+1}p^{k+1}}{1 - zq^j p^k}$$

be the elliptic gamma function of [32], [33]. Then an elliptic generalization of $Z(q,0)$ - which allows to take the limit to unimodular q – is given by

$$Z(q,p,0) = q^{-\frac{1}{12}}p^{-\frac{1}{12}} \oint \frac{dz}{2\pi i z} \frac{q^{-\left(\frac{\log(-z)}{2(\log(q)+\log(p))}\right)^2} p^{-\left(\frac{\log(-z)}{2(\log(q)+\log(p))}\right)^2}}{\Gamma\left(\frac{\log(-z)}{\log(q)+\log(p)};q^2,p^2\right) \Gamma\left(-\frac{\log(-z)}{\log(q)+\log(p)};q^2,p^2\right)}.$$

It is an open question for future research if this partition function can be related to a fermion system on the noncommutative torus in the limit of unimodular q.

As in the classical case of commutative elliptic curves, we can use the elliptic partition function to define Gromov-Witten invariants. Concretely, in the classical case the definition of the partition function as

$$Z\left(q,\lambda\right) = \exp\left(\sum_{g=1}^{\infty} \lambda^{2g-2} F_g\left(q\right)\right)$$

implies that we can calculate the Gromov-Witten invariants F_g as

$$F_g = \frac{1}{(2g-2)!} \left.\frac{\partial^{2g-2}\log(Z)}{\partial\lambda^{2g-2}}\right|_{\lambda=0}.$$

We can now use this equation, applied to the elliptic partition function from above, as a definition of elliptic Gromov-Witten invariants $F_g\left(q,p\right)$. The limit to a single unimodular parameter can be used as a definition of Gromov-Witten invariants for noncommutative elliptic curves.

Again, it is an open question what these invariants measure on the noncommutative torus. Definitely, they are not invariants of the smooth noncommutative torus since their commutative limit is a symplectic invariant. If they are in any sense nice invariants, they should go along with deformations of Floer homology and the Fukaya category of classical elliptic curves. Since the elliptic gamma function is a much more complicated and considerably richer object than the classical Jacobi theta function, we expect that the noncommutative Gromov-Witten invariants should be related to deformations of the Fukaya category of an elliptic curve to an algebraically even more complicated object. Without being able, at present, to answer the open questions sketched here, we will give an argument in the next section that exotic deformations of the Fukaya category of an elliptic curve should, indeed, exist.

4. Exotic deformations of the Fukaya category

This section contains material on work in progress. The work of [4] and [35] suggests that the Fukaya category of $E_{t,\tau}$ is closely related to the category of projective modules over a noncommutative torus (where the deformation of the algebra of functions is induced from the choice of symplectic structure involved in the definition of $E_{t,\tau}$). Let us assume this to be true in the sequel. We will argue that the moduli space of the Fukaya category (and consequently the moduli space of its bounded derived version), which is an object of central interest in homological mirror symmetry (see [17]), should possibly be larger than usually assumed (i.e. larger than the so called extended moduli space of [17], [40] for the mirror symmetric dual).

Consider the noncommutative torus T_Θ^2 (see [6] for the definition). As mentioned above, we assume that the category $\mathbf{Proj}(T_\Theta^2)$ of projective modules over the algebra of functions on T_Θ^2 is closely related to the Fukaya category of some elliptic curve (see [26] for a detailed discussion of this relationship). We now make a special choice of object M of $\mathbf{Proj}(T_\Theta^2)$ by assuming that M is a Heisenberg module (see e.g. [27]). As shown in [28] one can attach a two dimensional noncommutative gauge theory to M. While M has to be viewed as the noncommutative counterpart of a vector bundle on T_Θ^2, i.e. it refers to the classical gauge theory, it is also discussed in [28] how to quantize this theory. In conclusion, starting from M and turning on the deformation parameter $\hbar$, we arrive at a two dimensional quantized gauge theory (see also [7] for related work). As in the commutative case (see [5]), it was shown in [28] that this gauge theory is a cohomological field theory. In the limit $\hbar \to 0$ this field theory reduces to the usual cohomological field theory on an elliptic curve $E_{t,\tau}$ which is studied in [5] and determines the Gromov-Witten invariants of $E_{t,\tau}$. An algebraic model for this cohomological field theory is given by the Fukaya category $\mathbf{Fuk}(E_{t,\tau})$ of $E_{t,\tau}$. It is therefore natural to assume that the quantized cohomological field theory gained from M can algebraically be described by some deformation $\mathbf{F}_M$ of $\mathbf{Fuk}(E_{t,\tau})$. Now define the following 2-category $\widehat{F}(T_\Theta^2)$: For any Heisenberg module M take $\mathbf{F}_M$ as an object. Since it should be possible to reproduce M from $\mathbf{F}_M$ in the limit $\hbar \to 0$, we assume that we can treat all the other non-Heisenberg modules of $\mathbf{Proj}(T_\Theta^2)$ as trivial categories in some sense. We then define the 1-morphisms as functors and the 2-morphisms as natural transformations in the obvious way. This argument suggests that the moduli space of $\mathbf{Fuk}(E_{t,\tau})$ should be larger than usually assumed and should possibly include deformations of $\mathbf{Fuk}(E_{t,\tau})$ into 2-categories (observe that by considering the limit $\hbar \to 0$, for which $\mathbf{F}_M \to M$, $\widehat{F}(T_\Theta^2)$ is a continuous deformation of $\mathbf{Proj}(T_\Theta^2)$).

Remark 4.1. *The argument above shows that $\widehat{F}(T_\Theta^2)$ should be at least a 2-category. Since noncommutative deformations sometimes relate to higher categorical structures, one can a priori not exclude that $\mathbf{F}_M$ might itself be a 2-category (or a bicategory) and $\widehat{F}(T_\Theta^2)$ might therefore be an even higher category. We will give an argument below that this should, indeed, be the case.*

The view we have suggested here is that $\mathbf{F}_M$ appears as a categorical generalization of the module M while $\widehat{F}(T_\Theta^2)$ should be seen as a (higher) categorical generalization of $\mathbf{Proj}(T_\Theta^2)$. It is natural to assume then that there might exist even higher categorical deformations of these objects, i.e. we suggest to construct a 2-category deformation $\widehat{F}_M$ of $\mathbf{F}_M$ and from this – by replacing $\mathbf{F}_M$ with $\widehat{F}_M$ in the construction of $\widehat{F}(T_\Theta^2)$ – a deformation $\widehat{F}^{(2)}(T_\Theta^2)$ of $\widehat{F}(T_\Theta^2)$ which is at least a 3-category. Proceeding iteratively in this way, we could arrive at an n-category $\widehat{F}^{(n-1)}(T_\Theta^2)$.

So, the construction above suggests that there might even exist n-category deformations of $\mathbf{Fuk}(E_{t,\tau})$ beyond the 2-category deformation $\widehat{F}(T_\Theta^2)$, further enlarging the moduli space.

Let us now discuss the structural properties of the cohomological field theory on T_Θ^2, studied in [28], in more detail. As is shown there, this field theory has a BRST supercharge Q with respect to which it is a cohomological field theory. It follows from the general properties of the BRST-complex of a cohomological field theory that the operator product expansion of the theory should determine the structure of an A_∞-algebra. This A_∞-algebra should – as in the commutative case – be the formal completion of an A_∞-category (by introducing a formal value zero for undefined products), which in the limit where T_Θ^2 is send to the commutative elliptic curve $E_{t,\tau}$ becomes the A_∞-category determined by the usual cohomological field theory of [5] on $E_{t,\tau}$.

In the next step, we have to consider the inherent Morita equivalence on the cohomological field theory studied in [28] (for a general overview on Morita equivalence for noncommutative tori and its consequences we refer the reader to [30]). At the classical level, Morita equivalence means that the Heisenberg module M is actually a bimodule for the algebras of functions on the Morita equivalent noncommutative tori T_Θ^2 and $T_{\tilde\Theta}^2$. But there is strong evidence that Morita equivalence extends to a duality between the full quantized field theories defined from M on T_Θ^2 and $T_{\tilde\Theta}^2$, respectively (see [1], [3], [38]). This suggests that the abstract A_∞-category, related to the noncommutative cohomological field theory, should have different – dual – realizations on T_Θ^2 and $T_{\tilde\Theta}^2$.

Assume a precise notion of field theoretic realization of the A_∞-category would be available. From the above discussion, one has to conclude then that what one should actually consider from the quantum field theory point of view is all realizations of the A_∞-category (this is very much in accordance with the results gained in [13] in a different setting; in the noncommutative case this viewpoint seems unavoidable, due to Morita equivalence). One might be tempted to introduce the following 2-category as the correct algebraic framework, therefore: Objects should be the realizations of the A_∞-category and 1- and 2-morphisms should be given by the obvious functors and natural transformations. But this is not what one really wants from the physics perspective: The morphisms of the realization of the A_∞-category should correspond to the fields (ghost-fields, anti-ghost-fields) of the cohomological field theory. Given two realizations, we do not only want to know *which* fields are mapped onto each other (as would be the information given by a functor) but also *how* they are mapped to each other, i.e. we would like to know the precise transformation rule of the fields. As a consequence, the correct algebraic structure should not be that of a 2-category but that of an fc-multicategory as introduced in [21], [22]: The objects and the horizontal morphisms should be that of the realization of the A_∞-category while the vertical morphisms and the 2-cells should give the above mentioned transformation rules. Since an A_∞-category is not a proper category, one can not expect to get a double category, here, but has to assume the full structure of an fc-multicategory. In conclusion, we make the following conjecture:

Conjecture 4.2. *The algebraic structure, defined by the operator product expansion of the cohomological field theory on T_Θ^2 discussed in [28], should be given – correctly taking into account Morita equivalence – by an fc-multicategory.*

Returning to the deformation $\widehat{F}\left(T_\Theta^2\right)$ of **Fuk** $(E_{t,\tau})$, we therefore conjecture that $\widehat{F}\left(T_\Theta^2\right)$ should be given as the 2-category of all the fc-multicategories attached to the projective modules M in the way just described (actually, the dependence should now be not on T_Θ^2 but on the Morita equivalence class of T_Θ^2).

Remark 4.3. *It follows immediately that* $\widehat{F}\left(T_\Theta^2\right)$ *should have much more structure than that of a simple 2-category since every object is not just a category but an fc-multicategory. We suspect that* $\widehat{F}\left(T_\Theta^2\right)$ *gives an example of a next higher categorical analogue of an fc-multicategory, especially, it should be a generalization of a 3-category.*

There is also a more mathematical argument suggesting the same conclusion: If mirror symmetry generalizes to the noncommutative setting, one would expect that there is a dual description of $\widehat{F}\left(T_\Theta^2\right)$ as something like a category of coherent sheaves on a noncommutative torus. In the commutative case, the ring of classical theta functions gives a basis for global sections of line bundles. The noncommutative elliptic curve should then relate to a deformed ring of theta functions. This is indeed the case but the product of quantum theta functions is only partially defined, i.e. quantum theta functions do no longer form a ring but a linear category (see [23]). Vector bundles over a noncommutative elliptic curve (in the classical case corresponding to projective modules over the algebra of functions) should therefore correspond to categories with a module like structure. Consequently, the category of vector bundles and anything like a derived category of vector bundles (or derived category of coherent sheaves) should have the structure of a bicategory. Since in the classical case higher Massey products, turning the category of coherent sheaves into an A_∞-category, have to be taken into account to make homological mirror symmetry work (see [25]), we actually have to expect an A_∞-version of a bicategory (we can probably not expect the strictified version of a bicategory, called a 2-category, to appear on the nose, here). Again, Morita equivalence and the precise transformations induced by it on such A_∞-bicategories have to be taken into account for noncommutative elliptic curves and we finally arrive at the conclusion that the noncommutative analogue of the category of coherent sheaves - and therefore, if mirror symmetry holds, also $\widehat{F}\left(T_\Theta^2\right)$ - should be a generalization of a 3-category in the sense of a next higher categorical analogue of an fc-multicategory.

Future work is intended to deal with the question if these exotic and very rich deformations of the Fukaya category of an elliptic curve are in any way related to the elliptic deformation of the Hurwitz partition function, as introduced above.

5. Conclusion and outlook

The relation of the Fukaya category on X to a category of modules on a noncommutative deformation of X is argued in [15] to hold also for certain holomorphic symplectic manifolds X, especially for $X = T^4$. This means that also four dimensional noncommutative gauge theory might have a relation to deformations of Fukaya categories. This would mean that for noncommutative field theories the algebraic structure of two and four dimensional theories could be much more similar than in the commutative case (where e.g. no generalization of the full vertex algebra structure of two dimensional conformal field theory to higher dimensions has been found, so far). Observe that in this case the noncommutative field theories – since they are supposed to define deformations of cohomology theories – should be well-defined, especially they should be renormalizable. This is very different e.g. from deriving noncommutative field theories from open strings ending on D-branes with B-field backgrounds where the noncommutative field theories appear only as effective limits. We think that this indicates that the quantum field theories behind deformations of Fukaya categories should provide an extremely interesting testing ground for the subject of noncommutative quantum field theory.

Acknowledgements

I would like to thank H. Grosse, M. Kreuzer, and J. V. Stokman for discussions on or related to the material of this paper. I would like to thank the organizers (B. Fauser, J. Tolksdorf, and E. Zeidler) of the workshop *Recent developments in quantum field theory* at the MPI Leipzig for providing a productive and highly stimulating atmosphere. Finally, I would like to thank an unknown referee who made several suggestions improving the value of the paper.

References

[1] L. Alvarez-Gaume, J. L. F. Barbon, *Morita duality and large N limits*, Nucl.Phys. B623 (2002), 165–200.

[2] H. Awata, M. Fukuma, Y. Matsuo, S. Odake, *Representation theory of the* $W_{1+\infty}$ *algebra*, Prog. Theor. Phys. Suppl. 118 (1995), 343–374.

[3] J. Ambjorn, Y. M. Makeenko, J. Nishimura, R. J. Szabo, *Lattice gauge fields and discrete noncommutative Yang-Mills theory*, JHEP 0005 (2000), 023.

[4] P. Bressler, Y. Soibelman, *Mirror symmetry and deformation quantization*, hep-th/0202128.

[5] S. Cordes, G. Moore, S. Ramgoolam, *Lectures on 2d Yang-Mills theory, equivariant cohomology and topological field theories*, Nucl. Phys. Proc. Suppl. 41 (1995), 184–244.

[6] A. Connes, *Noncommutative geometry*, Academic Press, San Diego 1994.

[7] A. Connes, M. A. Rieffel, *Yang-Mills for noncommutative two-tori*, in: *Operator algebras and mathematical physics*, Iowa City 1985, Contemp. Math. Vol. 62, AMS, Providence, RI (1987), 237–266.

[8] R. Dijkgraaf, *Mirror symmetry and elliptic curves*, in *The moduli space of curves*, Proceedings of the Texel Island Meeting, April 1994, Birkhäuser Verlag, Basel 1995.

[9] R. Dijkgraaf, *Chiral deformations of conformal field theories*, Nucl. Phys. B493 (1997), 588–612.

[10] M. R. Douglas, *Conformal field theory techniques in large N Yang-Mills theory*, hep-th/9311130.

[11] V. Ginzburg, *Lectures on noncommutative geometry*, math/0506603.

[12] V. Ginzburg, *Calabi-Yau algebras*, math/0612139.

[13] M. Herbst, K. Hori, *B-type D-branes in toric Calabi-Yau varieties*, in *Homological Mirror Symmetry: Proceedings of the workshop at the Erwin Schrödinger Institute, Vienna, June 2006* (A. Kapustin, M. Kreuzer, K.-G. Schlesinger, eds.), to appear.

[14] A. Hurwitz, *Über die Anzahl der Riemannschen Flächen mit gegebenen Verzweigungspunkten*, Math. Ann. 55 (1902), 53.

[15] A. Kapustin, *A-branes and noncommutative geometry*, hep-th/0502212.

[16] M. Kaneko, D. Zagier, *A generalized Jacobi theta function and quasimodular forms*, in *The moduli space of curves*, Proceedings of the Texel Island Meeting, April 1994, Birkhäuser Verlag, Basel 1995.

[17] M. Kontsevich, *Homological algebra of mirror symmetry*, Proceedings ICM Zürich 1994, Birkhäuser Verlag, Basel 1995, 120–139.

[18] M. Kontsevich, Y. Soibelman, *Homological mirror symmetry and torus fibrations*, in: *Symplectic geometry and mirror symmetry*, Seoul 2000, World Scientific, River Edge, NJ 2001, 203–263.

[19] M. Kontsevich, Y. Soibelman, *Notes on A-infinity algebras, A-infinity categories and non-commutative geometry. I*, math/0606241, in *Homological Mirror Symmetry: Proceedings of the workshop at the Erwin Schrödinger Institute, Vienna, June 2006* (A. Kapustin, M. Kreuzer, K.-G. Schlesinger, eds.), to appear.

[20] A. Kapustin, E. Witten, *Electric-magnetic duality and the geometric Langlands program*, hep-th/0604151.

[21] T. Leinster, *fc-multicategories*, math/9903004.

[22] T. Leinster, *Higher operads, higher categories*, London Mathematical Society Lecture Notes Series 298, Cambridge University Press, Cambridge 2004.

[23] Y. I. Manin, *Theta functions, quantum tori and Heisenberg groups*, math/0011197.

[24] V. Mathai, J. Rosenberg, *T-duality for torus bundles with H-fluxes via noncommutative topology*, Comm. Math. Phys. 253 (2005), 705–721.

[25] A. Polishchuk, *Massey and Fukaya products on elliptic curves*, Adv. Theor. Math. Phys. 4 (2000), 1187–1207.

[26] A. Polishchuk, A. Schwarz, *Categories of holomorphic vector bundles on noncommutative two-tori*, Comm. Math. Phys. 236 (2003), 135–159.

[27] L. D. Paniak, R. J. Szabo, *Lectures on two-dimensional noncommutative gauge theory 1: Classical aspects*, hep-th/0302195.

[28] L. D. Paniak, R. J. Szabo, *Lectures on two-dimensional noncommutative gauge theory 2: Quantization*, hep-th/0304268.

[29] A. Polishchuk, E. Zaslow, *Categorical mirror symmetry: the elliptic curve*, Adv. Theor. Math. Phys. 2(2) (1998), 443–470.

[30] M. A. Rieffel, *Noncommutative tori - a case study of noncommutative differentiable manifolds*, in *Geometric and topological invariants of elliptic operators*, Brunswick, ME, 1988, Contemp. Math. vol. 105, AMS, Providence, RI, 1990, 191–211.

[31] R. Rudd, *The string partition function for QCD on the torus*, hep-th/9407176.

[32] S. N. M. Ruijsenaars, *First order analytic difference equations and integrable quantum systems*, J. Math. Phys. 38 (1997), no.2, 1069–1146.

[33] S. N. M. Ruijsenaars, *Special functions defined by analytic difference equations*, in J. Bustoz, M. E. H. Ismail, S. K. Suslov (eds.), *Special functions 2000: Current perspective and future directions*, NATO Sci. Ser. II Math. Phys Chem. 30, Kluwer, Dordrecht 2001.

[34] K.-G. Schlesinger, *Some remarks on mirror symmetry and noncommutative elliptic curves*, hep-th/0706.3817, submitted.

[35] Y. Soibelman, *Quantum tori, mirror symmetry and deformation theory*, Lett. Math. Phys. 56 (2001), 99–125.

[36] J. V. Stokman, *Askey-Wilson functions and quantum groups*, math.QA/0301330.

[37] Y. Soibelman, V. Vologodsky, *Non-commutative compactifications and elliptic curves*, math/0205117.

[38] N. Seiberg, E. Witten, *String theory and noncommutative geometry*, JHEP 9909 (1999), 032.

[39] A. Strominger, S.-T. Yau, E. Zaslow, *Mirror symmetry is T-duality*, Nucl. Phys. B479 (1996), 243-259.

[40] E. Witten, *Mirror manifolds and topological field theory*, in: S. T. Yau, *Essays on mirror manifolds*, International Press, Hong Kong 1991.

Karl-Georg Schlesinger
Institute for Theoretical Physics
Vienna University of Technology
Wiedner Hauptstrasse 8-10/136
AT–1040 Vienna
e-mail: kgschles@esi.ac.at

Quantum Field Theory
B. Fauser, J. Tolksdorf and E. Zeidler, Eds., 95–116
© 2009 Birkhäuser Verlag Basel/Switzerland

Witten's Volume Formula, Cohomological Pairings of Moduli Space of Flat Connections and Applications of Multiple Zeta Functions

Partha Guha

Abstract. We use Witten's volume formula to calculate the cohomological pairings of the moduli space of flat SU(3) connections. The cohomological pairings of moduli space of flat SU(2) connections are known from the work of Thaddeus-Witten-Donaldson, but for higher holonomy groups these pairings are largely unknown. We make some progress on these problems, and show that the pairings can be expressed in terms of multiple zeta functions.

Mathematics Subject Classification (2000). Primary 58G26; Secondary 14D20; 32G13; 58D27; 81T13.

Keywords. vector bundle, flat connection, Verlinde formula, cohomological pairing, multiple zeta values.

1. Introduction

This article emerges from the recently obtained connection between quantum field theory and algebraic geometry and it is devoted to the study of some cohomological properties of the moduli space of flat $SU(3)$ connections over a Riemann surface. Roughly speaking, one can study the cohomological pairings in three different ways: the first method was due to Thaddeus [33], the second one by Witten [39, 40], Donaldson [13] also proposed another method. The most up-to-date one was proposed by Jeffrey and Kirwan [25, 26]. In an interesting paper Earl and Kirwan [14] studied a generalization of the ring structure of the cohomology for the $SU(2)$ and $SU(3)$ moduli spaces. They addressed the general case $n > 3$, by constructing a complete set of generators for the ideal of relations, by suitably generalising the Mumford relations.

The moduli space $M(n,d)$ of semistable rank n degree d holomorphic vector bundles with fixed determinant on a compact Riemann surface Σ is a smooth Kähler manifold when n and d are coprime [2, 3, 12, 31]. Jeffrey and Kirwan [25, 26] gave full details of a mathematically rigorous proof of certain formulas for intersection pairings in the cohomology of moduli space $M(n,d)$ with complex coefficients. These formulas have been found by Witten by formally applying his version of non-Abelian localization to the infinite-dimensional space $\mathcal{A}$ of all U(n)-connections on Σ and the group of gauge transformations. Jeffrey and Kirwan used a non-Abelian localization technique to a certain finite-dimensional extended moduli space from which a moduli space $M(n,d)$ of semistable rank n degree d may be obtained by ordinary symplectic reduction. In this way they obtained Witten's formulas. It has been known (see [24]) that a moduli space of flat connections on principal G-bundles over Σ is a Marsden-Weinstein symplectic quotient of a finite-dimensional symplectic manifold [17] by a G-action.

We use Verlinde's formula [33, 36] of conformal field theory and complex geometry. Verlinde's formula gives the dimension of the space of conformal blocks in the WZW model on a Riemann surface. Verlinde's [36] result on the diagonalization of the fusion algebra gives a compact formula for the dimension of the space of conformal blocks. This formula coincides with the dimension of $H^0(\mathcal{M}_G, L^{\otimes k})$, where $\mathcal{M}_G$ is the moduli space of flat G bundles over the Riemann surface Σ_g of genus g and L is the generator of $Pic(\mathcal{M}_G)$. The formula for the dimension of these spaces, which is independent of the Riemann surface Σ, was proven by Tsuchiya, Ueno and Yamada [34]. The Verlinde formula has given rise to a great deal of excitement and new mathematics of infinite-dimensional variety (it is an ind-scheme) [9, 15, 28]. An ind-scheme is a directed system of schemes, that is, an ind-object of the category of schemes.

We can write the celebrated formula of Verlinde as

$$\dim H^0(\mathcal{M}_G, L^{\otimes k}) = \sum_\alpha \frac{1}{S_{0,\alpha}^{2g-2}} \, . \tag{1.1}$$

Here, α runs over the representatives of G which are the highest weights of integrable representations of the corresponding affine group $\widehat{G}$ at level k and $S_{\alpha,\gamma}$ is a matrix arising from the modular transformation of the character of the affine group $\widehat{G}$ at level k. If $\chi_\alpha(\tau)$ is the character of $\widehat{G}$ at level k with highest weight α, then S is defined by the formula

$$\chi_\alpha(-1/\tau) = \sum_\beta S_{\alpha\beta}\chi_\beta(\tau).$$

As an example we see that when $G = SU(2)$, then

$$S_{ij} = \left(\frac{2}{k+2}\right)^{1/2} \sin\frac{\pi(i+1)(j+1)}{k+2}.$$

Hence we obtain

$$\dim H^0(\mathcal{M}, L^{\otimes k}) = (\frac{k+2}{2})^{g-1} \sum_{j=0}^{k} (\frac{1}{\sin\frac{\pi(j+1)}{k+2}})^{2g-2}. \tag{1.2}$$

The volume of the moduli space is obtained from the Verlinde formula (1.1) and is given by

$$\mathrm{Vol}^F(\mathcal{M}) = \lim_{k\to\infty} k^{-n} \dim H^0(\mathcal{M}, L^{\otimes k}). \tag{1.3}$$

The Hirzebruch-Riemann-Roch formula is

$$\dim H^0(\mathcal{M}, L^{\otimes k}) = \langle \exp(kc_1(L)) \cdot Td(\mathcal{M}), \mathcal{M} \rangle,$$

where $Td(\mathcal{M})$ denotes the Todd class. For large k, this yields (for $G = SU(2)$)

$$\dim H^0(\mathcal{M}, L^{\otimes k}) \sim \frac{k^{3g-3}}{(3g-3)!} \langle c_1(L)^{3g-3}, \mathcal{M} \rangle.$$

Since $c_1(L)$ is represented by the symplectic form ω in de Rham cohomology, $\frac{\langle c_1(L)^{3g-3}, \mathcal{M} \rangle}{(3g-3)!}$ coincides with the volume of the moduli space $\mathrm{Vol}(\mathcal{M})$.

Incidentally, Witten gave a volume formula in [39] for the moduli space of flat connections for general G. It is given by

$$\mathrm{Vol}(\mathcal{M}) = \frac{\sharp Z(G).(\mathrm{Vol}(G))^{2g-2}}{(2\pi)^{\dim \mathcal{M}}} \sum_{\alpha} \frac{1}{(\dim \alpha)^{2g-2}} \tag{1.4}$$

where α runs over all the irreducible representations of G. Here, $\sharp Z(G)$ is the number of elements in the center of G.

In principle, Witten's volume formula is applicable to any G. But unfortunately some computational difficulties arise when $G = SU(n)$ for $n \geq 3$. In this case one faces the problem to find the matrix $S_{\alpha\beta}$ from the modular transformation of the Weyl-Kac character formula [19, 23].

In this article we obtain the volume formula for $G = SU(3)$ by computing the matrix $S_{\alpha\beta}$:

$$S_{0\lambda} = \frac{8}{\sqrt{6}(k+3)} \sin\frac{\pi\lambda_1}{k+3} \sin\frac{\pi\lambda_2}{k+3} \sin\frac{\pi(\lambda_1+\lambda_2)}{k+3}. \tag{1.5}$$

The computation of the volume formula is a *two step process*. At first we obtain the Verlinde formula for the moduli space of $SU(3)$ of flat connections by substituting the value of $S_{\alpha\beta}$ in (1.1). We obtain

Proposition 1.1.

$$\dim H^0(\mathcal{M}, L^{\otimes k}) = \frac{(k+3)^{2g-2}6^{g-1}}{2^{6g-6}} \sum_{\lambda_1,\lambda_2} (\frac{1}{\sin\frac{\pi\lambda_1}{k+3} \sin\frac{\pi\lambda_2}{k+3} \sin\frac{\pi(\lambda_1+\lambda_2)}{k+3}})^{2g-2}.$$

In the next step, using the above formula and Witten's prescription for large k limit, we obtain the volume of the moduli space of flat $SU(3)$ connections; this yields

Proposition 1.2.

$$\text{Vol}(\mathcal{M})_{SU(3)} = 3\frac{6^{g-1}}{(2\pi)^{6g-6}} \sum_{n_1,n_2}^{\infty} n_1^{-(2g-2)} n_2^{-(2g-2)} (n_1 + n_2)^{-(2g-2)}.$$

Here, a multiple zeta function appears in the volume of flat $SU(3)$ connections $\text{Vol}(\mathcal{M})_{SU(3)}$:

$$\zeta_g(A, 2g - 2) = \sum_{n_1,n_2}^{\infty} \frac{1}{n_1^{2g-2} n_2^{2g-2} (n_1 + n_2)^{2g-2}}. \tag{1.6}$$

It is a member of a family of a much larger class of zeta functions, known as multiple zeta functions. The Euler-Zagier multiple zeta functions are nested generalizations of the Riemann zeta function [37, 38]. They are defined as

$$\zeta_k(s_1, \dots, s_k) = \sum_{0<n_1<\cdots<n_k} n_1^{-s_1} \cdots n_k^{-s_k}.$$

Here, $s_1, \cdots, s_k \in \mathbb{Z}$, $s_1 \geq 2$, $s_j \geq 1$ for $2 \leq j \leq k$. For $k = 1$ this reduces to Riemann's zeta function. We call k the length or depth of $\underline{s}$ and $|\underline{s}| = \sum s_j$ the weight of $\underline{s}$.

Unlike as for the Riemann zeta function one could determine several algebraic relations between the multiple zeta values (MZV). One type of such relations appears when one multiplies two such series. In fact, one gets a linear combination of MZV. A simple example is stated below:

$$\zeta(s)\zeta(s') = \sum_{n\geq 1} \frac{1}{n^s} \sum_{m\geq 1} \frac{1}{m^{s'}}$$

$$= \sum_{n>m} + \sum_{n<m} + \sum_{n=m} = \zeta(s, s') + \zeta(s', s) + \zeta(s + s').$$

This is a quadratic relation among zeta values. In general the quadratic relation is given as

$$\zeta(\underline{s})\zeta(\underline{s}') = \sum_{\sigma} \underline{\sigma}.$$

This manipulation leads to the definition of a product called stuffle product [6]. This is a formal sum defined recursively by

$$aP * bQ = a(P * bQ) + b(aP * Q) + (a + b)(P * Q).$$

A simple example is

$$\zeta(s)^2 = 2\zeta(s, s) + \zeta(2s).$$

Then for $s = 2$, $\zeta(2) = \pi^2/6$ and $\zeta(4) = \pi^4/90$, we obtain

$$\zeta(2,2) = \sum_{m>n\geq 1} (mn)^{-2} = \frac{\pi^4}{120}.$$

Another example is $\zeta(2)\zeta(3) = \zeta(2,3) + \zeta(3,2) + \zeta(5)$.

Remark. In the same way as the stuffle product arises in the reorganization of multiple sums, multiple integrals lead to the definition of the shuffle product of words over an alphabet (with two letters) $X = \{x_0, x_1\}$. The words are given as $X^* = \{x_0^{a_1} x_1^{b_1} \cdots x_0^{a_k} x_1^{b_k}\}$. This product is defined by the same formula as the stuffle product except that the last term in the sum is omitted. The algebraic relations between multiple polylogarithms

$$Li_{(s_1 \cdots s_k)}(z) = \sum_{n_1 > n_2 > \cdots n_k \geq 1} \frac{z^{n_1}}{n_1^{s_1} \cdots n_k^{s_k}} \qquad |z| < 1 \; \forall s_j \geq 1$$

is generated by the shuffle relation. In fact, these multiple polylogarithms can be expressed as iterated Chen integrals, and from this representation one obtains the shuffle relations (for an example, see [7]).

In our case, Zagier [41, 42] gave a formula for calculating the values of this particular multiple zeta function. This result is also derived using the stuffle product. The *key formula* to compute our volume form is given by

$$\sum_{m,n}^{\infty} \frac{1}{m^s n^s (m+n)^s} = \frac{4}{3} \sum_{0 \leq r \leq s;\, r \, even} \binom{2s-r-1}{s-1} \zeta(r)\zeta(3s-r). \qquad (1.7)$$

Witten's volume formula can be extended to the moduli space of vector bundles with marked points $z_1, z_2, \ldots, z_p \in \Sigma_g$. We associate to each marked point z_i an irreducible representation Γ of $G_{\mathbf{C}}$. If λ is the highest weight of Γ, then $(\lambda, \alpha_{max}) \leq k$ where α_{max} is the highest root and $(\ ,\ )$ is the basic inner product (see appendix, [23]): alternatively λ is in the fundamental domain of the action of the affine Weyl group at level $k + h$ on the Cartan subalgebra (Lie algebra of the maximal torus). We sum over representations Γ for which, if λ is the highest weight of Γ_λ, the representation of dimension $(n + 1)$ and all the marked points are labeled by Γ_{n_r}. We associate a complex vector space to each labeled Riemann surface.

The generalized Verlinde formula for a group G in the presence of marked points [39] is

$$\dim H^0(\mathcal{M}_G, L^{\otimes k} \otimes \bigotimes_i \Gamma_{n_i}) = \sum_{j=0}^{k} \frac{1}{S_{0,j}^{2g-2+p}} \prod_{i=1}^{p} S_{n_i,j} \qquad (1.8)$$

and the vector space $H^0(\mathcal{M}_G, L^{\otimes k} \otimes \bigotimes_i \Gamma_{n_i})$ is independent of the details of the positions of the marked points.

Similarly, the volume of the generalized moduli space can be obtained from this generalized Verlinde formula (1.8) by extracting the term at the large k limit (1.3).

The volume formula for $G = SU(2)$ with marked points is given as

$$\text{Vol}^F(\mathcal{M}_t) = 2.\frac{1}{2^{g-1}\pi^{2g-2+p}} \sum_{n=1}^{p} \frac{\prod_{i=1}^{p} sin(\pi n t_i)}{n^{2g-2+p}}. \tag{1.9}$$

Our strategy is to compute the volume using the Verlinde formula in the large k limit rather than using Witten's volume formula directly, and this will be our recipe to find the volume of the moduli space.

Unlike the $SU(2)$ case, we obtain the volume formula for the moduli space of flat $SU(3)$ connections in terms of multiple zeta functions or double Bernoulli numbers [4, 5].

We obtain the volume formula of the moduli space of flat $SU(3)$ connections over one marked point Riemann surface.

Proposition 1.3.

$$\text{Vol}^F(\mathcal{M}_t) = \frac{3.6^{g-1}}{2^{6g-6}\pi^{6g-3}} \sum \frac{\sin \pi n_1 t_1. \sin \pi n_2 t_2. \sin \pi (n_1 + n_2)(t_1 + t_2)}{n_1^{2g-1} n_2^{2g-1}(n_1 + n_2)^{2g-1}}$$

t_1 and t_2 are restricted to

$$0 < t_i < 1.$$

When we expand the sine terms, we obtain a comprehensive volume formula for finding the intersection pairings of moduli space.

Witten's idea is based on the symplectic volume of the moduli space of flat connections. The moduli space $\mathcal{M}$ of flat connections of any semi-simple group G is a symplectic variety with a symplectic form ω. The volume of the moduli space of flat $SU(n)$ connections is

$$\text{Vol}^S(\mathcal{M}) = \frac{1}{r!} \int_{\mathcal{M}} \omega^r,$$

where $r = (n^2 - 1)(g - 1) = (g - 1)\dim G$ is the dimension of the moduli space.

Witten showed in [39] that the Reidemeister torsion of a Riemann surface, equipped with a flat connection, determines a natural volume form on the moduli space of flat connections, which agrees with the symplectic volume. Given a chain complex $C_\bullet$ that computes $H_*(\Sigma, ad(E))$, we define the torsion $\tau(C_\bullet)$ as a vector in

$$(\det H_0(\Sigma, ad(E)))^{-1} \otimes \det H_1(\Sigma, ad(E)) \otimes \det H_2(\Sigma, ad(E)))^{-1}.$$

For an irreducible flat connection:

$$H_0(\Sigma, ad(E)) = H_2(\Sigma, ad(E)) = 0.$$

So $\tau(C_\bullet)$ defines a vector in $\det H_1(\Sigma, \mathrm{ad}(E))$. Witten [39] gave the actual road map to compute the volume of $\mathcal{M}$.

Motivation. The result of this paper first appeared in [20]. Apparently one would ask why do we need another paper to study cohomological pairings, when Jeffrey and Kirwan [25, 26] gave full details of a mathematically rigorous proof of Witten's formulas for intersection numbers in the moduli spaces of flat connections. Indeed, the knowledge of the volume formula in principle allows us to calculate the full list of cohomology pairings for the moduli space of arbitrary rank. The main difficulty comes from computation. Our article is an explicit example of the computation of intersection pairings in the cohomology of the moduli space of flat $SU(3)$ connections. This involves the computation of multiple zeta functions and hence it is fairly difficult to compute intersection pairings for higher rank vector bundles. Our approach is based on the volume of the moduli spaces of parabolic bundles prescribed by Witten. The Jeffrey-Kirwan formulation yields formulas for all intersection numbers, whereas our approach yields formulas for the intersection numbers of restricted cases. For example, we exclude some cases that could yield the intersection numbers of some algebraic cycles in the moduli spaces.

2. Background about moduli space

Let us give a quick description of a moduli space [3]. Let Σ_g be a compact Riemann surface of genus g. Let E be the G bundle over Σ_g — here G can be any compact Lie group. For simplicity we shall work with the special case $G = SU(n)$. Let us consider the space of flat G connections over a Riemann surface Σ_g. We consider the space $\mathrm{Hom}(\pi_(\Sigma_g), G)/G$, which parameterizes the conjugacy classes of homomorphisms

$$\pi_1(\Sigma_g) \longrightarrow G. \qquad (2.1)$$

Now, $\pi_1(\Sigma_g)$ has generators $A_1, A_2, \cdots, A_g, B_1, B_2, \cdots, B_g$ that satisfy

$$\prod_{i=1}^{g} [A_i, B_i] = 1. \qquad (2.2)$$

It follows that $H^1(\Sigma_g, G)$ is the quotient by G of the subset of G^{2g} lying over 1 in the map $G^g \times G^g \to G$ given by $\prod[A_i, B_i]$. This shows clearly that $H^1(\Sigma_g, G)$ is a compact Hausdorff space.

We fix our structure group $G = SU(n)$. Let us consider a point $x \in \Sigma_g$. Suppose we cut out a small disc D around the point x. We fix the holonomy of the connection around the disc D to be $\exp(2\pi i p/n)$, where p and n are coprime to each other. Actually this holonomy $\exp(2\pi i p/n)$ around the point x ensures the irreducibility of the connection.

Consider a map

$$f_g : SU(n)^g \times SU(n)^g \longmapsto SU(n)$$

defined by

$$(A_1, B_1, \ldots, A_g, B_g) \longmapsto \prod_i^g A_i B_i A_i^{-1} B_i^{-1}.$$

In particular we select the subspace $W_g = f_g^{-1}(\exp(2\pi i p/n)$ of $SU(n)^{2g}$.

A point, say x, in the space $SU(n)^g \times SU(n)^g$ is considered to be reducible if there exists a matrix T in $SU(n)$ such that $(TA_iT^{-1}, TB_iT^{-1} \ldots)$ are all diagonal. When $n > 2$, we should include also those points where there exist matrices that can be simultaneously block diagonalized, for example, in the case of $SU(3)$ this would go into $S(U(2) \times U(1))$. If x is a reducible point of $SU(n)^{2g}$, then $f_g(x) = I$, so the connections take values in the Abelian subgroup of $SU(n)$.

Now it follows that the diagonal conjugation action of $SU(n)/Z(G) = PU(n)$ (where $Z(G)$ is the center of $SU(n)$) on $SU(n)^{2g}$ clearly preserves W_g and also by Schur's Lemma that the restriction of the action is free. Hence, the quotient $W_g/PU(n)$ is a smooth compact Hausdorff space, it is a manifold of dimension $2(\mathbf{g} - 1) \dim \mathbf{G}$ at all irreducible points (i.e. when the image of $\pi_1(\Sigma)$ generates G).

We can give an *equivalent description* of this moduli space in the holomorphic way (see [2, 21, 22]). The space of connections $\mathcal{A}$ over E is an affine space modeled on $\Omega^1(\Sigma_g, \mathrm{ad}\, E)$, such that the tangent space of $\mathcal{A}$ at any point is canonically identified with $\Omega^1(\Sigma_g, \mathrm{ad}\, E)$. Let us consider a decomposition of

$$\Omega^1(\Sigma_g, \mathrm{ad}\, E) \otimes \mathbf{C} = \Omega^{1,0}(\Sigma_g, \mathrm{ad}\, E^C) \oplus \Omega^{0,1}(\Sigma_g, \mathrm{ad}\, E^C).$$

If we consider an isomorphism between $\Omega^1(\Sigma_g, \mathrm{ad}\, E)$ and $\Omega^{0,1}(\Sigma_g, \mathrm{ad}\, E^C)$, we obtain a complex structure on the modeled space of $\mathcal{A}$ and hence also on $\mathcal{A}$. We say that $\mathcal{A}$ is the space of $\bar{\partial}$ operators on E^C. In the holomorphic picture we must restrict to a stable bundle [30] in order to obtain a smooth moduli space. A holomorphic vector bundle E is semi-stable over a Riemann surface, if for all sub-bundles F it satisfies

$$\frac{degF}{rankF} \leq \frac{degE}{rankE}.$$

Here, "degree" stands for the value of the first Chern class. The vector bundle E is a stable bundle if this inequality is strict. When the degree and the rank are coprime, then all the semi-stable bundles are stable. In this holomorphic picture the moduli space is interpreted as the space of gauge equivalence classes of stable vector bundles, i.e. $\mathcal{M}(\Sigma, G) = \mathcal{A}^S/G^C$, where $\mathrm{Aut}(E^C) = G^C$ acts on $\mathcal{A}^S$ with the constant scalars as the only isotropy group. The celebrated theorem of Narasimhan and Seshadri [31] connects both the pictures and it states that stable bundles arising from the representations of π_1 give irreducible representations.

Theorem 2.1 (Narasimhan-Seshadri). [31] *A holomorphic vector bundle of rank n is stable, if and only if it arises from an irreducible projective unitary representation of the fundamental group. Moreover, isomorphic bundles correspond to equivalent representations.*

The more general moduli spaces of flat $SU(n)$ connections over punctured Riemann surfaces have been studied by Mehta and Seshadri [29]. In the presence of a marked point on Σ_g we associate a conjugacy class of $SU(n)$ to it:

$$\Gamma \sim \begin{pmatrix} e^{2\pi i \gamma_1 / n} & 0 & \cdots & 0 \\ 0 & e^{2\pi i \gamma_2 / n} & \cdots & 0 \\ \cdots & \cdots & \cdots & \cdots \\ 0 & 0 & \cdots & e^{2\pi i \gamma_n / n} \end{pmatrix}$$

for all $0 < \gamma_i < 1$, where $\sum_{i=1}^{n} \gamma_i = 0$. The holonomy around this marked point takes its value in this conjugacy class. In presence of the marked points $z_1, z_2, \ldots, z_p$ we associate a set of conjugacy classes Γ_i of $SU(n)$. Consider a homomorphism

$$\pi_1(\Sigma_g - (z_1 \cup z_2 \cup \ldots \cup z_p)) \longrightarrow G$$

such that the loop around each z_i takes values in Γ_i, and the moduli space is the quotient by G of the fibre over 1 in the multiplication map

$$\Gamma_1 \times \Gamma_2 \times \cdots \times \Gamma_p \longrightarrow SU(n).$$

In other words, when we factor out the conjugacy, then we obtain the moduli space of parabolic bundles with weight $(\gamma_1, \ldots, \gamma_n)$. The dimension of the generalized moduli space [3] is

$$2(g-1)\dim G + \sum_{j=1}^{p} \dim \Gamma_j.$$

This moduli space of parabolic bundles over the punctured Riemann surface can be given a holomorphic picture, too. Mehta and Seshadri [29] have given the notion of stability in this case. This involves assigning weights given by the eigenvalues of Γ_i at each marked point.

3. Volume of the moduli space of $SU(2)$ flat connections

Let us quickly recapitulate the known case. We recall that $S_{\alpha\beta}$ is obtained from the modular transformation induced on the characters of level k. Let $\chi_\alpha(\tau)$ be the character of the affine group $\widehat{G}$, then by the modular transformation $\tau \to \frac{1}{\tau}$ [11, 27], we obtain matrix $S_{\alpha\beta}$, where α is the highest weight.

$$\chi_\alpha\left(-\frac{1}{\tau}\right) = \sum_\beta S_{\alpha\beta} \chi_\beta(\tau).$$

The key to construct this character is from the Weyl-Kac formula (for example [11, 27]). We define the character of the representation $L(\lambda)$ to be the function

$$ch_\lambda(t) = tr_{L(\lambda)} \exp(t)$$

where $t \in \hat{t}$ and $\hat{t}$ is the Cartan subalgebra of $\widehat{sl_n}$. The Weyl-Kac character formula is given by

$$tr_{L(\lambda)} \exp(t) = \frac{\sum sign(w) \exp(w(\lambda + \rho)|t)}{\sum sign(w) \exp(w(\rho)|t)}$$

where summations run over w in the Weyl group. The Weyl-Kac character formula is essentially the same as the Weyl character formula, but it differs in two minor ways, i.e. besides the usual root vectors, we also describe states by the number operator and the c-number term. Then k is the eigenvalue of the number operator.

The affine Weyl group W_{aff} is the semi-direct product of the ordinary Weyl group and the translation T_λ given by the co-root $\lambda^\vee$ of the highest root λ.

$$S_{ij} = (\frac{2}{k+2})^{\frac{1}{2}} \frac{\sin \pi(i+1)(j+1)}{k+2}.$$

Using Verlinde's formula, we obtain

$$\dim H^0(\mathcal{M}, L^k) = \sum_j (\frac{1}{S_{0,j}})^{2g-2}$$

$$R.H.S = \sum_j (\frac{1}{S_{0,j}})^{2g-2}$$

$$= (\frac{k+2}{2})^{g-1} \sum_{j=0}^{k} (\frac{1}{\frac{\sin \pi(j+1)}{k+2}})^{2g-2}.$$

Since our goal is to obtain a formula for the volume of the moduli space $\mathcal{M}$, we need to extract a term proportional to $k^{\dim_C \mathcal{M}} = k^{3g-3}$ for $k \longrightarrow \infty$. The two regions, namely $j \ll k$ and $k - j \ll k$, make equal contributions. In order to see this we use asymptotic analysis.

3.1. Asymptotic analysis and computation of volume of moduli spaces
We want to show that this is asymptotic to

$$\frac{2k^{3g-3}}{2^{g-1}\pi^{2g-2}} \cdot \sum_{r=1}^{\infty} \frac{1}{r^{2g-2}} \qquad \text{as} \quad k \longrightarrow \infty.$$

We assume that $g \geq 2$ and write $n = 2g - 2 \geq 2$. We also replace k by $l = k+2$, so the sum is:

$$\Sigma_l = (\frac{l}{2})^{n/2} \sum_{j=1}^{l-1} (\frac{1}{\sin(\pi j/l)})^n.$$

We divide this sum into the combination from $j \leq l/2$, $j \geq l/2$: these are essentially the same, hence it suffices to treat the first one. For $\epsilon > 0$, we write

$$\sum_{j=1}^{l/2}(\frac{1}{\sin \pi j/l})^n = \sum_{j=1}^{[\epsilon l]}(\frac{1}{\sin \pi j/l})^n + \sum_{[\epsilon l]+1}^{l/2}(\frac{1}{\sin \pi j/l})^n$$

$$= S + T, \quad \text{say}.$$

We want to compare the sum S with $S' = \sum_{j=1}^{\infty}(\frac{l}{\pi j})^n$. The difference $S - S'$ arises from two factors – approximating the sin function by its derivative and changing the range of summation. For the first case we have, for small ϵ and $j/l < \epsilon$,

$$\pi j/l \geq \sin(\pi j/l) \geq \pi j/l - 1/6(\pi j/l)^3.$$

This implies that, for some constant C,

$$(\frac{l}{\pi j})^n \leq (\frac{1}{\sin(\pi j/l)})^n \leq (\frac{k^n}{\pi j})^n(1 + C(\frac{j}{l})^2).$$

So

$$|\sum_{j=1}^{[\epsilon l]}(\frac{1}{\sin(\frac{\pi j}{l})})^n - \sum_{j=1}^{[\epsilon l]}(\frac{l}{\pi j})^n| \leq C\sum_{j=1}^{\epsilon l}\frac{l^{n-2}}{j^{n-2}} \leq C'l^{n-1},$$

for some C' (since these are $O(l)$ terms in the sum).
For the second factor:

$$\sum_{j=1}^{\infty}(\frac{l}{\pi j})^n - \sum_{j=1}^{[\epsilon l]}(\frac{l}{\pi j})^n = \sum_{[\epsilon l]+1}^{\infty}(\frac{l}{\pi j})^n = O(l^{n-1})$$

by comparing with the integral $\int_{\epsilon l}^{\infty} x^{-n} dx$. Hence, we see that $S - S'$ is $O(l^{n-1})$. Finally consider the other term T:

$$T = \sum_{[\epsilon l]+1}^{l/2}(\frac{1}{\sin(\pi j/l)})^n.$$

In this sum

$$\sin(\pi j/l) \geq \delta(\epsilon)$$

say so, for fixed ϵ, $T = O(l)$ (the number of terms in the sum). Putting all of this together, we see that

$$\Sigma_l = (\frac{l}{2})^{n/2}.(2\sum_{j=1}^{\infty}(\frac{l}{\pi j})^n + O(l^{n-1}),$$

which gives the required result.

Hence, for large k we obtain

$$\dim H^0(\mathcal{M}, L^k) \sim 2(\frac{k+2}{2})^{g-1} \sum_{j=0}^{k} (\frac{k+2}{\pi(j+1)})^{2g-2}$$

$$= 2\frac{k^{3g-3}}{2^{g-1}\pi^{2g-2}} \sum_{n=1}^{\infty} \frac{1}{n^{2g-2}}.$$

This finally yields

$$\dim H^0(\mathcal{M}, L^k) = 2\frac{k^{3g-3}}{2^{g-1}\pi^{2g-2}}\zeta(2g-2). \tag{3.1}$$

From the algebraic geometry point of view this dimension can be expressed via the Hirzebruch-Riemann-Roch theorem [18]

$$\dim H^0(\mathcal{M}, L^k) = \langle \exp(kc_1(L)).Td(\mathcal{M}), \mathcal{M} \rangle$$

and for large k,

$$\dim H^0(\mathcal{M}, L^k) \sim \frac{k^{3g-3}}{(3g-3)!}\langle c_1(L)^{3g-3}, \mathcal{M} \rangle. \tag{3.2}$$

Now $c_1(L)$ is represented by the symplectic form ω in de Rham cohomology. Hence

$$\frac{\langle c_1(L)^{3g-3}, \mathcal{M} \rangle}{(3g-3)!}$$

coincides with $\mathrm{Vol}(\mathcal{M})$. Equating the expressions (3.1) and (3.2), we obtain

$$\mathrm{Vol}(\mathcal{M}) = 2\frac{1}{(2\pi^2)^{g-1}} \sum_{n=1}^{\infty} n^{-(2g-2)}$$

$$= 2\frac{\zeta(2g-2)}{(2\pi^2)^{g-1}}$$

This is known as Witten's volume formula [39] of the moduli space of flat $SU(2)$ connections.

4. Volume of the moduli space of flat $SU(3)$ connections

We first recall some definitions of affine $\widehat{SU(3)}$ characters (for example [11, 27]). The affine $\widehat{SU(3)}$ characters are labeled by a highest weight $\Lambda = \lambda_1\Lambda_1 + \lambda_2\Lambda_2$ where the Λ_i's are the fundamental weights and the set of components $\{\lambda_i\}$ contains non-negative integers. If the height of affine $\widehat{SU(3)}$ is $n = k + 3$ with level $k \geq 0$, the highest weights corresponding to unitary representations satisfy $\lambda_1 + \lambda_2 \leq k$. Thus, there are $\frac{(k+1)(k+2)}{2} = \frac{(n-1)(n-2)}{2}$ independent affine characters. To see these more explicitly, let us consider shifted weight $\lambda = \Lambda + \Lambda_1 + \Lambda_2 = p_1\Lambda_1 + p_2\Lambda_2$. Unitarity of the representations implies that λ belongs to the fundamental domain $\mathcal{W}$,

$$\mathcal{W} = \{\lambda = p_1\Lambda_1 + p_2\Lambda_2, \quad p_i \geq 1 \quad \text{and} \quad p_1 + p_2 \leq n - 1\},$$

where the Λ_i's are the fundamental weights and the set of components $\{p_i\}$ are truncated by the level k. $\mathcal{W}$ is known as the Weyl alcove (see for example [8, 16, 27]).

Our starting point will be the Weyl-Kac character for $\widehat{SU(3)}$. We obtain the matrix $S_{\alpha\beta}$ (see details in [27]) from the modular transformations $\tau \longrightarrow -\frac{1}{\tau}$ of χ. The S matrix of $\widehat{SU(3)}$ is given below.

$$S_{0\lambda} = \frac{8}{\sqrt{6(k+3)}} \sin \frac{\pi\lambda_1}{k+3} \sin \frac{\pi\lambda_2}{k+3} \sin \frac{\pi(\lambda_1+\lambda_2)}{k+3}, \qquad (4.1)$$

which can be also derived from Weyl-Kac factorized form [11]:

$$\phi_\lambda = \frac{\sin \frac{\pi\lambda_1}{k+3} \sin \frac{\pi\lambda_2}{k+3} \sin \frac{\pi(\lambda_1+\lambda_2)}{k+3}}{\sin^2 \frac{\pi}{k+3} \sin \frac{2\pi}{k+3}}$$

after normalization.

Note that $k+3$ is the shifting of level k, and the shifting will be exactly equal to the Coxeter number of the group G. The Coxeter number of $SU(n)$ is n. Substituting the modular transformation $S_{0\lambda}$ in (4.1), we obtain the Verlinde formula for the moduli space of flat $SU(3)$ connections.

$$\dim H^0(\mathcal{M}, L^k) = \frac{(k+3)^{2g-2}6^{g-1}}{2^{7g-7}} \sum_{\lambda_1,\lambda_2} \left(\frac{1}{\sin \frac{\pi\lambda_1}{k+3} \sin \frac{\pi\lambda_2}{k+3} \sin \frac{\pi(\lambda_1+\lambda_2)}{k+3}} \right)^{2g-2}. \qquad (4.2)$$

Here the summation satisfies $\lambda_1 + \lambda_2 \leq k + 2$.

To find the volume our goal is again to extract the term proportional to

$$k^{\dim_c \mathcal{M}} = k^{8g-8}$$

for $k \longrightarrow \infty$.

Like in the $SU(2)$ case, here the contribution for large k comes from 3 different regions. Finally, we obtain

$$\dim H^0(\mathcal{M}, L^k) \sim 3 \frac{k^{8g-8}}{(2\pi)^{6g-6}2^{g-1}} 6^{g-1} \sum_{\lambda_1=1,\lambda_2=1}^{\infty} \frac{1}{\lambda_1^{2g-2}} \frac{1}{\lambda_2^{2g-2}} \frac{1}{(\lambda_1+\lambda_2)^{2g-2}}$$

$$\dim H^0(\mathcal{M}, L^k) \sim 3 \frac{k^{8g-8}}{(2\pi)^{6g-6}2^{g-1}} 6^{g-1} \zeta_g(2g-2),$$

where the generalized zeta function

$$\zeta_g(2g-2) = \sum n_1^{-(2g-2)} n_2^{-(2g-2)} (n_1 + n_2)^{-(2g-2)} \qquad (4.3)$$

can be expressed in terms of double Bernoulli numbers [4, 5, 32] or multiple zeta functions [41, 42].

Hence, using the Riemann-Roch formula, for large $k \longrightarrow \infty$ we obtain

$$\dim H^0(\mathcal{M}, L^k) \sim \frac{k^{8g-8}}{(8g-8)!} \langle c_1(L)^{8g-8}, \mathcal{M} \rangle.$$

Again, $c_1(L)$ is represented by the symplectic form ω and the term

$$\frac{\langle c_1(L)^{8g-8}, \mathcal{M}\rangle}{(8g-8)!}$$

coincides with $\text{Vol}(\mathcal{M})$. Hence we obtain

$$\text{Vol}(\mathcal{M})_{SU(3)} = 3\frac{6^{g-1}}{(2\pi)^{6g-6}2^{9-1}}\zeta_g(2g-2).$$

Finally, using the formula [42] of the multiple zeta function

$$\sum_{m,n}^{\infty}\frac{1}{m^s n^s (m+n)^s} = \frac{4}{3}\sum_{0\leq r\leq s;r\,even}\left(\begin{array}{c}2s-r-1\\s-1\end{array}\right)\zeta(r)\zeta(3s-r)$$

we obtain following examples.

Example : For $g = 2$ we know the value of the zeta function from Zagier [41, 42]: $\zeta_2(2) = (2\pi)^6/7!\,36$. So the volume is

$$\text{Vol}(\mathcal{M}) = 3.\frac{6}{(2\pi)^6.2}.\frac{(2\pi)^6}{7!\,36} = 1/4.7!$$

This is the first generalization of Witten's result [39] for the moduli space of $SU(2)$ flat connections to the moduli space of flat $SU(3)$ connections.

5. Cohomological pairings of the moduli space

This is the central theme of the whole article. Our goal here is to find out the cohomological pairings of the moduli space of flat $SU(3)$ connections on the Riemann surface. Our recipe to find the volume will be to use a generalized Verlinde formula (for the marked point case) (1.8) in the large k-limit. This volume formula contains all the information of certain cohomological pairings.

5.1. Review of Donaldson-Thaddeus-Witten's work on $SU(2)$ moduli space

Let $\mathcal{M}_1$ be the moduli space of flat $SU(2)$ connections. For a rational number $0 < t < 1$, we consider $\mathcal{M}_t$ to be the moduli space of flat connections on $\Sigma_g - x$, such that monodromy around x is in the conjugacy classes of $SU(2)$

$$T = \left(\begin{array}{cc}\exp(i\pi t) & 0\\0 & exp(-i\pi t)\end{array}\right).$$

One can show that for t close to 1, $\mathcal{M}_t$ is a CP^1 bundle over the moduli space $\mathcal{M}_1$.

$$\mathbf{CP}^1 \longrightarrow \mathcal{M}_t$$
$$\downarrow$$
$$\mathcal{M}_1$$

For $\mathcal{M}_t$, we still have a natural symplectic structure ω, but the periods of ω are no longer integers. Then ω is expressed by $a + th$ in $H^2(\mathcal{M}_t)$ generated by $a \in H^2(\mathcal{M}_1)$ and $h \in H^2(\mathcal{M}_t)$ takes value 1 on the fibre. Hence, for small t its symplectic volume will be

$$\text{Vol}^S(\mathcal{M}_t) = \langle \frac{1}{(3g-2)!}(a + th)^{3g-2}, [\mathcal{M}_t]\rangle.$$

Using the relation $h^2 = b \in H^4[\mathcal{M}_1]$, we can expand the above expression:

$$\frac{1}{(3g-2)!} \sum_{j=0}^{(\frac{3g-2}{2})} \binom{3g-2}{2j+1} t^{2j+1} a^{3g-3-2j} b^j [\mathcal{M}_1]. \tag{5.1}$$

On the other hand we use Witten's prescription [39] to obtain the volume of the moduli space of flat $SU(2)$ connections over the Riemann surface with p-marked points from the generalized Verlinde formula (1.8) in the large k limit.

$$\text{Vol}^F(\mathcal{M}_t) = 2.\frac{1}{2^{g-1}\pi^{2g-2+p}} \sum_{n=1}^{p} \frac{\prod_{i=1}^{p} sin(\pi n t_i)}{n^{2g-2+p}}.$$

This volume for the one marked point case is

$$\text{Vol}(\mathcal{M}_t) = \frac{2}{2^{g-1}\pi^{2g-1}} \sum_{n=1}^{\infty} \frac{sin(n\pi t)}{n^{2g-1}}. \tag{5.2}$$

Equating the two expressions (5.1) and (5.2), one obtains the pairing in terms of Bernoulli numbers.

$$\langle a^m b^n, [\mathcal{M}]\rangle = (-1)^g \frac{m!}{(g-1-m)!} 2^{1-g}(2^{g-1-m} - 2)B_{m-g+1}$$

where $m = 3g - 3 - 2j$ and we have used

$$\zeta(2k) = \frac{(-1)^{k+1}(2\pi)^{2k}}{2.(2k)!} B_{2k}.$$

This exactly coincides with Thaddeus' formula [33] which is verified by Donaldson [13] using topological gluing techniques extracted from the Verlinde algebra.

After the demonstration of the known case, we shall give our result in the remaining part of the article.

5.2. Cohomological pairings for $SU(3)$ connections

Our goal is to obtain the cohomological pairings for the moduli space of flat $SU(3)$ connections.

To begin with, let $\mathcal{M}_t$ be the moduli space of flat $SU(3)$ connections over a Riemann surface $\Sigma_g - x$ having one marked point x such that the holonomy around x is characterized by two rational numbers t_1, t_2 satisfying $0 < t_1 < 1$

and $0 < t_2 < 1$. The prescribed holonomy around x takes values in the conjugacy classes of $SU(3)$

$$\Theta \sim \begin{pmatrix} e^{2\pi i t_1/3} & 0 & 0 \\ 0 & e^{2\pi i t_2/3} & 0 \\ 0 & 0 & e^{-2\pi i(t_1+t_2)/3} \end{pmatrix}.$$

Then, for small values of t, $\mathcal{M}_t$ is the bundle over the ordinary smooth moduli space and the flag manifold is the fibre on it. It can be represented by

$$\mathcal{F} \longrightarrow \mathcal{M}_t$$
$$\downarrow$$
$$\mathcal{M}_1$$

In other words, the fibre is a flag manifold

$$\mathcal{F} = \frac{SU(3)}{U(1) \times U(1)} = \frac{SL(3, C)}{B^+},$$

where B^+ is the Borel subgroup of $SL(3, C)$. We now give a brief description of the flag manifold from the classic Bott and Tu [10].

5.2.1. Flag manifolds and cohomology. We define a *flag* in a complex vector space V of dimension n as a sequence of subspaces

$$V_1 \subset V_2 \subset \ldots \ldots \subset V_n, \quad \dim_{\mathbf{C}} V_i = i.$$

Let $Fl(V)$ be the collection of all flags in V. Any flag can be carried into any other flag in V by an element of the general linear group $GL(n, C)$, and the stabilizer of a flag is the Borel subgroup B^+ of the upper triangular matrices. Then, the set $Fl(V)$ is isomorphic to the coset space $GL(n, C)/B^+$. The quotient of any smooth manifold by the free action of a compact Lie group is again a smooth manifold. Hence, $Fl(V)$ is a manifold and it is called the flag manifold of V.

Similarly, we can construct a flag structure on bundles. Let $\pi : E \longrightarrow M$ be a C^∞ complex vector bundle of rank n over a manifold M. The associated flag bundle $Fl(E)$ is obtained from E by replacing each fibre E_p by the flag manifold $Fl(E_p)$, the local trivialization

$$\phi_\alpha : E|_{U_\alpha} \simeq U_\alpha \times \mathbf{C}^n$$

induces a natural trivialization

$$Fl(E)|_{U_\alpha} \simeq U_\alpha \times Fl(\mathbf{C}^n).$$

Since $GL(n, C)$ acts on $Fl(\mathbf{C}^n)$ we may take the transition function of $Fl(E)$ to be those of E.

Let us discuss a few things about split manifolds. Given a map $\sigma : Fl(E) \longrightarrow M$ we can define a split manifold as follows:

1. the pull back of E to $F(E)$ splits into a direct sum of line bundles

$$\sigma^{-1}E = L_1 \oplus \ldots \ldots \oplus L_n.$$

2. σ^* embeds $H^*(M)$ in $H^*(Fl(E))$.

The split manifold $Fl(E)$ is obtained by a sequence of $n-1$ projectivization. We shall now apply all these to obtain cohomology rings of flag manifolds.

Proposition 5.1. *The associated flag bundle $Fl(E)$ of a vector bundle is the split manifold.*

Proof: Given in Bott [10] (chapter 4).   □

If E is a rank n complex vector bundle over M, then the cohomology ring of its projectivization is

$$H^*(P(E)) = H^*(M)[c_1, \ldots, c_n, d_1, \ldots, d_n]/\{C(S)C(Q) = \pi^*C(E)\},$$

where $c_1, \ldots, c_n$ are the Chern classes of the universal subbundle S and $d_1, \ldots, d_n$ are the classes of the universal quotient bundle Q. Also $C(S)$ and $C(Q)$ denote the total Chern classes of S and Q. respectively. The flag manifold is obtained from a sequence of $(n-1)$ projectivizations

$$H^*(Fl(E))$$
$$= H^*(M)[C(S_1), \ldots, C(S_{n-1}), C(Q_1, \ldots, Q_{n-1})]/C(S_1)\ldots C(S_{n-1})C(Q_{n-1})$$
$$= C(E).$$

If

$$h_i = C_1(S_i), \quad i = 1 \ldots n-1$$
$$h_n = C(Q_{n-1}),$$

then we have

$$H^*(Fl(E)) = H^*(M)[h_1, \ldots, h_n]/(\prod_{i=1}^{n}(1 + h_i)) = C(E)).$$

In order to obtain the cohomology ring of the flag manifold F [10], we have to consider a trivial bundle over a point.

$$H^*(F) = R[h_1, \ldots, h_n]/(\prod_{i=1}^{n}(1 + h_i) = 1).$$

For the special case $n = 3$, we obtain

$$H^*(F) = R[h_1, h_2, h_3]/(\prod_{i=1}^{3}(1 + h_i) = 1).$$

5.2.2. Computation of the intersection pairings. We are going to apply our previous scheme. Since $\mathcal{M}_t$ is a bundle over $\mathcal{M}_1$, we can pull back the cohomology from the base manifold $\mathcal{M}_1$. In fact, it is not hard to see that the symplectic form ω represents the class

$$a + t_1 h_1 + t_2 h_2 \in H^2(\mathcal{M}_t),$$

where $a \in H^2(\mathcal{M}_1)$ and $h_i \in H^2(\mathcal{M}_t)$. So from this symplectic form the volume will be the following

$$\text{Vol}^S(\mathcal{M}_t) = \langle \frac{1}{(8g-5)!}(a + t_1 h_1 + t_2 h_2)^{8g-5}, [\mathcal{M}_t] \rangle.$$

When we expand this expression, then we obtain the following results.

Proposition 5.2.

$$\text{Vol}^S(\mathcal{M}_t) = \sum_{k,l} \frac{1}{(8g-5-k-l)!k!l!} t_1^k t_2^l \langle a^{8g-5-k-l} h_1^k h_2^l, [\mathcal{M}_t] \rangle.$$

But to get the exact pairing, we have to use the knowledge of Witten's volume function for small t_1, t_2, and also we use the following identities viz.

$$h_1^2 = b \in H^4; \quad h_2^2 = c \quad \in H^4; \quad -h_1 h_2 = d \quad \in H^4$$
$$h_1^2 h_2 = e \quad \in H^6; \quad -h_2^2 h_1 = f \quad \in H^6.$$

Note that $h_1^2 h_2 = -h_1 h_2^2$ is a fundamental class of the Flag manifold, these are top cohomology modules. The key lemma for obtaining the cohomology ring over the moduli space $\mathcal{M}_1$ follows from the Leray-Hirsch theorem [10].

Theorem 5.3. *(Leray-Hirsch) Let $\mathcal{E}$ be a fibre bundle over a manifold M with fibre $\mathcal{F}$. Assume M has finite good cover and suppose there are global cohomology classes $e_1, e_2, \ldots, e_r$ on $\mathcal{E}$ which – when restricted to each fibre – freely generate the cohomology of the fibre. Then $H^*(\mathcal{E})$ is a free module over $H^*(M)$ with basis $\{e_1, e_2, \ldots, e_r\}$, i.e.*

$$H^*(\mathcal{E}) \cong H^*(M) \otimes \mathbf{R}[e_1, e_2, \ldots, e_r]$$
$$\cong H^*(M) \otimes H^*(\mathcal{F}).$$

Now we formulate an important statement.

Lemma 5.4. *The fundamental classes of $\mathcal{M}_t$ are the product of the fundamental classes of the moduli space without marked point $\mathcal{M}_1$ and the fundamental classes of the flag manifold.*

We use the same recipe, i.e., extracting the volume from the generalized Verlinde formula (1.8), to find the volume of the moduli space.

If we feed the value of $S_{\alpha\beta}$ of $\widehat{SU(3)}$ obtained from the modular transformation of the Weyl-Kac character into the Verlinde formula (1.8) and repeat the derivation as in the previous section, we obtain the torsion volume

$$\text{Vol}^F(\mathcal{M}_t) = \frac{3.6^{g-1}}{2^{7g-7}\pi^{6g-3}} \sum \frac{\sin \pi n_1 t_1 \sin \pi n_2 t_2 \sin \pi (n_1+n_2)(t_1+t_2)}{n_1^{2g-1} n_2^{2g-1}(n_1+n_2)^{2g-1}}.$$

This is the generalization of Witten's volume formula for the moduli space of flat $SU(3)$ connections. It is the volume of the moduli space of flat $SU(3)$ connections over a Riemann surface of genus g with one marked point.

After a tedious calculation which makes use of the Taylor expansion of $\sin \pi n_1 t_1$, $\sin \pi n_2 t_2$ and $\sin \pi (n_1 + n_2)(t_1 + t_2)$, the above expression for small t_1, t_2 gives us a comprehensive formula:

Proposition 5.5.

$$\text{Vol}(\mathcal{M}_t) = \frac{3.6^{g-1}}{2^{7g-7}\pi^{6g-3}} \sum_n \sum_j \times$$

$$\frac{(-1)^{j_1+j_2+j_3}\pi^{2j_1+2j_2+j_3}t_1^{2(j_1+j_3-j_4)+1}t_2^{2(j_2+j_4)+2}}{(2j_1+1)!(2j_2+1)!(2j_3-2j_4)!(2j_4+1)!n_1^{2g-2j_1-2}n_2^{2g-2j_2-2}(n_1+n_2)^{2g-2j_3-2}}.$$

This is the key formula for getting the cohomology pairings of the moduli space of flat $SU(3)$ connections, and is Witten's volume formula for moduli space of flat $SU(3)$ connections. This formula is too bigm but can be handled for some lower genus cases.

5.3. Concrete examples

It is clear that the two volumes of the moduli space, namely the symplectic volume $\text{Vol}^S(\mathcal{M}_t)$ and the volume from Verlinde's formula $Vol(\mathcal{M}_t)$ in claim (1.9) are equal.

Using this simple prescription we obtain *explicit examples* of the cohomological pairings of the moduli space of flat $SU(3)$ connections. This pairing is expressed in terms of a multiple zeta function [41, 42] or double Bernoulli numbers [4, 5]. Equating the powers of $t_1^k t_2^l$ we obtain explicit pairings.

1. We consider *genus* = 3. Thus we obtain

$$\langle a^{10} e^1 f^1 [\mathcal{M}] \rangle = \frac{3.7!5!3!.2^2.8.9.10}{2^8.(2\pi)^6}\zeta_{SU(3)}.$$

From Zagier's formula we now come to know that the value of this function is $(2\pi)^6./7!.36$. Hence

$$\langle a^{10} e^1 f^1 [M] \rangle = (10.9.3.6.2).(2^4.5!)/4.36.2^6 = 5.9.15 = 675.$$

2. Once again consider *genus* = 3. We obtain

$$\langle a^{10} f^2 [M] \rangle = \frac{10.9.8.6.7!6!.2^2}{(2\pi)^6.2^8}\zeta(2,A)_2$$

i.e.

$$\langle a^{10} f^2 [M] \rangle = 10.9.8.5!.2^2/2^8 = 1350.$$

Thus, we gave two explicit examples of pairings. Indeed it is really hard to compute any arbitrary higher genus pairings. We hope that our readers realize the degree of complications for further computations of intersection pairings.

Acknowledgement. First of all, the author would like to express his respectful gratitude to his thesis advisor, Professor Simon Donaldson, for numerous suggestions, guidance and encouragement. He is also grateful to Professors Nigel Hitchin, Peter Kronheimer and Michael Thaddeus for their valuable comments and suggestions. He is extremely grateful to Professor Don Zagier for his numerical help and for calculating the expression of the multiple zeta function.

References

[1] T. Apostol, *Introduction to Analytic Number theory.* Springer-Verlag. 1976.

[2] M.F. Atiyah and R. Bott, *The Yang-Mills equation over the Riemann surface.* Philos.Trans. R.Soc.London A308,523–615. 1982.

[3] M.F. Atiyah. *The Geometry and Physics of Knots.* Cambridge University Press. 1990.

[4] E.W. Barnes. *The theory of the double gamma function.* Phil.Trans. Roy-Soc.A 196, 265–388.1901.

[5] E.W. Barnes. *On the theory of the multiple gamma function.* Trans. Cambridge Phil.Soc 19,374–425. 1904.

[6] J.M. Borwein, D.M. Bradley and D.J. Broadhurst, *Evaluations of k-fold Euler/Zagier sums: a compendium of results for arbitrary k.* The Wilf Festschrift (Philadelphia, PA, 1996).

[7] D. Bowman and D.M. Bradley, *Multiple polylogarithms: a brief survey.* q-series with applications to combinatorics, number theory, and physics (Urbana, IL, 2000), 71–92, Contemp. Math., 291, Amer. Math. Soc., Providence, RI, 2001. Electron. J. Combin. 4 (1997), no. 2, Research Paper 5.

[8] R. Baston and M. Eastwood. *The Penrose Transform , its interaction with Representation theory.* Oxford University Press. 1989.

[9] A. Beauville and Y. Laszlo. *Conformal blocks and generalized theta divisor.* Comm. Math. Phys. 164, 385–419. 1994.

[10] R. Bott and L. Tu. *Differential forms in Algebraic topology.* G.T.M. 82 Springer-Verlag. 1982.

[11] P. Di Francesco. *Integrable lattice models, graphs and modular invariant conformal field theories.* Int. J. Mod. Phys. A 7, 407–500, 1992.

[12] S.K. Donaldson. *A new proof to Narasimhan Seshadri's theorem.* J. Diff. Geom. 18 , 269–277, 1983.

[13] S.K. Donaldson. *Gluing technique in the cohomology of Moduli space.* in Topological Methods in Modern Mathematics, J. Milnor 60th birthday festschrift. Publish or Perish. 1992.

[14] R. Earl and F. Kirwan, *Complete sets of relations in the cohomology rings of moduli spaces of holomorphic bundles and parabolic bundles over a Riemann surface.* Proc. London Math. Soc. (3) 89 (2004), no. 3, 570–622.

[15] G. Faltings. *A proof of Verlinde's formula.* J. Alg. Geom. 3, 347–374, 1994.

[16] W. Fulton and J. Harris. *Representation theory, A first Course.* Springer-Verlag. 1991.

[17] V. Guillemin and S. Sternberg. *Symplectic techniques in Physics*. Cambridge Univ. Press. 1984.

[18] P. Griffith and J. Harris. *Principles of Algebraic geometry*. Wiley Interscience. 1978.

[19] D. Gepner and E. Witten. *String theory on group manifolds*. Nucl. Phys. B278, 493–549. Appendix A. 1986.

[20] P. Guha, *Topics in the geometry of holomorphic vector bundles*, Thesis, Oxford University 1994.

[21] N.J. Hitchin. *The symplectic geometry of moduli spaces of connection and geometric quantization*. Prog. of Theo Phys. Suppl.No.102, 159–174. 1990.

[22] N.J. Hitchin. *Flat connections and geometric quantization*. Commn. Math. Phys. 131, 347. 1990.

[23] L. Jeffrey. *Chern-Simons-Witten invariants of lens spaces and the torus bundles, and the semiclassical approximation*. Commun. Math. Phys. 147, 563–604. 1992.

[24] L. Jeffrey. *Extended moduli spaces of flat connections on Riemann surfaces*. Math. Ann. 298, 667–698. 1994.

[25] L. Jeffrey and F. Kirwan.*Intersection numbers in moduli spaces of holomorphic bundles on a Riemann surface*. Elec. Res. Ann. AMS 1, 57–71. 1995;

[26] L. Jeffrey and F. Kirwan *Intersection pairings in moduli spaces of vector bundles of arbitrary rank over a Riemann surface*. Ann. of Math. (2) 148 (1998), no. 1, 109–196.

[27] V. Kac. *Infinite dimensional Lie algebras*. Cambridge University Press. 1990.

[28] S. Kumar, M.S. Narasimhan and A. Ramanathan. *Infinite Grassmannian and moduli spaces of G-bundles*. Math. Ann. 300, 41–75. 1994.

[29] V.B. Mehta and C.S. Seshadri. *Moduli of vector bundles on curves with parabolic structures*. Math. Ann. 248, 205–239. 1980.

[30] D. Mumford. *Projective invariants and projective structures and application*. Proc. Inst. Cong. Math. Institut Mittag-Leffler. 1962

[31] M.S. Narasimhan and C.S. Seshadri. *Stable and unitary vector bundle on a compact Riemann surface*. Ann. Math. 82, 540–567. 1965.

[32] T. Shintani. *On special values of zeta function of totally real algebraic number fields*. Proc. Int. Cong. of Maths. Acad. Sci. Fennica, Helsinki. 591–597. 1978.

[33] M. Thaddeus. *Conformal field theory and Cohomology of the Moduli space of stable bundles*. J.Diff.Geom. 35, 131–149. 1992.

[34] A. Tsuchiya, K. Ueno and Y. Yamada, *Conformal field theory on universal family of stable curves with gauge symmetry*. Advanced Studies in Pure Math. vol. 19, Princeton Univ. Press, 459–566. 1989.

[35] V.S. Varadarajan. *Lie groups, Lie Algebras and their Representations*. Springer-Verlag. 1984.

[36] E. Verlinde. *Fusion rules and modular transformations in 2D conformal field theory*. Nucl.Phys B 300, 360–370. 1988.

[37] M. Waldschmidt, Valeurs zeta multiples: une introduction. Journal de Théorie des Nombres de Bordeaux, 12.2 (2000), 581–595.

[38] M. Waldschmidt, Multiple Polylogarithms: An Introduction. (full text) 12p. Number Theory and Discrete Mathematics Hindustan Book Agency, 2002, 1–12 Editors: A.K. Agarwal, B.C. Berndt, C.F. Krattenthaler, Gary L. Mullen, K. Ramachandra and Michel Waldschmidt. Institut de Mathematiques de Jussieu, Prépublication 275 (janvier 2001), 11 p.

[39] E. Witten. *On Quantum gauge theories in two dimensions.* Commn. Math. Phys. 141, 153–209. 1991.

[40] E. Witten. *Two dimensional gauge theories revisited.* J. Geom. Phys. 9, 303–368. 1992.

[41] D.B. Zagier. *Private communication.*

[42] D.B. Zagier. *Values of Zeta functions and their Applications.* First European Congress of Mathematics, 1992, Vol.II, Edited by A. Joseph et. al. Birkhäuser Verlag, 1994.

Partha Guha
Max Planck Institute for Mathematics in the Sciences
Inselstrasse 22
DE–04103 Leipzig

and

S.N. Bose National Centre for Basic Sciences
JD Block, Sector - III, Salt Lake
IN–700098 Kolkata
e-mail: `partha.guha@mis.mpg.de` & `partha@bose.res.in`

Quantum Field Theory
B. Fauser, J. Tolksdorf and E. Zeidler, Eds., 117–135
© 2009 Birkhäuser Verlag Basel/Switzerland

Noncommutative Field Theories from a Deformation Point of View

Stefan Waldmann

Abstract. In this review we discuss the global geometry of noncommutative field theories from a deformation point of view: The space-times under consideration are deformations of classical space-time manifolds using star products. Then matter fields are encoded in deformation quantizations of vector bundles over the classical space-time. For gauge theories we establish a notion of deformation quantization of a principal fibre bundle and show how the deformation of associated vector bundles can be obtained.

Mathematics Subject Classification (2000). Primary 53D55; Secondary 58B34; 81T75.

Keywords. noncommutative field theory, deformation quantization, principal bundles, locally noncommutative space-time.

1. Introduction

Noncommutative geometry is commonly believed to be a reasonable candidate for the marriage of classical gravity theory in form of Einstein's general relativity on one hand and quantum theory on the other hand. Both theories are experimentally well-established within large regimes of energy and distance scales. However, from a more fundamental point of view, the coexistence of these two theories becomes inevitably inconsistent when one approaches the Planck scale where gravity itself gives significant quantum effects.

Since general relativity is ultimately the theory of the geometry of space-time it seems reasonable to use notions of 'quantum geometry' known under the term *noncommutative geometry* in the sense of Connes [11] to achieve appropriate formulations of what eventually should become quantum gravity. Of course, this ultimate goal has not yet been reached but techniques of noncommutative geometry have been used successfully to develop models of quantum field theories on quantum space-times being of interest of their own. Moreover, a deeper understanding of ordinary quantum field theories can be obtained by studying their counterparts on 'nearby' noncommutative space-times. On the other hand, people started to investigate experimental implications of a possible noncommutativity of space-time in future particle experiments.

Such a wide scale of applications and interests justifies a more *conceptual* discussion of noncommutative space-times and (quantum) field theories on them in order to clarify fundamental questions and generic features expected to be common to all examples.

In this review, we shall present such an approach from the point of view of deformation theory: noncommutative space-times are not studied by themselves but always with respect to a classical space-time, being suitably deformed into the noncommutative one. Clearly, this point of view can not cover all possible (and possibly interesting) noncommutative geometries but only a particular class. Moreover, we focus on *formal* deformations for technical reasons. It is simply the most easy approach where one can rely on the very powerful machinery of algebraic deformation theory. But it also gives hints on approaches beyond formal deformations: finding obstructions in the formal framework will indicate even more severe obstructions in any non-perturbative approach.

In the following, we discuss mainly two questions: first, what is the appropriate description of matter fields on deformed space-times and, second, what are the deformed analogues of principal bundles needed for the formulation of gauge theories. The motivation for these two questions should be clear.

The review is organized as follows: in Section 2, we recall some basic definitions and properties concerning deformation quantizations and star products needed for the set-up of noncommutative space-times. We discuss some fundamental examples as well as a new class of locally noncommutative space-times. Section 3 is devoted to the study of matter fields: we use the Serre-Swan theorem to relate matter fields to projective modules and discuss their deformation theory. Particular interest is put on the mass terms and their positivity properties. In Section 4 we establish the notion of deformation quantization of principal fibre bundles and discuss the existence and uniqueness results. Finally, in Section 5 we investigate the resulting commutant and formulate an appropriate notion of associated (vector) bundles. This way we make contact to the results of Section 3. The review is based on joint works with Henrique Bursztyn on one hand as well as with Martin Bordemann, Nikolai Neumaier and Stefan Wei on the other hand.

2. Noncommutative space-times

In order to implement uncertainty relations for measuring coordinates of events in space-time it has been proposed already very early to replace the commutative algebra of (coordinate) functions by some noncommutative algebra. In [17] a concrete model for a noncommutative Minkowski space-time was introduced with commutation relations of the form

$$[\hat{x}^\mu, \hat{x}^\nu] = \mathrm{i}\lambda\theta^{\mu\nu}, \qquad (2.1)$$

where λ plays the role of the deformation parameter and has the physical dimension of an area. Usually, this area will be interpreted as the Planck area. Moreover, θ is a real, antisymmetric tensor which in [17] and many following papers is assumed to be *constant*: in [17] this amounts to require that $\theta^{\mu\nu}$ belongs to the center of the new algebra of noncommutative coordinates. In fact, the approach in [17] was more subtle: the constants $\theta^{\mu\nu}$ are subject to a tensor transforming non-trivially under Lorentz transformations in such a way that the whole algebra is endowed with a representation of the Lorentz (in fact Poincaré) group by *-automorphisms. This was achieved by adding the whole Lorentz orbit of a constant tensor $\theta^{\mu\nu}$ with rank 4. Unfortunately, in the vast majority of the follow-up papers this feature of Lorentz invariance has been neglected. In the following we shall ignore symmetries, too, as in the end we focus on generic space-times without symmetries anyway. However, if there is a reasonably large symmetry group present, one can always perform the construction parallel to [17] and add the orbit of the Poisson tensor under the symmetry group.

Instead of constructing an abstract algebra where commutation relations like (2.1) are fulfilled, it is convenient to use a 'symbol calculus' and encode (2.1) already for the classical coordinate functions by changing the multiplication law instead. For functions f and g on the classical Minkowski space-time one defines the Weyl-Moyal star product by

$$ f \star g = \mu \circ e^{\frac{i\lambda}{2} \theta^{\mu\nu} \frac{\partial}{\partial x^\mu} \otimes \frac{\partial}{\partial x^\nu}} (f \otimes g), \qquad (2.2) $$

where $\mu(f \otimes g) = fg$ denotes the undeformed, pointwise product. Then (2.1) holds for the classical coordinate functions with respect to the $\star$-commutator.

Clearly, one has to be slightly more careful with expressions like (2.2): in order to make sense out of the infinite differentiations the functions f and g first should be C^∞. But then the exponential series does not converge in general whence a more sophisticated analysis is required. Though this can be done in a completely satisfying way for this particular example, we shall not enter this discussion here but consider (2.2) as a formal power series in the deformation parameter λ. Then $\star$ becomes an associative $\mathbb{C}[[\lambda]]$-bilinear product for $C^\infty(\mathbb{R}^4)[[\lambda]]$, i. e. a star product in the sense of [3]. It should be noted that the interpretation of (2.2) as formal series in λ is physically problematic: λ is the Planck area and hence a physically measurable and non-zero quantity. Thus our point of view only postpones the convergence problem and can be seen as a perturbative approach.

With this example in mind, one arrives at several conceptual questions: The first is that Minkowski space-time is clearly not a very realistic background when one wants to consider quantum effects of 'hard' gravity. Here already classically nontrivial curvature and even nontrivial topology may arise. Thus one is forced to consider more general and probably even generic Lorentz manifolds instead. Fortunately, deformation quantization provides a well-established and successful mathematical framework for this geometric situation.

Recall that a star product on a manifold M is an associative $\mathbb{C}[[\lambda]]$-bilinear multiplication $\star$ for $f, g \in C^\infty(M)[[\lambda]]$ of the form

$$f \star g = \sum_{r=0}^{\infty} \lambda^r C_r(f, g), \qquad (2.3)$$

where $C_0(f, g) = fg$ is the undeformed, pointwise multiplication and the C_r are bidifferential operators. Usually, one requires $1 \star f = f = f \star 1$ for all f. It is easy to see that $\{f, g\} = \frac{1}{i}(C_1(f, g) - C_1(g, f))$ defines a Poisson bracket on M. Conversely, and this is the highly nontrivial part, any Poisson bracket $\{f, g\} = \theta(\mathrm{d}f, \mathrm{d}g)$, where

$$\theta \in \Gamma^\infty(\Lambda^2 TM), \quad [\![\theta, \theta]\!] = 0 \qquad (2.4)$$

is the corresponding Poisson tensor, can be quantized into a star product [14, 29]. Beside these existence results one has a very good understanding of the classification of such star products [20, 29, 31], see also [15, 19] for recent reviews and [40] for an introduction.

With this geometric interpretation the Weyl-Moyal star product on Minkowski space-time turns out to be a deformation quantization of the *constant* Poisson structure

$$\theta = \frac{1}{2} \theta^{\mu\nu} \frac{\partial}{\partial x^\mu} \wedge \frac{\partial}{\partial x^\nu}. \qquad (2.5)$$

On a generic space-time M there is typically *no* transitive action of isometries which would justify the notion of a 'constant' bivector field. Thus a star product $\star$ on M is much more complicated than (2.2) in general: already the first order term is a (nontrivial) Poisson structure and for the higher order terms one has to invoke the (unfortunately rather inexplicit) existence theorems.

Thus answering the first question by using general star products raises the second: what is the physical role of a Poisson structure on space-time? While on Minkowski space-time with constant θ we can view the finite number of coefficients $\theta^{\mu\nu} \in \mathbb{R}$ as *parameters* of the theory this is certainly no longer reasonable in the more realistic geometric framework: there is an infinity of Poisson structures on each manifold whence an interpretation as 'parameter' yields a meaningless theory. Instead, θ has to be considered as a *field* itself, obeying its own dynamics compatible with the constraint of the Jacobi identity $[\![\theta, \theta]\!] = 0$. Unfortunately, up to now a reasonable 'field equation' justified by first principles seems to be missing.

This raises a third conceptual question, namely why should there be any Poisson structure on M and what are possible experimental implications? In particular, the original idea of introducing a noncommutative structure was to implement uncertainty relations forbidding the precise localization of events. The common believe is that such quantum effects should only play a role when approaching the Planck scale. Now it turns out that the quantum field theories put on such a noncommutative Minkowski space-time (or their Euclidian counterparts) suffer all from quite unphysical properties: Typically, the noncommutativity enters in long-distance/low-energy features contradicting our daily life experience. Certainly, a last word is not said but there might be a simple explanation why such effects should be expected: the *global* θ (constant or not) yields global effects on M. This was the starting point of a more refined notion of noncommutative space-times advocated in [1, 23] as *locally* noncommutative space-times. Roughly speaking, without entering the technical details, it is not M which should become noncommutative but TM. Here the tangent bundle is interpreted as the bundle of all normal charts on M and for each normal chart with origin $p \in M$ one constructs its own star product $\star_p$. The crucial property is then that $\star_p$ is the pointwise, commutative product outside a (small) compact subset around p. This way, the long-distance behavior (with respect to the reference point p) is classical while close to p there is a possibly even very strong noncommutativity. In some sense, this is an implementation of an idea of Julius Wess, proposing that the transition from classical geometry to quantum geometry should be understood as a kind of phase transition taking place at very small distances [42]. Of course, the conceptual question about the physical origin of the corresponding Poisson structure on TM as well as the convergence problem still persists also in this approach.

Ignoring these questions about the nature of θ, we shall assume in the following that we are given a star product $\star$ on a manifold M which can be either space-time itself or its tangent bundle in the locally noncommutative case. Then we address the question how to formulate reasonable field theories on $(M, \star)$. Here we shall focus on *classical* field theories which still need to be quantized later on. On the other hand, we seek for a *geometric* formulation not relying on particular assumptions about the underlying classical space-time.

3. Matter fields and deformed vector bundles

In this section we review some results from [6, 9, 35, 39].

In classical field theories both bosonic and fermionic matter fields are given by sections of appropriate vector bundles. For convenience, we choose the vector bundles to be complex as also the function algebra $C^\infty(M)$ consists of complex-valued functions. However, the real case can be treated completely analogously. Thus let $E \longrightarrow M$ be a complex vector bundle over M. Then the E-valued fields are the (smooth) sections $\Gamma^\infty(E)$ which form a module over $C^\infty(M)$ by pointwise multiplication. Thanks to the commutativity of $C^\infty(M)$ we have the freedom to choose this module structure to be a right module structure for later convenience.

It is a crucial feature of vector bundles that $\Gamma^\infty(E)$ is actually a finitely generated and projective module:

Theorem 3.1 (Serre-Swan). *The sections $\Gamma^\infty(E)$ of a vector bundle $E \longrightarrow M$ are a finitely generated and projective $C^\infty(M)$-module. Conversely, any such module arises this way up to isomorphism.*

Recall that a right module $\mathcal{E}_\mathcal{A}$ over an algebra $\mathcal{A}$ is called finitely generated and projective if there exists an idempotent $e^2 = e \in M_n(\mathcal{A})$ such that $\mathcal{E}_\mathcal{A} \cong e\mathcal{A}^n$ as right $\mathcal{A}$-modules. More geometrically speaking, for any vector bundle $E \longrightarrow M$ there is another vector bundle $F \longrightarrow M$ such that their Whitney sum $E \oplus F$

is isomorphic to a trivial vector bundle $M \times \mathbb{C}^n \longrightarrow M$. Note that the Serre-Swan theorem has many incarnations, e.g. the original version was formulated for compact Hausdorff spaces and continuous sections/functions. Note also that for our situation no compactness assumption is necessary (though it drastically simplifies the proof) as manifolds are assumed to be second countable.

Remark 3.2. The Serre-Swan theorem is the main motivation for noncommutative geometry to consider finitely generated and projective modules over a not necessarily commutative algebra $\mathcal{A}$ as 'vector bundles' over the (noncommutative) space described by $\mathcal{A}$ in general.

For physical applications in field theory one usually has more structure on E than just a bare vector bundle. In particular, for a Lagrangian formulation a 'mass term' in the Lagrangian is needed. Geometrically such a mass term corresponds to a Hermitian fibre metric h on E. One can view a Hermitian fibre metric as a map

$$h : \Gamma^\infty(E) \times \Gamma^\infty(E) \longrightarrow C^\infty(M), \qquad (3.1)$$

which is $\mathbb{C}$-linear in the second argument and satisfies $\overline{h(\phi, \psi)} = h(\psi, \phi)$, $h(\phi, \psi f) = h(\phi, \psi)f$ as well as

$$h(\phi, \phi) \geq 0 \qquad (3.2)$$

for $\phi, \psi \in \Gamma^\infty(E)$ and $f \in C^\infty(M)$. The pointwise non-degeneracy of h is equivalent to the property that

$$\Gamma^\infty(E) \ni \phi \;\mapsto\; h(\phi, \cdot) \in \Gamma^\infty(E^*) \qquad (3.3)$$

is an anti-linear module isomorphism. Note that the sections of the dual vector bundle $E^* \longrightarrow M$ coincide with the dual module, i.e. we have $\Gamma^\infty(E^*) = \mathrm{Hom}_{C^\infty(M)}(\Gamma^\infty(E), C^\infty(M))$.

In order to encode now the positivity (3.2) in a more algebraic way suitable for deformation theory, we have to consider the following class of algebras: First, we use a ring of the form $\mathsf{C} = \mathsf{R}(\mathrm{i})$ with $\mathrm{i}^2 = -1$ for the scalars where R is an *ordered ring*. This includes both $\mathbb{R}$ and $\mathbb{R}[[\lambda]]$, where positive elements in $\mathbb{R}[[\lambda]]$ are defined by

$$a = \sum_{r=r_0}^{\infty} \lambda^r a_r > 0 \quad \text{if} \quad a_{r_0} > 0. \qquad (3.4)$$

In fact, this way $\mathbb{R}[[\lambda]]$ becomes an ordered ring whenever R is ordered. More physically speaking, the ordering of $\mathbb{R}[[\lambda]]$ refers to a kind of asymptotic positivity. Then the algebras in question should be *-algebras over C: Indeed, $C^\infty(M)$ is a *-algebra over $\mathbb{C}$ where the *-involution is the pointwise complex conjugation. For the deformed algebras $(C^\infty(M)[[\lambda]], \star)$ we require that the star product is Hermitian, i.e.

$$\overline{f \star g} = \overline{g} \star \overline{f} \qquad (3.5)$$

for all $f, g \in C^\infty(M)[[\lambda]]$. For a real Poisson structure θ this can be achieved by a suitable choice of $\star$.

For such a *-algebra we can now speak of positive functionals and positive elements [7] by mimicking the usual definitions from operator algebras, see e.g. [33] for the case of (unbounded) operator algebras and [38] for a detailed comparison.

Definition 3.3. Let $\mathcal{A}$ be a *-algebra over $\mathsf{C} = \mathsf{R}(\mathrm{i})$. A C-linear functional $\omega : \mathcal{A} \longrightarrow \mathsf{C}$ is called positive if $\omega(a^*a) \geq 0$ for all $a \in \mathcal{A}$. An element $a \in \mathcal{A}$ is called positive if $\omega(a) \geq 0$ for all positive functionals ω.

We denote the convex cone of positive elements in $\mathcal{A}$ by $\mathcal{A}^+$. It is an easy exercise to show that for $\mathcal{A} = C^\infty(M)$ the positive functionals are the compactly supported Borel measures and $\mathcal{A}^+$ consists of functions f with $f(x) \geq 0$ for all $x \in M$.

Using this notion of positive elements and motivated by [30], the algebraic formulation of a fibre metric is now as follows [6,9]:

Definition 3.4. Let $\mathcal{E}_\mathcal{A}$ be a right $\mathcal{A}$-module. Then an inner product $\langle \cdot, \cdot \rangle$ on $\mathcal{E}_\mathcal{A}$ is a map

$$\langle \cdot, \cdot \rangle : \mathcal{E}_\mathcal{A} \times \mathcal{E}_\mathcal{A} \longrightarrow \mathcal{A}, \tag{3.6}$$

which is C-linear in the second argument and satisfies $\langle x, y \rangle = \langle y, x \rangle^*$, $\langle x, y \cdot a \rangle = \langle x, y \rangle a$, and $\langle x, y \rangle = 0$ for all y implies $x = 0$. The inner product is called strongly non-degenerate if in addition

$$\mathcal{E}_\mathcal{A} \ni x \mapsto \langle x, \cdot \rangle \in \mathcal{E}^* = \mathsf{Hom}_A(\mathcal{E}_\mathcal{A}, \mathcal{A}) \tag{3.7}$$

is bijective. It is called completely positive if for all $n \in \mathbb{N}$ and $x_1, \ldots, x_n \in \mathcal{E}_\mathcal{A}$ one has $(\langle x_i, x_j \rangle) \in M_n(\mathcal{A})^+$.

Clearly, a Hermitian fibre metric on a complex vector bundle endows $\Gamma^\infty(E)$ with a completely positive, strongly non-degenerate inner product in the sense of Definition 3.4.

With the above definition in mind we can now formulate the following deformation problem [6]:

Definition 3.5. Let $\star$ be a Hermitian star product on M and $E \longrightarrow M$ a complex vector bundle with fibre metric h.

1. A deformation quantization $\bullet$ of E is a right module structure $\bullet$ for $\Gamma^\infty(E)[[\lambda]]$ with respect to $\star$ of the form

$$\phi \bullet f = \sum_{r=0}^{\infty} \lambda^r R_r(\phi, f) \tag{3.8}$$

   with bidifferential operators R_r and $R_0(\phi, f) = \phi f$.
2. For a given deformation quantization $\bullet$ of E a deformation quantization of h is a completely positive inner product $\boldsymbol{h}$ for $(\Gamma^\infty(E)[[\lambda]], \bullet)$ of the form

$$\boldsymbol{h}(\phi, \psi) = \sum_{r=0}^{\infty} \lambda^r \boldsymbol{h}_r(\phi, \psi) \tag{3.9}$$

   with (sesquilinear) bidifferential operators $\boldsymbol{h}_r$ and $\boldsymbol{h}_0 = h$.

In addition, we call two deformations $\bullet$ and $\tilde{\bullet}$ *equivalent* if there exists a formal series of differential operators

$$T = \mathsf{id} + \sum_{r=1}^{\infty} \lambda^r T_t : \Gamma^\infty(E)[[\lambda]] \longrightarrow \Gamma^\infty(E)[[\lambda]], \qquad (3.10)$$

such that

$$T(\phi \bullet f) = T(\phi) \tilde{\bullet} f. \qquad (3.11)$$

With other words, T is a module isomorphism starting with the identity in order λ^0 such that T is not visible in the classical/commutative limit. Conversely, starting with one deformation $\bullet$ and a T like in (3.10), one obtains another equivalent deformation $\tilde{\bullet}$ by defining $\tilde{\bullet}$ via (3.11). Similarly, we define two deformations $\boldsymbol{h}$ and $\tilde{\boldsymbol{h}}$ to be *isometric* if there exists a self-equivalence U with

$$\boldsymbol{h}(\phi, \psi) = \tilde{\boldsymbol{h}}(U(\phi), U(\psi)). \qquad (3.12)$$

The relevance of the above notions for noncommutative field theories should now be clear: for a classical matter field theory modeled on $E \longrightarrow M$ we obtain the corresponding noncommutative field theory by choosing a deformation $\bullet$ (if it exists!) together with a deformation $\boldsymbol{h}$ (if it exists!) in order to write down noncommutative Lagrangians involving expressions like $\mathcal{L}(\phi) = \boldsymbol{h}(\phi, \phi) + \cdots$.

Note that naive expressions like $\overline{\phi} \star \phi$ do not make sense geometrically, even on the classical level: sections of a vector bundle can not be 'multiplied' without the extra structure of a fibre metric h unless the bundle is trivial *and* trivialized. In this particular case we can of course use the canonical fibre metric coming from the canonical inner product on $\mathbb{C}^n$. We refer to [35, 39] for a further discussion.

We can now state the main results of this section, see [6,9] for detailed proofs:

Theorem 3.6. *For any star product $\star$ on M and any vector bundle $E \longrightarrow M$ there exists a deformation quantization $\bullet$ with respect to $\star$ which is unique up to equivalence.*

Theorem 3.7. *For any Hermitian star product $\star$ on M and any fibre metric h on $E \longrightarrow M$ and any deformation quantization $\bullet$ of E there exists a deformation quantization $\boldsymbol{h}$ of h which is unique up to isometry.*

The first theorem relies heavily on the Serre-Swan theorem and the fact that algebraic K_0-theory is stable under formal deformations [32]. In fact, projections and hence projective modules can always be deformed in an essentially unique way. The second statement follows for much more general deformed algebras than only for star products, see [9].

Remark 3.8.

1. In case M is symplectic, one has even a rather explicit Fedosov-like construction for $\bullet$ and $\boldsymbol{h}$ in terms of connections, see [36].
2. It turns out that also $\Gamma^\infty(\mathsf{End}(E))$ becomes deformed into an associative algebra $(\Gamma^\infty(\mathsf{End}(E))[[\lambda]], \star')$ such that $\Gamma^\infty(E)[[\lambda]]$ becomes a Morita equivalence

bimodule between the two deformed algebras $\star$ and $\star'$. Together with the deformation $\boldsymbol{h}$ of h one obtains even a strong Morita equivalence bimodule [9].

3. Note also that the results of the two theorems are more than just the 'analogy' used in the more general framework of noncommutative geometry: we have here a precise link between the noncommutative geometries and their classical/commutative limits via deformation. For general noncommutative geometries it is not even clear what a classical/commutative limit is.

4. Deformed principal bundles

This section contains a review of results obtained in [5] as well as in [41].

In all fundamental theories of particle physics the field theories involve gauge fields. Geometrically, their formulation is based on the use of a principal bundle $\mathrm{pr} : P \longrightarrow M$ with structure group G, i.e. P is endowed with a (right) action of G which is proper and free whence the quotient $P/G = M$ is again a smooth manifold. All the matter fields are then obtained as sections of associated vector bundles by choosing an appropriate representation of G.

In the noncommutative framework there are several approaches to gauge theories: for particular structure groups and representations notions of gauge theories have been developed by Jurco, Schupp, Wess and coworkers [24–28]. Here the focus was mainly on local considerations and the associated bundles but not on the principal bundle directly. Conversely, there is a purely algebraic and intrinsically global formulation of Hopf-Galois extensions where not only the base manifold M is allowed to be noncommutative but even the structure group is replaced by a general Hopf algebra, see e.g. [12] and references therein for the relation of Hopf-Galois theory to noncommutative gauge field theories. However, as we shall see below, in this framework which a priori does not refer to any sort of deformation, in general only very particular Poisson structures on M can be used. Finally, in [37] a local approach to principal $\mathrm{Gl}(n, \mathbb{C})$ or $\mathrm{U}(n)$ bundles was implicitly used via deformed transition matrices.

We are now seeking for a definition of a deformation quantization of a principal bundle P for a generic structure Lie group G, arbitrary M and arbitrary star product $\star$ on M without further assumptions on P. In particular, the formulation should be intrinsically global.

The idea is to consider the classical algebra homomorphism

$$\mathrm{pr}^* : C^\infty(M) \longrightarrow C^\infty(P) \tag{4.1}$$

and try to find a reasonable deformation of pr^*. The first idea would be to find a star product $\star_P$ on P with a deformation $\boldsymbol{\mathrm{pr}^*} = \sum_{r=0}^{\infty} \lambda^r \boldsymbol{\mathrm{pr}^*}_r$ of $\boldsymbol{\mathrm{pr}^*}_0 = \mathrm{pr}^*$ into an algebra homomorphism

$$\boldsymbol{\mathrm{pr}^*}(f \star g) = \boldsymbol{\mathrm{pr}^*}(f) \star_P \boldsymbol{\mathrm{pr}^*}(g) \tag{4.2}$$

with respect to the two star products $\star$ and $\star_P$. In some sense this would be the first (but not the only) requirement for a Hopf-Galois extension. In fact, the first order of (4.2) implies that the *classical* projection map pr is a Poisson map with respect to the Poisson structures induced by $\star$ on M and $\star_P$ on P. The following example shows that in general there are obstructions to achieve (4.2) already on the classical level:

Example. Consider the Hopf fibration pr $: S^3 \longrightarrow S^2$ (which is a nontrivial principal S^1-bundle over S^2) and equip S^2 with the canonical symplectic Poisson structure. Then there exists *no* Poisson structure on S^3 such that pr becomes a Poisson map. Indeed, if there would be such a Poisson structure then necessarily all symplectic leaves would be two-dimensional as symplectic leaves are mapped into symplectic leaves and S^2 is already symplectic. Fixing one symplectic leaf in S^3 one checks that pr restricted to this leaf is still surjective and thus provides a covering of S^2. But S^2 is simply connected whence the symplectic leaf is itself a S^2. This would yield a section of the nontrivial principal bundle pr $: S^3 \longrightarrow S^2$, a contradiction.

Remark 4.1. Note that there are prominent examples of Hopf-Galois extensions using quantum spheres, see e.g. [21] and references therein. The above example shows that when taking the semi-classical limit of these q-deformations one obtains Poisson structures on S^2 which are certainly not symplectic. Note that this was a crucial feature in the above example. A further investigation of these examples is work in progress.

The above example shows that the first idea of deforming the projection map into an algebra homomorphism leads to hard obstructions in general, even though there are interesting classes of examples where the obstructions are absent. However, as we are interested in an approach not making too much assumptions in the beginning, we abandon this first idea. The next weaker requirement would be to deform pr* not into an algebra homomorphism but only turning $C^\infty(P)$ into a *bimodule*. This would have the advantage that there is no Poisson structure on P needed. However, a more subtle analysis shows that again for the Hopf fibration such a bimodule structure is impossible if one uses a star product on S^2 coming from the symplectic Poisson structure. Thus we are left with a *module structure*: for later convenience we choose a right module structure and state the following definition [5]:

Definition 4.2. Let pr $: P \circlearrowleft G \longrightarrow M$ be a principal G-bundle over M and $\star$ a star product on M. A deformation quantization of P is a right $\star$-module structure $\bullet$ for $C^\infty(P)[[\lambda]]$ of the form

$$F \bullet f = F \mathrm{pr}^* f + \sum_{r=1}^\infty \lambda^r \varrho_r(F, f), \tag{4.3}$$

where $\varrho_r : C^\infty(P) \times C^\infty(M) \longrightarrow C^\infty(P)$ is a bidifferential operator (along pr) for all $r \geq 1$, such that in addition one has the G-equivariance

$$g^*(F \bullet f) = g^* F \bullet f \tag{4.4}$$

for all $F \in C^\infty(P)[[\lambda]]$, $f \in C^\infty(M)[[\lambda]]$ and $g \in G$.

Note that as G acts on P from the right, the pull-backs with the actions of $g \in G$ provide a left action on $C^\infty(P)$ in (4.4). Then this condition means that the G-action commutes with the module multiplications.

Note that the module property $F \bullet (f \star g) = (F \bullet f) \bullet g$ implies that the constant function 1 acts as identity. Indeed, since $1 \star 1 = 1$ the action of 1 via $\bullet$ is a projection. However, in zeroth order the map $F \mapsto F \bullet 1$ is just the identity and hence invertible. But the only invertible projection is the identity map itself. Thus

$$F \bullet 1 = F \tag{4.5}$$

for all $F \in C^\infty(P)[[\lambda]]$, so the module structure $\bullet$ is necessarily unital.

Finally, we call two deformation quantizations $\bullet$ and $\tilde{\bullet}$ *equivalent*, if there exists a G-equivariant equivalence transformation between them, i.e. a formal series of differential operators $T = \mathrm{id} + \sum_{r=1}^{\infty} \lambda^r T_r$ on $C^\infty(P)[[\lambda]]$ such that

$$T(F \bullet f) = T(F) \tilde{\bullet} f \quad \text{and} \quad g^* T = T g^* \tag{4.6}$$

for all $F \in C^\infty(P)[[\lambda]]$, $f \in C^\infty(M)[[\lambda]]$ and $g \in G$.

We shall now discuss the existence and classification of such module structures. For warming up we consider the situation of a *trivial* principal fibre bundle:

Example. Let $P = M \times G$ be the trivial (and trivialized) principal G-bundle over M with the obvious projection. For any star product $\star$ on M we can now extend $\star$ to $C^\infty(M \times G)[[\lambda]]$ by simply acting only on the M-coordinates in the Cartesian product. Here we use the fact that we can canonically extend multidifferential operators on M to $M \times G$. Clearly, all algebraic properties are preserved whence in this case we even get a star product $\star_P = \star \otimes \mu$ with the undeformed multiplication μ for the G-coordinates. In particular, $C^\infty(M \times G)[[\lambda]]$ becomes a right module with respect to $\star$. So locally there are no obstructions even for the strongest requirement (4.2) and hence also for (4.3).

The problem of finding $\bullet$ is a global question whence we can not rely on local considerations directly. The most naive way to construct a $\bullet$ is an order-by-order construction: In general, one has to expect obstructions in each order which we shall now compute explicitly. This is a completely standard approach from the very first days of algebraic deformation theory [13, 18] and will in general only yield the result that there are possible obstructions: in this case one needs more refined arguments to ensure existence of deformations whence the order-by-order argument in general is rather useless. In our situation, however, it turns out that we are surprisingly lucky.

The following argument applies essentially to arbitrary algebras and module deformations and should be considered to be folklore. Suppose we have already found $\varrho_0 = \mathrm{pr}^*$, ϱ_1, ..., ϱ_k such that

$$F \bullet^{(k)} f = F \mathrm{pr}^* f + \sum_{r=1}^{k} \lambda^r \varrho_r(F, f) \tag{4.7}$$

is a module structure up to order λ^k and each ϱ_r fulfils the G-equivariance condition. Then in order to find ϱ_{k+1} such that $\bullet^{(k+1)} = \bullet^{(k)} + \lambda^{k+1}\varrho_{k+1}$ is a module structure up to order λ^{k+1} we have to satisfy

$$\varrho_{k+1}(F, f)\mathrm{pr}^* g - \varrho_{k+1}(F, fg) + \varrho_{k+1}(F\mathrm{pr}^* f, g)$$

$$= \sum_{r=1}^{k} \left(\varrho_r(F, C_{k+1-r}(f, g)) - \varrho_r(\varrho_{k+1-r}(F, f), g) \right) = R_k(F, f, g), \qquad (4.8)$$

for all $F \in C^\infty(P)[[\lambda]]$ and $f, g \in C^\infty(M)[[\lambda]]$. Here C_r denotes the r-th cochain of the star product $\star$ as in (2.3). In order to interpret this equation we consider the ϱ_r as maps

$$\varrho_r : C^\infty(M) \ni f \mapsto \varrho_r(\cdot, f) \in \mathrm{Diffop}(P) \qquad (4.9)$$

and similarly

$$R_k : C^\infty(M) \times C^\infty(M) \ni (f, g) \mapsto R_k(\cdot, f, g) \in \mathrm{Diffop}(P). \qquad (4.10)$$

Viewing $\mathrm{Diffop}(P)$ as $C^\infty(M)$-bimodule via pr^* in the usual way, we can now reinterpret (4.8) as equation between a Hochschild one-cochain ϱ_{k+1} and a Hochschild two-cochain R_k

$$\delta\varrho_{k+1} = R_k \qquad (4.11)$$

in the Hochschild (sub-)complex $\mathrm{HC}^\bullet_{\mathrm{diff}}(C^\infty(M), \mathrm{Diffop}(P))$ consisting of *differential* cochains taking values in the bimodule $\mathrm{Diffop}(P)$. Here δ is the usual Hochschild differential. Using the assumption that the $\varrho_0, \ldots, \varrho_k$ have been chosen such that $\bullet^{(k)}$ is a module structure up to order λ^k it is a standard argument to show

$$\delta R_k = 0. \qquad (4.12)$$

Thus the necessary condition for (4.11) is always fulfilled by construction whence (4.11) is a cohomological condition: The equation (4.11) has solutions if and only if the class of R_k in the second Hochschild cohomology $\mathrm{HH}^2_{\mathrm{diff}}(C^\infty(M), \mathrm{Diffop}(P))$ is trivial.

In fact, we have also to take care of the G-equivariance of ϱ_{k+1}. If all the $\varrho_0, \ldots, \varrho_k$ satisfy the G-equivariance then it is easy to see that also R_k has the G-equivariance property. Thus we have to consider yet another subcomplex of the differential Hochschild complex, namely

$$\mathrm{HC}^\bullet_{\mathrm{diff}}(C^\infty(M), \mathrm{Diffop}(P)^G) \subseteq \mathrm{HC}^\bullet_{\mathrm{diff}}(C^\infty(M), \mathrm{Diffop}(P)). \qquad (4.13)$$

Thus the obstruction for (4.11) to have a G-equivariant solution is the Hochschild cohomology class

$$[R_k] \in \mathrm{HH}^2_{\mathrm{diff}}(C^\infty(M), \mathrm{Diffop}(P)^G). \qquad (4.14)$$

A completely analogous order-by-order construction shows that also the obstructions for equivalence of two deformations $\bullet$ and $\tilde{\bullet}$ can be formulated using the differential Hochschild complex of $C^\infty(M)$ with values in $\mathrm{Diffop}(P)^G$. Now the obstruction lies in the first cohomology $\mathrm{HH}^1_{\mathrm{diff}}(C^\infty(M), \mathrm{Diffop}(P)^G)$.

The following (nontrivial) theorem solves the problem of existence and uniqueness of deformation quantizations now in a trivial way [5]:

Theorem 4.3. *Let* $\mathrm{pr} : P \longrightarrow M$ *be a surjective submersion.*

1. *We have*

$$\mathrm{HH}^k_{\mathrm{diff}}(C^\infty(M), \mathrm{Diffop}(P)) = \begin{cases} \mathrm{Diffop}_{\mathrm{ver}}(P) & \text{for } k = 0 \\ \{0\} & \text{for } k \geq 1. \end{cases} \quad (4.15)$$

2. *If in addition* $\mathrm{pr} : P \circlearrowleft G \longrightarrow M$ *is a principal G-bundle then we have*

$$\mathrm{HH}^k_{\mathrm{diff}}(C^\infty(M), \mathrm{Diffop}(P)^G) = \begin{cases} \mathrm{Diffop}_{\mathrm{ver}}(P)^G & \text{for } k = 0 \\ \{0\} & \text{for } k \geq 1. \end{cases} \quad (4.16)$$

The main idea is to proceed in three steps: first one shows that one can localize the problem to a bundle chart. For the local situation one can use the explicit homotopies from [4] to show that the cohomology is acyclic. This is the most nontrivial part. By a suitable partition of unity one can glue things together to end up with the global statement. For a detailed proof we refer to [5].

From this theorem and the previous considerations we obtain immediately the following result [5]:

Corollary 4.4. *For every principal G-bundle* $\mathrm{pr} : P \circlearrowleft G \longrightarrow M$ *and any star product* $\star$ *on M there exists a deformation quantization* $\bullet$ *which is unique up to equivalence.*

In particular, the deformation for the trivial bundle as in Example 4 is the unique one up to equivalence.

Remark 4.5.

1. It should be noted that the use of Theorem 4.3 gives existence and uniqueness but no explicit construction of deformation quantizations of principal bundles. Here the cohomological method is not sufficient even though in [5] rather explicit homotopies were constructed which allow to determine further properties of $\bullet$.

2. In the more particular case of a symplectic Poisson structure on M, Weiss used in his thesis [41] a variant of Fedosov's construction which gives a much more geometric and explicit approach: there is a well-motivated geometric input, namely a symplectic covariant derivative on M as usual for Fedosov's star products and a principal connection on P. Out of this the module multiplication $\bullet$ is constructed by a recursive procedure. The dependence of $\bullet$ on the principal connection should be interpreted as a global version of the Seiberg-Witten map [34], now of course in a much more general framework for arbitrary principal bundles, see also [2, 24, 25].

3. For the general Poisson case a more geometric construction is still missing. However, it seems to be very promising to combine global formality theorems like the one in [16] or the approach in [10] with the construction [41]. These possibilities will be investigated in future works.

5. The commutant and associated bundles

Theorem 4.3 gives in addition to the existence and uniqueness of deformation quantizations of P also a description of the *differential commutant* of the right multiplications by functions on M via $\bullet$: we are interested in those formal series $D = \sum_{r=0}^{\infty} \lambda^r D_r \in \mathrm{Diffop}(P)[[\lambda]]$ of differential operators with the property

$$D(F \bullet f) = D(F) \bullet f \tag{5.1}$$

for all $F \in C^{\infty}(P)[[\lambda]]$ and $f \in C^{\infty}(M)[[\lambda]]$. In particular, if $D_0 = \mathrm{id}$ then (5.1) gives a *self-equivalence*. Clearly, the differential commutant

$$\mathcal{K} = \big\{ D \in \mathrm{Diffop}(P)[[\lambda]] \mid D \text{ satisfies } (5.1) \big\} \subseteq \mathrm{Diffop}(P)[[\lambda]] \tag{5.2}$$

is a subalgebra of $\mathrm{Diffop}(P)[[\lambda]]$ over $\mathbb{C}[[\lambda]]$.

Note that there are other operators on $C^{\infty}(P)[[\lambda]]$ which commute with all right multiplications, namely the highly non-local pull-backs g^* with $g \in G$. This was just part of the Definition 4.2 of a deformation quantization of a principal bundle. However, in this section we shall concentrate on the differential operators with (5.1) only.

Before describing the commutant it is illustrative to consider the classical situation. Here the commutant is simply given by the vertical differential operators

$$\mathrm{Diffop}_{\mathrm{ver}}(P) = \big\{ D \in \mathrm{Diffop}(P) \mid D(F\mathrm{pr}^* f) = D(F)\mathrm{pr}^* f \big\} \tag{5.3}$$

by the very definition of vertical differential operators. Alternatively, the commutant is the zeroth Hochschild cohomology. More interesting is now the next statement which gives a quantization of the classical commutant, see [5].

Theorem 5.1. *There exists a* $\mathbb{C}[[\lambda]]$-*linear bijection*

$$\varrho' : \mathrm{Diffop}_{\mathrm{ver}}(P)[[\lambda]] \longrightarrow \mathcal{K} \subseteq \mathrm{Diffop}(P)[[\lambda]] \tag{5.4}$$

of the form

$$\varrho' = \mathrm{id} + \sum_{r=1}^{\infty} \lambda^r \varrho_r' \tag{5.5}$$

which is G-*equivariant, i.e.*

$$g^* \varrho' = \varrho' g^* \tag{5.6}$$

for all $g \in G$. *The choice of such a* ϱ' *induces an associative deformation* $\star'$ *of* $\mathrm{Diffop}_{\mathrm{ver}}(P)[[\lambda]]$ *which is uniquely determined by* $\star$ *up to equivalence. Finally,* ϱ' *induces a left* $(\mathrm{Diffop}_{\mathrm{ver}}(P)[[\lambda]], \star')$-*module structure* $\bullet'$ *on* $C^{\infty}(P)[[\lambda]]$ *via*

$$D \bullet' F = \varrho'(D)F. \tag{5.7}$$

The proof relies on an adapted symbol calculus for the differential operators $\mathrm{Diffop}(P)$: using an appropriate G-invariant covariant derivative ∇^P on P which preserves the vertical distribution and a principal connection on P one can induce a G-equivariant splitting of the differential operators $\mathrm{Diffop}(P)$ into the vertical differential operators and those differential operators which differentiate at least once in horizontal directions. Note that this complementary subspace has no intrinsic meaning but depends on the choice of ∇^P and the principal connection. A recursive construction gives the corrections terms $\varrho'_r(D)$ for a given $D \in \mathrm{Diffop}_{\mathrm{ver}}(P)$, heavily using the fact that the *first* Hochschild cohomology $\mathrm{HH}^1_{\mathrm{diff}}(C^\infty(M), \mathrm{Diffop}(P))$ vanishes. Since the commutant itself is an associative algebra the remaining statements follow.

Corollary 5.2. *For the above choice of ϱ' the resulting deformation $\star'$ as well as the module structure are G-invariant, i.e. we have*

$$g^*(D \star' \tilde{D}) = g^* D \star' g^* \tilde{D} \quad and \quad g^*(D \bullet' F) = g^* D \bullet' g^* F \qquad (5.8)$$

for all $D, \tilde{D} \in \mathrm{Diffop}_{\mathrm{ver}}(P)[[\lambda]]$ and $F \in C^\infty(P)[[\lambda]]$.

This follows immediately from the G-equivariance of $\bullet$ and the G-equivariance of ϱ'.

Remark 5.3. A simple induction shows that the commutant of $(\mathrm{Diffop}_{\mathrm{ver}}(P)[[\lambda]], \star')$ inside all differential operators $\mathrm{Diffop}(P)[[\lambda]]$ is again $(C^\infty(M)[[\lambda]], \star)$, where both algebras act by $\bullet'$ and $\bullet$, respectively. This way $C^\infty(P)[[\lambda]]$ becomes a $(\star', \star)$-bimodule such that the two algebras acting from left and right are mutual commutants inside all differential operators. Though this resembles already much of a Morita context, it is easy to see that $C^\infty(P)[[\lambda]]$ is *not* a Morita equivalence bimodule, e.g it is not finitely generated and projective. However, as we shall see later, there is still a close relation to Morita theory to be expected.

Remark 5.4. Note that classically $\mathrm{pr}^* : C^\infty(M) \longrightarrow \mathrm{Diffop}(P)$ is an algebra homomorphism, too. Thus the questions raised at the beginning of Section 4 can now be rephrased as follows: for a *bimodule deformation* of $C^\infty(P)$ into a bimodule over $C^\infty(M)[[\lambda]]$ equipped with possibly two different star products for the left and right action, one has to deform pr^* into a map

$$\mathbf{pr}^* : C^\infty(M)[[\lambda]] \longrightarrow (\mathrm{Diffop}_{\mathrm{ver}}(P)[[\lambda]], \star') \qquad (5.9)$$

such that the image is a subalgebra. In this case, we can induce a new product $\star'_M$ also for $C^\infty(M)[[\lambda]]$ making $C^\infty(P)[[\lambda]]$ a bimodule for the two, possibly different, star product algebras $(C^\infty(M)[[\lambda]], \star'_M)$ from the left and $(C^\infty(M)[[\lambda]], \star)$ from the right. Note that this is the only way to achieve it since $\star'$ is uniquely determined by $\star$. Thus it is clear that we have to expect obstructions in the general case as there might be no subalgebra of $(\mathrm{Diffop}_{\mathrm{ver}}(P)[[\lambda]], \star')$ which is in bijection to $C^\infty(M)[[\lambda]]$. Even if this might be the case, the resulting product $\star'_M$ might be inequivalent to $\star$. Note however, that we have now a very precise framework for the question whether pr^* can be deformed into a bimodule structure.

Remark 5.5. As a last remark we note that changing $\star$ to an equivalent $\tilde{\star}$ via an equivalence transformation Φ yields a corresponding right module structure $\tilde{\bullet}$ by

$$F \tilde{\bullet} f = F \bullet \Phi(f), \qquad (5.10)$$

which is still unique up to equivalence by Theorem 4.3. It follows that the commutants are *equal* (for this particular choice of $\tilde{\bullet}$) whence the induced deformations

$\star'$ and $\tilde{\star}'$ coincide. An equivalent choice of $\tilde{\bullet}$ would result in an equivalent $\tilde{\star}'$. This shows that we obtain a well-defined map

$$\mathrm{Def}(C^\infty(M)) \longrightarrow \mathrm{Def}(\mathrm{Diffop}_{\mathrm{ver}}(P)) \tag{5.11}$$

for the sets of equivalence classes of associative deformations. In fact, the resulting deformations $\star'$ are even G-invariant, whence the above map takes values in the smaller class of G-invariant deformations $\mathrm{Def}_G(\mathrm{Diffop}_{\mathrm{ver}}(P))$.

To make contact with the deformed vector bundles from Section 3 we consider now the association process. Recall that on the classical level one starts with a (continuous) representation π of G on a finite-dimensional vector space V. Then the associated vector bundle is

$$E = P \times_G V \longrightarrow M, \tag{5.12}$$

where the fibred product is defined via the equivalence relation $(p \cdot g, v) \sim (p, \pi(g)v)$ as usual. As the action of G on P is proper and free, E is a smooth manifold again and, in fact, a vector bundle over M with typical fibre V. Rather tautologically, any vector bundle is obtained like this by association from its own frame bundle. For the sections of E one has the canonical identifications

$$\Gamma^\infty(E) \cong C^\infty(P,V)^G \tag{5.13}$$

as right $C^\infty(M)$-modules, where the G-action of $C^\infty(P,V)$ is the obvious one.

After this preparation it is clear how to proceed in the deformed case. From the G-equivariance of $\bullet$ we see that

$$\Gamma^\infty(E)[[\lambda]] \cong C^\infty(P,V)^G[[\lambda]] \subseteq C^\infty(P,V)[[\lambda]] \tag{5.14}$$

is a $\star$-submodule with respect to the restricted module multiplication $\bullet$. It induces a right $\star$-module structure for $\Gamma^\infty(E)[[\lambda]]$ which we still denote by $\bullet$. This way we recover the deformed vector bundle as in Section 3.

Moreover, we see that the $\mathsf{End}(V)$-valued differential operators $\mathrm{Diffop}(P) \otimes \mathsf{End}(V)$ canonically act on $C^\infty(P,V)$ whence $((\mathrm{Diffop}_{\mathrm{ver}}(P) \otimes \mathsf{End}(V))[[\lambda]], \star')$ acts via $\bullet'$ on $C^\infty(P,V)[[\lambda]]$ in such a way that the action commutes with the $\bullet$-multiplications from the right. By the G-invariance of $\star'$ we see that the invariant elements $(\mathrm{Diffop}_{\mathrm{ver}}(P) \otimes \mathsf{End}(V))^G[[\lambda]]$ form a $\star'$-subalgebra which preserves (via $\bullet'$) the $\bullet$-submodule $C^\infty(P,V)^G[[\lambda]]$. Thus we obtain an algebra homomorphism

$$((\mathrm{Diffop}_{\mathrm{ver}}(P) \otimes \mathsf{End}(V))^G[[\lambda]], \star') \longrightarrow (\Gamma^\infty(\mathsf{End}(E))[[\lambda]], \star') \tag{5.15}$$

where $\star'$ on the left hand side is the deformation from Remark 3.8, part 2.

We conclude this section with some remarks and open questions:

Remark 5.6.

1. The universal enveloping algebra valued gauge fields of [24,25] can now easily be understood. For two vertical *vector fields* $\xi, \eta \in \mathrm{Diffop}_{\mathrm{ver}}(P)$ we have an action on $C^\infty(P)[[\lambda]]$ via $\bullet'$-left multiplication. In zeroth order this is just the usual Lie derivative $\mathscr{L}_\xi$. Now the module structure says that

$$\xi \bullet' (\eta \bullet' F) - \eta \bullet' (\xi \bullet' F) = ([\xi, \eta]_{\star'}) \bullet' F \tag{5.16}$$

for all $F \in C^\infty(P)[[\lambda]]$. Here $[\xi, \eta]_{\star'} = \xi \star' \eta - \eta \star' \xi \in \text{Diffop}_{\text{ver}}(P)[[\lambda]]$ is the $\star'$-commutator. In general, this commutator is a formal series of vertical differential operators but not necessarily a vector field any more. Note that (5.16) holds already on the level of the principal bundle.

2. For noncommutative gauge field theories we still need a good notion of gauge fields, i.e. connection one-forms, and their curvatures within our global approach. Though there are several suggestions from e.g. [27] a conceptually clear picture seems still to be missing.

3. In a future project we plan to investigate the precise relationship between $(\text{Diffop}_{\text{ver}}(P)[[\lambda]], \star')$ and the Morita theory of star products [6–8]. Here (5.15) already suggests that one can re-construct all algebras Morita equivalent to $(C^\infty(M)[[\lambda]], \star)$ out of $\star'$.

Acknowledgement

It is a pleasure for me to thank the organizers Bertfried Fauser, Jürgen Tolksdorf and Eberhard Zeidler for their invitation to the very stimulating conference "Recent Developments in Quantum Field Theory". Moreover, I would like to thank Rainer Matthes for valuable discussions on Hopf-Galois extensions, Jim Stasheff for pointing out reference [13], and Stefan Weiß for many comments on the first draft of the manuscript.

References

[1] Bahns, D., Waldmann, S., *Locally Noncommutative Space-Times*. Rev. Math. Phys. **19** (2007), 273–305.

[2] Barnich, G., Brandt, F., Grigoriev, M., *Seiberg-Witten maps and noncommutative Yang-Mills theories for arbitrary gauge groups*. J. High Energy Phys. .8 (2002), 23.

[3] Bayen, F., Flato, M., Frønsdal, C., Lichnerowicz, A., Sternheimer, D., *Deformation Theory and Quantization*. Ann. Phys. **111** (1978), 61–151.

[4] Bordemann, M., Ginot, G., Halbout, G., Herbig, H.-C., Waldmann, S., *Formalité G_∞ adaptee et star-représentations sur des sous-variétés coïsotropes*. Preprint **math.QA/0504276** (2005), 56 pages. Extended version of the previous preprint math/0309321.

[5] Bordemann, M., Neumaier, N., Waldmann, S., Weiss, S., *Deformation quantization of surjective submersions and principal fibre bundles*. Preprint (2007). In preparation.

[6] Bursztyn, H., Waldmann, S., *Deformation Quantization of Hermitian Vector Bundles*. Lett. Math. Phys. **53** (2000), 349–365.

[7] Bursztyn, H., Waldmann, S., *Algebraic Rieffel Induction, Formal Morita Equivalence and Applications to Deformation Quantization*. J. Geom. Phys. **37** (2001), 307–364.

[8] Bursztyn, H., Waldmann, S., *The characteristic classes of Morita equivalent star products on symplectic manifolds*. Commun. Math. Phys. **228** (2002), 103–121.

[9] Bursztyn, H., Waldmann, S., *Completely positive inner products and strong Morita equivalence*. Pacific J. Math. **222** (2005), 201–236.

[10] Cattaneo, A. S., Felder, G., Tomassini, L., *From local to global deformation quantization of Poisson manifolds.* Duke Math. J. **115**.2 (2002), 329–352.

[11] Connes, A., *Noncommutative Geometry.* Academic Press, San Diego, New York, London, 1994.

[12] Dabrowski, L., Grosse, H., Hajac, P. M., *Strong Connections and Chern-Connes Pairing in the Hopf-Galois Theory.* Commun. Math. Phys. **220** (2001), 301–331.

[13] Donald, J. D., Flanigan, F. J., *Deformations of algebra modules.* J. Algebra **31** (1974), 245–256.

[14] DeWilde, M., Lecomte, P. B. A., *Existence of Star-Products and of Formal Deformations of the Poisson Lie Algebra of Arbitrary Symplectic Manifolds.* Lett. Math. Phys. **7** (1983), 487–496.

[15] Dito, G., Sternheimer, D., *Deformation quantization: genesis, developments and metamorphoses.* In: Halbout, G. (eds.), *Deformation quantization.* [22], 9–54.

[16] Dolgushev, V. A., *Covariant and equivariant formality theorems.* Adv. Math. **191** (2005), 147–177.

[17] Doplicher, S., Fredenhagen, K., Roberts, J. E., *The Quantum Structure of Spacetime at the Planck Scale and Quantum Fields.* Commun. Math. Phys. **172** (1995), 187–220.

[18] Gerstenhaber, M., *On the Deformation of Rings and Algebras.* Ann. Math. **79** (1964), 59–103.

[19] Gutt, S., *Variations on deformation quantization.* In: Dito, G., Sternheimer, D. (eds.), *Conférence Moshé Flato 1999. Quantization, Deformations, and Symmetries, Mathematical Physics Studies* no. **21**, 217–254. Kluwer Academic Publishers, Dordrecht, Boston, London, 2000.

[20] Gutt, S., Rawnsley, J., *Equivalence of star products on a symplectic manifold; an introduction to Deligne's Čech cohomology classes.* J. Geom. Phys. **29** (1999), 347–392.

[21] Hajac, P. M., Matthes, R., Szymański, W., *Chern numbers for two families of noncommutative Hopf fibrations.* C. R. Math. Acad. Sci. Paris **336**.11 (2003), 925–930.

[22] Halbout, G. (eds.), *Deformation Quantization*, vol. 1 in *IRMA Lectures in Mathematics and Theoretical Physics.* Walter de Gruyter, Berlin, New York, 2002.

[23] Heller, J. G., Neumaier, N., Waldmann, S., *A C^*-Algebraic Model for Locally Noncommutative Spacetimes.* Lett. Math. Phys. **80** (2007), 257–272.

[24] Jurčo, B., Möller, L., Schraml, S., Schupp, P., Wess, J., *Construction of non-Abelian gauge theories on noncommutative spaces.* Eur. Phys. J. **C21** (2001), 383–388.

[25] Jurčo, B., Schraml, S., Schupp, P., Wess, J., *Enveloping algebra-valued gauge transformations for non-abelian gauge groups on non-commutative spaces.* Eur. Phys. J. C Part. Fields **17**.3 (2000), 521–526.

[26] Jurčo, B., Schupp, P., *Noncommutative Yang-Mills from equivalence of star products.* Eur. Phys. J. **C14** (2000), 367–370.

[27] Jurčo, B., Schupp, P., Wess, J., *Noncommutative gauge theory for Poisson manifolds.* Nucl. Phys. **B584** (2000), 784–794.

[28] Jurčo, B., Schupp, P., Wess, J., *Nonabelian noncommutative gauge theory via noncommutative extra dimensions.* Nucl.Phys. **B 604** (2001), 148–180.

[29] Kontsevich, M., *Deformation Quantization of Poisson manifolds*. Lett. Math. Phys. **66** (2003), 157–216.

[30] Lance, E. C., *Hilbert C*-modules. A Toolkit for Operator algebraists*, vol. 210 in *London Mathematical Society Lecture Note Series*. Cambridge University Press, Cambridge, 1995.

[31] Nest, R., Tsygan, B., *Algebraic Index Theorem*. Commun. Math. Phys. **172** (1995), 223–262.

[32] Rosenberg, J., *Rigidity of K-theory under deformation quantization*. Preprint q-alg/9607021 (July 1996).

[33] Schmüdgen, K., *Unbounded Operator Algebras and Representation Theory*, vol. 37 in *Operator Theory: Advances and Applications*. Birkhäuser Verlag, 1990.

[34] Seiberg, N., Witten, E., *String Theory and Noncommutative Geometry*. J. High. Energy Phys. **09** (1999), 032.

[35] Waldmann, S., *Deformation of Hermitian Vector Bundles and Non-Commutative Field Theory*. In: Maeda, Y., Watamura, S. (eds.), *Noncommutative Geometry and String Theory*, vol. 144 in *Prog. Theo. Phys. Suppl.*, 167–175. Yukawa Institute for Theoretical Physics, 2001. Proceedings of the International Workshop on Noncommutative Geometry and String Theory.

[36] Waldmann, S., *Morita equivalence of Fedosov star products and deformed Hermitian vector bundles*. Lett. Math. Phys. **60** (2002), 157–170.

[37] Waldmann, S., *On the representation theory of deformation quantization*. In: Halbout, G. (eds.), *Deformation quantization*. [22], 107–133.

[38] Waldmann, S., *The Picard Groupoid in Deformation Quantization*. Lett. Math. Phys. **69** (2004), 223–235.

[39] Waldmann, S., *Morita Equivalence, Picard Groupoids and Noncommutative Field Theories*. In: Carow-Watamura, U., Maeda, Y., Watamura, S. (eds.), *Quantum Field Theory and Noncommutative Geometry*, vol. 662 in *Lect. Notes Phys.*, 143–155. Springer-Verlag, 2005.

[40] Waldmann, S., *Poisson-Geometrie und Deformationsquantisierung. Eine Einführung*. Springer-Verlag, 2007.

[41] Weiß, S., *Nichtkommutative Eichtheorien und Deformationsquantisierung von Hauptfaserbündeln*. Master thesis, Fakultät für Mathematik und Physik, Physikalisches Institut, Albert-Ludwigs-Universität, Freiburg, 2006. Available at http://idefix.physik.uni-freiburg.de/~weiss/.

[42] Wess, J., *Privat communication*. Discussion on noncommutativity as phase transition.

Stefan Waldmann
Fakultät für Mathematik und Physik
Albert-Ludwigs-Universität Freiburg
Physikalisches Institut
Hermann Herder Strasse 3
DE–79104 Freiburg
e-mail: Stefan.Waldmann@physik.uni-freiburg.de

Quantum Field Theory
B. Fauser, J. Tolksdorf and E. Zeidler, Eds., 137–154
© 2009 Birkhäuser Verlag Basel/Switzerland

Renormalization of Gauge Fields using Hopf Algebras

Walter D. van Suijlekom

Abstract. We describe the Hopf algebra structure of Feynman graphs for non-Abelian gauge theories and prove compatibility of the so-called Slavnov–Taylor identities with the coproduct. When these identities are taken into account, the coproduct closes on the Green's functions, which thus generate a Hopf subalgebra.

Mathematics Subject Classification (2000). Primary 81T15; Secondary 81T18; 81T13; 16W30.

Keywords. Feynman diagrams, perturbative quantum field theory, renormalization, Greens functions, 1-particle irreducible (1PI) graphs, Yang-Mills theories.

1. Introduction

Quantum field theories have been widely accepted in the physics community, mainly because of their well-tested predictions. One of the famous numbers predicted by quantum electrodynamics is the electromagnetic moment of the electron which has been tested up to a previously unencountered precision.

Unfortunately, quantum field theories are perceived with some suspicion by mathematicians. This is mainly due to the appearance of divergences when naively computing probability amplitudes. These *infinities* have to be dealt with properly by an apparently obscure process called renormalization.

Nevertheless, mathematical interest has been changing lately in favor of quantum field theories, the general philosophy being that such a physically accurate theory should have some underlying mathematically rigorous description. One of these interests is in the process of renormalization, and has been studied in the context of Hopf algebras [6, 3]. Of course, the process of renormalization was already quite rigorously defined by physicists in the early second half of the previous century. However, the structure of a coproduct describing how to subtract divergence really clarified the process.

One could argue though that since the elements in the Hopf algebra are individual Feynman graphs, it is a bit unphysical. Rather, one would like to describe the renormalization process on the level of the 1PI Green's functions, since these correspond to actual physical processes. Especially for (non-Abelian) gauge theories, the graph-by-graph approach of for instance the BPHZ-procedure is usually replaced by more powerful methods based on BRST-symmetry and the Zinn-Justin equation (and its far reaching generalization: the Batalin-Vilkovisky formalism). They all involve the 1PI Green's functions or even the full effective action that is generated by them.

The drawback of these latter methods, is that they rely heavily on functional integrals and are therefore completely formal. One of the advantages of BPHZ-renormalization is that if one accepts the perturbative series of Green's function in terms of Feynman graphs as a starting point, the procedure is completely rigorous. Of course, this allowed the procedure to be described by a mathematical structure such as a Hopf algebra.

In this article, we prove some of the results on Green's functions starting with the Hopf algebra of Feynman graphs for non-Abelian gauge theories. We derive the existence of Hopf subalgebras generated by the 1PI Green's functions. We do this by showing that the coproduct takes a closed form on these Green's functions, thereby relying heavily on a formula that we have previously derived [14]. Already in [1] Hopf subalgebras were given for any connected graded Hopf algebra as solutions to Dyson-Schwinger equations. It turned out that there was a close relation with Hochschild cohomology. It was argued by Kreimer in [8, 7] that – for the case of non-Abelian gauge theories – the existence of Hopf subalgebras follows from the validity of the Slavnov–Taylor identities *inside* the Hopf algebra of (QCD) Feynman graphs. We now fully prove this claim by applying a formula for the coproduct on Green's functions that we have derived before in [14]. In fact, that formula allowed us to prove compatibility of the Slavnov–Taylor identities with the Hopf algebra structure.

This paper is organized as follows. In Section 2, we start by giving some background from physics. Of course, this can only be a quick *lifting of the curtain* and is meant as a motivation for the present work. In Section 3, we make precise our setup by defining the Hopf algebra of Feynman graphs and introduce several combinatorial factors associated to such graphs. We put the process of renormalization in the context of a Birkhoff decomposition.

Section 4 contains the derivation of the Hopf algebra structure at the level of Green's functions, rather then the individual Feynman graphs. We will encounter the crucial role that is played by the so-called Slavnov–Taylor identities.

2. Preliminaries on perturbative quantum field theory

We start by giving some background from physics and try to explain the origin of
Feynman graphs in the perturbative approach to quantum field theory.

We understand *probability amplitudes for physical processes as formal expansions in Feynman amplitudes*, thereby avoiding the use of path integrals. We make
this more explicit by some examples taken from physics.

Example 1. *The interaction of the photon with the electron in quantum electrodynamics (QED) is described by the following expansion,*

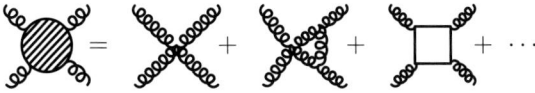

*Here all graphs appear that can be built from the vertex that connects a wiggly line
(the photon) to two straight lines (the electron).*

Example 2. *The quartic gluon self-interaction in quantum chromo dynamics is
given by*

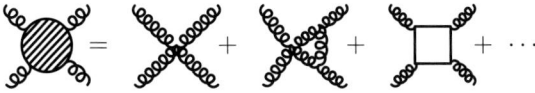

*This expansion involves the gluon vertex of valence 3 and 4 (wiggly lines), as well
as the quark-gluon interaction (involving two straight lines)*

We shall call these expansions **Green's functions**. Of course, this names originates from the theory of partial differential equations and the zeroth order terms
in the above expansions are in fact Green's functions in the usual sense. We use the
notation $G^{\prec}$ and $G^{\times}$ for the Green's function, indicating the external structure
of the graphs in the above two expansions, respectively.

From these expansions, physicists can actually derive numbers, giving the
probability amplitudes mentioned above. The rules of this game are known as
the Feynman rules; we briefly list them for the case of quantum electrodynamics.
Feynman rules for non-Abelian gauge theories can be found in most standard
textbooks on quantum field theory (see for instance [2]).

Assigning a momentum k to each edge of a graph, we have:

$$\text{~~~}_{k} = \frac{1}{k^2 + i\epsilon}\left(-\delta_{\mu\nu} + \frac{k_\mu k_\nu}{k^2 + i\epsilon}(1-\xi)\right)$$

$$\text{———}_{k} = \frac{1}{\gamma^\mu k_\mu + m}$$

$$\text{———} = -ie\gamma^\mu \delta(k_1 + k_2 + k_3).$$

Here, e is the electron charge, m the electron mass and γ^μ are 4×4 Dirac gamma matrices; they satisfy $\gamma^\mu\gamma^\nu + \gamma^\nu\gamma^\mu = -2\delta^{\mu\nu}$. Also, ϵ is an infrared regulator and $\xi \in \mathbb{R}$ is the so-called gauge fixing parameter. In addition to the above assignments, one integrates the above internal momenta k (for each internal edge) over $\mathbb{R}^4$.

Example 3. *Consider the following electron self-energy graph*

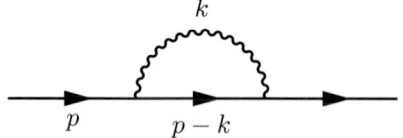

According to the Feynman rules, the amplitude for this graph is

$$U(\Gamma) = \int d^4k \; (e\gamma^\mu)\frac{1}{\gamma^\kappa(p_\kappa + k_\kappa) + m}(e\gamma^\nu)\left(-\frac{\delta_{\mu\nu}}{k^2 + i\epsilon} + \frac{k_\mu k_\nu}{(k^2 + i\epsilon)^2}(1 - \xi)\right) \quad (1)$$

with summation over repeated indices understood.

The alert reader may have noted that the above improper integral is actually not well-defined. This is the typical situation – happening for most graphs – and are the famous divergences in perturbative quantum field theory. This apparent failure can be resolved, leading eventually to spectacularly accurate predictions in physics.

The theory that proposes a solution to these divergences is called *renormalization*. This process consists of two steps. Firstly, one introduces a *regularization parameter* that controls the divergences. For instance, in *dimensional regularization* one integrates in $4 + z$ dimensions instead of in 4, with z a complex number.

Adopting certain rules[1] for this integration in complex dimensions, one obtains for instance for the above integral (1):

$$U(\Gamma)(z) \sim \Gamma(z)\text{Pol}(p),$$

where the Γ on the left-hand-side is the graph and the Γ on the right-hand-side is the gamma function from complex analysis. Moreover, $\text{Pol}(p)$ is a polynomial in the external momentum p. The previous divergence has been translated into a pole of the gamma function at $z = 0$ and we have thus obtained a control on the divergence.

The second step in the process of renormalization is *subtraction*. We let T be the projection onto the pole part of Laurent series in z, i.e.,

$$T\left[\sum_{n=-\infty}^{\infty} a_n z^n\right] = \sum_{n<0} a_n z^n.$$

More generally, we have a projection on the divergent part in the regularizing parameter. This is the origin of the study of Rota-Baxter algebras in the setting of quantum field theories [5]. We will however restrict ourselves to dimensional regularization, which is a well suited regularization for gauge theories. For the above graph Γ, we define the **renormalized amplitude** $R(\Gamma)$ by simply subtracting the divergent part, that is, $R(\Gamma) = U(\Gamma) - T[U(\Gamma)]$. Clearly, the result is finite for $z \to 0$. More generally, a graph Γ might have subgraphs $\gamma \subset \Gamma$ which lead to sub-divergences in $U(\Gamma)$. The so-called **BPHZ-procedure** (after its inventors Bogoliubov, Parasiuk, Hepp and Zimmermann) provides a way to deal with those sub-divergences in a recursive manner. It gives for the **renormalized amplitude**:

$$R(\Gamma) = U(\Gamma) + C(\Gamma) + \sum_{\gamma \subset \Gamma} C(\gamma)U(\Gamma/\gamma) \tag{2a}$$

where C is the so-called **counterterm** defined recursively by

$$C(\Gamma) = -T\left[U(\Gamma) + \sum_{\gamma \subset \Gamma} C(\gamma)U(\Gamma/\gamma)\right]. \tag{2b}$$

The two sums here are over all subgraphs in a certain class; we will make this more precise in the next section.

2.1. Gauge theories

We now focus on a special class of quantum field theories – quantum gauge theories – which are of particular interest for real physical processes. Without going into details on what classical gauge field theories are, we focus on the consequences on the quantum side of the presence of a classical gauge symmetry. Such a gauge symmetry acts (locally) on the classical fields by **gauge transformations** and these

---

[1]Essentially, one only needs the rule that the formula familiar in integer dimension $\int d^D e^{-\pi\lambda k^2} = \lambda^{D/2}$ holds for complex dimension D as well. Indeed, using Schwinger parameters, or, equivalently, the Laplace transform, one can write $1/k^2$ as the integral over $s > 0$ of e^{-sk^2}.

transformations form a group, the gauge group. This group is typically infinite dimensional, since it consists of functions on space-time taking values in a Lie group. For quantum electrodynamics this Lie group is Abelian and just $U(1)$, for quantum chromo dynamics – the theory of gluons and quarks – it is $SU(3)$.

When (perturbatively) quantizing the gauge theory, one is confronted with this extra infinity. A way to handle it is by *fixing the gauge*, in other words, choosing an orbit under the action of the gauge group. All this can be made quite precise in *BRST-quantization*. Although in this process the gauge symmetry completely disappears, certain identities between Green's functions appear. This is a purely 'quantum property' and therefore interesting to study. In addition, being identities between full Green's functions, it is interesting with a view towards nonperturbative quantum field theory.

For quantum electrodynamics, the identities are simple and linear in the Green's functions:

$$U\left(G \overset{\prec}{}\right) = U\left(G {-}\right). \tag{3}$$

These are known as **Ward identities** since they were first derived by Ward in [15]. The apparent mismatch between the number of external lines on the left and right-hand-side is resolved because the vertex graphs are considered at *zero momentum transfer*. This means that the momentum on the photon line is evaluated at $p = 0$.

For non-Abelian gauge theories such as quantum chromo dynamics (QCD), the identities are quadratic in the fields and read:

$$
\begin{aligned}
U\left(G \overset{\prec}{}\right) U\left(G \overset{\prec}{}\right) &= U\left(G {\times}\right) U\left(G {-}\right); \\
U\left(G \overset{\prec}{}\right) U\left(G {\cdots}\right) &= U\left(G {\times}\right) U\left(G {}\right); \\
U\left(G \overset{\prec}{}\right) U\left(G \overset{\prec}{}\right) &= U\left(G {\times}\right) U\left(G {\cdots}\right).
\end{aligned}
\tag{4}
$$

The dotted and straight line here corresponds to the ghost and quark, respectively. After their inventors, they are called the **Slavnov–Taylor identities** [11, 12].

The importance of these identities lie in the fact that they are compatible with renormalization under the condition that gauge invariance is compatible with the regularization procedure. In fact, it turns out that dimensional regularization satisfies this requirement, see for instance Section 13.1 of [9]. As a consequence, the Slavnov-Taylor identities hold after replacing U by R or C in the above formula. For instance, in the case of quantum electrodynamics one obtains the identity $Z_1 = Z_2$ actually derived by Ward, where $Z_1 = C(G \overset{\prec}{})$ and $Z_2 = C(G {-})$. For quantum chromo dynamics on the other hand, one derives the formulae

$$
\frac{Z \overset{\prec}{}}{Z - \sqrt{Z {\cdots}}} = \frac{Z {\cdots}}{Z {\cdots} \sqrt{Z {\cdots}}} = \frac{Z \overset{\prec}{}}{\left(Z {\cdots}\right)^{3/2}} = \frac{\sqrt{Z {\times}}}{Z {\cdots}},
\tag{5}
$$

where the notation is as above: $Z^r := C(G^r)$. The above formula can be readily obtained from the above Slavnov–Taylor identities (4) afterreplacing U by C .

They are the key to proving renormalizability of non-Abelian gauge theories, let us try to sketch this argument.

First of all, the different interactions that are present in the theory can be weighted by a coupling constant. For example, in QCD there are four different interactions: gluon-quark, gluon-ghost, cubic and quartic gluon self-interaction. All of these come with their own coupling constants and gauge invariance (or rather, BRST-invariance) requires them to be identical. In the process of renormalization, the coupling constants are actually not constant and depend on the energy scale. This is the *running of the coupling constant* and is the origin of the renormalization group describing how they change. For QCD, the four coupling constants $g_{0,\,\text{gluon-quark}}, g_{0,\,\text{gluon-ghost}}, g_{0,\,\text{cubic}}, g_{0,\,\text{quartic}}$ are expressed in terms of the original coupling constant g as

$$g_{0,\,\text{gluon-quark}} = \frac{Z^{\,\text{gluon-quark}}}{Z - \sqrt{Z^{\,\text{gluon}}}}g, \qquad g_{0,\,\text{gluon-ghost}} = \frac{Z^{\,\text{gluon-ghost}}}{Z - \sqrt{Z^{\,\text{gluon}}}}g,$$

$$g_{0,\,\text{cubic}} = \frac{Z^{\,\text{cubic}}}{\left(Z^{\,\text{gluon}}\right)^{3/2}}g, \qquad g_{0,\,\text{quartic}} = \frac{\sqrt{Z^{\,\text{quartic}}}}{Z^{\,\text{gluon}}}g. \tag{6}$$

We see that the Slavnov–Taylor identities guarantee that *the four coupling constants remain equal* after renormalization.

The above compatibility of renormalization with the Slavnov–Taylor identities is usually derived using the Zinn-Justin equation (or the more general BV-formalism) relying heavily on path integral techniques. Our goal in the next sections is to derive this result taking the formal expansion of the Green's functions in Feynman graphs as a starting point. We will work in the setting of the Connes-Kreimer Hopf algebra of renormalization.

3. The Hopf algebra of Feynman graphs

We suppose that we have defined a (renormalizable) quantum field theory and specified the possible interactions between different types of particles. We indicate the interactions by vertices and the propagation of particles by lines. This leads us to define a set $R = R_V \cup R_E$ of vertices and edges; for QED we have

$$R_V = \{\ \text{⤬}\ \}; \qquad R_E = \{\ \text{———}\ ,\ \text{∿∿}\ \},$$

whereas for QCD we have,

$$R_V = \{\ \text{⤬}\ ,\ \text{⤙}\ ,\ \text{⤏}\ ,\ \text{✕}\ \}; \qquad R_E = \{\ \text{———}\ ,\ \cdots\cdots\ ,\ \text{⦵⦵⦵}\ \}.$$

We stress for what follows that it is not necessary to define the set explicitly.

A **Feynman graph** is a graph built from vertices in R_V and edges in R_E. Naturally, we demand edges to be connected to vertices in a compatible way, respecting the type of vertex and edge. As opposed to the usual definition in graph theory, Feynman graphs have no external vertices, they only have external lines. We assume those lines to carry a labelling.

An **automorphism** of a Feynman graph is a graph automorphism leaving the external lines fixed and respects the types of vertices and edges. This definition is motivated by the fact that the external lines correspond physically to particles prepared for some collision experiment – the interior of the graph – and those lines are thus fixed. The order of the group of automorphisms $\mathrm{Aut}(\Gamma)$ of a graph Γ is called its **symmetry factor** and denoted by $\mathrm{Sym}(\Gamma)$. Let us give two examples:

$$\mathrm{Sym}(\text{\scriptsize figure}) = 2; \qquad \mathrm{Sym}(\text{\scriptsize figure}) = 1.$$

For disconnected graphs, the symmetry factor is given recursively as follows. Let Γ' be a connected graph; we set

$$\mathrm{Sym}(\Gamma\,\Gamma') = (n(\Gamma,\Gamma') + 1)\mathrm{Sym}(\Gamma)\mathrm{Sym}(\Gamma'), \qquad (7)$$

with $n(\Gamma,\Gamma')$ the number of connected components of Γ that are isomorphic to Γ'.

We define the **residue** $\mathrm{res}(\Gamma)$ of a graph Γ as the vertex or edge the graph reduces to after collapsing all its internal vertices and edges to a point. For example, we have:

$$\mathrm{res}\left(\text{\scriptsize figure}\right) = \text{\scriptsize figure} \qquad \text{and} \qquad \mathrm{res}\left(\text{\scriptsize figure}\right) = \text{——}.$$

Henceforth, we will *restrict to graphs with residue in R*; these are the relevant graphs to be considered for the purpose of renormalization.

For later use, we introduce another combinatorial quantity, which is the **number of insertion places** $\Gamma \mid \gamma$ for the graph γ in Γ. It is defined as the number of elements in the set of vertices and internal edges of Γ of the form $\mathrm{res}(\gamma) \in R$. For disconnected graphs $\gamma = \gamma_1 \cup \cdots \cup \gamma_n$, the number $\Gamma \mid \gamma$ counts the number of $n-tuples$ of disjoint insertion places of the type $\mathrm{res}(\gamma_1), \cdots, \mathrm{res}(\gamma_n)$.

We exemplify this quantity by

$$\text{\scriptsize figure} \Big| \text{\scriptsize figure} = 2 \quad \text{whereas} \quad \text{\scriptsize figure} \Big| \text{\scriptsize figure} = 6.$$

Here, one allows multiple insertions of edge graphs (i.e. a graph with residue in R_E) on the same edge; the underlying philosophy is that insertion of an edge graph creates a new edge.

For the definition of the Hopf algebra of Feynman graphs [3], we restrict to **one-particle irreducible** (1PI) Feynman graphs. These are graphs that are not trees and cannot be disconnected by cutting a single internal edge. For example, all graphs in this paper are one-particle irreducible, *except* the following which is one-particle reducible:

Connes and Kreimer then defined the following Hopf algebra. We refer to the appendix for a quick review on Hopf algebras.

Definition 4. *The Hopf algebra H of Feynman graphs is the free commutative $\mathbb{Q}$-algebra generated by all 1PI Feynman graphs, with counit $\epsilon(\Gamma) = 0$ unless $\Gamma = \emptyset$, in which case $\epsilon(\emptyset) = 1$, coproduct,*

$$\Delta(\Gamma) = \Gamma \otimes 1 + 1 \otimes \Gamma + \sum_{\gamma \subsetneq \Gamma} \gamma \otimes \Gamma/\gamma,$$

where the sum is over disjoint unions of subgraphs with residue in R. The antipode is given recursively by,

$$S(\Gamma) = -\Gamma - \sum_{\gamma \subsetneq \Gamma} S(\gamma)\Gamma/\gamma. \tag{8}$$

Two examples of this coproduct, taken from QED, are:

$$\Delta\left(\text{⤳⟨⟩⤳}\right) = \text{⤳⟨⟩⤳} \otimes 1 + 1 \otimes \text{⤳⟨⟩⤳} + 2 \,\text{⤳⟨}\, \otimes\, \text{⤳○⤳},$$

$$\Delta\left(\text{⤳⟨⟩⤳}\right) = \text{⤳⟨⟩⤳} \otimes 1 + 1 \otimes \text{⤳⟨⟩⤳} + 2\,\text{⤳⟨}\, \otimes\, \text{⤳○⤳}$$

$$+ 2\,\text{⤳⟨}\, \otimes\, \text{⤳⟨⟩⤳} + \text{⤳⟨ ⤳⟨}\, \otimes\, \text{⤳○⤳}.$$

The above Hopf algebra is an example of a connected graded Hopf algebra, i.e. $H = \oplus_{n \in \mathbb{N}} H^n$, $H^0 = \mathbb{C}$ and

$$H^k H^l \subset H^{k+l}; \qquad \Delta(H^n) = \sum_{k=0}^{n} H^k \otimes H^{n-k}.$$

Indeed, the Hopf algebra of Feynman graphs is graded by the **loop number** $L(\Gamma)$ of a graph Γ; then H^0 consists of rational multiples of the empty graph, which is the unit in H, so that $H^0 = \mathbb{Q}1$.

Remark 5. *One can enhance the Feynman graphs with an external structure. This involves the external momenta on the external lines and can be formulated mathematically by distributions, see for instance [3]. The case of quantum electrodynamics has been worked out in detail in [13].*

3.1. Renormalization as a Birkhoff decomposition

We now demonstrate how to obtain Equation (2) for the renormalized amplitude and the counterterm for a graph as a Birkhoff decomposition in the group of characters of H. Let us first recall the definition of a Birkhoff decomposition.

We let $l : C \to G$ be a loop with values in an arbitrary complex Lie group G, defined on a smooth simple curve $C \subset \mathbb{P}_1(\mathbb{C})$. Let $C_\pm$ be the two complements of C in $\mathbb{P}_1(\mathbb{C})$, with $\infty \in C_-$. A **Birkhoff decomposition** of l is a factorization of the form

$$l(z) = l_-(z)^{-1} l_+(z); \qquad (z \in C),$$

where $l_\pm$ are (boundary values of) two holomorphic maps on $C_\pm$, respectively, with values in G. This decomposition gives *a natural way to extract finite values from a divergent expression.* Indeed, although $l(z)$ might not holomorphically extend to C_+, $l_+(z)$ is clearly finite as $z \to 0$.

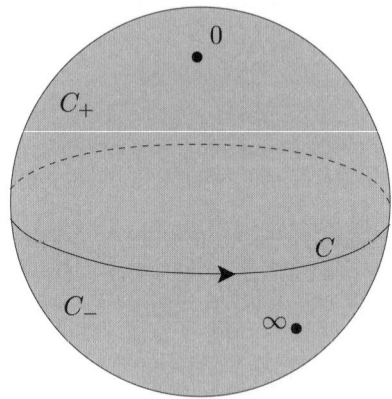

We now look at the group $G(K) = \mathrm{Hom}_{\mathbb{Q}}(H, K)$ of K-valued characters of a connected graded commutative Hopf algebra H, where K is the field of convergent Laurent series in z.[2] The product, inverse and unit in the group $G(K)$ are defined by the respective equations:

$$\phi * \psi(X) = \langle \phi \otimes \psi, \Delta(X) \rangle,$$
$$\phi^{-1}(X) = \phi(S(X)),$$
$$e(X) = \epsilon(X),$$

for $\phi, \psi \in G(K)$. We claim that a map $\phi \in G(K)$ is in one-to-one correspondence with loops l on an infinitesimal circle around $z = 0$ and values in $G(\mathbb{Q}) = \mathrm{Hom}_{\mathbb{Q}}(H, \mathbb{Q})$. Indeed, the correspondence is given by

$$\phi(X)(z) = l(z)(X),$$

---

[2]In the language of algebraic geometry, there is an affine group scheme G represented by H in the category of commutative algebras. In other words, $G = \mathrm{Hom}_{\mathbb{Q}}(H, \ . \ )$ and $G(K)$ are the K-points of the group scheme.

and to give a Birkhoff decomposition for l is thus equivalent to giving a factorization $\phi = \phi_-^{-1} * \phi_+$ in $G(K)$. It turns out that for graded connected commutative Hopf algebras such a factorization exists.

Theorem 6 (Connes–Kreimer [3]). *Let H be a graded connected commutative Hopf algebra. The Birkhoff decomposition of $l : C \to G$ (given by an algebra map $\phi : H \to K$) exists and is given dually by*

$$\phi_-(X) = \epsilon(X) - T\left[m(\phi_- \otimes \phi)(1 \otimes (1 - \epsilon)\Delta(X)\right]$$

and $\phi_+ = \phi_- * \phi$.

The graded connected property of H assures that the recursive definition of ϕ_- actually makes sense. In the case of the Hopf algebra of Feynman graphs defined above, the factorization takes the following form:

$$\phi_-(\Gamma) = -T\left[\phi(\Gamma) + \sum_{\gamma \subsetneq \Gamma} \phi_-(\gamma)\phi(\Gamma/\gamma)\right]$$

$$\phi_+(\Gamma) = \phi(\Gamma) + \phi_-(\Gamma) + \sum_{\gamma \subsetneq \Gamma} \phi_-(\gamma)\phi(\Gamma/\gamma)$$

The key point is now that the Feynman rules actually define an algebra map $U : H \to K$ by assigning to each graph Γ the regularized Feynman rules $U(\Gamma)$, which are Laurent series in z. When compared with Equations (2) one concludes that the algebra maps U_+ and U_- in the Birkhoff factorization of U are precisely the renormalized amplitude R and the counterterm C, respectively. Summarizing, we can write the BPHZ-renormalization as the Birkhoff decomposition $U = C^{-1}*R$ of the map $U : H \to K$ dictated by the Feynman rules.

Although the above construction gives a very nice geometrical description of the process of renormalization, it is a bit unphysical in that it relies on individual graphs. Rather, as mentioned before, in physics the probability amplitudes are computed from the full expansion of Green's functions. Individual graphs do not correspond to physical processes and therefore a natural question to pose is how the Hopf algebra structure behaves at the level of the Green's functions. We will see in the next section that they generate Hopf subalgebras, i.e. the coproduct closes on Green's functions. In proving this, the Slavnov–Taylor identities turn out to play an essential role.

4. The Hopf algebra of Green's functions

For a vertex or edge $r \in R$ we define the **1PI Green's function** by

$$G^r = 1 \pm \sum_{\mathrm{res}(\Gamma)=r} \frac{\Gamma}{\mathrm{Sym}(\Gamma)}, \tag{9}$$

where the sign is $+$ if r is a vertex and $-$ if it is an edge. The restriction of the sum to graphs Γ at loop order $L(\Gamma) = L$ is denoted by G_L^r.

Proposition 7 ([14]). *The coproduct takes the following form on the 1PI Green's functions:*

$$\Delta(G^r) = \sum_{\gamma} \sum_{\mathrm{res}(\Gamma)=r} \frac{\Gamma \mid \gamma}{\mathrm{Sym}(\gamma)\mathrm{Sym}(\Gamma)} \gamma \otimes \Gamma,$$

with the sum over γ over all disjoint unions of 1PI graphs.

The sketch of the proof is as follows. First, one writes the coproduct Δ as a sum of maps Δ_γ where these maps only detects subgraphs isomorphic to γ. One then proves the above formula for Δ_γ with γ a 1PI graph using simply the orbit-stabilizer theorem for the automorphism group of graphs. Finally, writing $\Delta_{\gamma\gamma'}$ in terms of Δ_γ and $\Delta_{\gamma'}$ one proceeds by induction to derive the above expression.

One observes that the coproduct does not seem to close on Green's functions due to the appearance of the combinatorial factor $\Gamma \mid \gamma$. Let us try to elucidate this and compute these factors explicitly.

Let $m_{\Gamma,r}$ be the number of vertices/internal edges of type r appearing in Γ, for $r \in R$. Moreover, let $n_{\gamma,r}$ be the number of connected components of γ with residue r. Since insertion of a vertex graph (i.e. with residue in R_V) on a vertex v in Γ prevents a subsequent insertion at v of a vertex graph with the same residue, whereas insertion of an edge graph (i.e. with residue in R_E) creates two new edges and hence two insertion places for a subsequent edge graph, we find the following expression,

$$\Gamma \mid \gamma = \prod_{v \in R_V} n_{\gamma,v}! \binom{m_{\Gamma,v}}{n_{\gamma,v}} \prod_{e \in R_E} n_{\gamma,e}! \binom{m_{\Gamma,e} + n_{\gamma,e} - 1}{n_{\gamma,e}}.$$

Indeed, the binomial coefficients arise for each vertex v since we are choosing $n_{\gamma,v}$ out of $m_{\Gamma,v}$ whereas for an edge e we choose $n_{\gamma,e}$ out of $m_{\Gamma,e}$ *with repetition*.

We claim that this counting enhances our formula to the following

$$\Delta(G^r) = \sum_{\mathrm{res}(\Gamma)=r} \prod_{v \in R_V} (G^v)^{m_{\Gamma,v}} \prod_{e \in R_E} (G^e)^{-m_{\Gamma,e}} \otimes \frac{\Gamma}{\mathrm{Sym}(\Gamma)}. \tag{10}$$

Before proving this, we explain the meaning of the inverse of Green's functions in our Hopf algebra. Since any Green's function starts with the identity, we can surely write its inverse formally as a geometric series. Recall that the Hopf algebra is graded by loop number. Hence, the inverse of a Green's function at a fixed loop order is in fact well-defined; it is given by restricting the above formal series expansion to this loop order. In the following, also rational powers of Green's functions will appear; they will be understood in like manner.

Proof of Eq. (10). Let us simplify a little and consider a scalar field theory with just one type of vertex and edge, i.e. $R = \{ \prec, \; - \}$. We consider the sum

$$\sum_{\gamma} \frac{\Gamma \mid \gamma}{\mathrm{Sym}(\gamma)} \gamma = \sum_{\gamma_v} \frac{n_{\gamma,v}!}{\mathrm{Sym}(\gamma_v)} \binom{m_{\Gamma,v}}{n_{\gamma,v}} \gamma_v \sum_{\gamma_e} \frac{n_{\gamma,e}!}{\mathrm{Sym}(\gamma_e)} \binom{m_{\Gamma,e} + n_{\gamma,v} - 1}{n_{\gamma,v}} \gamma_e,$$

naturally split into a sum over vertex and edge graphs. We have also inserted the above combinatorial expression for the number of insertion places. Next, we write $\gamma_v = \gamma_v'\gamma_v''$ and try factorize the sum over γ_v into a sum over γ_v' (connected) and γ_v''. Some care should be taken here regarding the combinatorial factors but let us ignore them for the moment. In fact, if we fix the number of connected components $h^0(\gamma_v)$ of γ_v in the sum to be n_V we can write

$$\sum_{h^0(\gamma_v)=n_V} n_V! \frac{\gamma_v}{\mathrm{Sym}(\gamma_v)} = \sum_{h^0(\gamma_v)=n_V} \left(\sum_{\substack{\gamma_v',\gamma_v'' \\ \gamma_v'\gamma_v''\simeq\gamma_v}} \frac{n(\gamma_v'',\gamma_v')+1}{n_V} \right) n_V! \frac{\gamma_v}{\mathrm{Sym}(\gamma_v)},$$

with γ_v' a connected graph. Here, we have simply inserted 1,

$$\sum_{\substack{\gamma_v',\gamma_v'' \\ \gamma_v'\gamma_v''\simeq\gamma_v}} \frac{n(\gamma_v'',\gamma_v')+1}{n_V} = \sum_{\gamma_v'} \frac{n(\gamma_v,\gamma_v')}{n_V} = 1,$$

which follows directly from the definition of $n(\gamma_v,\gamma_v')$ as the number of connected components of γ_v isomorphic to γ_v'. Now, by definition $\mathrm{Sym}(\gamma_v'\gamma_v'') = (n(\gamma_v'',\gamma_v') + 1)\mathrm{Sym}(\gamma_v')\mathrm{Sym}(\gamma_v'')$ for a connected graph γ_v' so that we obtain for the above sum

$$\sum_{\gamma_v'} \frac{\gamma_v'}{\mathrm{Sym}(\gamma_v')} \sum_{h^0(\gamma_v'')=n_V-1} (n_V-1)! \frac{\gamma_v''}{\mathrm{Sym}(\gamma_v'')} = \cdots = (G^v - 1)^{n_V},$$

by applying the same argument n_V times. Recall also the definition of the Green's function G^v from Eq. (9). A similar argument applies to the edge graphs, leading to a contribution $(1 - G^e)^{n_E}$, with n_E the number of connected components of γ_e. When summing over n_V and n_E, taking also into account the combinatorial factors, we obtain:

$$\sum_{n_V=0}^{\infty} \binom{m_{\Gamma,v}}{n_V}(G^v - 1)^{n_V} \sum_{n_E=0}^{\infty} \binom{m_{\Gamma,e} + n_E - 1}{n_E}(1 - G^e)^{n_E} = (G^v)^{m_{\Gamma,v}}(G^e)^{-m_{\Gamma,e}}.$$

The extension to the general setting where the set R contains different types of vertices and edges is straightforward. □

An additional counting of the number of edges and numbers of vertices in Γ gives the following relations:

$$2m_{\Gamma,e} + N_e(\mathrm{res}(\Gamma)) = \sum_{v\in R_V} N_e(v)m_{\Gamma,v}$$

where $N_e(r)$ is the number of lines (of type e) attached to $r \in R$. For instance $N_e(\text{◂})$ equals 2 if e is an electron line and 1 if e is a photon line. One checks the above equality by noting that the left-hand-side counts the number of internal half lines plus the external lines which are connected to the vertices that appear at the right-hand-side, taken into account their valence.

With this formula, we can write Eq. (10) as

$$\Delta(G^r) = \prod_e (G^e)^{N_e(r)/2} \sum_{\text{res}(\Gamma)=r} \prod_v \left(\frac{G^v}{\prod_e (G^e)^{N_e(v)/2}} \right)^{m_{\Gamma,v}} \otimes \frac{\Gamma}{\text{Sym}(\Gamma)}. \quad (11)$$

This is still not completely satisfactory since it involves the number of vertices in Γ which prevents us from separating the summation of Γ from the other terms. We introduce the following notation for the fraction of Green's functions above:

$$X_v = \left(\frac{G^v}{\prod_e (G^e)^{N_e(v)/2}} \right)^{1/(N(v)-2)} \quad (12)$$

with $N(v)$ the total number of edges attached to v. Before we state our main theorem, let us motivate the definition of these elements in the case of QCD.

Example 8. *In QCD, there are four vertices and the corresponding elements X_v are given by,*

$$X_{\text{⫞⟨}} = \frac{G^{\text{⫞⟨}}}{G - \sqrt{G^{\text{⟋⟍}}}}, \qquad X_{\text{⫞⫞}} = \frac{G^{\text{⫞⫞}}}{G\ \sqrt{G^{\text{⟋⟍}}}},$$

$$X_{\text{⫞⟨}} = \frac{G^{\text{⟨}}}{\left(G^{\text{⟋⟍}} \right)^{3/2}}, \qquad X_{\text{✕}} = \frac{\sqrt{G}^{\text{✕}}}{G^{\text{⟋⟍}}}.$$

The combinations of the Green's functions are identical to those appearing in formulas (5). Indeed, as we will see in a moment, setting them equal in H is compatible with the coproduct.

Although motivated by the study of the Slavnov–Taylor identities in non-Abelian gauge theories, the following result holds in complete generality.

Theorem 9. *The ideal $I = \langle X_v - X_{v'} \rangle_{v' \in R_V}$ is a Hopf ideal, i.e.*

$$\Delta(I) \subset I \otimes H + H \otimes I, \qquad \epsilon(I) = 0, \qquad S(I) \subset I.$$

Proof. Let us write the above Eq. (11) in terms of the X_v's:

$$\Delta(G^r) = \prod_e (G^e)^{\frac{1}{2}N_e(r)} \sum_{\text{res}(\Gamma)=r} \prod_{v'} (X_{v'})^{(N(v')-2)m_{\Gamma,v'}} \otimes \frac{\Gamma}{\text{Sym}(\Gamma)}.$$

In this expression, $X_{v'}$ appears with a certain power, say s, and we can replace $(X_{v'})^s$ by $(X_v)^s$ as long as we add the term $(X_{v'})^s - (X_v)^s$. This latter term can be factorized as $X_{v'} - X_v$ times a certain polynomial in X_v and $X_{v'}$ and thus corresponds to an element in I. As a result, we can replace all $X_{v'}$'s by X_v for some fixed v modulo addition of terms in $I \otimes H$.

The second step uses the following equality between vertices and edges:

$$\sum_{v' \in R_V} (N(v') - 2)m_{\Gamma,v'} = 2L + N(r) - 2 \quad (13)$$

in terms of the loop number L and residue r of Γ. The equality follows by an easy induction on the number of internal lines of Γ (cf. [14]). Finally, one can separate the sum over Γ at a fixed loop order to obtain

$$\Delta(G^r) = \prod_e (G^e)^{\frac{1}{2}N_e(r)} \sum_{L=0}^{\infty} (X_v)^{2L+N(r)-2} \otimes G_L^r, \tag{14}$$

understood modulo terms in $I \otimes H$. From this one derives that $\Delta(X_v - X_{v'})$ lies in $I \otimes H + H \otimes I$ as follows. Let us first find a more convenient choice of generators of I. By induction, one can show that

$$X_v - X_{v'} = \left(X_v^{(N(v')-2))(N(v)-2)} - X_v^{(N(v')-2)(N(v)-2)} \right) \mathrm{Pol}(X_v, X_v'),$$

where Pol is a (formally) invertible series in X_v and $X_{v'}$. In fact, it starts with a nonzero term of order zero. By multiplying out both denominators in the X_v and $X_{v'}$, we arrive at the following set of (equivalent) generators of I

$$(G^v)^{N(v')-2} \prod_e (G^e)^{(N(v')-2)N_e(v)/2} - \left(G^{v'} \right)^{N(v)-2} \prod_e (G^e)^{(N(v)-2)N_e(v')/2}$$

with $v, v' \in R_V$. A little computation shows that the first leg of the tensor product in the coproduct on these two terms coincide, using Eq. (14). As a consequence, one can combine these terms to obtain an element in $H \otimes I$ modulo the aforementioned terms in $I \otimes H$ needed to arrive at (14). $\qquad\square$

As a consequence, we can work on the **quotient Hopf algebra** $\widetilde{H} = H/I$. Suppose we work in the case of a non-Abelian gauge theory such as QCD, with the condition that the regularization procedure is compatible with gauge invariance such as dimensional regularization (see also [10]). In such a case, the map $U : H \to K$ defined by the (regularized) Feynman rules vanishes on the ideal I because of the Slavnov–Taylor identities. Hence, it factors through an algebra map from $\widetilde{H}$ to the field K. Since $\widetilde{H}$ is still a commutative connected Hopf algebra, there is a Birkhoff decomposition $U = C^{-1} * R$ as before *with C and R algebra maps from $\widetilde{H}$ to K*. This is the crucial point, because it implies that both C and R vanish automatically on I. In other words, both the counterterms and the renormalized amplitudes satisfy the Slavnov–Taylor identities. In particular, the $C(X_v)$'s are the terms appearing in Eq. (5) which coincide because $C(I) = 0$. Note also that in $\widetilde{H}$ expression (14) holds so that the coproduct closes on Green's functions, i.e. they generate Hopf subalgebras.

As a corollary to this, we can derive a generalization of Dyson's formula originally derived for QED [4]. It provides a relation between the renormalized Green's function written in terms of the coupling constant g and the unrenormalized Green's function written in terms of the bare coupling constant defined by $g_0 = C(X_v)g$ for some $v \in R_V$.

Corollary 10 (Dyson's formula). *The following analogue of Dyson's formula for QED holds in general,*

$$R(G^r)(g) = \prod_e (Z^e)^{N_e(r)/2} U(G^r)(g_0),$$

where $Z_e = C(G^e)$.

Proof. This follows from an application of $R = C * U$ to G^r using Eq. (14) while counting the number of times the coupling constant g appears when applying the Feynman rules to a graph with residue r and loop number L. In fact, this number is $\sum_v (N(v) - 2)m_{\Gamma,v}$ which is also $2L + N(r) - 2$ as noted before. $\square$

Appendix A. Hopf algebras

For convenience, let us briefly recall the definition of a (commutative) Hopf algebra. It is the dual object to a group and, in fact, there is a one-to-one correspondence between groups and commutative Hopf algebras.

Let G be a group with product, inverse and identity element. We consider the algebra of representative functions $H = \mathcal{F}(G)$. This class of functions is such that $\mathcal{F}(G \times G) \simeq \mathcal{F}(G) \otimes \mathcal{F}(G)$. For instance, if G is a (complex) matrix group, then $\mathcal{F}(G)$ could be the algebra generated by the coordinate functions x_{ij} so that $x_{ij}(g) = g_{ij} \in \mathbb{C}$ are just the (i,j)'th entries of the matrix g.

Let us see what happens with the product, inverse and identity of the group on the level of the algebra $H = \mathcal{F}(G)$. The multiplication of the group can be seen as a map $G \times G \to G$, given by $(g,h) \to gh$. Since dualization reverses arrows, this becomes a map $\Delta : H \to H \otimes H$ called the *coproduct* and given for $f \in H$ by

$$\Delta(f)(g,h) = f(gh).$$

The property of associativity on G becomes *coassociativity* on H:

$$(\Delta \otimes \mathrm{id}) \circ \Delta = (\mathrm{id} \otimes \Delta) \circ \Delta, \tag{A1}$$

stating simplify that $f((gh)k) = f(g(hk))$.

The unit $e \in G$ gives rise to a *counit*, as a map $\epsilon : H \to \mathbb{C}$, given by $\epsilon(f) = f(e)$ and the property $eg = ge = g$ becomes on the algebra level

$$(\mathrm{id} \otimes \epsilon) \circ \Delta = \mathrm{id} = (\epsilon \otimes \mathrm{id}) \circ \Delta, \tag{A2}$$

which reads explicitly $f(ge) = f(eg) = f(g)$.

The inverse map $g \mapsto g^{-1}$, becomes the *antipode* $S : H \to H$, defined by $S(f)(g) = f(g^{-1})$. The property $gg^{-1} = g^{-1}g = e$, becomes on the algebra level:

$$m(S \otimes \mathrm{id}) \circ \Delta = m(\mathrm{id} \otimes S) \circ \Delta = 1_H \epsilon, \tag{A3}$$

where $m : H \otimes H \to H$ denotes pointwise multiplication of functions in H.

From this example, we can now abstract the conditions that define a general Hopf algebra.

Definition 11. *A Hopf algebra H is an algebra H, together with two algebra maps $\Delta : H \otimes H \to H$ (coproduct), $\epsilon : H \to \mathbb{C}$ (counit), and a bijective $\mathbb{C}$-linear map $S : H \to H$ (antipode), such that equations (A1)–(A3) are satisfied.*

If the Hopf algebra H is commutative, we can conversely construct a (complex) group from it as follows. Consider the collection G of multiplicative linear maps from H to $\mathbb{C}$. We will show that G is a group. Indeed, we have the *convolution product* between two such maps ϕ, ψ defined as the dual of the coproduct: $(\phi * \psi)(X) = (\phi \otimes \psi)(\Delta(X))$ for $X \in H$. One can easily check that coassociativity of the coproduct (Eq. (A1)) implies associativity of the convolution product: $(\phi * \psi) * \chi = \phi * (\psi * \chi)$. Naturally, the counit defines the unit e by $e(X) = \epsilon(X)$. Clearly $e * \phi = \phi = \phi * e$ follows at once from Eq. (A2). Finally, the inverse is constructed from the antipode by setting $\phi^{-1}(X) = \phi(S(X))$ for which the relations $\phi^{-1} * \phi = \phi * \phi^{-1} = e$ follow directly from Equation (A3).

With the above explicit correspondence between groups and commutative Hopf algebras, one can translate practically all concepts in group theory to Hopf algebras. For instance, a subgroup $G' \subset G$ corresponds to a *Hopf ideal* $I \subset \mathcal{F}(G)$ in that $\mathcal{F}(G') \simeq \mathcal{F}(G)/I$ and vice versa. The conditions for being a subgroup can then be translated to give the following three conditions defining a Hopf ideal I in a commutative Hopf algebra H

$$\Delta(I) \subset I \otimes H + H \otimes I, \qquad \epsilon(I) = 0, \qquad S(I) \subset I.$$

References

[1] C. Bergbauer and D. Kreimer. Hopf algebras in renormalization theory: Locality and Dyson-Schwinger equations from Hochschild cohomology. *IRMA Lect. Math. Theor. Phys.* 10 (2006) 133–164.

[2] J. Collins. *Renormalization.* Cambridge University Press, 1984.

[3] A. Connes and D. Kreimer. Renormalization in quantum field theory and the Riemann- Hilbert problem. I: The Hopf algebra structure of graphs and the main theorem. *Comm. Math. Phys.* 210 (2000) 249–273.

[4] F. J. Dyson. The S matrix in quantum electrodynamics. *Phys. Rev.* 75 (1949) 1736–1755.

[5] K. Ebrahimi-Fard and L. Guo. Rota-Baxter algebras in renormalization of perturbative quantum field theory. *Fields Inst. Commun.* 50 (2007) 47–105.

[6] D. Kreimer. On the Hopf algebra structure of perturbative quantum field theories. *Adv. Theor. Math. Phys.* 2 (1998) 303–334.

[7] D. Kreimer. Dyson–Schwinger equations: From Hopf algebras to number theory. hep-th/0609004.

[8] D. Kreimer. Anatomy of a gauge theory. *Ann. Phys.* 321 (2006) 2757–2781.

[9] G. 't Hooft and M. J. G. Veltman. Diagrammar. *CERN yellow report.* 73 (1973) 1–114.

[10] D. V. Prokhorenko. Renormalization of gauge theories and the hopf algebra of diagrams. arXiv:0705.3906 [hep-th].

[11] A. A. Slavnov. Ward identities in gauge theories. *Theor. Math. Phys.* 10 (1972) 99–107.

[12] J. C. Taylor. Ward identities and charge renormalization of the yang-mills field. *Nucl. Phys.* B33 (1971) 436–444.

[13] W. D. van Suijlekom. The Hopf algebra of Feynman graphs in QED. *Lett. Math. Phys.* 77 (2006) 265–281.

[14] W. D. van Suijlekom. Renormalization of gauge fields: A Hopf algebra approach. *Commun. Math. Phys.* 276 (2007) 773–798.

[15] J. C. Ward. An identity in quantum electrodynamics. *Phys. Rev.* 78 (1950) 182.

Walter D. van Suijlekom
Institute for Mathematics, Astrophysics and Particle Physics
Faculty of Science, Radboud Universiteit Nijmegen
Toernooiveld 1
NL–6525 ED Nijmegen
e-mail: `waltervs@math.ru.nl`

Quantum Field Theory
B. Fauser, J. Tolksdorf and E. Zeidler, Eds., 155–162
© 2009 Birkhäuser Verlag Basel/Switzerland

Not so Non-Renormalizable Gravity

Dirk Kreimer

Abstract. We review recent ideas [1] how gravity might turn out to be a renormalizable theory after all.

Mathematics Subject Classification (2000). Primary 81T15; Secondary 81T13; 83C47.

Keywords. gravitation, renormalization, Feynman rules, Dyson-Schwinger equations, gauge theory, perturbative methods.

1. Introduction

Renormalizable perturbative quantum field theories are embarrassingly successful in describing observed physics. Whilst their mathematical structure is still a challenge albeit an entertaining one, they are testimony to some of the finest achievements in our understanding of nature. The physical laws as far as they are insensitive to the surrounding geometry seems completely described by such theories. Alas, if we incorporate gravity, and want to quantize it, we seem at a loss.

In this talk, we report on some recent work [1] which might give hope. Our main purpose is to review the basic idea and to put it into context.

As in [1], we will proceed by a comparison of the structure of a renormalizable theory, quantum electrodynamics in four dimensions, and gravity.

It is the role of the Hochschild cohomology [2] in those two different situations, which leads to surprising new insights. We will discuss them at an elementary level for the situation of pure gravity. We also allow, in the spirit of the workshop for the freedom to muse about conceptual consequences at the end.

Talk given at the "Max Planck Institute for Mathematics in the Natural Sciences", Leipzig. Work supported in parts by grant NSF-DMS/0603781. Author supported by CNRS.

2. The structure of Dyson–Schwinger Equations in QED$_4$

2.1. The Green functions

Quantum electrodynamics in four dimensions of space-time (QED$_4$) is described in its short-distance behavior by four Green functions

$$G^{\bar{\psi}\gamma\cdot\partial\psi}, G^{m\bar{\psi}\psi}, G^{\bar{\psi}\gamma\cdot A\psi}, G^{\frac{1}{4}F^2}, \tag{1}$$

corresponding to the four monomials in its Lagrangian

$$L = \bar{\psi}\gamma\cdot\partial\psi - \bar{\psi}m\psi - \bar{\psi}\gamma\cdot A\psi - \frac{1}{4}F^2. \tag{2}$$

Here, $G^i = G^i(\alpha, L)$, with α the fine structure constant and $L = \ln q^2/\mu^2$, so that we work in a MOM (momentum) scheme, subtract at $q^2 = \mu^2$, project the vertex function to its scalar formfactor $G^{\bar{\psi}\gamma\cdot\partial\psi}$ with UV divergences evaluated at zero photon momentum. Similarly the other Green functions are normalized as to be the multiplicative quantum corrections to the tree level monomials above, in momentum space.

In perturbation theory, the degree of divergence of a graph Γ with f external fermion lines and m external photon lines in D dimensions is

$$\omega_D(\Gamma) = \frac{3}{2}f + m - D - (D-4)(|\Gamma|-1) \Rightarrow \omega_4(\Gamma) = \frac{3}{2}f + m - 4. \tag{3}$$

This is independent of the loop number for QED$_4$, $D = 4$, and is a sole function of the number and type of external legs. $\omega_D(\Gamma)$ determines the number of derivatives with respect to masses or external momenta needed to render a graph logarithmically divergent, and hence identifies the top-level residues, which drive the iteration of Feynman integrals according to the quantum equations of motion [3].

We define these four Green functions as an evaluation by renormalized Feynman rules of a series of one-particle irreducible (1PI) Feynman graphs $\Gamma \in \mathcal{FG}_i$. These series are determined as a fixpoint of the following system in Hochschild cohomology.

$$X^{\bar{\psi}\gamma\cdot\partial\psi} = 1 - \sum_{k=1}^{\infty} \alpha^k B_+^{\bar{\psi}\gamma\cdot\partial\psi,k}(X^{\bar{\psi}\gamma\cdot\partial\psi}Q^{2k}(\alpha)), \tag{4}$$

$$X^{\bar{\psi}\gamma\cdot A\psi} = 1 + \sum_{k=1}^{\infty} \alpha^k B_+^{\bar{\psi}\gamma\cdot A\psi,k}(X^{\bar{\psi}\gamma\cdot A\psi}Q^{2k}(\alpha)), \tag{5}$$

$$X^{\bar{\psi}m\psi} = 1 - \sum_{k=1}^{\infty} \alpha^k B_+^{\bar{\psi}m\psi,k}(X^{\bar{\psi}m\psi}Q^{2k}(\alpha)), \tag{6}$$

$$X^{\frac{1}{4}F^2} = 1 - \sum_{k=1}^{\infty} \alpha^k B_+^{\frac{1}{4}F^2,k}(X^{\frac{1}{4}F^2}Q^{2k}(\alpha)). \tag{7}$$

Here,

$$B_+^{i,k} = \sum_{|\gamma|=k, \Delta'(\gamma)=0, \gamma \in \mathcal{FG}_i} B_+^{\gamma}, \; \forall i \in \mathcal{R}_{\text{QED}}, \tag{8}$$

is a sum over all Hopf algebra primitive graphs with given loop number k and contributing to superficially divergent amplitude i, and

$$B_+^{\gamma}(X) = \sum_{\Gamma \in <\Gamma>} \frac{\mathbf{bij}(\gamma, X, \Gamma)}{|X|_v} \frac{1}{\mathrm{maxf}(\Gamma)} \frac{1}{(\gamma|X)} \Gamma, \tag{9}$$

where $\mathrm{maxf}(\Gamma)$ is the number of maximal forests of Γ, $|X|_\gamma$ is the number of distinct graphs obtainable by permuting edges of X, $\mathbf{bij}(\gamma, X, \Gamma)$ is the number of bijections of external edges of X with an insertion place in γ such that the result is Γ, and finally $(\gamma|X)$ is the number of insertion places for X in γ [4], and

$$\mathcal{R}_{\text{QED}} = \{\bar{\psi}\gamma \cdot \partial\psi, \bar{\psi}\gamma \cdot A\psi, m\bar{\psi}\psi, \frac{1}{4}F^2\}. \tag{10}$$

Also, we let

$$Q = \frac{X^{\bar{\psi}\gamma \cdot A\psi}}{X^{\bar{\psi}\gamma \cdot \partial\psi}\sqrt{X^{\frac{1}{4}F^2}}}. \tag{11}$$

The resulting maps $B_+^{i,K}$ are Hochschild closed

$$bB_+^{i,K} = 0 \tag{12}$$

in the sense of [5]. We have in fact

$$\Delta(B_+^{\gamma}(X)) = \sum_{\Gamma} n_{\Gamma, X, \gamma}\Gamma \tag{13}$$

where $n_{\Gamma, X, \gamma}$ can be determined from (9,12).

Furthermore, one can choose a basis of primitives γ [3] such that their Mellin transforms $M_\gamma(\rho)$ have the form

$$M_\gamma(\rho) = \int \iota_\gamma(k_i; q)_{|_{q^2=\mu^2}} \prod_{s=1}^{|\gamma|} \frac{[k_s^2/\mu^2]^{-\rho/|\gamma|}d^4k_i}{(2\pi)^4} \text{ for } 1 > \Re(\rho) > 0, \tag{14}$$

where the integrand ι_γ is a function of internal momenta k_i and an external momentum q, subtracted at $q^2 = \mu^2$.

The Dyson Schwinger equations then take the form

$$G_R^i(\alpha, L) = 1 \pm \lim_{\rho \to 0} \left[\sum_k \alpha^k \sum_{|\gamma|=k} G_r^i(\alpha, \partial_\rho)\mathcal{Q}(\alpha, \partial_\rho)M_\gamma(\rho) \left[\left(\frac{q^2}{\mu^2}\right)^{-\rho} - 1 \right] \right] \tag{15}$$

where

$$\Phi_R(X^i) = G_R^i(\alpha, L), \tag{16}$$

and

$$\Phi_R(Q) = \mathcal{Q}(\alpha, L), \tag{17}$$

is the invariant charge, all calculated with renormalized Feynman rules in the MOM scheme.

2.2. Gauge theoretic aspects

Using Ward identities, we can reduce the set $\mathcal{R}_{\mathrm{QED}} = \{\bar{\psi}\gamma\cdot\partial\psi, m\bar{\psi}\psi, \bar{\psi}\gamma\cdot A\psi, \frac{1}{4}F^2\}$ to three elements upon identifying $G^{\bar{\psi}\gamma\cdot\partial\psi} = G^{\bar{\psi}\gamma\cdot A\psi}$. Using the Baker–Johnson–Willey gauge [6] we can furthermore trivialize

$$G^{\bar{\psi}\gamma\cdot\partial\psi} \;=\; G^{\bar{\psi}\gamma\cdot A\psi} = 1. \tag{18}$$

Using their work again [7], we have that $m\bar{\psi}\psi$ can be ignored in $\mathcal{R}_{\mathrm{QED}}$.

We are hence left with the determination of a single gauge-independent Green function $G^{\frac{1}{4}F^2}$, which in the MOM scheme takes the form

$$G^{\frac{1}{4}F^2}(\alpha, L) \;=\; 1 - \sum_{k=1}^{\infty} \gamma_k(\alpha)L^k, \tag{19}$$

and the renormalization group determines [8]

$$\gamma_k(\alpha) \;=\; \frac{1}{k}\gamma_1(\alpha)(1 - \alpha\partial_\alpha)\gamma_{k-1}(\alpha). \tag{20}$$

Here, $\gamma_1(\alpha) = 2\psi(\alpha)/\alpha$, where $\psi(\alpha)$ is the MOM scheme β-function of QED, which is indeed half of the anomalous dimension γ_1 of the photon field in that scheme.

One can show that $\gamma_1(\alpha)$ as a perturbative series ($\gamma_1(\alpha) = \sum_{j=1}^{\infty} \gamma_{1,j}\alpha^j$) is Gevrey–1 and that the series $\sum_{j=1}^{\infty} \gamma_{1,j}\alpha^j/j!$ has a finite radius of convergence, with a bound involving the lowest order contribution of the β-function and the one-instanton action [3].

Furthermore, $\gamma_1(\alpha)$ fulfils [3]

$$\gamma_1(\alpha) \;=\; P(\alpha) - \gamma_1(\alpha)(1 - \alpha\partial_\alpha)\gamma_1(\alpha), \tag{21}$$

an equation which has been studied in numerical detail recently [9], with more of its analytic structure to be exhibited there. In this equation, $P(\alpha)$ is obtained from the primitives of the Hopf algebra

$$P(\alpha) \;=\; \sum_{\gamma} \alpha^{|\gamma|} \lim_{\rho \to 0} \rho M_\gamma(\rho), \tag{22}$$

and $P(\alpha)$ is known perturbatively as a fifth order polynomial [10] and its asymptotics have been conjectured long ago [11].

This finishes our summary of QED$_4$ as a typical renormalizable theory.

2.3. Non-Abelian gauge theory

The above approach to Green functions remains valid for a non-Abelian gauge theory with the definition of a single invariant charge $\mathcal{Q}(\alpha, L)$ being the crucial requirement. This can be consistently done, [4], upon recognizing that the celebrated

Slavnov–Taylor identities for the couplings fulfil

$$\frac{S_R^\phi(X^{\bar\psi\gamma\cdot A\psi})}{S_R^\phi(X^{\bar\psi\gamma\cdot\partial\psi})} = \frac{S_R^\phi(X^{AA\partial A})}{S_R^\phi(X^{\partial A\partial A})} = \frac{S_R^\phi(X^{AAAA})}{S_R^\phi(X^{AA\partial A})} = \frac{S_R^\phi(X^{\bar\phi A\cdot\partial\phi})}{S_R^\phi(X^{\phi\Box\phi})} \tag{23}$$

for the set of amplitudes

$$\mathcal{R}_{\mathrm{QCD}} = \{DADA, \bar\psi\gamma\cdot\partial\psi, \bar\phi\Box\phi, AADA, AAAA, \bar\phi A\cdot\partial\phi, \bar\psi\gamma\cdot A\psi\}, \tag{24}$$

needing renormalization in QCD.

This allows to define a Hochschild cohomology on the sum of graphs at a given loop order, and hence to obtain multiplicative renormalization in this language from the resulting coideals in the Hopf algebra [4, 12].[1]

Note that the structure of the sub-Hopf algebras underlying this approach [5, 8] implies that the elements $X^i(\alpha)$ close under the coproduct. A general classification of related sub-Hopf algebras has been recently obtained by Loic Foissy [13]. He considers only the case that the lowest order Hochschild cocycle is present in the combinatorial Dyson–Schwinger equations, but his study is rather complete when augmented by the results of [2].

3. Gravity

We consider pure gravity understood as a theory based on a graviton propagator and n-graviton couplings as vertices. A fuller discussion incorporating ghosts and matter fields is referred to future work.

3.1. Summary of some results obtained for quantum gravity

We summarize here some results published in [1].

Corollary 1. *Let* $|\Gamma| = k$. *Then* $\omega(\Gamma) = -2(|\Gamma| + 1)$.

This is a significant change from the behavior of a renormalizable theory: in the renormalizable case, each graph contributing to the same amplitude i has the same powercounting degree regardless of the loop number. Here, we have the dual situation: the loop number determines the powercounting degree, regardless of the amplitude.

Theorem 2. *The set* $d_{\omega(\Gamma)}$ *contains no primitive element beyond one loop.*

The set $d_{\omega(\Gamma)}$ is determined as a set of dotted graphs, with dots representing $\omega(\Gamma)$ derivatives with respect to masses or external momenta such that the corresponding integrand ι_Γ is overall log-divergent. Whilst in a renormalizable theory, we find for each amplitude in the finite set $\mathcal{R}$ primitives at each loop order in $d_{\omega(\Gamma)}$, here we have an infinite set $\mathcal{R}$, but only one primitive in it.

---

[1]Eds. note: See also W. van Suijlekom's contribution in this book.

Proposition 3. *The relations*

$$\frac{X^{n+1}}{X^n} = \frac{X^n}{X^{n-1}}, \ n \geq 3, \tag{25}$$

define a sub-Hopf algebra with Hochschild closed one-cocycles $B_+^{1,n}$.

Here, X^n is the sum of all graphs with n external graviton lines. One indeed finds that the combinatorial Dyson–Schwinger equations for gravity provide a sub-Hopf algebra upon requiring these relations, in straightforward generalization of the situation in a non-Abelian gauge theory.

3.2. Comments

3.2.1. Gauss-Bonnet.
The Gauss–Bonnet theorem ensures here, in the form

$$0 = \int_{\mathrm{M}} \sqrt{g} \left(R_{\mu\nu\rho\sigma} R^{\mu\nu\rho\sigma} - 4R_{\alpha\beta} R^{\alpha\beta} + R^2 \right), \tag{26}$$

the vanishing of the one-loop renormalization constants. This does not imply the vanishing of the two-loop renormalization constants as their one-loop subdivergences are off-shell. But it implies that the two-loop counter term has only a first order pole by the scattering type formula, in agreement with the vanishing of $\phi_{\text{off-shell}}(\gamma)\phi_{\text{on-shell}}(\Gamma/\gamma)$. Here, γ, Γ/γ is the decomposition of Γ into one-loop graphs and $\phi_{\text{on/off-shell}}$ denotes suitable Feynman rules.

3.2.2. Two-loop counterterm.
Also, the universality of the two-loop counterterm suggests that indeed

$$\frac{Z^{\mathrm{gr}}{}_{n+1}}{Z^{\mathrm{gr}}{}_n} = \frac{Z^{\mathrm{gr}}{}_n}{Z^{\mathrm{gr}}{}_{n-1}}, \ \text{with } Z^{\mathrm{gr}}{}_n = S_R^\phi(X^n), \tag{27}$$

holds for off-shell counterterms. In particular, if we compute in a space of constant curvature and conformally reduced gravity, which maintains many striking features of asymptotic safe gravity [14, 15], the above identities should hold for suitably defined characters: indeed, in such circumstances we can renormalize using a graviton propagator, which is effectively massive with the mass $\sqrt{R/6}$ provided by the constant curvature R, and hence can renormalize at zero external momentum. Using the KLT relations [16, 17], this reduces the above identities to a (cumbersome) combinatorial exercise on one-loop graphs to be worked out in the future.

Continuing this line of thought one expects that a single quantity, the β-function of gravity, exhibits short-distance singularities. If this expectation bears out, it certainly is in nice conceptually agreement with the expectation that in theories where gravity has a vanishing β-function, gravity is indeed a finite theory [18].

3.2.3. Other instances of gravity powercounting.

The appearance of Feynman rules such that the powercounting of vertex amplitudes in $\mathcal{R}_V$ cancels the powercounting of propagator amplitudes in $\mathcal{R}_E$, $\mathcal{R} = \mathcal{R}_V \cup \mathcal{R}_E$, is not restricted to gravity. It indeed appears for example also in the field theoretic description of bulk materials like glass, which were recently described at tree-level as a field theory [19], and whose renormalization will have powercounting properties similar to the present discussion.

Acknowledgements

It is a pleasure to thank Bertfried Fauser, Jürgen Tolksdorf and Eberhard Zeidler for inviting me to this workshop.

References

[1] D. Kreimer, *A remark on quantum gravity*, Annals Phys. **323** (2008) 49 [arXiv:0705.3897 [hep-th]].

[2] C. Bergbauer and D. Kreimer, *Hopf algebras in renormalization theory: Locality and Dyson-Schwinger equations from Hochschild cohomology*, IRMA Lect. Math. Theor. Phys. **10** (2006) 133 [arXiv:hep-th/0506190].

[3] D. Kreimer and K. Yeats, *An etude in non-linear Dyson-Schwinger equations*, Nucl. Phys. Proc. Suppl. **160** (2006) 116 [arXiv:hep-th/0605096].

[4] D. Kreimer, *Anatomy of a gauge theory*, Annals Phys. **321** (2006) 2757 [arXiv:hep-th/0509135].

[5] D. Kreimer, *Dyson–Schwinger Equations: From Hopf algebras to Number Theory*, in *Universality and Renormalization*, I. Binder, D. Kreimer, eds., Fields Inst Comm. **50** (2007) 225, AMS.

[6] K. Johnson, M. Baker and R. Willey, *Selfenergy Of The Electron*, Phys. Rev. **136** (1964) B1111.

[7] M. Baker and K. Johnson, *Asymptotic form of the electron propagator and the selfmass of the electron*, Phys. Rev. D **3** (1971) 2516.

[8] D. Kreimer and K. Yeats, *Recursion and growth estimates in renormalizable quantum field theory*, Commun. Math. Phys. (2008) DOI:10.1007/s00220-008-0431-7; arXiv:hep-th/0612179.

[9] D. Kreimer, K. Yeats, G. Van Baalen, D. Uminsky, in preparation, see also Karen Yeats homepage at
http://math.bu.edu/people/kayeats/papers/picturetalkshort4up.pdf

[10] S. G. Gorishnii, A. L. Kataev, S. A. Larin and L. R. Surguladze, *The Analytical four loop corrections to the QED Beta function in the MS scheme and to the QED psi function: Total reevaluation*, Phys. Lett. B **256** (1991) 81.

[11] C. Itzykson, G. Parisi and J. B. Zuber, *Asymptotic Estimates In Quantum Electrodynamics*, Phys. Rev. D **16** (1977) 996;
R. Balian, C. Itzykson, G. Parisi and J. B. Zuber, *Asymptotic Estimates In Quantum Electrodynamics. 2*, Phys. Rev. D **17** (1978) 1041.

[12] W. D. van Suijlekom, *Renormalization of gauge fields: A Hopf algebra approach,* Commun. Math. Phys. **276** (2007) 773 [arXiv:hep-th/0610137].

[13] L. Foissy, *Faa di Bruno subalgebras of the Hopf algebra of planar trees from combinatorial Dyson–Schwinger equations,* arXiv:0707.1204 [math.RA].

[14] M. Reuter, *Nonperturbative Evolution Equation for Quantum Gravity,* Phys. Rev. D **57** (1998) 971 [arXiv:hep-th/9605030];
O. Lauscher and M. Reuter, *Ultraviolet fixed point and generalized flow equation of quantum gravity,* Phys. Rev. D **65** (2002) 025013 [arXiv:hep-th/0108040];
M. Niedermaier, *The asymptotic safety scenario in quantum gravity: An introduction,* arXiv:gr-qc/0610018.

[15] M. Reuter, H. Weyer, *Background Independence and Asymptotic Safety in Conformally Reduced Gravity,* arXiv:0801.3287 [hep-th].

[16] H. Kawai, D. C. Lewellen, and S. H. Tye, "A relation between tree amplitudes of closed and open strings", Nucl. Phys. B, 269, 1-23, (1986).

[17] Z. Bern, D. C. Dunbar and T. Shimada, *String based methods in perturbative gravity,* Phys. Lett. B **312** (1993) 277 [arXiv:hep-th/9307001];
Z. Bern, L. J. Dixon, D. C. Dunbar, M. Perelstein and J. S. Rozowsky, *On the relationship between Yang-Mills theory and gravity and its implication for ultraviolet divergences,* Nucl. Phys. B **530** (1998) 401 [arXiv:hep-th/9802162];
Z. Bern and A. K. Grant, *Perturbative gravity from QCD amplitudes,* Phys. Lett. B **457** (1999) 23 [arXiv:hep-th/9904026].

[18] L. Dixon, *Is N=8 supergravity finite?,* available via http://www.slac.stanford.edu/~lance/Neq8.ppt.

[19] A. Velenich, C. Chamon, L. Cugliandolo, D. Kreimer, *On the Brownian gas: a field theory with a Poissonian ground state,* preprint.

Dirk Kreimer
Institut des Hautes Études Scientifiques (IHES)
35 route de Chartres
FR–91440 Bures-sur-Yvette
http://www.ihes.fr
and Boston University
http://math.bu.edu
e-mail: kreimer@ihes.fr

Quantum Field Theory
B. Fauser, J. Tolksdorf and E. Zeidler, Eds., 163–175
© 2009 Birkhäuser Verlag Basel/Switzerland

The Structure of Green Functions in Quantum Field Theory with a General State

Christian Brouder

Abstract. In quantum field theory the Green function is usually calculated as the expectation value of the time-ordered product of fields over the vacuum. In some cases, especially in degenerate systems, expectation values over general states are required. The corresponding Green functions are essentially more complex than in the vacuum, because they cannot be written in terms of standard Feynman diagrams. Here a method is proposed to determine the structure of these Green functions and to derive nonperturbative equations for them. The main idea is to transform the cumulants describing correlations into interaction terms.

Mathematics Subject Classification (2000). Primary 81T99; Secondary 81V70; 81T10.

Keywords. Nonequilibrium quantum field theory, initial correlations, structure of Green functions.

1. Introduction

High-energy physics uses quantum field theory mainly to describe scattering experiments through the S-matrix. In solid-state or molecular physics we are rather interested in the value of physical observables such as the charge and current densities inside the sample or the response to an external perturbation. At the quantum field theory (QFT) level, these quantities are calculated as expectation values of Heisenberg operators. For example, the current density for a system in a state $|\Phi\rangle$ is $\langle \Phi | \mathbf{J}(x) | \Phi \rangle$, where $|\Phi\rangle$ and $\mathbf{J}(x)$ are written in the Heisenberg picture.

The first QFT calculation of Heisenberg operators was made by Dyson in two difficult papers [1, 2] that were completely ignored. At about the same time, Gell-Mann and Low discovered that, when the initial state of the system is non-degenerate, the expectation value of a Heisenberg operators can be obtained by a relatively simple formula [3]. The Gell-Mann and Low formula has been immensely successful and is a key element of the many-body theory of condensed matter [4, 5].

Its main advantage over the formalism developed by Dyson is that all the standard tools of QFT can be used without change.

However, it was soon realized that the assumption of a nondegenerate initial state is not always valid. As a matter of fact, the problem of what happens when the initial state is not trivial is so natural that it was discussed in many fields of physics: statistical physics [6], many-body physics [7], solid-state physics [8], atomic physics [9], quantum field theory and nuclear physics [10, 11]. As a consequence, the theory developed to solve this problem received several names such as nonequilibrium quantum field theory (or quantum statistical mechanics) with initial correlations (or with cumulants, or for open shells, or for degenerate systems). It is also called the closed-time path or the (Schwinger-)Keldysh approach for an arbitrary initial density matrix.

It should be stressed that the problem of the quantum field theory of a degenerate system is not only of academic interest. For instance, many strongly-correlated systems contain open-shell transition metal ions which are degenerate by symmetry. This degeneracy makes the system very sensitive to external perturbation and, therefore, quite useful for the design of functional materials.

The elaboration of a QFT for degenerate systems took a long time. It started with Symanzik [12] and Schwinger [13] and made slow progress because the combinatorial complexity is much higher than with standard QFT. To illustrate this crucial point, it is important to consider an example. According to Wick's theorem, the time-ordered product of free fields can be written in terms of normal order products:

$$T\varphi(x_1)\ldots\varphi(x_4) = \;:\varphi(x_1)\ldots\varphi(x_4): + \sum_{ijkl} :\varphi(x_i)\varphi(x_j): G_0(x_k,x_l)$$
$$+ \sum_{ijkl} :\varphi(x_k)\varphi(x_l): G_0(x_i,x_j) + \sum_{ijkl} G_0(x_i,x_j)G_0(x_k,x_l),$$

where the quadruple of indices (i,j,k,l) runs over $(1,2,3,4)$, $(1,3,2,4)$ and $(1,4,2,3)$. The expectation value of this expression over the vacuum gives the familiar result $\sum_{ijkl} G_0(x_i,x_j)G_0(x_k,x_l)$. However, when the initial state $|\psi\rangle$ is not the vacuum (as in solid-state physics), we obtain

$$\langle\psi|T\varphi(x_1)\ldots\varphi(x_4)|\psi\rangle = \langle\psi|:\varphi(x_1)\ldots\varphi(x_4):|\psi\rangle + \sum_{ijkl} \rho_2(x_i,x_j)G_0(x_k,x_l)$$
$$+ \sum_{ijkl} \rho_2(x_k,x_l)G_0(x_i,x_j) + \sum_{ijkl} G_0(x_i,x_j)G_0(x_k,x_l),$$

where $\rho_2(x,y) = \langle\psi|:\varphi(x)\varphi(y):|\psi\rangle$. If we assume, for notational convenience, that the expectation value of the normal product of an odd number of field operators is zero, the fourth cumulant $\rho_4(x_1,\ldots,x_4)$ is defined by the equation

$$\langle\psi|:\varphi(x_1)\ldots\varphi(x_4):|\psi\rangle = \rho_4(x_1,\ldots,x_4) + \sum_{ijkl} \rho_2(x_k,x_l)\rho_2(x_i,x_j).$$

If we put $g = G_0 + \rho_2$, the free four-point Green function becomes

$$\langle\psi|T\varphi(x_1)\ldots\varphi(x_4)|\psi\rangle \;\; = \;\; \rho_4(x_1,\ldots,x_4) + \sum_{ijkl} g(x_i, x_j)g(x_k, x_l).$$

When $\rho_4 = 0$, the expression is the same as over the vacuum, except for the fact that the free Feynman propagator G_0 is replaced by g. When this substitution is valid, standard QFT can be applied without major change and the structure of the interacting Green functions is not modified. For fermionic systems described by a quadratic Hamiltonian H_0, this happens when the ground state is nondegenerate, so that $|\psi\rangle$ is a Slater determinant. When $\rho_4 \neq 0$, the expression becomes essentially different because the cumulant ρ_4 appears as a sort of free Feynman propagator with four legs. In general, the expectation value of a time-ordered product of n free fields involves ρ_k with $k \leq n$.

In other words, the perturbative expansion of the Green functions can no longer be written as a sum of standard Feynman diagrams. Generalized Feynman diagrams have to be used, involving free Feynman propagators with any number of legs [6, 7, 14].

Because of this additional complexity, the structure of the Green functions for degenerate systems is almost completely unknown. The only result available is the equivalent of the Dyson equation for the one-body Green function $G(x, y)$ [7]

$$G \;\; = \;\; (1 - A)^{-1}(G_0 + C)(1 - B)^{-1}(1 + \Sigma G),$$

where A, B, C and Σ are sums of one-particle irreducible diagrams. When the initial state is nondegenerate, $A = B = C = 0$ and the Dyson equation $G = G_0 + G_0 \Sigma G$ is recovered.

In the present paper, a formal method is presented to determine the structure of Green functions for degenerate systems. The main idea is to use external sources that transform the additional propagators ρ_n into *interaction terms*. This brings the problem back into the standard QFT scheme, where many structural results are available.

2. Expectation value of Heisenberg operators

Let us consider a physical observable $A(t)$, for instance the charge density or the local magnetic field. In the Heisenberg picture, this observable is represented by the operator $A_H(t)$ and the value of its observable when the system is in the state $|\Phi_H\rangle$ is given by the expectation value $\langle A(t)\rangle = \langle\Phi_H|A_H(t)|\Phi_H\rangle$.

Going over to the interaction picture, we write the Hamiltonian of the system as the sum of a free and an interaction parts: $H(t) = H_0 + H_I(t)$, we define the time evolution operator $U(t, t') = T\big(\exp(-i\int_{t'}^{t} H_I(t)dt)\big)$ and we assume that the state $|\Phi_H\rangle$ can be obtained as the adiabatic evolution of an eigenstate $|\Phi_0\rangle$ of H_0. The expectation value of A becomes

$$\langle A(t)\rangle \;\; = \;\; \langle\Phi_0|U(-\infty, t)A(t)U(t, -\infty)|\Phi_0\rangle,$$

where $A(t)$ on the right hand side is the operator representing the observable in the interaction picture.expectation value!in interaction picture The identity $1 = U(t,t')U(t',t)$ and the definition $S = U(\infty,-\infty)$ enable us to derive the basic expression for the expectation value of an observable in the interaction picture:

$$\langle A(t)\rangle = \langle\Phi_0|S^\dagger T(A(t)S)|\Phi_0\rangle. \tag{2.1}$$

When $|\Phi_0\rangle$ is nondegenerate, this expression can be further simplified into the Gell-Mann and Low formula

$$\langle\Phi|A(t)|\Phi\rangle = \frac{\langle\Phi_0|T(A(t)S)|\Phi_0\rangle}{\langle\Phi_0|S|\Phi_0\rangle}.$$

If the system is in a mixed state, as is the case for a degenerate system by Lüders' principle, the expectation value becomes

$$\langle A(t)\rangle = \sum_n p_n\langle\Phi_n|S^\dagger T(A(t)S)|\Phi_n\rangle,$$

where p_n is the probability to find the system in the eigenstate $|\Phi_n\rangle$. It will be convenient to use more general mixed states $\sum_{mn} w_{mn}|\Phi_m\rangle\langle\Phi_n|$, where w_{mn} is a density matrix (i.e. a nonnegative Hermitian matrix with unit trace). Such a mixed state corresponds to a linear form ω defined by its value over an operator O:

$$\omega(O) = \sum_{mn} w_{mn}\langle\Phi_n|O|\Phi_m\rangle.$$

Then, the expectation value of $A(t)$ becomes

$$\langle A(t)\rangle = \omega\big(S^\dagger T(A(t)S)\big). \tag{2.2}$$

3. QFT with a general state

In all practical cases, the operator representing the observable $A(t)$ in the interaction picture is a polynomial in φ and its derivatives. Its expectation value (2.2) can be expressed in terms of Green functions that are conveniently calculated by a formal trick due to Symanzik [12] and Schwinger [13], and reinterpreted by Keldysh [15].

The first step is to define an S-matrix in the presence of an external current j as $S(j) = T\big(e^{-i\int H^{\text{int}}(t)dt+i\int j(x)\varphi(x)dx}\big)$, where H^{int} is the interaction Hamiltonian in the interaction picture. The interaction Hamiltonian is then written in terms of a Hamiltonian density $V(x)$, so that $\int H^{\text{int}}(t)dt = \int V(x)dx$ and the generating function of the interacting Green functions is defined by $Z(j_+,j_-) = \omega\big(S^\dagger(j_-)S(j_+)\big)$. The interacting Green functions can then be obtained as functional derivatives of Z with respect to the external currents j_+ and j_-. For example

$$\langle T(\varphi(x)\varphi(y))\rangle = -\frac{\delta^2 Z(j_+,j_-)}{\delta j_+(x)\delta j_+(y)}, \quad\text{and}\quad \langle\varphi(x)\varphi(y)\rangle = \frac{\delta^2 Z(j_+,j_-)}{\delta j_-(x)\delta j_+(y)}.$$

As in standard QFT, the connected Green functions are generated by $\log Z$.

In the functional method [16, 17], the generating function Z of the interacting system is written as $Z = e^{-iD} Z_0$, where D is the interaction in terms of functional derivatives

$$D = \int V\left(\frac{-i\delta}{\delta j_+(x)}\right) - V\left(\frac{i\delta}{\delta j_-(x)}\right) dx,$$

and where $Z_0(j_+, j_-) = \omega\big(S_0^\dagger(j_-)S_0(j_+)\big)$, with $S_0(j) = T\big(e^{i\int j(x)\varphi(x)dx}\big)$. Note that $Z_0(j_+, j_-)$ is the generating function of the free Green functions.

A straightforward calculation [17] leads to

$$Z^0(j_+, j_-) = e^{-1/2 \int \mathbf{j}(x)G_0'(x,y)\mathbf{j}(y)dxdy} e^{\rho'(j_+ - j_-)},$$

where $\mathbf{j} = (j_+, j_-)$ is the source vector,

$$G_0'(x, y) = \begin{pmatrix} \langle 0|T\big(\phi(x)\phi(y)\big)|0\rangle & -\langle 0|\phi(y)\phi(x)|0\rangle \\ -\langle 0|\phi(x)\phi(y)|0\rangle & \langle 0|\bar{T}\big(\phi(x)\phi(y)\big)|0\rangle \end{pmatrix}, \qquad (3.1)$$

is a free Green function (with $\bar{T}$ the anti-time ordering operator) and

$$e^{\rho'(j)} = \omega\big(:e^{i\int j(x)\varphi(x)dx}:\big) \qquad (3.2)$$

defines the generating function $\rho'(j)$ of the cumulants of the initial state ω.

The free Green function G_0' describes the dynamics generated by the free Hamiltonian H_0. It can also be written in terms of advanced and retarded Green functions [13].

The idea of describing a state by its cumulants was introduced in QFT by Fujita [6] and Hall [7]. It was recently rediscovered in nuclear physics [10, 11] and in quantum chemistry [18].

The next step is to modify the definition of the free Green function. The cumulant function is Taylor expanded

$$\rho'(j) = \sum_{n=2}^{\infty} \frac{1}{n!} \int dx_1 \dots dx_n \rho_n(x_1, \dots, x_n) j(x_1) \dots j(x_n).$$

The expansion starts at $n = 2$ because $\omega(1) = 1$ and the linear term can be removed by shifting the field φ. The bilinear term $\rho_2(x, y)$ is included into the free Green function by defining

$$G_0(x, y) = G_0'(x, y) + \rho_2(x, y) \begin{pmatrix} 1 & -1 \\ -1 & 1 \end{pmatrix},$$

and the corresponding cumulant function becomes

$$\rho(j) = \rho'(j) - (1/2) \int dxdy j(x)\rho_2(x, y)j(y)$$

$$= \sum_{n=3}^{\infty} \frac{1}{n!} \int dx_1 \dots dx_n \rho_n(x_1, \dots, x_n) j(x_1) \dots j(x_n).$$

Remark 3.1. There are several good reasons to use G_0 and ρ instead of G_0' and ρ': (i) This modification is exactly what is done in solid-state physics when the free Green function includes a sum over occupied states [19]; (ii) At a fundamental level, G_0 and ρ have a more intrinsic meaning than G_0' and ρ' because they do not depend on the state $|0\rangle$ chosen as the vacuum; (iii) An important theorem of quantum field theory [20] states that, under quite general conditions, $\rho_n(x_1, \ldots, x_n)$ is a smooth function of its arguments when $n > 2$, so that G_0 gathers all possible singular terms (a related result was obtained by Tikhodeev [21]); (iv) A state for which $\rho(j) = 0$ is called a quasi-free state [22], quasi-free states are very convenient in practice because the rules of standard QFT can be used without basic changes. Thus, the additional complications arise precisely when ρ (and not ρ') is not zero.

4. Nonperturbative equations

To size up the combinatorial complexity due to the presence of a non-zero ρ, we present the diagrammatic expansion of the one-body Green function $G(x, y)$ for the φ^3 theory to second order in perturbation theory. For this illustrative purpose, it will be enough to say that the cumulant $\rho_n(x_1, \ldots, x_n)$ is pictured as a white vertex with n edges attached to it, the other vertex of the edge is associated with one of the points $x_1, \ldots, x_n$. For example, $\rho_4(x_1, \ldots, x_4)$ is represented by the diagram In this diagram, the white dot does not stand for a spacetime point, it

$$\rho_4(x_1, x_2, x_3, x_4) \quad = \quad \begin{array}{c} x_1 \quad\quad x_2 \\ \times \\ x_3 \quad\quad x_4 \end{array}$$

just indicates that the points x_1 to x_4 are arguments of a common cumulant. If we restrict the calculation to the case when $\rho_n = 0$ if n is odd, we obtain the following expansion In standard QFT, only the first and last diagrams of the right hand side

$$G(x, y) \quad = \quad \underset{x \quad\quad y}{\bullet\!\!-\!\!-\!\!-\!\!\bullet} + \text{(diagrams)} + \cdots$$

are present. In the general case when all $\rho_n \neq 0$, the number of diagrams is still much larger.

4.1. Generalized Dyson equation

As mentioned in the introduction, the only known result concerning the structure of Green functions with a general state was derived by Hall for the one-body Green function $G(x,y)$ [7]

$$G = (1-A)^{-1}(G_0 + C)(1-B)^{-1}(1+\Sigma G).$$

In diagrammatic terms the quantities A, B, C and Σ are sums of one-particle irreducible diagrams. If we take our example of the Green function of φ^3 theory up to second order, we find

$$A = \quad + \quad + \quad + \ldots$$

$$B = \quad + \quad + \quad + \ldots$$

$$C = \quad + \quad + \quad + \quad + \quad + \quad + \quad + \quad + \ldots$$

$$\Sigma = \quad + \quad + \quad + \quad + \ldots$$

In standard QFT, we have $A = B = C = 0$ and the diagrammatic representation of Σ contains much less terms. However, the difference with standard QFT is not only limited to the number of diagrams. The definition (3.2) of the cumulant function, and the fact that the free field φ is a solution of the Klein-Gordon equation imply that ρ_n is a solution of the Klein-Gordon equation in each of its variables. Thus, $A(x,y)$, $B(x,y)$ and $C(x,y)$ are solutions of the Klein-Gordon equation for x and y. As a consequence, applying the Klein-Gordon operator to the Green function gives us $(\Box+m^2)G = (1-B)^{-1}(1+\Sigma G)$. In other words, applying the Klein-Gordon operator kills a large number of terms of G. This is in stark contrast with standard QFT, where $(\Box+m^2)G = 1+\Sigma G$ and amputating a Green function does not modify its structure. This important difference makes some tools of standard QFT (e.g. amputated diagrams or Legendre transformation) invalid in the presence of a general state.

All those difficulties explain the scarcity of results available in non-perturbative QFT with a general state. Apart from Hall's work [7], the only non-perturbative results are Tikhodeev's cancellation theorems [23, 24] and the equation of motion for the Green functions [25].

In the next section, we present a simple trick to derive the structure of Green functions with a general state.

4.2. Quadrupling the sources

We first determine the main formal difference between standard QFT and QFT with a general state. In both cases, the generating function of the Green functions can be written $Z = \mathrm{e}^{-iD}Z_0$, where D describes the interaction and Z_0 the initial state. In the presence of a general state, the interaction D is simple but Z_0 is made non standard by the cumulant factor e^ρ. The idea of the solution is to transfer the cumulant function ρ from Z_0 to D, because powerful functional methods were developed to deal with general interactions D. These methods were first proposed by Dominicis and Englert [26] and greatly expanded by the Soviet school [27, 28, 29, 30, 31, 32, 33, 34, 35].

This transfer from the initial state to the interaction can be done easily by introducing two additional external sources k_+ and k_- and using the identity

$$\mathrm{e}^{\rho(j_+ - j_-)} = \mathrm{e}^{\rho(-i\frac{\delta}{\delta k_+} - i\frac{\delta}{\delta k_-})}\mathrm{e}^{i\int (j_+(x)k_+(x) - j_-(x)k_-(x))\mathrm{d}x}\Big|_{k_+ = k_- = 0}.$$

The term involving ρ can now be transferred from Z_0 to D by defining the new generating function

$$\bar{Z}(j_\pm, k_\pm) \;=\; \mathrm{e}^{-i\bar{D}}\bar{Z}_0(j_\pm, k_\pm),$$

where the modified interaction is

$$\bar{D} \;=\; \int V\Big(\frac{-i\delta}{\delta j_+(x)}\Big) - V\Big(\frac{i\delta}{\delta j_-(x)}\Big)\mathrm{d}x - i\rho(-i\frac{\delta}{\delta k_+} - i\frac{\delta}{\delta k_-}),$$

and the modified free generating function is

$$\bar{Z}_0(j_\pm, k_\pm) \;=\; \mathrm{e}^{-1/2\int \mathbf{J}(x)\bar{G}_0(x,y)\mathbf{J}(y)\mathrm{d}x\mathrm{d}y},$$

with $\mathbf{J} = (j_+, j_-, k_+, k_-)$. The modified free Green function $\bar{G}_0$ is now a 4x4 matrix that can be written as a 2x2 matrix of 2x2 matrices

$$\bar{G}_0 \;=\; \begin{pmatrix} G_0 & -i\mathbf{1} \\ -i\mathbf{1} & 0 \end{pmatrix}.$$

In contrast to the standard case, the free Green function $\bar{G}_0$ is invertible

$$\bar{G}_0^{-1} \;=\; \begin{pmatrix} 0 & i\mathbf{1} \\ i\mathbf{1} & G_0 \end{pmatrix},$$

and it is again possible to use amputated diagrams and Legendre transformations. The free generating function $\bar{Z}_0$ is the exponential of a function that is bilinear in the sources, and all the standard structural tools of QFT are available again. We illustrate this by recovering Hall's analogue of the Dyson equation.

4.3. An algebraic proof of Hall's equation

The free generating function $\bar{Z}_0$ has a standard form and the Dyson equation holds again: $\bar{G} = \bar{G}_0 + \bar{G}_0\bar{\Sigma}\bar{G}$, where $\bar{G}$ is the 4x4 one-body Green function obtained from the generating function $\bar{Z}$ and $\bar{\Sigma}$ is the corresponding self-energy. Each 4x4 matrix is written as a 2x2 matrix of 2x2 matrices. For example

$$\bar{G} = \begin{pmatrix} \bar{G}_{11} & \bar{G}_{12} \\ \bar{G}_{21} & \bar{G}_{22} \end{pmatrix}.$$

We want to determine the structure of the 2x2 Green function G, which is equal to $\bar{G}_{11}$ when $k_+ = k_- = 0$.

The upper-left component of the Dyson equation for $\bar{G}$ is

$$\bar{G}_{11} = G_0 + (G_0\bar{\Sigma}_{11} - i\bar{\Sigma}_{21})\bar{G}_{11} + (G_0\bar{\Sigma}_{12} - i\bar{\Sigma}_{22})\bar{G}_{21}. \tag{4.1}$$

The lower-left component gives us $\bar{G}_{21} = -i(1 + i\bar{\Sigma}_{12})^{-1}(1 + \bar{\Sigma}_{11}\bar{G}_{11})$. If we introduce this expression for $\bar{G}_{21}$ into equation (4.1), rearrange a bit and use the operator identity $1 + O(1 - O)^{-1} = (1 - O)^{-1}$, we obtain

$$(1 + i\bar{\Sigma}_{21})\bar{G}_{11} = (G_0 - \bar{\Sigma}_{22})(1 + i\bar{\Sigma}_{12})^{-1}(1 + \bar{\Sigma}_{11}\bar{G}_{11}).$$

Hall's equation is recovered by identifying $A = -i\bar{\Sigma}_{21}$, $B = -i\bar{\Sigma}_{12}$ and $C = -\bar{\Sigma}_{22}$, where the right hand side is taken at $k_+ = k_- = 0$. Note that Hall's equation is now obtained after a few lines of algebra instead of a subtle analysis of the graphical structure of the diagrams.

With the same approach, all the nonperturbative methods used in solid-state physics, such as the GW approximation [36] and the Bethe-Salpeter equation [37, 38], can be transposed to the case of a general initial state. This will be presented in a forthcoming publication.

5. Determination of the ground state

QFT with a general state was studied because the initial eigenstate of a quantum system is sometimes degenerate. However, it remains to determine which density matrix ω_{mn} of the free Hamiltonian leads to the ground state of the interacting system.

A solution to this problem was inspired by quantum chemistry methods [39]. A number of eigenstates $|\Phi_n\rangle$ of H_0 are chosen, for example the complete list of degenerate eigenstates corresponding to a given energy. These eigenstates span the so-called *model space* and the ground state of the interacting system is assumed to belong to the adiabatic evolution of the model space. This model space generates, for each density matrix, a linear form ω as described in equation (2.2). The problem boils down to the determination of the density matrix ω_{mn} that minimizes the energy of the interacting system.

This minimization leads to an effective Hamiltonian and the proper density matrix is obtained by diagonalizing the effective Hamiltonian. This type of method is typical of atomic and molecular physics [40]. However, the effective Hamiltonian can now be determined by powerful non-perturbative Green function methods. Therefore, the present approach leads to a sort of unification of quantum chemistry and QFT: it contains standard QFT when the dimension of the model space is one, it contains standard quantum chemistry (more precisely many-body perturbation theory) when the Green functions are expanded perturbatively.

Therefore, the present approach might help developing some new nonperturbative methods in quantum chemistry. On the other hand, quantum chemistry has accumulated an impressive body of results. The physics Nobel-prize winner Kenneth Wilson stated that [41] "Ab initio quantum chemistry is an emerging computational area that is fifty years ahead of lattice gauge theory." Therefore, the experience gained in quantum chemistry can be used to solve some of the remaining problems of the present approach, such as the removal of the secular terms [14] to all order.

6. Conclusion

The present paper sketched a new method to determine the Green functions of quantum field theory with a general state. The main idea is to transform the cumulant function describing the initial state into an interaction term. As a consequence, the cumulants become dressed by the interaction, providing a much better description of the correlation in the system.

An alternative method would be to work at the operator level, as was done recently by Dütsch and Fredenhagen [42], and to take the expectation value at the end of the calculation. This would have the obvious advantage of dealing with a fully rigorous theory. However, we would loose the non-perturbative aspects of the present approach.

Although this approach seems promising, much remains to be done before it can be applied to realistic systems: (i) our description is purely formal; (ii) the degenerate initial eigenstates lead to secular terms that must be removed [14]; (iii) renormalization must be included, although this will probably not be very different from the standard case, because all the singularities of the free system are restricted to G_0.

Interesting connections can be made with other problems. For example, the cancellation theorem [23] seems to be interpretable as a consequence of the unitarity of the S-matrix. It would extend Veltman's largest time equation [43] to the case of spacetime points with equal time. Another exciting track would be a connection with noncommutative geometry. Keldysh [15] noticed that the doubling of sources could be replaced by a doubling of spacetime points. In other words, $j_\pm(x)$ becomes $j(x_\pm)$, where $x_\pm$ are two copies of the spacetime point x: time travels from the past to the future for x_+ and in the other direction for x_-. Sivasubramanian and coll. [44] have proposed to interpret this doubling of spacetime points in terms of noncommutative geometry. It would be interesting to follow this track for our quadrupling of spacetime points.

From the practical point of view, the main applications of our scheme will be for the calculation of strongly-correlated systems, in particular for the optical response of some materials, such as gemstones, that remain beyond the reach of the standard tools of contemporary solid-state physics.

After the completion of this work, we came across a little known article by Sergey Fanchenko, where the cumulants are used to define an effective action [45]. His paper is also interesting because it gives a path integral formulation of quantum field theory with a general state. His approach and the one of the present paper provide complementary tools to attack nonperturbative problems of quantum field theory with a general state.

Acknowledgement

I thank Alessandra Frabetti, Frédéric Patras, Sergey Fanchenko and Pierre Cartier for very useful discussions.

References

[1] F. J. Dyson. *Heisenberg operators in quantum electrodynamics. I.* Phys. Rev. **82** (1951), 428–39.

[2] F. J. Dyson. *Heisenberg operators in quantum electrodynamics. II.* Phys. Rev. **83** (1951), 608–27.

[3] M. Gell-Mann and F. Low. *Bound states in quantum field theory.* Phys. Rev. **84** (1951), 350–4.

[4] A. L. Fetter and J. D. Walecka. *Quantum Theory of Many-Particle Systems.* McGraw-Hill, Boston (1971).

[5] E. K. U. Gross, E. Runge and O. Heinonen. *Many-Particle Theory.* Adam Hilger, Bristol (1991).

[6] S. Fujita. *Introduction to Non Equilibrium Quantum Statistical Mechanics.* Saunders, Philadelphia (1966).

[7] A. G. Hall. *Non-equilibrium Green functions: Generalized Wick's theorem and diagrammatic perturbation theory with initial correlations.* J. Phys. A: Math. Gen. **8** (1975), 214–25.

[8] D. M. Esterling and R. V. Lange. *Degenerate mass operator perturbation theory in the Hubbard model.* Rev. Mod. Phys. **40** (1968), 796–9.

[9] I. Lindgren, B. Åsén, S. Salomonson and A.-M. Mårtensson-Pendrill. *QED procedure applied to the quasidegenerate fine-structure levels of He-like ions.* Phys. Rev. A **64** (2001), 062505.

[10] P. A. Henning. *On the treatment of initial correlations in quantum field theory of non-equilibrium states.* Nucl. Phys. B **337** (1990), 547–68.

[11] R. Fauser and H. H. Wolter. *Non-equilibrium quantum field theory and perturbation theory.* Nucl. Phys. A **584** (1995), 604–620.

[12] K. Symanzik. *On the many-body structure of Green's functions in quantum field theory.* J. Math. Phys. **1** (1960), 249–73.

[13] J. Schwinger. *Brownian motion of a quantum oscillator.* J. Math. Phys. **2** (1961), 407–32.

[14] Y. A. Kukharenko and S. G. Tikhodeev. *Diagram technique in the theory of relaxation processes.* Soviet Phys. JETP **56** (1982), 831–8.

[15] L. V. Keldysh. *Diagram technique for nonequilibrium processes.* Soviet Phys. JETP **20** (1965), 1018–26.

[16] J. Schwinger. *On the Green's functions of quantized fields. I.* Proc. Nat. Acad. Sci. **37** (1951), 452–5.

[17] K.-C. Chou, Z.-B. Su, B.-L. Hao and L. Yu. *Equilibrium and nonequilibrium formalisms made unified.* Phys. Repts. **118** (1985), 1–131.

[18] W. Kutzelnigg and D. Mukherjee. *Cumulant expansion of the reduced density matrices.* J. Chem. Phys. **110** (1999), 2800–9.

[19] C. Brouder. *Matrix elements of many-body operators and density correlations.* Phys. Rev. A **72** (2005), 032720.

[20] S. Hollands and W. Ruan. *The state space of perturbative quantum field theory in curved spacetimes.* Ann. Inst. Henri Poincaré **3** (2003), 635–57.

[21] S. G. Thikhodeev. *Relations between many-body Green's functions and correlation functions.* Sov. Phys. Doklady **27** (1982), 492–3.

[22] B. S. Kay and R. M. Wald. *Theorems on the uniqueness and thermal properties of stationary, nonsingular, quasifree states on spacetimes with a bifurcate Killing horizon.* Phys. Repts. **207** (1991), 49–136.

[23] S. G. Thikhodeev. *On the vanishing of correlation-connected diagrams in Hall's diagram technique.* Sov. Phys. Doklady **27** (1982), 624–5.

[24] P. Danielewicz. *Quantum theory of nonequilibrium processes, I.* Ann. Phys. **152** (1984), 239–304.

[25] C. Brouder. *Green function hierarchy for open shells.* Euro. Phys. Lett. **71** (2005), 556–62.

[26] C. de Dominicis and F. Englert. *Potential-correlation function duality in statistical physics.* J. Math. Phys. **8** (1967), 2143–6.

[27] A. N. Vasil'ev and A. K. Kazanskii. *Legendre transforms of the generating functionals in quantum field theory.* Theor. Math. Phys. **12** (1972), 875–87.

[28] A. N. Vasil'ev and A. K. Kazanskii. *Equations of motion for a Legendre transform of arbitrary order.* Theor. Math. Phys. **14** (1973), 215–226.

[29] A. N. Vasil'ev, A. K. Kazanskii and Y. M. Pis'mak. *Diagrammatic analysis of the fourth Legendre transform.* Theor. Math. Phys. **20** (1974), 754–62.

[30] A. N. Vasil'ev, A. K. Kazanskii and Y. M. Pis'mak. *Equations for higher Legendre transforms in terms of 1-irreducible vertices.* Theor. Math. Phys. **19** (1974), 443–53.

[31] A. N. Vasil'ev. *The field theoretic renormalization group in critical behavior theory and stochastic dynamics.* Chapman and Hall/CRC, New York (2004).

[32] Y. M. Pis'mak. *Proof of the 3-irreducibility of the third Legendre transform.* Theor. Math. Phys. **18** (1974), 211–8.

[33] Y. M. Pis'mak. *Combinatorial analysis of the overlapping problem for vertices with more than four legs.* Theor. Math. Phys. **24** (1975), 649–58.

[34] Y. M. Pis'mak. *Combinatorial analysis of the overlapping problem for vertices with more than four legs: II Higher Legendre transforms.* Theor. Math. Phys. **24** (1975), 755–67.

[35] Y. M. Pis'mak. *n-particle problem in quantum field theory and the functional Legendre transforms*. Int. J. Mod. Phys. **7** (1992), 2793–808.

[36] F. Aryasetiawan and O. Gunnarson. *The GW method*. Rep. Prog. Phys. **61** (1998), 237–312.

[37] S. Albrecht, L. Reining, R. D. Sole and G. Onida. *Ab initio calculation of excitonic effects in the optical spectra of semiconductors*. Phys. Rev. Lett. **80** (1998), 4510–3.

[38] L. X. Benedict, E. L. Shirley and R. B. Bohn. *Optical absorption of insulators and the electron-hole interaction: An ab initio calculation*. Phys. Rev. Lett. **80** (1998), 4514–7.

[39] C. Brouder. *Many-body approach to crystal field theory*. Phys. Stat. Sol. (c) **2** (2005), 472–5.

[40] I. Lindgren and J. Morrison. *Atomic Many-Body Theory*. Second edition. Springer-Verlag, (1986).

[41] K. G. Wilson. *Ab initio quantum chemistry: A source of ideas for lattice gauge theorists*. Nucl. Phys. Suppl. **17** (1990), 82–92.

[42] M. Dütsch and K. Fredenhagen. *Causal perturbation theory in terms of retarded products, and a proof of the Action Ward Identity*. Rev. Math. Phys. **16** (2004), 1291–348.

[43] M. Veltman. *Unitarity and causality in a renormalizable field theory with unstable particles*. Physica **29** (1963), 186–207.

[44] S. Sivasubramanian, Y. N. Srivastava, G. Vitiello and A. Widom. *Quantum dissipation induced noncommutative geometry*. Phys. Lett. A **311** (2003), 97–105.

[45] S. S. Fanchenko. *Generalized diagram technique of nonequilibrium processes*. Theor. Math. Phys. **55** (1983), 406–9.

Christian Brouder
Institut de Minéralogie et de Physique des Milieux Condensés
CNRS UMR7590, Université Paris 6
140 rue de Lourmel
FR–75015 Paris
e-mail: `christian.brouder@impmc.jussieu.fr`

Quantum Field Theory
B. Fauser, J. Tolksdorf and E. Zeidler, Eds., 177–196
© 2009 Birkhäuser Verlag Basel/Switzerland

The Quantum Action Principle in the Framework of Causal Perturbation Theory

Ferdinand Brennecke and Michael Dütsch

Abstract. In perturbative quantum field theory the maintenance of classical symmetries is quite often investigated by means of algebraic renormalization, which is based on the Quantum Action Principle. We formulate and prove this principle in a new framework, in causal perturbation theory with localized interactions. Throughout this work a universal formulation of symmetries is used: the Master Ward Identity.

Mathematics Subject Classification (2000). Primary 81T15; Secondary 70S10; 81T50.

Keywords. Perturbative renormalization, quantum action principle, master Ward identity, symmetries, causal perturbation theory, local fields.

1. Introduction

The main problem in perturbative renormalization is to prove that symmetries of the underlying classical theory can be maintained in the process of renormalization. In traditional renormalization theory this is done by 'algebraic renormalization' [26]. This method relies on the '*Quantum Action Principle*' (QAP), which is due to Lowenstein [23] and Lam [22]. This principle states that the most general violation of an identity expressing a relevant symmetry ('Ward identity') can be expressed by the insertion of a local field with appropriately bounded mass dimension. Proceeding in a proper field formalism[1] by induction on the order of $\hbar$, this knowledge about the structure of violations of Ward identities and often cohomological results are used to remove these violations by finite renormalizations. For example, this method has been used to prove BRST-symmetry of Yang-Mills gauge theories [2, 3, 31, 17, 1].

---

[1] By 'proper field formalism' we mean the description of a perturbative QFT in terms of the generating functional of the 1-particle irreducible diagrams.

Traditionally, algebraic renormalization is formulated in terms of a renormalization method in which the interaction is not localized (i.e. $S_{\mathrm{int}} = \int dx \, \mathcal{L}_{\mathrm{int}}(x)$, where $\mathcal{L}_{\mathrm{int}}$ is a polynomial in the basic fields with constant coefficients), for example the BPHZ momentum space subtraction procedure [32, 23, 22] or the pole subtractions of dimensionally regularized integrals [5]. In [25] it is pointed out (without proof) that the QAP is a general theorem in perturbative QFT for non-localized interactions, i.e. it holds in any renormalization scheme.[2]

However, for the generalization of perturbative QFT to general globally hyperbolic *curved space-times*, it is advantageous to work with *localized interactions* (i.e. $S_{\mathrm{int}} = \int dx \, \sum_{n \geq 1} (g(x))^n \, \mathcal{L}_{\mathrm{int},n}(x)$, where g is a test function with compact support) and to use a renormalization method which proceeds in configuration space and in which the locality and causality of perturbative QFT is clearly visible [8, 18, 19]. It is *causal perturbation theory* (CPT) [4, 15, 14] which is distinguished by these criteria.

Since it is the framework of *algebraic QFT* [16] in which the problems specific for curved space-times (which mainly rely on the absence of translation invariance) can best be treated, our main goal is the perturbative construction of the net of local algebras of interacting fields ('perturbative algebraic QFT'). Using the formulation of causality in CPT, it was possible to show that for this construction it is sufficient to work with *localized* interactions [8, 12]. Hence, a main argument against localized interactions, namely that a space or time dependence of the coupling constants has not been observed in experiments, does not concern perturbative algebraic QFT. Because of the localization of the interactions, the construction of the local algebras of interacting fields is not plagued by infrared divergences, the latter appear only in the construction of physical states.

Due to these facts it is desirable to transfer the techniques of algebraic renormalization to CPT, that is to formulate the $\hbar$-expansion, a proper field formalism and the QAP in the framework of CPT. For the $\hbar$-expansion the difficulty is that CPT is a construction of the perturbation series by induction on the coupling constant, a problem solved in [11, 12]. A formulation of the QAP in the framework of CPT has partially been given in [11] and in [27]; but for symmetries relying on a variation of the fields (as e.g. BRST-symmetry) an appropriate formulation and a proof were missing up to the appearance of the paper [6]. In the latter, also a proper field formalism and algebraic renormalization are developed in the framework of CPT.

In this paper we concisely review main results of that work [6], putting the focus on the QAP. To be closer to the conventional treatment of perturbative QFT in Minkowski space and to simplify the formalism, we work with the Wightman 2-point function instead of a Hadamard function.[3] Compared with [6], we formulate some topics alternatively, in particular we introduce the proper field formalism

---

[2]Causal perturbation theory, with the adiabatic limit carried out, is included in that statement.
[3]In [6] smoothness in the mass m is required for $m \geq 0$ which excludes the Wightman 2-point function.

without using arguments relying on Wick's theorem and the corresponding diagrammatic interpretation. In addition we prove a somewhat stronger version of the QAP.

The validity of the QAP is very general. Therefore, we investigate a universal formulation of Ward identities: the *Master Ward Identity* (MWI) [9, 13]. This identity can be derived in the framework of classical field theory simply from the fact that classical fields can be multiplied pointwise. Since this is impossible for quantum fields (due to their distributional character), the MWI is a highly nontrivial renormalization condition, which cannot be fulfilled in general, the well known anomalies of perturbative QFT are the obstructions.

2. The off-shell Master Ward Identity in classical field theory

For algebraic renormalization it is of crucial importance that the considered Ward identities hold true in classical field theory. Therefore, in this section, we derive the off-shell MWI in the classical framework. The formalism of classical field theory, which we are going to introduce, will be used also in perturbative QFT, since the latter will be obtained by deformation of the classical Poisson algebra (Sect. 3) [11, 12, 13, 14].

For simplicity we study the model of a real scalar field φ on d dimensional Minkowski space $\mathbb{M}$, $d > 2$. The field φ and partial derivatives $\partial^a \varphi$ ($a \in \mathbb{N}_0^d$) are evaluation functionals on the configuration space $\mathcal{C} \equiv C^\infty(\mathbb{M}, \mathbb{R})$: $(\partial^a \varphi)(x)(h) = \partial^a h(x)$. Let $\mathcal{F}$ be the space of all functionals

$$F(\varphi) : \mathcal{C} \longrightarrow \mathbb{C}, \quad F(\varphi)(h) = F(h) , \tag{2.1}$$

which are localized polynomials in φ:

$$F(\varphi) = \sum_{n=0}^{N} \int dx_1 \ldots dx_n \, \varphi(x_1) \cdots \varphi(x_n) f_n(x_1, \ldots, x_n) , \tag{2.2}$$

where $N < \infty$ and the f_n's are $\mathbb{C}$-valued distributions with compact support, which are symmetric under permutations of the arguments and whose wave front sets satisfy the condition

$$\mathrm{WF}(f_n) \cap \left(\mathbb{M}^n \times (\overline{V}_+^n \cup \overline{V}_-^n) \right) = \emptyset \tag{2.3}$$

and $f_0 \in \mathbb{C}$. ($\overline{V}_\pm$ denotes the closure of the forward/backward light-cone.) Endowed with the *classical product* $(F_1 \cdot F_2)(h) := F_1(h) \cdot F_2(h)$, the space $\mathcal{F}$ becomes a commutative algebra. By the support of a functional $F \in \mathcal{F}$ we mean the support of $\frac{\delta F}{\delta \varphi}$.

The space of *local functionals* $\mathcal{F}_{\mathrm{loc}} \subset \mathcal{F}$ is defined as

$$\mathcal{F}_{\mathrm{loc}} \overset{\mathrm{def}}{=} \left\{ \int dx \sum_{i=1}^{N} A_i(x) h_i(x) \equiv \sum_{i=1}^{N} A_i(h_i) \, \middle| \, A_i \in \mathcal{P}, \, h_i \in \mathcal{D}(\mathbb{M}) \right\} , \tag{2.4}$$

where $\mathcal{P}$ is the linear space of all polynomials of the field φ and its partial derivatives:

$$\mathcal{P} := \bigvee \left\{ \partial^a \varphi \, | \, a \in \mathbb{N}_0^d \right\} . \qquad (2.5)$$

We consider action functionals of the form $S_{\text{tot}} = S_0 + \lambda S$ where $S_0 \stackrel{\text{def}}{=} \int dx \frac{1}{2}(\partial_\mu \varphi \partial^\mu \varphi - m^2 \varphi^2)$ is the free action, λ a real parameter and $S \in \mathcal{F}$ some compactly supported interaction, which may be *non-local*. The retarded Green function $\Delta_{S_{\text{tot}}}^{\text{ret}}$ corresponding to the action S_{tot}, is defined by

$$\int dy \, \Delta_{S_{\text{tot}}}^{\text{ret}}(x,y) \frac{\delta^2 S_{\text{tot}}}{\delta\varphi(y)\delta\varphi(z)} = \delta(x-z) = \int dy \, \frac{\delta^2 S_{\text{tot}}}{\delta\varphi(x)\delta\varphi(y)} \Delta_{S_{\text{tot}}}^{\text{ret}}(y,z) \qquad (2.6)$$

and $\Delta_{S_{\text{tot}}}^{\text{ret}}(x,y) = 0$ for x sufficiently early. In the following we consider only actions S_{tot} for which the retarded Green function exists and is unique in the sense of formal power series in λ.

To introduce the perturbative expansion around the free theory and to define the Peierls bracket, we define retarded wave operators which map solutions of the free theory to solutions of the interacting theory [13]. However, we define them as maps on the space $\mathcal{C}$ of all field configurations ('*off-shell formalism*') and not only on the space of free solutions:

Definition 2.1. A retarded wave operator is a family of maps $(r_{S_0+S,S_0})_{S\in\mathcal{F}}$ from $\mathcal{C}$ into itself with the properties

  (i) $r_{S_0+S,S_0}(f)(x) = f(x)$ for x sufficiently early
 (ii) $\frac{\delta(S_0+S)}{\delta\varphi} \circ r_{S_0+S,S_0} = \frac{\delta S_0}{\delta\varphi}$.

The following Lemma is proved in [6].

Lemma 2.2. *The retarded wave operator* $(r_{S_0+S,S_0})_{S\in\mathcal{F}}$ *exists and is unique and invertible in the sense of formal power series in the interaction* S.

Motivated by the interaction picture known from QFT, we introduce retarded fields: the classical retarded field to the interaction S and corresponding to the functional $F \in \mathcal{F}$ is defined by

$$F_S^{\text{cl}} \stackrel{\text{def}}{=} F \circ r_{S_0+S,S_0} : \mathcal{C} \longrightarrow \mathbb{C}. \qquad (2.7)$$

The crucial factorization property,

$$(F \cdot G)_S^{\text{cl}} = F_S^{\text{cl}} \cdot G_S^{\text{cl}} , \qquad (2.8)$$

cannot be maintained in the process of quantization, because quantum fields are distributions. This is why many proofs of symmetries in classical field theory do not apply to QFT (cf. Sect. 5).

The perturbative expansion around the free theory is defined by expanding the retarded fields with respect to the interaction. The coefficients are given by the classical retarded product R_{cl} [13]:

$$R_{\text{cl}} : \mathbb{T}\mathcal{F} \otimes \mathcal{F} \to \mathcal{F} , \quad R_{\text{cl}}(S^{\otimes n}, F) \stackrel{\text{def}}{=} \frac{d^n}{d\lambda^n}\Big|_{\lambda=0} F \circ r_{S_0+\lambda S,S_0} , \qquad (2.9)$$

where $\mathbb{T}\mathcal{V} \stackrel{\text{def}}{=} \mathbb{C} \oplus \bigoplus_{n=1}^{\infty} \mathcal{V}^{\otimes n}$ denotes the tensor algebra corresponding to some vector space $\mathcal{V}$. For non-diagonal entries, $R_{\text{cl}}(\otimes_{j=1}^{n} S_j, F)$ is determined by linearity and symmetry under permutations of $S_1, ..., S_n$. Interacting fields can then be written as

$$F_S^{\text{cl}} \simeq \sum_{n=0}^{\infty} \frac{1}{n!} R_{\text{cl}}(S^{\otimes n}, F) \equiv R_{\text{cl}}(e_{\otimes}^{S}, F) \ . \tag{2.10}$$

The r.h.s. of $\simeq$ is interpreted as a *formal power series* (i.e. we do not care about convergence of the series).

By means of the retarded wave operator one can define an off-shell version [6] of the Peierls bracket associated to the action S [24], $\{\cdot, \cdot\}_S : \mathcal{F} \otimes \mathcal{F} \to \mathcal{F}$, and one verifies that this is indeed a Poisson bracket, i.e. that $\{\cdot, \cdot\}_S$ is linear, antisymmetric and satisfies the Leibniz rule and the Jacobi identity [13, 6].

Following [6], we are now going to derive the classical off-shell MWI from the factorization (2.8) and the definition of the retarded wave operators. Let $\mathcal{J}$ be the ideal generated by the free field equation,

$$\mathcal{J} \stackrel{\text{def}}{=} \left\{ \sum_{n=1}^{N} \int dx_1 \dots dx_n \, \varphi(x_1) \cdots \varphi(x_{n-1}) \frac{\delta S_0}{\delta \varphi(x_n)} f_n(x_1, \dots, x_n) \right\} \subset \mathcal{F} \ ,$$

with $N < \infty$ and the f_n's being defined as in (2.2). Obviously, every $A \in \mathcal{J}$ can be written as

$$A \stackrel{\text{def}}{=} \int dx \, Q(x) \frac{\delta S_0}{\delta \varphi(x)} \ , \tag{2.11}$$

where Q may be non-local. Given $A \in \mathcal{J}$ we introduce a corresponding derivation [13]

$$\delta_A \stackrel{\text{def}}{=} \int dx \, Q(x) \frac{\delta}{\delta \varphi(x)}. \tag{2.12}$$

Notice $F(\varphi + Q) - F(\varphi) = \delta_A F + \mathcal{O}(Q^2)$ (for $F \in \mathcal{F}$) that is, $\delta_A F$ can be interpreted as the variation of F under the infinitesimal field transformation $\varphi(x) \mapsto \varphi(x) + Q(x)$. From the definition of the retarded wave operators Def. 2.1 we obtain

$$\begin{aligned}
(A + \delta_A S) \circ r_{S_0+S, S_0} &= \int dx \, Q(x) \circ r_{S_0+S, S_0} \frac{\delta(S_0 + S)}{\delta \varphi(x)} \circ r_{S_0+S, S_0} \\
&= \int dx \, Q(x) \circ r_{S_0+S, S_0} \frac{\delta S_0}{\delta \varphi(x)} \ . \tag{2.13}
\end{aligned}$$

In terms of the perturbative expansion this relation reads

$$R_{\text{cl}}(e_{\otimes}^{S}, A + \delta_A S) = \int dx \, R_{\text{cl}}(e_{\otimes}^{S}, Q(x)) \frac{\delta S_0}{\delta \varphi(x)} \in \mathcal{J}. \tag{2.14}$$

This is the MWI written in the off-shell formalism. When restricted to the solutions of the free field equation, the right-hand side vanishes and we obtain the on-shell version of the MWI, as it was derived in [13]. For the simplest case $Q = 1$ the MWI reduces to the off-shell version of the (interacting) field equation

$$R_{\mathrm{cl}}\left(e_\otimes^S, \frac{\delta(S_0 + S)}{\delta\varphi(x)}\right) = \frac{\delta S_0}{\delta\varphi(x)} . \tag{2.15}$$

3. Causal perturbation theory

Following [14], we quantize perturbative classical fields by deforming the underlying free theory as a function of $\hbar$: we replace $\mathcal{F}$ by $\mathcal{F}[[\hbar]]$ (i.e. all functionals are formal power series in $\hbar$) and deform the classical product into the $\star$-product, $\star : \mathcal{F} \times \mathcal{F} \to \mathcal{F}$ (for simplicity we write $\mathcal{F}$ for $\mathcal{F}[[\hbar]]$):

$$(F \star G)(\varphi) \overset{\mathrm{def}}{=} \sum_{n=0}^{\infty} \frac{\hbar^n}{n!} \int dx_1 \ldots dx_n dy_1 \ldots dy_n \frac{\delta^n F}{\delta\varphi(x_1) \cdots \delta\varphi(x_n)}$$

$$\cdot \prod_{i=1}^{n} \Delta_m^+(x_i - y_i) \frac{\delta^n G}{\delta\varphi(y_1) \cdots \delta\varphi(y_n)} . \tag{3.1}$$

The $\star$-product is still associative but non-commutative.

In contrast to the classical retarded field F_S^{cl} (2.7), one assumes in perturbative QFT that the interaction S and the field F are *local* functionals. For an interacting quantum field F_S one makes the Ansatz of a *formal power series* in the interaction S:

$$F_S = \sum_{n=0}^{\infty} \frac{1}{n!} R_{n,1}(S^{\otimes n}, F) \equiv R(e_\otimes^S, F) . \tag{3.2}$$

The 'retarded product' $R_{n,1}$ is a **linear** map, from $\mathcal{F}_{\mathrm{loc}}^{\otimes n} \otimes \mathcal{F}_{\mathrm{loc}}$ into $\mathcal{F}$ which is **symmetric in the first n variables**. We interpret $R(A_1(x_1), \ldots; A_n(x_n)), A_1, \ldots, A_n \in \mathcal{P}$, as $\mathcal{F}$-valued distributions on $\mathcal{D}(\mathbb{M}^n)$, which are defined by: $\int dx\, h(x)\, R(\ldots, A(x), \ldots)$ $:= R(\ldots \otimes A(h) \otimes \ldots)\ \forall h \in \mathcal{D}(\mathbb{M})$.

Since the retarded products depend only on the functionals (and not on how the latter are written as smeared fields (2.4)), they must satisfy the **Action Ward Identity (AWI)** [14, 29, 30]:

$$\partial_\mu^x R_{n-1,1}(\ldots A_k(x) \ldots) = R_{n-1,1}(\ldots, \partial_\mu A_k(x), \ldots) . \tag{3.3}$$

Interacting fields are defined by the following axioms [14], which are motivated by their validity in classical field theory. The **basic axioms** are the initial condition $R_{0,1}(1, F) = F$ and

Causality: $F_{G+H} = F_G$ if $\mathrm{supp}\,(\frac{\delta F}{\delta\varphi}) \cap (\mathrm{supp}\,(\frac{\delta H}{\delta\varphi}) + \bar{V}_+) = \emptyset$ y;

GLZ Relation: $F_G \star H_G - H_G \star F_G = \frac{d}{d\lambda}\big|_{\lambda=0} (F_{G+\lambda H} - H_{G+\lambda F})$.

Using only these requirements, the retarded products $R_{n,1}$ can be constructed by induction on n (cf. [28]). However, in each inductive step one is free to add a local functional, which corresponds to the usual renormalization ambiguity. This ambiguity is reduced by imposing **renormalization conditions** as further axioms, see below.

Mostly, perturbative QFT is formulated in terms of the **time ordered product** ('T-product') $T : \mathbb{TF}_{\text{loc}} \to \mathcal{F}$, which is a **linear** and **totally symmetric** map. Compared with the R-product, the T-product has the advantage of being totally symmetric and the disadvantage that its classical limit does not exist [11]. R- and T-products are related by Bogoliubov's formula:

$$R(e^S_\otimes, F) = \frac{\hbar}{i} \mathbf{S}(S)^{-1} \star \frac{d}{d\tau}\Big|_{\tau=0} \mathbf{S}(S + \tau F) , \tag{3.4}$$

where

$$\mathbf{S}(S) \equiv T(e^{iS/\hbar}_\otimes) \equiv \sum_{n=0}^{\infty} \frac{i^n}{n!\hbar^n} T_n(S^{\otimes n}) . \tag{3.5}$$

The basic axioms for retarded products translate into the following basic axioms for T-products: the initial conditions $T_0(1) = 1$, $T_1(F) = F$ and **causal factorization**:

$$T_n(A_1(x_1), ..., A_n(x_n)) =$$
$$T_k(A_1(x_1), ..., A_k(x_k)) \star T_{n-k}(A_{k+1}(x_{k+1}), ..., A_n(x_n)) \tag{3.6}$$

if $\{x_1, ..., x_k\} \cap (\{x_{k+1}, ..., x_n\} + \bar{V}_-) = \emptyset$. There is no axiom corresponding to the GLZ Relation. The latter can be interpreted as 'integrability condition' for the 'vector potential' $R(e^S_\otimes, F)$, that is it ensures the existence of the 'potential' $\mathbf{S}(S)$ fulfilling (3.4); for details see [7] and Proposition 2 in [10].

For this paper the following **renormalization conditions** are relevant (besides the MWI).

Translation Invariance: The group $(\mathbb{R}^d, +)$ of space and time translations has an obvious automorphic action β on $\mathcal{F}$, which is determined by $\beta_a \varphi(x) = \varphi(x + a)$, $a \in \mathbb{R}^d$. We require

$$\beta_a \mathbf{S}(S) = \mathbf{S}(\beta_a S) , \quad \forall a \in \mathbb{R}^d . \tag{3.7}$$

Field Independence: $\frac{\delta T}{\delta \varphi(x)} = 0$. This axiom implies the causal Wick expansion of [15] as follows [14]: since $T(\otimes_{j=1}^n F_j) \in \mathcal{F}$ is polynomial in φ, it has a finite Taylor expansion in φ. By using Field Independence, this expansion can be written as

$$T_n(A_1(x_1), \cdots, A_n(x_n)) = \sum_{l_1,...,l_n} \frac{1}{l_1! \cdots l_n!}$$

$$\cdot T_n\Big(\cdots, \sum_{a_{i1}...a_{il_i}} \frac{\partial^{l_i} A_i}{\partial(\partial^{a_{i1}}\varphi) \cdots \partial(\partial^{a_{il_i}}\varphi)}(x_i), \cdots \Big)\Big|_{\varphi=0} \prod_{i=1}^{n} \prod_{j_i=1}^{l_i} \partial^{a_{ij_i}} \varphi(x_i) \tag{3.8}$$

with multi-indices $a_{ij_i} \in \mathbb{N}_0^d$.

Scaling: This requirement uses the *mass dimension of a monomial* in $\mathcal{P}$, which is defined by the conditions

$$\dim(\partial^a \varphi) = \frac{d-2}{2} + |a| \quad \text{and} \quad \dim(A_1 A_2) = \dim(A_1) + \dim(A_2) \tag{3.9}$$

for all monomials $A_1, A_2 \in \mathcal{P}$. The mass dimension of a *polynomial* in $\mathcal{P}$ is the maximum of the mass dimensions of the contributing monomials. We denote by $\mathcal{P}_{\mathrm{hom}}$ the set of all field polynomials which are homogeneous in the mass dimension.

The axiom **Scaling Degree** requires that 'renormalization may not make the interacting fields more singular' (in the UV-region). Usually this is formulated in terms of Steinmann's *scaling degree* [28]:

$$\mathrm{sd}(f) \overset{\mathrm{def}}{=} \inf\{\delta \in \mathbb{R} \mid \lim_{\rho \downarrow 0} \rho^{\delta} f(\rho x) = 0\}, \quad f \in \mathcal{D}'(\mathbb{R}^k) \text{ or } f \in \mathcal{D}'(\mathbb{R}^k \setminus \{0\}). \tag{3.10}$$

Namely, one requires

$$\mathrm{sd}\Big(T(A_1, ..., A_n)|_{\varphi=0}(x_1 - x_n, ...)\Big) \leq \sum_{j=1}^{n} \dim(A_j), \quad \forall A_j \in \mathcal{P}_{\mathrm{hom}}, \tag{3.11}$$

where Translation Invariance is assumed. Notice that this condition restricts *all* coefficients in the causal Wick expansion (3.8).

In the inductive construction of the sequence $(R_{n-1,1})_{n\in\mathbb{N}}$ or $(T_n)_{n\in\mathbb{N}}$, respectively, the problem of renormalization appears as the extension of the coefficients in the causal Wick expansion (which are $\mathbb{C}[[\hbar]]$-valued distributions) from $\mathcal{D}(\mathbb{R}^{d(n-1)} \setminus \{0\})$ to $\mathcal{D}(\mathbb{R}^{d(n-1)})$. This extension has to be done in the sense of formal power series in $\hbar$, that is individually in each order in $\hbar$. With that it holds

$$\lim_{\hbar \to 0} R = R_{\mathrm{cl}}. \tag{3.12}$$

In [14] it is shown that there **exists** a T-product which fulfils all axioms. The **non-uniqueness** of solutions is characterized by the 'Main Theorem'; for a complete version see [14].

4. Proper vertices

A main motivation for introducing proper vertices is to select that part of a T-product for which renormalization is non-trivial (cf. [21]). This is the contribution of all 1-particle-irreducible (1PI) subdiagrams. This selection can be done as follows: first one eliminates all disconnected diagrams. Then, one interprets each connected diagram as tree diagram with *non-local* vertices ('proper vertices') given by the 1PI-subdiagrams. The proper vertices can be interpreted as the 'quantum part' of the Feynman diagrams. Since renormalization is unique and trivial for tree diagrams, Ward identities can equivalently be formulated in terms of proper vertices (Sect. 5.1).

Essentially we follow this procedure, however, we avoid to argue in terms of diagrams, i.e. to use Wick's Theorem. It has been shown in [6] that with our definition (4.6) of the vertex functional Γ the 'proper interaction' $\Gamma(e_{\otimes}^S)$ corresponds to the sum of all 1PI-diagrams of $T(e_{\otimes}^{iS/\hbar})$.

The connected part T^c of a time-ordered T can be defined recursively by [11]

$$T^c_n(\otimes^n_{j=1}F_j) \overset{\text{def}}{=} T_n(\otimes^n_{j=1}F_j) - \sum_{|P|\geq 2}\prod_{J\in P} T^c_{|J|}(\otimes_{j\in J}F_j) \ . \tag{4.1}$$

It follows that T and T^c are related by the linked cluster theorem:

$$T(e^{iF}_\otimes) = \exp_\bullet(T^c(e^{iF}_\otimes)) \ , \tag{4.2}$$

where $\exp_\bullet$ denotes the exponential function with respect to the classical product.

For $F \in \mathcal{F}_{\text{loc}}$ the connected tree part $T^c_{\text{tree},n}(F^{\otimes n})$ can be defined as follows [11]: since $T^c_n = \mathcal{O}(\hbar^{n-1})$, the limit

$$\hbar^{-(n-1)}\, T^c_{\text{tree},n} \overset{\text{def}}{=} \lim_{\hbar\to 0} \hbar^{-(n-1)}\, T^c_n \tag{4.3}$$

exists. This definition reflects the well known statements that T^c_{tree} is the 'classical part' of T^c and that connected loop diagrams are of higher orders in $\hbar$.

Since proper vertices are non-local, we need the connected tree part $T^c_{\text{tree}}(\otimes^n_{j=1}F_j)$ for non-local entries $F_j \in \mathcal{F}$. This can be defined recursively [6]:

$$T^c_{\text{tree}}(\otimes^{n+1}_{j=1}F_j) = \sum_{k=1}^n \int dx_1...dx_k\, dy_1...dy_k\, \frac{\delta^k F_{n+1}}{\delta\varphi(x_1)...\delta\varphi(x_k)}$$

$$\cdot \prod_{j=1}^k \Delta^F_m(x_j - y_j)\, \frac{1}{k!} \sum_{I_1 \sqcup ... \sqcup I_k = \{1,...,n\}} \frac{\delta}{\delta\varphi(y_1)} T^c_{\text{tree}}(\otimes_{j\in I_1}F_j) \cdot ...$$

$$\cdot \frac{\delta}{\delta\varphi(y_k)} T^c_{\text{tree}}(\otimes_{j\in I_k}F_j) \ , \tag{4.4}$$

where $I_j \neq \emptyset \ \forall j$, $\sqcup$ means the disjoint union and Δ^F_m is the Feynman propagator for mass m. (Note that in the sum over $I_1, ..., I_k$ the order of $I_1, ..., I_k$ is distinguished and, hence, there is a factor $\frac{1}{k!}$.) For local entries the two definitions (4.3) and (4.4) of T^c_{tree} agree, as explained in [6].

The 'vertex functional' Γ is defined by the following proposition [6]:

Proposition 4.1. *There exists a* **totally symmetric** *and* **linear** *map*

$$\Gamma \ : \ \mathbb{T}\mathcal{F}_{\text{loc}} \to \mathcal{F} \tag{4.5}$$

which is uniquely determined by

$$T^c(e^{iS/\hbar}_\otimes) = T^c_{\text{tree}}\left(e^{i\Gamma(e^S_\otimes)/\hbar}_\otimes\right) \ . \tag{4.6}$$

To zeroth and first order in S we obtain

$$\Gamma(1) = 0 \ , \quad \Gamma(S) = S \ . \tag{4.7}$$

Since T^c, T^c_{tree} and Γ are linear and totally symmetric, the defining relation (4.6) implies

$$T^c(e^{iS/\hbar}_\otimes \otimes F) = T^c_{\text{tree}}\left(e^{i\Gamma(e^S_\otimes)/\hbar}_\otimes \otimes \Gamma(e^S_\otimes \otimes F)\right) \ . \tag{4.8}$$

To prove the proposition, one constructs $\Gamma(\otimes_{j=1}^n F_j)$ by induction on n, using (4.6) and the requirements total symmetry and linearity:

$$\Gamma(\otimes_{j=1}^n F_j) = (i/\hbar)^{n-1} T^c(\otimes_{j=1}^n F_j) - \sum_{|P|\geq 2} (i/\hbar)^{|P|-1} T_{\text{tree}}^c\left(\bigotimes_{J \in P} \Gamma(\otimes_{j\in J} F_j)\right),$$

(4.9)

where P is a partition of $\{1, ..., n\}$ in $|P|$ subsets J.

From this recursion relation and from $T_n^c - T_{\text{tree},n}^c = \mathcal{O}(\hbar^n)$ we inductively conclude

$$\Gamma(e_\otimes^S) = S + \mathcal{O}(\hbar), \quad \Gamma(e_\otimes^S \otimes F) = F + \mathcal{O}(\hbar) \quad \text{if} \quad F, S \sim \hbar^0. \quad (4.10)$$

Motivated by this relation and (4.6) we call $\Gamma(e_\otimes^S)$ the 'proper interaction' corresponding to the classical interaction S.

The validity of renormalization conditions for T implies corresponding properties of Γ, as worked out in [6].

Analogously to the conventions for R- and T-products we sometimes write $\int dx\, g(x)\, \Gamma(A(x) \otimes F_2...)$ for $\Gamma(A(g) \otimes F_2...)$ ($A \in \mathcal{P}$, $g \in \mathcal{D}(\mathbb{M})$). Since Γ depends only on the *functionals*, it fulfils the AWI: $\partial_x^\mu \Gamma(A(x) \otimes F_2...) = \Gamma(\partial^\mu A(x) \otimes F_2...)$.

5. The Quantum Action Principle

5.1. Formulation of the Master Ward Identity in terms of proper vertices

The classical MWI was derived for arbitrary interaction $S \in \mathcal{F}$ and arbitrary $A \in \mathcal{J}$. For *local* functionals $S \in \mathcal{F}_{\text{loc}}$ and

$$A = \int dx\, h(x)\, Q(x) \frac{\delta S_0}{\delta \varphi(x)} \in \mathcal{J} \cap \mathcal{F}_{\text{loc}}, \quad h \in \mathcal{D}(\mathbb{M}), \quad Q \in \mathcal{P}, \quad (5.1)$$

it can be transferred formally into perturbative QFT (by the replacement $R_{\text{cl}} \to R$), where it serves as an additional, highly non-trivial renormalization condition:

$$R(e_\otimes^S, A + \delta_A S) = \int dy\, h(y) R(e_\otimes^S, Q(y)) \frac{\delta S_0}{\delta \varphi(y)}. \quad (5.2)$$

Since the MWI holds true in classical field theory (i.e. for connected tree diagrams, see below) it is possible to express this renormalization condition in terms of the 'quantum part' (described by the loop diagrams) - that is in terms of proper vertices. We do this in several steps:

Proof of the MWI for T_{tree}^c (connected tree diagrams). Since this is an alternative formulation of the *classical* MWI, we still include *non-local functionals* $S \in \mathcal{F}$, $A = \int dx\, Q(x) \frac{\delta S_0}{\delta \varphi(x)} \in \mathcal{J}$, as in Sect. 2. The classical field equation (2.15) can be expressed in terms of T_{tree}^c:

$$T_{\text{tree}}^c\left(e_\otimes^{iS/\hbar} \otimes \frac{\delta(S_0 + S)}{\delta \varphi(x)}\right) = \frac{\delta S_0}{\delta \varphi(x)}. \quad (5.3)$$

The only difference between R_{cl} and T^c_{tree} is that the retarded propagator $\Delta^{\mathrm{ret}}(y)(\neq \Delta^{\mathrm{ret}}(-y))$ is replaced by the Feynman propagator $\Delta^F(y)(= \Delta^F(-y))$, the combinatorics of the diagrams remains the same. Hence, the factorization of classical fields (2.8),

$$R_{\mathrm{cl}}\left(e^S_\otimes, F \cdot G\right) = R_{\mathrm{cl}}\left(e^S_\otimes, F\right) \cdot R_{\mathrm{cl}}\left(e^S_\otimes, G\right) \tag{5.4}$$

holds true also for T^c_{tree}:

$$T^c_{\mathrm{tree}}\left(e^{iS/\hbar}_\otimes \otimes FG\right) = T^c_{\mathrm{tree}}\left(e^{iS/\hbar}_\otimes \otimes F\right) \cdot T^c_{\mathrm{tree}}\left(e^{iS/\hbar}_\otimes \otimes G\right) . \tag{5.5}$$

We now multiply the field equation for T^c_{tree} with $T^c_{\mathrm{tree}}(e^{iS/\hbar}_\otimes \otimes Q(x))$ and integrate over x. This yields the MWI for T^c_{tree}:

$$T^c_{\mathrm{tree}}\left(e^{iS/\hbar}_\otimes \otimes (A + \delta_A S)\right) = \int dx\, T^c_{\mathrm{tree}}\left(e^{iS/\hbar}_\otimes \otimes Q(x)\right) \cdot \frac{\delta S_0}{\delta\varphi(x)} . \tag{5.6}$$

Translation of the (quantum) MWI from R into T^c. Using Bogoliubov's formula (3.4) and the identity

$$(F \star G) \cdot \frac{\delta S_0}{\delta\varphi} = F \star \left(G \cdot \frac{\delta S_0}{\delta\varphi}\right) \qquad \forall F, G \in \mathcal{F} \tag{5.7}$$

(which relies on $(\Box + m^2)\Delta^+_m = 0$), the MWI in terms of R-products (5.2) can be translated into T-products:

$$T\left(e^{iS/\hbar}_\otimes \otimes (A + \delta_A S)\right) = \int dy\, h(y)\, T\left(e^{iS/\hbar}_\otimes \otimes Q(y)\right)\frac{\delta S_0}{\delta\varphi(y)} , \quad h \in \mathcal{D}(\mathbb{M}), Q \in \mathcal{P} . \tag{5.8}$$

To translate it further into T^c we note that the linked cluster formula (4.2) implies

$$T^c\left(e^{iF}_\otimes \otimes G\right) = T\left(e^{iF}_\otimes\right)^{-1} \cdot T\left(e^{iF}_\otimes \otimes G\right) , \tag{5.9}$$

where the inverse is meant with respect to the classical product. It exists because $T\left(e^{iF}_\otimes\right)$ is a formal power series of the form $T\left(e^{iF}_\otimes\right) = 1 + \mathcal{O}(F)$. With that we conclude that the MWI can equivalently be written in terms of T^c by replacing T by T^c on both sides of (5.8).

Translation of the MWI from T^c into Γ. Applying (4.6) on both sides of the MWI in terms of T^c we obtain

$$\int dy\, h(y)\, T^c_{\mathrm{tree}}\left(e^{i\Gamma(e^S_\otimes)/\hbar}_\otimes \otimes \Gamma\left(e^S_\otimes \otimes Q(y)\frac{\delta(S_0 + S)}{\delta\varphi(y)}\right)\right)$$

$$= \int dy\, h(y) T^c_{\mathrm{tree}}\left(e^{i\Gamma(e^S_\otimes)/\hbar}_\otimes \otimes \Gamma\left(e^S_\otimes \otimes Q(y)\right)\right)\frac{\delta S_0}{\delta\varphi(y)}$$

$$= \int dy\, h(y)\, T^c_{\mathrm{tree}}\left(e^{i\Gamma(e^S_\otimes)/\hbar}_\otimes \otimes \Gamma(e^S_\otimes \otimes Q(y))\frac{\delta(S_0 + \Gamma(e^S_\otimes))}{\delta\varphi(y)}\right) ,$$

where we have used the classical MWI in terms of T^c_{tree} (5.6). It follows

$$\Gamma(e^S_\otimes \otimes Q(y))\frac{\delta(S_0 + \Gamma(e^S_\otimes))}{\delta\varphi(y)} = \Gamma\left(e^S_\otimes \otimes Q(y)\frac{\delta(S_0 + S)}{\delta\varphi(y)}\right) . \tag{5.10}$$

The various formulations of the MWI, in terms of R-products (5.2), T-products (5.8), T^c-products and in terms of proper vertices (5.10), they all are equivalent.

Remark 5.1. The off-shell field equation

$$T\left(e_\otimes^{iS/\hbar} \otimes \frac{\delta(S_0 + S)}{\delta\varphi(y)}\right) = \frac{\delta S_0}{\delta\varphi(y)} \cdot T\left(e_\otimes^{iS/\hbar}\right), \quad \forall S , \tag{5.11}$$

is a further renormalization condition, which can equivalently be expressed by

$$\Gamma(e_\otimes^S \otimes \varphi(y)) = \varphi(y) , \quad \forall S , \tag{5.12}$$

as shown in [6]. For a T-product satisfying this condition and for $Q = D\varphi$ (where D is a polynomial in partial derivatives) the QAP simplifies to

$$D\varphi(y) \frac{\delta(S_0 + \Gamma(e_\otimes^S))}{\delta\varphi(y)} = \Gamma\left(e_\otimes^S \otimes D\varphi(y) \frac{\delta(S_0 + S)}{\delta\varphi(y)}\right) . \tag{5.13}$$

5.2. The anomalous Master Ward Identity - Quantum Action Principle

The QAP is a statement about the structure of all possible violations of Ward identities. In our framework the main statement of the QAP is that any term violating the MWI can be expressed as $\Gamma(e_\otimes^S \otimes \Delta)$, where Δ is local (in a stronger sense than only $\Delta \in \mathcal{F}_{\mathrm{loc}}$) and $\Delta = \mathcal{O}(\hbar)$ and the mass dimension of Δ is bounded in a suitable way.

Theorem 5.2 (Quantum Action Principle). *(a) Let Γ be the vertex functional belonging to a time ordered product satisfying the basic axioms and Translation Invariance (3.7). Then there exists a unique sequence of* **linear** *maps $(\Delta^n)_{n\in\mathbb{N}}$,*

$$\Delta^n : \mathcal{P}^{\otimes(n+1)} \to \mathcal{D}'(\mathbb{M}, \mathcal{F}_{\mathrm{loc}}) , \quad \otimes_{j=1}^n L_j \otimes Q \mapsto \Delta^n(\otimes_{j=1}^n L_j(x_j); Q(y)) \tag{5.14}$$

($\mathcal{D}'(\mathbb{M}, \mathcal{F}_{\mathrm{loc}})$ is the space of $\mathcal{F}_{\mathrm{loc}}$-valued distributions on $\mathcal{D}(\mathbb{M})$), which are symmetric in the first n factors,

$$\Delta^n(\otimes_{j=1}^n L_{\pi j}(x_{\pi j}); Q(y)) = \Delta^n(\otimes_{j=1}^n L_j(x_j); Q(y)) \tag{5.15}$$

for all permutations π, and which are implicitly defined by the 'anomalous MWI'

$$\Gamma(e_\otimes^S \otimes Q(y)) \frac{\delta(S_0 + \Gamma(e_\otimes^S))}{\delta\varphi(y)} = \Gamma\left(e_\otimes^S \otimes \left(Q(y) \frac{\delta(S_0 + S)}{\delta\varphi(y)} + \Delta(L; Q)(g; y)\right)\right) , \tag{5.16}$$

where $S = L(g)$ ($L \in \mathcal{P}$, $g \in \mathcal{D}(\mathbb{M})$) and

$$\Delta(L; Q)(g; y) := \sum_{n=0}^\infty \frac{1}{n!} \int dx_1...dx_n \prod_{j=1}^n g(x_j) \Delta^n(\otimes_{j=1}^n L(x_j); Q(y)) . \tag{5.17}$$

As a consequence of (5.16) the maps Δ^n have the following properties:

(i) $\Delta^0 = 0$;

(ii) *locality: there exist* **linear** *maps* $P_a^n : \mathcal{P}^{\otimes(n+1)} \to \mathcal{P}$ *(where a runs through a finite subset of $(\mathbb{N}_0^d)^n$), which are symmetric in the first n factors, such that Δ^n can be written as*

$$\Delta^n(\otimes_{j=1}^n L_j(x_j); Q(y)) = \sum_{a \in (\mathbb{N}_0^d)^n} \partial^a \delta(x_1 - y, ..., x_n - y) \, P_a^n(\otimes_{j=1}^n L_j; Q)(y) . \quad (5.18)$$

(iii) $\Delta^n(\otimes_{j=1}^n L_j(x_j); Q(y)) = \mathcal{O}(\hbar) \quad \forall n > 0 \text{ if } L_j \sim \hbar^0, \ Q \sim \hbar^0 .$

(b) If the time ordered product satisfies the renormalization conditions Field Independence and Scaling Degree (3.11), then each term on the r.h.s. of (5.18) fulfils

$$|a| + \dim(P_a^n(\otimes_{j=1}^n L_j; Q)) \leq \sum_{j=1}^n \dim(L_j) + \dim(Q) + \frac{d+2}{2} - d\,n . \quad (5.19)$$

For a renormalizable interaction (that is $\dim(L) \leq d$) this implies

$$|a| + \dim(P_a^n(L^{\otimes n}; Q)) \leq \dim(Q) + \frac{d+2}{2} . \quad (5.20)$$

Note that (5.16) differs from the MWI (5.10) only by the local term $\Delta(L; Q)(g; y)$, which clearly depends on the chosen normalization of the time ordered product. Therefore, $\Delta(L; Q)(g; y) = 0$ is a sufficient condition for the validity of the MWI for Q and $S = L(g)$; it is also necessary due to the uniqueness of the maps Δ^n.

Proof. (a) Proceeding as in Sect. 5.1, the defining relation (5.16) can equivalently be written in terms of T-products:

$$T\left(e_\otimes^{iS/\hbar} \otimes \left(Q(y)\frac{\delta(S_0 + S)}{\delta\varphi(y)} + \Delta(L; Q)(g; y)\right)\right) = T\left(e_\otimes^{iS/\hbar} \otimes Q(y)\right)\frac{\delta S_0}{\delta\varphi(y)} . \quad (5.21)$$

To n-th order in g this equation reads

$$\Delta^n(L^{\otimes n}; Q(y))(g^{\otimes n}) = T\left((iS/\hbar)^{\otimes n} \otimes Q(y)\right) \cdot \frac{\delta S_0}{\delta\varphi(y)} - T\left((iS/\hbar)^{\otimes n} \otimes Q(y)\frac{\delta S_0}{\delta\varphi(y)}\right)$$

$$- n\,T\left((iS/\hbar)^{\otimes n-1} \otimes Q(y)\frac{\delta S}{\delta\varphi(y)}\right) - \sum_{l=0}^{n-1} \binom{n}{l} T\left((iS/\hbar)^{\otimes n-l} \otimes \Delta^l(L^{\otimes l}; Q(y))(g^{\otimes l})\right).$$

$$(5.22)$$

Taking linearity and symmetry (5.15) into account we extend this relation to non-diagonal entries and write it in terms of the distributional kernels

$$\Delta^n(\otimes_{j=1}^n L_j(x_j); Q(y)) = \left(\frac{i}{\hbar}\right)^n T\left(\otimes_{j=1}^n L_j(x_j) \otimes Q(y)\right) \cdot \frac{\delta S_0}{\delta\varphi(y)}$$

$$-\left(\frac{i}{\hbar}\right)^n T\left(\otimes_{j=1}^n L_j(x_j) \otimes Q(y) \cdot \frac{\delta S_0}{\delta\varphi(y)}\right)$$

$$-\sum_{l=1}^n \left(\frac{i}{\hbar}\right)^{n-1} T\left(\otimes_{j(\neq l)} L_j(x_j) \otimes Q(y) \sum_a (\partial^a \delta)(x_l - y) \frac{\partial L_l}{\partial(\partial^a \varphi)}(x_l)\right)$$

$$-\sum_{I \subset \{1,\dots,n\}, \, I^c \neq \emptyset} \left(\frac{i}{\hbar}\right)^{|I^c|} T\left(\otimes_{i \in I^c} L_i(x_i) \otimes \Delta^{|I|}(\otimes_{j \in I} L_j(x_j); Q(y))\right) \qquad (5.23)$$

This relation gives a unique inductive construction of the sequence $(\Delta^n)_{n \in \mathbb{N}}$ (if the distribution on the r.h.s. of (5.23) takes values in $\mathcal{F}_{\text{loc}}$) and it gives also the initial value $\Delta^0 = 0$. Obviously, the so obtained maps $\Delta^n : \mathcal{P}^{\otimes(n+1)} \to \mathcal{D}'(\mathbb{M}, \mathcal{F}_{\text{loc}})$ are linear and symmetric (5.15).

The main task is to prove that $\Delta^n(\otimes_{j=1}^n L_j; Q)$ (which is defined inductively by (5.23)) satisfies locality (5.18); the latter implies that $\Delta^n(\otimes_{j=1}^n L_j; Q)$ takes values in $\mathcal{F}_{\text{loc}}$. For this purpose we first prove

$$\text{supp } \Delta^n(\otimes_{j=1}^n L_j; Q) \subset \mathbb{D}_{n+1} \overset{\text{def}}{=} \{(x_1, \dots, x_{n+1}) \in \mathbb{M}^{n+1} \, | \, x_1 = \cdots = x_{n+1}\},$$
$$(5.24)$$

that is we show that the r.h.s. of (5.23) vanishes for $(x_1, \dots, x_n, y) \notin \mathbb{D}_{n+1}$. For such a configuration there exists a $K \subset \{1, \dots, n\}$ with $K^c := \{1, \dots, n\} \setminus K \neq \emptyset$ and either $(\{x_k \, | \, k \in K^c\} + \bar{V}_+) \cap (\{x_j \, | \, j \in K\} \cup \{y\}) = \emptyset$ or $(\{x_k \, | \, k \in K^c\} + \bar{V}_-) \cap (\{x_j \, | \, j \in K\} \cup \{y\}) = \emptyset$. We treat the first case, the second case is completely analogous. Using causal factorization of the T-products (3.6) and locality (5.24) of the inductively known $\Delta^{|I|}$, $|I| < n$, we write the r.h.s. of (5.23) as

$$\left(\frac{i}{\hbar}\right)^n \left(T\left(\otimes_{j \in K^c} L_j(x_j)\right) \star T\left(\otimes_{i \in K} L_i(x_i) \otimes Q(y)\right)\right) \frac{\delta S_0}{\delta\varphi(y)}$$

$$-T\left(\otimes_{j \in K^c} L_j(x_j)\right) \star \left[\left(\frac{i}{\hbar}\right)^n T\left(\otimes_{i \in K} L_i(x_i) \otimes Q(y) \frac{\delta S_0}{\delta\varphi(y)}\right)\right.$$

$$+\left(\frac{i}{\hbar}\right)^{n-1} \sum_{l \in K} T\left(\otimes_{i \in K, i \neq l} L_i(x_i) \otimes Q(y) \sum_a (\partial^a \delta)(x_l - y) \frac{\partial L_l}{\partial(\partial^a \varphi)}(x_l)\right)$$

$$+\left.\left(\frac{i}{\hbar}\right)^{|K^c|+|K\setminus I|} \sum_{I \subset K} T\left(\otimes_{i \in K\setminus I} L_i(x_i) \otimes \Delta^{|I|}(\otimes_{s \in I} L_s(x_s); Q(y))\right)\right]. \qquad (5.25)$$

Using (5.7) this can be written in the form $T(\otimes_{j \in K^c} L_j(x_j)) \star (\dots)$. The second factor vanishes due to the validity of (5.23) in order $|K|$. This proves (5.24).

$\Delta^n(\otimes_{j=1}^n L_j; Q)$ is, according to its inductive definition (5.23), a distribution on $\mathcal{D}(\mathbb{M}^{n+1})$ which takes values in $\mathcal{F}$. Hence, it is of the form

$$\Delta^n(\otimes_{j=1}^n L_j(x_j); Q(y)) = \sum_k \int dz_1...dz_k$$

$$f_k^n(\otimes_{j=1}^n L_j \otimes Q)(x_1, ..., x_n, y, z_1, ..., z_k) \, \varphi(z_1)...\varphi(z_k) \, , \qquad (5.26)$$

where $f_k^n(\otimes_{j=1}^n L_j \otimes Q)(x_1, ..., x_n, y, z_1, ..., z_k) \in \mathcal{D}'(\mathbb{M}^{n+k+1})$ has the following properties:

- it depends linearly on $(\otimes_{j=1}^n L_j \otimes Q)$;
- it is invariant under permutations of the pairs $(L_1, x_1), ..., (L_n, x_n)$.
- The distribution
$\int dx_1...dx_n dy \, f_k^n(\otimes_{j=1}^n L_j \otimes Q)(x_1, ..., x_n, y, z_1, ..., z_k) \, h(x_1, ..., x_n, y) \in \mathcal{D}'(\mathbb{M}^k)$ is symmetric under permutations of $z_1, ..., z_k$ and satisfies the wave front set condition (2.3), for all $h \in \mathcal{D}(\mathbb{M}^{n+1})$.
- From (5.23) we see that Translation Invariance of the T-product (3.7) implies the same property for Δ^n:

$$\beta_a \, \Delta^n(\otimes_{j=1}^n L_j(x_j); Q(y)) = \Delta^n(\otimes_{j=1}^n L_j(x_j + a); Q(y + a)) \, . \qquad (5.27)$$

Therefore, the distributions $f_k^n(\otimes_{j=1}^n L_j \otimes Q)$ depend only on the relative coordinates.

Due to (5.24) the support of $f_k^n(\otimes_{j=1}^n L_j \otimes Q)$ is contained in $\mathbb{D}_{n+1} \times \mathbb{M}^k$; but, to obtain the assertion (5.18), we have to show $\operatorname{supp} f_k^n(\otimes_{j=1}^n L_j \otimes Q) \subset \mathbb{D}_{n+k+1}$. For this purpose we take into account that

$$\frac{\delta \, T(\otimes_{j=1}^l A_j(x_j))}{\delta\varphi(z)} = 0 \quad \text{if} \quad z \neq x_j \,\, \forall j = 1, ..., l \, . \qquad (5.28)$$

This relation can be shown as follows: for the restriction of the time ordered product to $\mathcal{D}(\mathbb{M}^l \setminus \mathbb{D}_l)$ this property is obtained inductively by causal factorization (3.6). That (5.28) is maintained in the extension of the T-product to $\mathcal{D}(\mathbb{M}^l)$ can be derived from

$$[T(\otimes_{j=1}^l A_j(x_j)), \, \varphi(z)]_\star = 0 \quad \text{if} \quad (x_j - z)^2 < 0 \quad \forall j = 1, ..., l \, , \qquad (5.29)$$

which is a consequence of the causal factorization of $T(\varphi(z) \otimes \otimes_{j=1}^l A_j(x_j))$ (cf. Sect. 3 of [15]).

Applying (5.28) to the T-products on the r.h.s. of (5.23) and using (5.24), we conclude

$$\operatorname{supp} \frac{\delta \, \Delta^n(\otimes_{j=1}^n L_j; Q)}{\delta\varphi} \subset \mathbb{D}_{n+2} \, . \qquad (5.30)$$

It follows that the distributions $f_k^n(\otimes_{j=1}^n L_j \otimes Q)$ (5.26) have support on the total diagonal $\mathbb{D}_{n+k+1}$. Taking additionally Translation Invariance into account, we conclude that these distributions are of the form

$$f_k^n(\otimes_{j=1}^n L_j \otimes Q)(x_1, ..., x_n, y, z_1, ..., z_k) = \sum_{a,b} C_{a,b}(\otimes_{j=1}^n L_j \otimes Q)$$

$$\partial^a \delta(x_1 - y, ..., x_n - y)\, \partial^b \delta(z_1 - y, ..., z_k - y)\,, \tag{5.31}$$

where the coefficients $C_{a,b}(\otimes_{j=1}^n L_j \otimes Q) \in \mathbb{C}$ depend linearly on $(\otimes_{j=1}^n L_j \otimes Q)$ and are symmetric in the first n factors. Inserting (5.31) into (5.26) we obtain (5.18), the corresponding maps P_a^n having the asserted properties.

The important property (iii) is obtained by taking the classical limit $\hbar \to 0$ of the anomalous MWI (5.16): using (4.10) it results $\lim_{\hbar \to 0} \Delta(L; Q)(g; y) = 0$.

(b) The statement (5.19) is a modified version of Proposition 10(ii) in [6]. It follows from the formulas ([6]-5.32-33) and ([6]-5.46-47) of that paper. Namely, by using the causal Wick expansion of Δ^n (which follows from the Field Independence of the T-product) and (5.24) it is derived in ([6]-5.32-33) that Δ^n is of the form

$$\Delta^n(\otimes_{j=1}^n L_j(x_j); Q(y)) = \sum_{\mathbf{l,a,b}} C_{\mathbf{a},b}^{\mathbf{l}}(\partial^b \delta)(x_1 - y, ..., x_n - y)$$

$$\cdot \prod_{i=1}^n \prod_{j_i=1}^{l_i} \left(\partial^{a_{ij_i}} \varphi(x_i)\right) \cdot \prod_{j=1}^{l} \partial^{a_j} \varphi(y)$$

$$= \sum_{\mathbf{l,a,b}} \sum_{d \leq b} \tilde{C}_{\mathbf{a},b,d}^{\mathbf{l}}(\partial^d \delta)(x_1 - y, ..., x_n - y)$$

$$\cdot \prod_{i=1}^n \left(\partial^{b_i - d_i} \prod_{j_i=1}^{l_i} \left(\partial^{a_{ij_i}} \varphi(y)\right)\right) \cdot \prod_{j=1}^{l} \partial^{a_j} \varphi(y)\,, \tag{5.32}$$

where $\mathbf{l} \equiv (l_1, ..., l_n; l)$, $\mathbf{a} \equiv (a_{11}, ..., a_{1l_1}, ..., a_{n1}, ..., a_{nl_n}; a_1...a_l)$ and $C_{\mathbf{a},b}^{\mathbf{l}}$, $\tilde{C}_{\mathbf{a},b,d}^{\mathbf{l}}$ are numerical coefficients which depend also on $(L_1, ..., L_n, Q)$. Since the T-product satisfies the axiom Scaling Degree the range of b is bounded by ([6]-5.46). The l.h.s. of (5.19) is given by

$$|d| + |b - d| + \sum_{i=1}^n \sum_{j_i=1}^{l_i} \left(|a_{ij_i}| + \frac{d-2}{2}\right) + \sum_{j=1}^{l} \left(|a_j| + \frac{d-2}{2}\right)\,, \tag{5.33}$$

which agrees with the l.h.s. of ([6]-5.47). Hence, it is bounded by the r.h.s. of ([6]-5.47).   □

Remark 5.3. Since the T-product $T(F^{\otimes n})$ depends only on the (local) functional F and not on how F is written as $F = \sum_k \int dx\, g_k(x)\, P_k(x)$ $(g_k \in \mathcal{D}(\mathbb{M})$, $P_k \in \mathcal{P})$, we conclude from (5.23) that we may express the violating term $\Delta(L; Q)(g; y)$ as follows: given $A = \int dx\, h(x)\, Q(x)\, \delta S_0/\delta\varphi(x)$ $(h \in \mathcal{D}(\mathbb{M})$, $Q \in \mathcal{P})$, there exists a linear and symmetric map $\Delta_A : \mathbb{T}\mathcal{F}_{\mathrm{loc}} \to \mathcal{F}_{\mathrm{loc}}$ which is uniquely determined by

$$\Delta_A(e_{\otimes}^{L(g)}) \stackrel{\text{def}}{=} \int dy\, h(y)\, \Delta(L; Q)(g; y)\,. \tag{5.34}$$

A glance at (5.23) shows that Δ_A depends linearly on A. The corresponding smeared out version of the QAP is given in [6].

We are now going to reformulate our version of the QAP (Theorem 5.2) in the form given in the literature. Motivated by (4.10), we interpret $\Gamma_{\text{tot}}(S_0, S) \overset{\text{def}}{=} S_0 + \Gamma(e_\otimes^S)$ as the *proper total action* associated with the classical action $S_{\text{tot}} = S_0 + S$. For $P \in C^\infty(\mathbb{M}, \mathcal{P})$ the 'insertion' of $P(x)$ into $\Gamma_{\text{tot}}(S_0, S)$ is denoted and defined by[4]

$$P(x) \cdot \Gamma_{\text{tot}}(S_0, S) \overset{\text{def}}{=} \frac{\delta}{\delta\rho(x)}\Big|_{\rho\equiv 0} \Gamma_{\text{tot}}\left(S_0, S + \int dx \rho(x) P(x)\right) = \Gamma\left(e_\otimes^S \otimes P(x)\right), \tag{5.35}$$

where $\rho \in \mathcal{D}(\mathbb{M})$ is an 'external field'. Setting $S' \overset{\text{def}}{=} S + \int dx \rho(x) Q(x)$ and introducing the local field

$$\Delta(x) \overset{\text{def}}{=} Q(x) \frac{\delta(S_0 + S)}{\delta\varphi(x)} + \Delta(L; Q)(g; x) \in C^\infty(\mathbb{M}, \mathcal{P}), \tag{5.36}$$

the anomalous MWI (5.16) can be rewritten as

$$\frac{\delta\Gamma_{\text{tot}}(S_0, S')}{\delta\rho(x)} \frac{\delta\Gamma_{\text{tot}}(S_0, S')}{\delta\varphi(x)}\Big|_{\rho\equiv 0} = \Delta(x) \cdot \Gamma_{\text{tot}}(S_0, S). \tag{5.37}$$

The $\hbar$-expansion of the right-hand side starts with

$$\Delta(x) \cdot \Gamma_{\text{tot}}(S_0, S) = Q(x) \frac{\delta(S_0 + S)}{\delta\varphi(x)} + \mathcal{O}(\hbar) \equiv \frac{\delta(S_0 + S')}{\delta\rho(x)} \frac{\delta(S_0 + S')}{\delta\varphi(x)}\Big|_{\rho=0} + \mathcal{O}(\hbar), \tag{5.38}$$

where (4.10) is used. To discuss the mass dimension of the local insertion Δ (5.36), we assume that there is an open region $\emptyset \neq \mathcal{U} \subset \mathbb{M}$ such that the test function g which switches the interaction is constant in $\mathcal{U}$: $g|_\mathcal{U} = $ constant. For $x \in \mathcal{U}$ the insertion $\Delta(x)$ is a field polynomial with constant coefficients. By $\dim(\Delta)$ we mean the mass dimension of this polynomial. For a renormalizable interaction Theorem 5.2(b) implies

$$\dim(\Delta) \leq \dim(Q) + \frac{d+2}{2} = \dim(Q) - \dim(\varphi) + d. \tag{5.39}$$

This version (5.37)-(5.39) of the QAP, which we have proved in the framework of CPT, formally agrees with the literature, namely with the 'QAP for nonlinear variations of the fields' (formulas (3.82)-(3.83) in [26]). This is the most important and most difficult case of the QAP.

As explained in (2.15), the MWI reduces for $Q = 1$ to the off-shell field equation. Setting $Q = 1$ in (5.37)-(5.39) and using $\Gamma(e_\otimes^S \otimes 1) = 1$, we obtain $\delta\Gamma_{\text{tot}}(S_0, S)/\delta\varphi(x) = \Delta(x) \cdot \Gamma_{\text{tot}}(S_0, S)$, where $\Delta(x) \cdot \Gamma_{\text{tot}}(S_0, S) = \delta(S_0 + S)/\delta\varphi(x) + \mathcal{O}(\hbar)$ and $\dim(\Delta) \leq d - \dim(\varphi)$, which formally agrees with formulas (3.80)–(3.81) in [26]. The latter are called there the 'QAP for the equations of motion', as expected from (2.15).

---

[4]The dot does not mean the classical product here!

Remark 5.4. An 'insertion' (5.35) being a rather technical notion, the violating term $\Gamma\left(e_\otimes^S \otimes \Delta(L; Q)(g; y)\right)$ in the anomalous MWI (5.16) can be much better interpreted by writing (5.16) in terms of R-products:

$$R\left(e_\otimes^S \otimes Q(y)\frac{\delta(S_0 + S)}{\delta\varphi(y)}\right) + R(e_\otimes^S \otimes \Delta(L; Q)(g; y)) = R(e_\otimes^S \otimes Q(y))\frac{\delta S_0}{\delta\varphi(y)} . \quad (5.40)$$

In this form, the violating term $R(e_\otimes^S \otimes \Delta(L; Q)(g; y))$ is the *interacting field* to the interaction S and belonging to the local field $\Delta(L; Q)(g; y)$.

6. Algebraic renormalization

In this section we sketch, for the *non-expert reader*, the crucial role of the QAP in algebraic renormalization. For shortness, we strongly simplify.

In algebraic renormalization one investigates, whether violations of Ward identities can be removed by finite renormalizations of the T-products. The results about the structure of the violating term given by the QAP are used as follows.

- Algebraic renormalization starts with the anomalous MWI (5.16), that is the result that the MWI can be violated only by an insertion term, i.e. a term of the form $\Gamma\left(e_\otimes^S \otimes \Delta\right)$ for some $\Delta \in \mathcal{F}_{\text{loc}}$, cf. (5.40).
- Algebraic renormalization proceeds by induction on the order of $\hbar$. To start the induction one uses that $\Delta \equiv \Delta(L; Q)(g; y)$ is of order $\mathcal{O}(\hbar)$.
- Because the finite renormalization terms, which one may add to a T-product, must be *local* (in the strong sense of (5.18)) and compatible with the axiom Scaling Degree, it is of crucial importance that $\Delta(L; Q)(g; y)$ satisfies locality (5.18) and the bound (5.19) on its mass dimension.

For many Ward identities it is possible to derive a *consistency equation* for $\Delta(L; Q)(g; y)$. Frequently this equation can be interpreted as the statement that $\Delta(L; Q)(g; y)$ is a *cocycle* in the cohomology generated by the corresponding symmetry transformation δ acting on some space $\mathcal{K} \subset \mathcal{F}_{\text{loc}}$. For example, δ is a nilpotent derivation (as the BRST-transformation[5]) or a family of derivations $(\delta_a)_{a=1,...,N}$ fulfilling a Lie algebra relation $[\delta_a, \delta_b] = f_{abc}\delta_c$.

If the cocycle $\Delta(L; Q)(g; y)$ is a *coboundary*, it is usually possible to remove this violating term by a finite renormalization. Hence, in this case, the solvability of the considered Ward identity amounts to the question whether this cohomology is trivial. For a renormalizable interaction the bound (5.19) on the mass dimension makes it possible to reduce the space $\mathcal{K}$ to a finite dimensional space, this simplifies the cohomological question enormously.

Many examples for this pattern are given in [26]. In the framework of CPT the QAP and its application in algebraic renormalization have been used to prove the Ward identities of the $O(N)$ scalar field model [6] (as a simple example to

---

[5]The cohomological structure of BRST-symmetry is much richer as mentioned here, see [17].

illustrate how algebraic renormalization works in CPT) and, much more relevant, BRST-symmetry of Yang-Mills fields in curved space-times [20].

Acknowledgement

We very much profitted from discussions with Klaus Fredenhagen. We are grateful also to Raymond Stora for valuable and detailed comments. While writing this paper, M.D. was a guest of the Institute for Theoretical Physics of the University of Göttingen, he thanks for hospitality.

References

[1] G. Barnich, F. Brandt and M. Henneaux, *Local BRST cohomology in gauge theories.* Phys. Rept. **338** (2000), 439–569.

[2] C. Becchi, A. Rouet and R. Stora, *Renormalization of the abelian Higgs-Kibble Model.* Commun. Math. Phys. **42** (1975), 127–162

[3] C. Becchi, A. Rouet and R. Stora, *Renormalization of gauge theories.* Annals Phys. **98** (1976), 287–321.

[4] N.N. Bogoliubov and D. V. Shirkov, *Introduction to the Theory of Quantized Fields.* Interscience Publishers, Inc., New York (1959)

[5] P. Breitenlohner and D. Maison, *Dimensional renormalization and the action principle.* Commun. Math. Phys. **52** (1977), 11–38

[6] F. Brennecke and M. Dütsch, *Removal of violations of the Master Ward Identity in perturbative QFT.* Rev. Math. Phys. ... (2008), 100–120.

[7] R. Brunetti, M. Dütsch and K. Fredenhagen, *Retarded products versus time-ordered products: a geometrical interpretation*, work in preparation

[8] R. Brunetti and K. Fredenhagen, *Microlocal analysis and interacting quantum field theories: Renormalization on physical backgrounds.* Commun. Math. Phys. **208** (2000), 623–661

[9] M. Dütsch and F.-M. Boas, *The Master Ward identity.* Rev. Math. Phys. **14** (2002), 977–1049.

[10] M. Dütsch and K. Fredenhagen, *A local (perturbative) construction of observables in gauge theories: The example of QED.* Commun. Math. Phys. **203** (1999), 71–105.

[11] M. Dütsch and K. Fredenhagen, *Algebraic quantum field theory, perturbation theory, and the loop expansion.* Commun. Math. Phys. **219** (2001), 5–30.

[12] M. Dütsch and K. Fredenhagen, *Perturbative algebraic field theory, and deformation quantization.* Fields Institute Communications **30** (2001), 151–160

[13] M. Dütsch and K. Fredenhagen, *The Master Ward identity and generalized Schwinger-Dyson equation in classical field theory.* Commun. Math. Phys. **243** (2003), 275–314.

[14] M. Dütsch and K. Fredenhagen, *Causal perturbation theory in terms of retarded products, and a proof of the Action Ward identity.* Rev. Math. Phys. **16** (2004), 1291–1348.

[15] H. Epstein and V. Glaser, *The role of locality in perturbation theory.* Ann. Inst. H. Poincaré **A19** (1973), 211.

[16] R. Haag, *Local Quantum Physics: Fields, Particles and Algebras.* 2nd edn., Springer-Verlag, (1996)

[17] M. Henneaux and C. Teitelboim, *Quantization of gauge systems.* Princeton University Press, Princeton, New Jersey (1992), 1–520.

[18] S. Hollands and R.M. Wald, *Existence of local covariant time-ordered products of quantum fields in curved spacetimes.* Commun. Math. Phys. **231** (2002), 309–345

[19] S. Hollands and R.M. Wald, *Conservation of the stress tensor in interacting quantum field theory in curved spacetimes.* Rev. Math. Phys. **17** (2005), 227–312

[20] S. Hollands, *Renormalized Quantum Yang-Mills Fields in Curved Spacetime.* gr-qc/0705.3340 (2007)

[21] G. Jona-Lasinio, *Relativistic Field Theories with Symmetry-Breaking Solutions.* Nuovo Cimento **34** (1964), 1790–1795

[22] Y.-M. P. Lam, *Perturbation Lagrangian theory for scalar fields: Ward-Takahasi identity and current algebra.* Phys. Rev. **D6** (1972), 2145–2161

[23] J. H. Lowenstein, *Differential vertex operations in Lagrangian field theory.* Commun. Math. Phys. **24** (1971), 1–21

[24] R.E. Peierls, *The commutation laws of relativistic field theory.* Proc. Roy. Soc. (London) **A214** (1952), 143.

[25] O. Piguet and A. Rouet,*Symmetries in perturbtaive quantum field theory.* Phys. Rept. **76** (1981), 1–77

[26] O. Piguet and S. P. Sorella, *Algebraic renormalization: Perturbative renormalization, symmetries and anomalies.* Lect. Notes Phys. **M28** (1995), 1–134.

[27] G. Pinter, *The Action Principle in Epstein Glaser Renormalization and Renormalization of the S Matrix of ϕ^4-Theory.* hep-th/9911063

[28] O. Steinmann, *Perturbative Expansion in axiomatic field theory.* Lect. Notes Phys. **11** (1971)

[29] R. Stora, *Pedagogical Experiments in Renormalized Perturbation Theory.* Contribution to Conference 'Theory of Renormalization and Regularization', Hesselberg, Germany, http://wwwthep.physik.uni-mainz.de/~scheck (2002)

[30] R. Stora, *The Wess Zumino consistency condition: a paradigm in renormalized perturbation theory.* Fortsch. Phys. **54** (2006), 175–182

[31] I. V. Tyutin, *Gauge invariance in field theory and statistical physics in operator formalism.* LEBEDEV-75-39

[32] W. Zimmermann, Commun. Math. Phys. **15** (1969), 208

Ferdinand Brennecke
Institut für Quantenelektronik
ETH Zürich
CH–8093 Zürich
e-mail: brennecke@phys.ethz.ch

Michael Dütsch
Institut für Theoretische Physik
Universität Zürich
CH–8057 Zürich
e-mail: duetsch@physik.unizh.ch

Quantum Field Theory
B. Fauser, J. Tolksdorf and E. Zeidler, Eds., 197–216

Plane Wave Geometry and Quantum Physics

Matthias Blau

Abstract. I explain how the Lewis–Riesenfeld exact treatment of the time-dependent quantum harmonic oscillator can be understood in terms of the geodesics and isometries of a plane wave metric, and I show how a curious equivalence between two classes of Yang-Mills actions can be traced back to the transformation relating plane waves in Rosen and Brinkmann coordinates.

Mathematics Subject Classification (2000). Primary 81T30; Secondary 83E30; 81T13; 83C35.

Keywords. Rosen and Brinkmann coordinates, plane wave metric, gravitational waves, matrix string theory, harmonic oscillator, Yang-Mills theory, Heisenberg isometry algebra, duality in string theory.

1. Introduction

The characteristic interplay of geometry and gauge theory in string theory has led to many new and exciting developments in recent years, in particular to progress in the understanding of certain strongly coupled quantum field theories. However, since string theorists were (regrettably) absent from the list of speakers at this conference dedicated to recent developments in quantum field theory, I decided to talk about a subject on the interface of geometry and quantum physics that is only loosely inspired by, and not strictly dependent upon, string theory.[1]

Thus, as an embryonic example of the interplay between geometry and quantum physics in string theory (an example that requires neither knowledge nor appreciation of string theory, but also does not do justice to the depth and richness of these ideas in the string theory context), I will explain the relation between

---

This work has been supported by the Swiss National Science Foundation and by the EU under contract MRTN-CT-2004-005104.
[1]In retrospect this was a wise choice, because of the hostile attitude towards string theory at this, in all other respects very charming and enjoyable, meeting, expressed in particular by some of the members of the senior pontificating classes.

some geometric properties of plane wave space-time metrics on the one hand and some corresponding statements about quantum (gauge) theories on the other.

In section 2, I briefly review some of the basic and entertaining features of the geometry of plane wave metrics. In particular, I emphasize the ubiquitous and multifaceted role of the time-dependent harmonic oscillator in this context, which appears in the geodesic equations, in the description of the Heisenberg isometry algebra of plane wave metrics, and in the coordinate transformation between the two standard (Rosen and Brinkmann) coordinate systems for these metrics.

The first application I will discuss is then naturally to the quantum theory of time-dependent harmonic oscillators (section 3). In general one can quantize these systems exactly using the powerful Lewis–Riesenfeld method of invariants. Embedding the problem of a time-dependent harmonic oscillator into the plane wave setting equips it with a rich geometric structure, and links the dynamics of the harmonic oscillator to the conserved charges associated with the isometries. I will show that this provides a natural geometric explanation of the entire Lewis–Riesenfeld procedure.

As a second application, I will discuss a curious equivalence between two a priori apparently quite different classes of Yang-Mills theories (section 4). Once again, it is the plane wave perspective which provides an explanation for this. Namely, I will show that this equivalence can be traced back to the coordinate transformation relating plane waves in Rosen and Brinkmann coordinates, and I add a few comments on what is the string theory context for these particular Yang-Mills actions.

Section 2 is extracted (and adapted to present purposes) from my unpublished lecture notes on plane waves and Penrose limits [1]. The material in section 3 is based on [2], and section 4 is based on currently unpublished material that will appear in [3].

2. A brief introduction to the geometry of plane wave metrics

2.1. Plane waves in Rosen and Brinkmann coordinates: heuristics

Usually gravitational plane wave solutions of general relativity are discussed in the context of the linearized theory. There one makes the Ansatz that the metric takes the form

$$g_{\mu\nu} = \eta_{\mu\nu} + h_{\mu\nu} \tag{1}$$

where $h_{\mu\nu}$ is treated as a small perturbation of the Minkowski background metric $\eta_{\mu\nu}$. To linear order in $h_{\mu\nu}$, the Einstein equations (necessarily) reduce to a wave equation. One finds that gravitational waves are transversally polarized. For example, a wave traveling in the (t, z)-direction distorts the metric only in the transverse directions, and a typical solution of the linearized Einstein equations is

$$ds^2 = -dt^2 + dz^2 + (\delta_{ij} + h_{ij}(z - t))dy^i dy^j \quad . \tag{2}$$

Note that in terms of lightcone coordinates $U = z - t$, $V = (z + t)/2$ this can be written as

$$ds^2 = 2dU\,dV + (\delta_{ij} + h_{ij}(U))dy^i dy^j \ . \tag{3}$$

We will now simply define a plane wave metric in general relativity to be a metric of the above form, dropping the assumption that h_{ij} be "small",

$$ds^2 = 2dU\,dV + g_{ij}(U)dy^i dy^j \ . \tag{4}$$

We will say that this is a plane wave metric in *Rosen coordinates*. This is not the coordinate system in which plane waves are usually discussed, among other reasons because typically in Rosen coordinates the metric exhibits spurious coordinate singularities.

Another way of introducing (or motivating the definition of) these plane wave metrics is to start with the $D = d + 2$ dimensional Minkowski metric written in lightcone coordinates,

$$ds^2 = \eta_{\mu\nu}dx^\mu dx^\nu = 2dudv + \delta_{ab}dx^a dx^b \ , \tag{5}$$

with $a = 1, \ldots, d$. To this metric one adds a term corresponding to a perturbation traveling at the speed of light in the v-direction,

$$ds^2 = 2dudv + A(u, x^a)(du)^2 + \delta_{ab}dx^a dx^b \ , \tag{6}$$

and requires that the effect of this term is to exert a linear (harmonic) force on test particles, leading to

$$ds^2 = 2dudv + A_{ab}(u)x^a x^b(du)^2 + \delta_{ab}dx^a dx^b \ . \tag{7}$$

This is the metric of a plane wave in *Brinkmann coordinates*. We will see below that the two classes of metrics described by (4) and (7) are indeed equivalent. Every metric of the form (4) can be brought to the form (7), and conversely every metric of the type (7) can be written, in more than one way, as in (4).

These exact gravitational plane wave solutions have been discussed in the context of four-dimensional general relativity for a long time (see e.g. [4] and [5]), even though they are not (and were never meant to be) phenomenologically realistic models of gravitational plane waves. The reason for this is that in the far-field gravitational waves are so weak that the linearized Einstein equations and their solutions are adequate to describe the physics, whereas the near-field strong gravitational effects responsible for the production of gravitational waves, for which the linearized equations are indeed insufficient, correspond to much more complicated solutions of the Einstein equations (describing e.g. two very massive stars orbiting around their common center of mass).

Rather, as the in some sense simplest non-trivial genuinely Lorentzian metrics, and as exact solutions of the full non-linear Einstein equations (see section 2.2), these plane wave metrics have always been extremelyuseful as a theoretical play-ground. It has also long been recognized that gravitational wave metrics provide potentially exact and exactly solvable string theory backgrounds, and this led to a certain amount of activity in this field in the early 1990s (see e.g. [6] for a review). More recently, the observations in [7, 8, 9] have led to a renewed surge in interest in the subject in the string theory community, in particular in connection with the remarkable BMN correspondence (Berenstein-Maldacena-Nastase) [10].

In the following, however, we will just be interested in certain aspects of the geometry of plane waves *per se*, and the role they play in elucidating certain properties of much simpler physical systems. The most basic aspects of the geometry of a space-time metric are revealed by studying its curvature, geodesics, and isometries. This is actually all we need, and we will now address these issues in turn.

2.2. Curvature of plane waves

While not strictly needed for the applications in sections 3 and 4, this brief discussion of the curvature of plane waves provides some useful insight into the geometry and physics of plane waves and the nature of Brinkmann coordinates.

Since the plane wave metric in Brinkmann coordinates (7) is so simple, it is straightforward to see that the only non-vanishing components of its Riemann curvature tensor are

$$R_{uaub} = -A_{ab} \ . \tag{8}$$

In particular, therefore, there is only one non-trivial component of the Ricci tensor,

$$R_{uu} = -\delta^{ab} A_{ab} \ , \tag{9}$$

and the Ricci scalar is zero.

Thus the metric is flat iff $A_{ab} = 0$. Moreover, we see that in Brinkmann coordinates the vacuum Einstein equations reduce to a simple algebraic condition on A_{ab} (regardless of its u-dependence), namely that it be traceless. The number of degrees of freedom of this traceless matrix $A_{ab}(u)$ correspond precisely to those of a transverse traceless symmetric tensor (a.k.a. a graviton). In four dimensions, the general vacuum plane wave solution thus has the form

$$ds^2 = 2dudv + [A(u)(x^2 - y^2) + 2B(u)xy]du^2 + dx^2 + dy^2 \tag{10}$$

for arbitrary functions $A(u)$ and $B(u)$. This reflects the two polarization states or degrees of freedom of a four-dimensional graviton. This family of exact solutions to the full non-linear Einstein equations would deserve to have text-book status but does not, to the best of my knowledge, appear in any of the standard introductory texts on general relativity.

2.3. Geodesics, lightcone gauge and harmonic oscillators

We now take a look at geodesics of a plane wave metric in Brinkmann coordinates, i.e. the solutions $x^\mu(\tau)$ to the geodesic equations

$$\ddot{x}^\mu(\tau) + \Gamma^\mu_{\nu\lambda}(x(\tau))\dot{x}^\nu(\tau)\dot{x}^\lambda(\tau) = 0 \ , \tag{11}$$

where an overdot denotes a derivative with respect to the affine parameter τ. Rather than determining the geodesic equations by first calculating all the non-zero Christoffel symbols, we make use of the fact that the geodesic equations can be obtained more efficiently, and in a way that allows us to directly make use of the symmetries of the problem, as the Euler-Lagrange equations of the Lagrangian

$$\mathcal{L} = \tfrac{1}{2} g_{\mu\nu} \dot{x}^\mu \dot{x}^\nu$$
$$= \dot{u}\dot{v} + \tfrac{1}{2} A_{ab}(u) x^a x^b \dot{u}^2 + \tfrac{1}{2} \dot{\vec{x}}^2 \ , \tag{12}$$

supplemented by the constraint $2\mathcal{L} = \epsilon$, where $\epsilon = 0$ ($\epsilon = -1$) for massless (massive) particles. Since nothing depends on v, the *lightcone momentum*

$$p_v = \frac{\partial \mathcal{L}}{\partial \dot{v}} = \dot{u} \tag{13}$$

is conserved. For $p_v = 0$, the particle obviously does not feel the curvature term A_{ab}, and the geodesics are straight lines. When $p_v \neq 0$, one has $u = p_v \tau + u_0$, and by an affine transformation of τ one can always choose the *lightcone gauge*

$$u = \tau \ . \tag{14}$$

Then the geodesic equations for the transverse coordinates are the Euler-Lagrange equations

$$\ddot{x}^a(\tau) = A_{ab}(\tau) x^b(\tau) \ . \tag{15}$$

These are the equation of motion of a non-relativistic *harmonic oscillator* with (possibly time-dependent) frequency matrix $\omega_{ab}^2(\tau) = -A_{ab}(\tau)$. The constraint $2\mathcal{L} = \epsilon$, or

$$2\dot{v}(\tau) + A_{ab}(\tau) x^a(\tau) x^b(\tau) + \delta_{ab} \dot{x}^a(\tau) \dot{x}^b(\tau) = \epsilon, \tag{16}$$

is then a first integral of the equation of motion for the remaining coordinate $v(\tau)$, and is readily integrated to give $v(\tau)$ in terms of the solutions to the harmonic oscillator equation for $x^a(\tau)$.

By definition the lightcone Hamiltonian is (minus!) the momentum p_u conjugate to u (in the lightcone gauge $u = \tau$),

$$H_{lc} = -p_u \ . \tag{17}$$

Using

$$p_u = g_{u\mu} \dot{x}^\mu = \dot{v} + A_{ab}(\tau) x^a x^b \tag{18}$$

and the constraint, one finds that the lightcone Hamiltonian is just (for $\epsilon \neq 0$ up to an irrelevant constant) the Hamiltonian of the above harmonic oscillator,

$$H_{lc} = \tfrac{1}{2} (\delta_{ab} \dot{x}^a \dot{x}^b - A_{ab}(\tau) x^a x^b) - \tfrac{1}{2}\epsilon \equiv H_{ho} - \tfrac{1}{2}\epsilon \ . \tag{19}$$

Note also that, in the lightcone gauge, the complete relativistic particle Lagrangian

$$\mathcal{L} = \dot{v} + \tfrac{1}{2} A_{ab}(\tau) x^a x^b + \tfrac{1}{2} \dot{\vec{x}}^2 = \mathcal{L}_{ho} + \dot{v} \tag{20}$$

differs from the harmonic oscillator Lagrangian only by a total time-derivative.

In summary, we note that in the lightcone gauge the equations of motion of a *relativistic particle* in the plane wave metric reduce to those of a *non-relativistic harmonic oscillator*. This harmonic oscillator equation plays a central role in the following and will reappear several times below in different contexts, e.g. when discussing the transformation from Rosen to Brinkmann coordinates, or when analyzing the isometries of a plane wave metric.

2.4. From Rosen to Brinkmann coordinates (and back)

I will now describe the relation between the plane wave metric in Brinkmann coordinates,

$$ds^2 = 2dudv + A_{ab}(u)x^a x^b du^2 + d\vec{x}^2 \ , \tag{21}$$

and in Rosen coordinates,

$$ds^2 = 2dUdV + g_{ij}(U)dy^i dy^j \ . \tag{22}$$

It is clear that, in order to transform the non-flat transverse metric in Rosen coordinates to the flat transverse metric in Brinkmann coordinates, one should change variables as

$$x^a = E^a_{\ i} y^i \ , \tag{23}$$

where $E^a_{\ i}$ is a vielbein for g_{ij} in the sense that

$$g_{ij} = E^a_{\ i} E^b_{\ j} \delta_{ab} \ . \tag{24}$$

Plugging this into the metric, one sees that this has the desired effect provided that E satisfies the symmetry condition

$$\dot{E}_{ai} E^i_{\ b} = \dot{E}_{bi} E^i_{\ a} \tag{25}$$

(such an E can always be found), and provided that one accompanies this by a shift in V. The upshot of this is that the change of variables

$$
\begin{aligned}
U &= u \\
V &= v + \tfrac{1}{2}\dot{E}_{ai} E^i_{\ b} x^a x^b \\
y^i &= E^i_{\ a} x^a \ ,
\end{aligned}
\tag{26}
$$

transforms the Rosen coordinate metric (22) into the Brinkmann form (21), with A_{ab} given by [2]

$$A_{ab} = \ddot{E}_{ai} E^i_{\ b} \ . \tag{27}$$

This can also be written as the *harmonic oscillator equation* (again!)

$$\ddot{E}_{ai} = A_{ab} E_{bi} \tag{28}$$

we had already encountered in the context of the geodesic equation.

In practice, once one knows that Rosen and Brinkmann coordinates are indeed just two distinct ways of describing the same class of metrics, one does not need to perform explicitly the coordinate transformation mapping one to the other. All one is interested in is the relation between $g_{ij}(U)$ and $A_{ab}(u)$, which is just the relation (8)

$$A_{ab} = -E^i_{\ a} E^j_{\ b} R_{UiUj} = -R_{uaub} \tag{29}$$

between the curvature tensor in Rosen and Brinkmann coordinates.

There is a lot of nice geometry lurking behind the transformation from Rosen to Brinkmann coordinates. For example, the symmetry condition (25) says that $E^i_{\ a}$ is a parallel transported co-frame along the null geodesic congruence

$(U = \tau, V, y^i = const.)$ [11, 12], and the coordinate transformation itself can be interpreted as passing from Rosen coordinates to inertial Fermi coordinates adapted to the null geodesic $(U = \tau, V = 0, y^i = 0)$ [13].

For a different perspective, and a prescription for how to go back from Brinkmann to Rosen coordinates, note that the index i on E_{ai} in (28) can be thought of as labelling d out of the $2d$ linearly independent solutions of the oscillator equation. The symmetry condition (25) can equivalently be written as

$$\dot{E}_{ai}E^i_{\ b} = \dot{E}_{bi}E^i_{\ a} \quad \Leftrightarrow \quad \dot{E}_{ai}E^a_{\ k} = \dot{E}_{ak}E^a_{\ i} \ , \tag{30}$$

and can now be interpreted as the condition that the *Wronskian* of the i'th and k'th solution

$$W(E_i, E_k) := \dot{E}_{ak}E^a_{\ i} - \dot{E}_{ai}E^a_{\ k} \tag{31}$$

is zero. Thus, given the metric in Brinkmann coordinates, one can construct the metric in Rosen coordinates by solving the oscillator (geodesic) equation, choosing a maximally commuting set of solutions to construct E_{ai}, and then determining g_{ij} algebraically from the E_{ai} from (24).

2.5. The Heisenberg isometry algebra of a generic plane wave

We now study the isometries of a generic plane wave metric. In Brinkmann coordinates, because of the explicit dependence of the metric on u and the transverse coordinates, only one isometry is manifest, namely that generated by the parallel (covariantly constant) and hence in particular Killing null vector $Z = \partial_v$. In Rosen coordinates, the metric depends neither on V nor on the transverse coordinates y^k, and one sees that in addition to $Z = \partial_V$ there are at least d more Killing vectors, namely the ∂_{y^k}. Together these form an Abelian translation algebra acting transitively on the null hypersurfaces of constant U.

However, this is not the whole story. Indeed, one particularly interesting and peculiar feature of plane wave space-times is the fact that they generically possess a *solvable* (rather than semi-simple) isometry algebra, namely a Heisenberg algebra, only part of which we have already seen above.

All Killing vectors X can be found in a systematic way by solving the Killing equations

$$L_X g_{\mu\nu} = \nabla_\mu X_\nu + \nabla_\nu X_\mu = 0 \ . \tag{32}$$

I will not do this here but simply present the results of this analysis in Brinkmann coordinates (see [2] for details). The upshot is that a generic $(2 + d)$-dimensional plane wave metric has a $(2d + 1)$-dimensional isometry algebra generated by the Killing vector $Z = \partial_v$ and the $2d$ Killing vectors

$$X(f_{(K)}) \equiv X_{(K)} = f_{(K)a}\partial_a - \dot{f}_{(K)a}x^a\partial_v \ . \tag{33}$$

Here the $f_{(K)a}$, $K = 1, \ldots, 2d$ are the $2d$ linearly independent solutions of the *harmonic oscillator equation* (yet again!)

$$\ddot{f}_a(u) = A_{ab}(u)f_b(u) \ . \tag{34}$$

These Killing vectors satisfy the algebra

$$[X_{(J)}, X_{(K)}] = -W(f_{(J)}, f_{(K)})Z \tag{35}$$
$$[X_{(J)}, Z] = 0 , \tag{36}$$

where the Wronskian $W(f_{(J)}, f_{(K)})$ is, exactly as in (31), given by

$$W(f_{(J)}, f_{(K)}) = \sum_a (f_{(J)a} \dot{f}_{(K)a} - f_{(K)a} \dot{f}_{(J)a}) . \tag{37}$$

It is of course constant (independent of u) as a consequence of the harmonic oscillator equation. This is already the Heisenberg algebra. To make this more explicit, one can make a convenient choice of basis for the solutions $f_{(J)}$ by splitting the $f_{(J)}$ into two sets of solutions

$$\{f_{(J)}\} \to \{q_{(a)}, p_{(a)}\} \tag{38}$$

characterized by the initial conditions

$$
\begin{aligned}
q_{(a)b}(u_0) = \delta_{ab} \quad & \dot{q}_{(a)b}(u_0) = 0 \\
p_{(a)b}(u_0) = 0 \quad & \dot{p}_{(a)b}(u_0) = \delta_{ab} .
\end{aligned}
\tag{39}
$$

Since the Wronskian of these functions is independent of u, it can be determined by evaluating it at $u = u_0$. Then one can immediately read off that

$$
\begin{aligned}
W(q_{(a)}, q_{(b)}) = W(p_{(a)}, p_{(b)}) &= 0 \\
W(q_{(a)}, p_{(b)}) &= \delta_{ab} .
\end{aligned}
\tag{40}
$$

Therefore the corresponding Killing vectors

$$Q_{(a)} = X(q_{(a)}) , \qquad P_{(a)} = X(p_{(a)}) \tag{41}$$

and Z satisfy the canonically normalized Heisenberg algebra

$$
\begin{aligned}
[Q_{(a)}, Z] = [P_{(a)}, Z] &= 0 \\
[Q_{(a)}, Q_{(b)}] = [P_{(a)}, P_{(b)}] &= 0 \\
[Q_{(a)}, P_{(b)}] &= -\delta_{ab} Z .
\end{aligned}
\tag{42}
$$

As we had noted before, in Rosen coordinates, the $(d+1)$ translational isometries in the V and y^k directions, generated by the Killing vectors $Z = \partial_V$ and $Q_{(k)} = \partial_{y^k}$, are manifest. One can check that the "missing" d Killing vectors $P_{(k)}$ are given by

$$P_{(k)} = -y^k \partial_V + \int^u du' \, g^{km}(u') \partial_{y^m} . \tag{43}$$

It is straightforward to verify that together they also generate the Heisenberg algebra (42).

These considerations also provide yet another perspective on the transformation from Brinkmann to Rosen coordinates, and the vanishing Wronskian condition discussed at the end of section 2.4. Indeed, passing from Brinkmann to Rosen coordinates can be interpreted as passing to coordinates in which half of the translational Heisenberg algebra symmetries are manifest. This is achieved by choosing the (transverse) coordinate lines to be the integral curves of these Killing vectors. This is of course only possible if these Killing vectors commute, i.e. the Wronskian of the corresponding solutions of the harmonic oscillator equation is zero, and results in a metric which is independent of the transverse coordinates, namely the plane wave metric in Rosen coordinates.

2.6. Geodesics, isometries, and conserved charges

We can now combine the results of the previous sections to determine the conserved charges carried by particles moving geodesically in the plane wave geometry. In general, given any Killing vector X, there is a corresponding conserved charge $C(X)$,

$$C(X) = g_{\mu\nu}X^\mu \dot{x}^\mu \ . \tag{44}$$

That $C(X)$ is indeed constant along the trajectory of the geodesic $x^\mu(\tau)$ can easily be verified by using the geodesic and Killing equations.

The conserved charge corresponding to the Killing vector $Z = \partial_v$, the central element of the Heisenberg algebra, is, none too surprisingly, nothing other than the conserved lightcone momentum p_v (13) of section 2.3,

$$C(Z) = g_{v\mu}\dot{x}^\mu = \dot{u} = p_v \ . \tag{45}$$

In addition to Z, for any solution f of the harmonic oscillator equation we have a Killing vector $X(f)$ (33),

$$X(f) = f_a\partial_a - \dot{f}_a x^a \partial_v \ . \tag{46}$$

The associated conserved charge is

$$C(X(f)) = f_a p^a - \dot{f}_a x^a \ . \tag{47}$$

(here we have used the, now more appropriate, phase space notation $p^a = \dot{x}^a$). This is rather trivially conserved (constant), since both f_a and x^a are solutions of the same ubiquitous harmonic oscillator equation and $C(X(f))$ is nothing other than their constant Wronskian,

$$C(X(f)) = W(f, x) \ . \tag{48}$$

Thus these somewhat tautological conserved charges are not helpful in integrating the geodesic or harmonic oscillator equations. Nevertheless, the very fact that they exist, and that they satisfy a (Poisson bracket) Heisenberg algebra, will turn out to be conceptually important in section 3. We will denote the conserved charges corresponding to the Killing vectors $Q_{(a)}$ and $P_{(a)}$ (41) by

$$C(Q_{(a)}) \equiv \mathcal{Q}_{(a)} \qquad C(P_{(a)}) \equiv \mathcal{P}_{(a)} \ . \tag{49}$$

The Poisson brackets among the charges $C(X(f))$ can be determined from the canonical Poisson brackets $\{x^a, p^b\} = \delta_{ab}$ to be

$$\{X(f_1), X(f_2)\} = \{f_{1a}p^a - \dot{f}_{1a}x^a, f_{2a}p^a - \dot{f}_{2a}x^a\} = W(f_1, f_2) \tag{50}$$

(note the usual sign flip with respect to the Lie bracket (35) of the corresponding vector fields). In particular, as a consequence of (40) the charges $\mathcal{Q}_{(a)}$ and $\mathcal{P}_{(a)}$ have the canonical Poisson brackets

$$\{\mathcal{Q}_{(a)}, \mathcal{P}_{(b)}\} = \delta_{ab} \ . \tag{51}$$

Generically, a plane wave metric has just this Heisenberg algebra of isometries which acts transitively on the null hyperplanes $u = const.$, with a simply transitive Abelian subalgebra. However, for special choices of $A_{ab}(u)$, there may of course be more Killing vectors. These could arise from internal symmetries of A_{ab}, giving more Killing vectors (and corresponding conserved angular momenta) in the transverse directions, as for an isotropic harmonic oscillator.

Of more interest is the fact that for particular $A_{ab}(u)$ there may be Killing vectors with a ∂_u-component. The existence of such a Killing vector renders the plane wave homogeneous (away form the fixed points of this extra Killing vector). These homogeneous plane waves have been completely classified in [2]. The simplest examples, and the only ones that we will consider here, are plane waves with a u-independent profile A_{ab},

$$ds^2 = 2dudv + A_{ab}x^a x^b du^2 + d\vec{x}^2 \ , \tag{52}$$

which obviously, since now nothing depends on u, have the extra Killing vector $X = \partial_u$.

The existence of the additional Killing vector $X = \partial_u$ extends the Heisenberg algebra to the harmonic oscillator algebra, with X playing the role of the number operator or harmonic oscillator Hamiltonian. Indeed, X and $Z = \partial_v$ obviously commute, and the commutator of X with one of the Killing vectors $X(f)$ is

$$[X, X(f)] = X(\dot{f}) \ . \tag{53}$$

Note that this is consistent, i.e. the right-hand-side is again a Killing vector, because when A_{ab} is constant and f satisfies the harmonic oscillator equation then so does its u-derivative $\dot{f}$. In terms of the basis (41) we have

$$\begin{aligned} [X, Q_{(a)}] &= P_{(a)} \\ [X, P_{(a)}] &= A_{ab}Q_{(b)} \ , \end{aligned} \tag{54}$$

which is the harmonic oscillator algebra.

Another way of understanding the relation between $X = \partial_u$ and the harmonic oscillator Hamiltonian is to look at the conserved charge associated with $X = \partial_u$,

$$C(\partial_u) = g_{u\mu}\dot{x}^\mu = p_u \ , \tag{55}$$

which we had already identified (up to a constant for non-null geodesics) as minus the harmonic oscillator Hamiltonian in section 2.3. This is of course indeed a conserved charge iff the Hamiltonian is time-independent, i.e. iff A_{ab} is constant.

2.7. Synopsis

In the above I have reviewed, in somewhat more detail than strictly necessary for the following, some of the interesting and entertaining aspects of the geometry of plane wave metrics. The only things that we will actually directly make use of below are, in section 3,

- the Heisenberg isometry algebra (section 2.5)
- and the existence of the corresponding conserved charges (section 2.6),

and, in section 4,

- the lightcone gauge geodesic Lagrangian (section 2.3)
- and the transformation from Rosen to Brinkmann coordinates (section 2.4).

3. The Lewis–Riesenfeld theory of the time-dependent quantum oscillator

3.1. Description of the problem

We will now discuss the quantum theory of a time-dependent harmonic oscillator (for simplicity in $d = 1$ dimension, but the discussion generalizes in an obvious way to $d > 1$), with Hamiltonian

$$H_{ho}(t) = \tfrac{1}{2}(p^2 + \omega(t)^2 x^2) \ . \tag{56}$$

The aim is to find the solutions of the time-dependent Schrödinger equation (in units with $\hbar = 1$)

$$i\partial_t|\psi(t)\rangle = \hat{H}_{ho}(t)|\psi(t)\rangle \ . \tag{57}$$

Standard textbook treatments of this problem employ the following strategy:

- When the Hamiltonian is time-independent, then the standard procedure is of course to reduce this problem to that of finding the stationary eigenstates $|\psi_n\rangle$ of $\hat{H}_{ho}$,

$$\hat{H}_{ho}|\psi_n\rangle = E_n|\psi_n\rangle \ , \tag{58}$$

  with $E_n = \omega(n + \tfrac{1}{2})$ etc., in terms of which the general solution to the time-dependent Schrödinger equation can then be written as

$$|\psi(t)\rangle = \sum_n c_n e^{-iE_n t}|\psi_n\rangle \ , \tag{59}$$

  where the c_n are constants.

- When the Hamiltonian is time-dependent, on the other hand, then in principle the solution is given by the time-ordered exponential of $\hat{H}_{ho}(t)$,

$$|\psi(t)\rangle = \left(\mathcal{T}e^{-i\int_{t_0}^t dt'\, \hat{H}_{ho}(t')}\right)|\psi(t_0)\rangle \ , \tag{60}$$

  but in practice this cannot be evaluated to get an exact solution. One thus needs to then invoke some kind of adiabatic approximation to perturbatively determine the solution (and then calculate transition and decay rates etc.).

What Lewis and Riesenfeld observed [14] is that, even in the time-dependent case, there is a procedure analogous to the one used in the time-independent case which allows one to explicitly find the exact solutions of the time-dependent Schödinger equation.

3.2. Outline of the Lewis–Riesenfeld procedure

The idea of [14] is to base the construction of the solutions of the Schrödinger equation not on the stationary eigenstates of the Hamiltonian (which does not make sense when the Hamiltonian depends explicitly on time) but on the eigenstates of another operator $\hat{I}$ which is an *invariant* of the system. This means that

$$\hat{I}(t, \hat{x}, \hat{p}) \equiv \hat{I}(t) \tag{61}$$

is a (typically explicitly time-dependent) operator satisfying

$$i\frac{d}{dt}\hat{I}(t) \equiv i\partial_t\hat{I}(t) + [\hat{I}(t), \hat{H}_{ho}(t)] = 0 \tag{62}$$

(when $\hat{H}_{ho}$ is time-independent, then one can of course just take $\hat{I} = \hat{H}_{ho}$). The Lewis–Riesenfeld procedure now consists of two parts:

1. The first is to show how one can construct all the solutions of the time-dependent Schrödinger equation for $\hat{H}_{ho}(t)$ from the spectrum and eigenstates of the invariant $\hat{I}(t)$.

2. The second is an algorithm which provides an invariant for any time-dependent harmonic oscillator, and which moreover has the feature that $\hat{I}(t)$ itself has the form of a *time-independent* harmonic oscillator (so that it is straightforward to determine the spectrum and eigenstates of $\hat{I}(t)$).

Issue (1) can be established by straightforward and relatively standard quantum mechanical manipulations. I will briefly recall these below but have nothing new to add to that part of the discussion. Issue (2), on the other hand, is usually established by a direct but rather brute-force calculation which does not appear to provide any conceptual insight into why invariants with the desired properties exist. I will show in section 3.3 that this conceptual insight is obtained by embedding the time-dependent harmonic oscillator into the plane wave setting.

To address (1), let us assume that an invariant $\hat{I}(t)$ satisfying (62) exists and that it is hermitian. We choose a complete set of eigenstates, labeled by the real eigenvalues λ of $\hat{I}(t)$,

$$\hat{I}(t)|\lambda\rangle = \lambda|\lambda\rangle \ . \tag{63}$$

It follows from (62) that the eigenvalues λ are time-independent, and that

$$\langle\lambda'|i\partial_t - \hat{H}_{ho}(t)|\lambda\rangle = 0 \tag{64}$$

for all $\lambda \neq \lambda'$. We would like this equation to be true also for the diagonal elements, in which case we would have already found the solutions of the time-dependent Schrödinger equation for $\hat{H}_{ho}(t)$. To accomplish this, we slightly modify the eigenfunctions by multiplying them by a time-dependent phase,

$$|\lambda\rangle \rightarrow e^{i\alpha_\lambda(t)}|\lambda\rangle \ . \tag{65}$$

It can be seen immediately that this phase factor does not change the off-diagonal matrix elements of $i\partial_t - \hat{H}_{ho}(t)$ (since the eigenstates are orthogonal). Requiring the validity of (64) also for $\lambda = \lambda'$ then leads to a first-order differential equation for $\alpha_\lambda(t)$,

$$\tfrac{d}{dt}\alpha_\lambda(t) = \langle\lambda|i\partial_t - \hat{H}_{ho}(t)|\lambda\rangle. \tag{66}$$

Solving this equation, the general solution to the time-dependent Schrödinger equation for $\hat{H}_{ho}(t)$ is, similarly to (59),

$$|\psi(t)\rangle = \sum_\lambda c_\lambda e^{i\alpha_\lambda(t)}|\lambda\rangle \;, \tag{67}$$

where the c_λ are constants.

This is an extremely neat way of solving exactly the quantum theory of the time-dependent harmonic oscillator. Its usefulness, however, depends on the ability to construct a suitable invariant $\hat{I}(t)$ which is such that (a) one can explicitly find its spectrum and eigenstates and (b) it is sufficiently closely related to $\hat{H}_{ho}(t)$ so that one can evaluate the diagonal matrix elements of $\hat{H}_{ho}(t)$ in the basis of eigenstates $|\lambda\rangle$ of the invariant $\hat{I}(t)$ (in order to determine the phases $\alpha_\lambda(t)$).

In a nutshell, this is achieved in [14] as follows (see also [2] for a detailed account with further comments on the procedure). Let $\sigma(t)$ be any solution to the non-linear differential equation

$$\ddot{\sigma}(t) + \omega(t)^2\sigma(t) = \sigma(t)^{-3} \;, \tag{68}$$

where $\omega(t)$ is the harmonic oscillator frequency. Then it can be checked by a straightforward but unenlightening calculation that

$$\hat{I}(t) = \tfrac{1}{2}(\hat{x}^2\sigma(t)^{-2} + (\sigma(t)\hat{p} - \dot{\sigma}(t)\hat{x})^2) \tag{69}$$

is an invariant in the sense of (62). As a first sanity check on this construction, note that for ω time-independent one can also choose $\sigma = \omega^{-1/2}$ to be constant, upon which the invariant becomes

$$\hat{I} = \tfrac{1}{2}(\omega\hat{x}^2 + \omega^{-1}\hat{p}^2) = \omega^{-1}\hat{H}_{ho} \;, \tag{70}$$

which is of course the privileged invariant of a time-independent system. In general, in terms of the hermitian conjugate operators

$$\hat{a} = \tfrac{1}{\sqrt{2}}(\hat{x}\sigma^{-1} + i(\sigma\hat{p} - \dot{\sigma}\hat{x})) \qquad \hat{a}^\dagger = \tfrac{1}{\sqrt{2}}(\hat{x}\sigma^{-1} - i(\sigma\hat{p} - \dot{\sigma}\hat{x})) \tag{71}$$

which satisfy the canonical commutation relations $[\hat{a}, \hat{a}^\dagger] = 1$, $\hat{I}(t)$ has the standard oscillator representation

$$\hat{I}(t) = \hat{a}^\dagger\hat{a} + \tfrac{1}{2} \tag{72}$$

of a time-independent harmonic oscillator, and the original Hamiltonian is a quadratic function of $\hat{a}$ and $\hat{a}^\dagger$,

$$\hat{H}_{ho}(t) = c(t)(\hat{a})^2 + c(t)^*(\hat{a}^\dagger)^2 + d(t)(\hat{a}^\dagger\hat{a} + \tfrac{1}{2}) \;, \tag{73}$$

where e.g. $d(t) = \frac{1}{2}(\omega(t)^2\sigma(t)^2 + \dot{\sigma}(t)^2 + \sigma(t)^{-2})$. This makes it straightforward to evaluate e.g. the diagonal matrix elements of $\hat{H}_{ho}(t)$ in the standard basis of eigenstates of $\hat{I}(t)$.

Finally, the general solution to (68) can be written in terms of any two linearly independent solutions f_1, f_2 to the classical harmonic oscillator equation for $H_{ho}(t)$ (this is (68) with zero on the rhs instead of the non-linear term). Normalizing their Wronskian to 1, the general solution $\sigma(t)$ is [14]

$$\sigma = \pm\left[c_1^2 f_1^2 + c_2^2 f_2^2 \pm 2(c_1^2 c_2^2 - 1)^{1/2} f_1 f_2\right]^{1/2} , \qquad (74)$$

where c_i are constants subject to the condition that the solution is real, and the signs can be chosen independently.

3.3. Deducing the procedure from the plane wave geometry

While the procedure outlined above provides a concrete (and in practice also very useful) algorithm to solve exactly the quantum theory of a time-dependent harmonic oscillator (and certain other time-dependent systems [14]), it remains somewhat unsatisfactory from a conceptual point of view. In particular, it is not clear from the construction

- why invariants with the desired properties exist in the first place;
- why solutions to the classical equations play a role in the construction of these quantum invariants;
- why one should solve the non-linear equation (68) if, in any case, in the end it all boils down to solutions of the ordinary linear classical harmonic oscillator equation appearing in (74).

Here is where insight is gained by realizing the harmonic oscillator equation as the geodesic equation in a plane wave metric. Recall that in section 2.5 we had found a Heisenberg isometry algebra which, in particular, includes the "hidden" symmetries generated by the Killing vector fields (33,46)

$$X(f) = f_a\partial_a - \dot{f}_a x^a \partial_v , \qquad (75)$$

where f is a solution of the classical harmonic oscillator equation, and the corresponding "hidden" conserved charges (47)

$$C(X(f)) = f_a p^a - \dot{f}_a x^a . \qquad (76)$$

In particular, we had obtained the conserved charges $\mathcal{Q}_{(a)}$ and $\mathcal{P}_{(a)}$ (49). These are linear in the phase space variables x^a and p^a, and thus we can unambiguously associate to them quantum operators

$$\mathcal{Q}_{(a)} \rightarrow \hat{\mathcal{Q}}_{(a)} \qquad \mathcal{P}_{(a)} \rightarrow \hat{\mathcal{P}}_{(a)} \qquad (77)$$

which, by construction, are invariants in the sense of (62),

$$\frac{d}{dt}\hat{\mathcal{Q}}_{(a)} = \frac{d}{dt}\hat{\mathcal{P}}_{(a)} = 0 , \qquad (78)$$

and which satisfy the canonical commutation relations (cf. (51))

$$[\hat{\mathcal{Q}}_{(a)}, \hat{\mathcal{P}}_{(b)}] = i\delta_{ab} \ . \tag{79}$$

Note that to "see" these invariants, one has to extend the harmonic oscillator configuration space not just by the time-direction $t = u$, but one also has to add yet another dimension, the null direction v.

The rest is now straightforward. Since $\hat{\mathcal{Q}}_{(a)}$ and $\hat{\mathcal{P}}_{(b)}$ are invariants, also any quadratic operator in these variables (with constant coefficients) is an invariant. In the one-dimensional case ($d = 1$), we can e.g. consider invariants of the form

$$\hat{I}(t) = \tfrac{1}{2M}\hat{\mathcal{P}}^2 + \tfrac{M\Omega^2}{2}\hat{\mathcal{Q}}^2 \ , \tag{80}$$

which we can write in terms of invariant creation and annihilation operators $\hat{\mathcal{A}}$ and $\hat{\mathcal{A}}^\dagger$ (constructed in the usual way from $\hat{\mathcal{Q}}$ and $\hat{\mathcal{P}}$) as

$$\hat{I}(t) = \Omega(\hat{\mathcal{A}}^\dagger\hat{\mathcal{A}} + \tfrac{1}{2}) \ . \tag{81}$$

Let us now compare this in detail with the results of section 3.2. First of all, to match with the (arbitrary choice of) normalization of the invariant (72), we choose $\Omega = 1$. Next we can identify what $\sigma(t)^2$ is by identifying it with the coefficient of $\hat{p}^2$ in the expansion of (80) in terms of $\hat{p}$ and $\hat{x}$. The upshot is that $\sigma(t)$ has precisely the form given in (74), with $c_1^2 = 1$ and $c_2^2 = 1/M$. Finally, one sees that the invariant oscillators $\hat{\mathcal{A}}$ and $\hat{\mathcal{A}}^\dagger$ are related to the oscillators $\hat{a}$ and $\hat{a}^\dagger$ by a unitary transformation which is precisely the unitary transformation that implements the phase transformation (65) on the eigenstates of the invariant.

We have thus come full circle. Starting with the conserved charges associated with the Heisenberg algebra Killing vectors, we have constructed quadratic quantum invariants and have reproduced all the details of the Lewis–Riesenfeld algorithm, including the phase factors $\alpha_\lambda(t)$. Constructing the Fock space in the usual way, one then obtains all the solutions (67) to the time-dependent Schrödinger equation.

4. A curious equivalence between two classes of Yang-Mills actions

4.1. Description of the problem

A prototypical non-Abelian Yang-Mills + scalar action in $n = p+1$ dimensions, obtained e.g. by the dimensional reduction of pure Yang-Mills theory (with standard Lagrangian $L \sim \mathrm{tr}\, F_{MN}F^{MN}$) in D space-time dimensions down to n dimensions, has the form

$$S_{YM} = \int d^n\sigma\, \mathrm{Tr}\left(-\tfrac{1}{4}g_{YM}^{-2}\eta^{\alpha\gamma}\eta^{\beta\delta}F_{\alpha\beta}F_{\gamma\delta} - \tfrac{1}{2}\eta^{\alpha\beta}D_\alpha\phi^a D_\beta\phi^a + \tfrac{1}{4}g_{YM}^2[\phi^a,\phi^b]^2\right) \ . \tag{82}$$

Here the ϕ^a, $a = 1, \ldots, D - n$, are hermitian scalar fields arising from the internal components of the gauge field and thus taking values in the adjoint representation of the gauge group, the covariant derivative is

$$D_\alpha \phi^a = \partial_\alpha \phi^a - i[A_\alpha, \phi^a] \ , \tag{83}$$

A_α is the gauge field, $F_{\alpha\beta}$ its curvature, g_{YM}^2 denotes the Yang-Mills coupling constant, Tr a Lie algebra trace, and in writing the above action I have suppressed all Lie algebra labels.

This basic action can of course be modified in various ways, e.g. by adding further fields (we will not do this), or by modifying the couplings of the scalar fields. We will consider two such modifications. The first one simply consists of adding (possibly time-dependent) mass terms for the scalars. Denoting the scalars in this model by X^a, the action reads

$$S_{BC} = \int d^n\sigma \, \mathrm{Tr} \left(-\tfrac{1}{4} g_{YM}^{-2} \eta^{\alpha\gamma}\eta^{\beta\delta} F_{\alpha\beta}F_{\gamma\delta} - \tfrac{1}{2}\eta^{\alpha\beta}\delta_{ab}D_\alpha X^a D_\beta X^b \right.$$
$$\left. +\tfrac{1}{4} g_{YM}^2 \delta_{ac}\delta_{bd}[X^a,X^b][X^c,X^d] + \tfrac{1}{2} A_{ab}(t) X^a X^b \right) \ , \tag{84}$$

with $A_{ab}(t)$ minus the mass-squared matrix.

The second class of actions arises from (82) by replacing the flat metric δ_{ab} on the scalar field space (suppressed in (82) but written out explicitly in (84)) by a time-dependent matrix $g_{ij}(t)$ of "coupling constants", but without adding any mass terms. Denoting the scalars in this model by Y^i, the action reads

$$S_{RC} = \int d^n\sigma \, \mathrm{Tr} \left(-\tfrac{1}{4} g_{YM}^{-2} \eta^{\alpha\gamma}\eta^{\beta\delta} F_{\alpha\beta}F_{\gamma\delta} - \tfrac{1}{2}\eta^{\alpha\beta} g_{ij}(t) D_\alpha Y^i D_\beta Y^j \right.$$
$$\left. +\tfrac{1}{4} g_{YM}^2 g_{ik}(t)g_{jl}(t)[Y^i,Y^j][Y^k,Y^l] \right) \ . \tag{85}$$

The reason for the subscripts $_{BC}$ and $_{RC}$ on the actions will, if not already obvious at this stage, become apparent below. In any case, the claim is now that these two, apparently rather different, classes of Yang-Mills actions are simply related by a certain linear field redefinition $Y^i = E^i_a X^a$ of the scalar fields,

$$S_{RC}[A_\alpha, Y^i = E^i_a X^a] = S_{BC}[A_\alpha, X^a] \ . \tag{86}$$

We could straightaway prove this by a brute-force calculation, but this would be rather unenlightening. Instead, we will first consider a much simpler classical mechanics toy model of this equivalence (section 4.2), and we will then be able to establish (86) with hardly any calculation at all (section 4.3). At the end, I will briefly indicate why one is led to consider actions of the type (84,85) in the first place, and why from that point of view one can a priori anticipate the validity of an identity like (86).

4.2. A classical mechanics toy model

As a warm-up exercise, consider the standard harmonic oscillator Lagrangian (now mysteriously labeled $_{bc}$)

$$L_{bc}(x) = \tfrac{1}{2}(\dot{x}^2 - \omega^2 x^2) \ , \tag{87}$$

and the (exotic) Lagrangian

$$L_{rc}(y) = \tfrac{1}{2}\sin^2 \omega t \, \dot{y}^2 \tag{88}$$

with a time-dependent kinetic term. Now consider the transformation

$$y = (\sin \omega t)^{-1} x \ . \tag{89}$$

Then one finds

$$
\begin{aligned}
L_{rc}(y) &= \tfrac{1}{2}(\dot{x}^2 + \omega^2 x^2 \cot^2 \omega t - 2\omega x \dot{x} \cot \omega t) \\
&= \tfrac{1}{2}(\dot{x}^2 - \omega^2 x^2) - \tfrac{d}{dt}(\tfrac{1}{2}\omega x^2 \cot \omega t) \\
&= L_{bc}(x) + \tfrac{d}{dt}(\dots)
\end{aligned}
\tag{90}
$$

Thus, up to a total time-derivative the linear transformation (89) transforms the exotic (and seemingly somewhat singular) Lagrangian (88) to the completely regular "massive" Lagrangian (87), and the corresponding actions are essentially identical. This should be thought of as the counterpart of the statement that the Rosen coordinate plane wave metric

$$ds^2 = 2dU\,dV + \sin^2 \omega U (dy)^2 \tag{91}$$

can, in Brinkmann coordinates, be written as

$$2dU\,dV + \sin^2 \omega U (dy)^2 = 2dudv - \omega^2 x^2 (du)^2 + (dx)^2 \ . \tag{92}$$

We can now generalize this in the following way. Consider the Lagrangian L_{bc} corresponding to the lightcone Hamiltonian (19) of a (massless, say) particle in a plane wave in Brinkmann coordinates (in the lightcone gauge $u = t$),

$$L_{bc}(x) = \tfrac{1}{2}(\delta_{ab}\dot{x}^a \dot{x}^b + A_{ab}(t)x^a x^b) \ , \tag{93}$$

and the corresponding Lagrangian in Rosen coordinates,

$$L_{rc}(y) = \tfrac{1}{2}g_{ij}(t)\dot{y}^i \dot{y}^j \ . \tag{94}$$

The claim is that these two Lagrangians are equal up to a total time-derivative. To see this, recall first of all the coordinate transformation (26)

$$
\begin{aligned}
y^i &= E^i{}_a x^a \\
V &= v + \tfrac{1}{2}\dot{E}_{ai} E^i{}_b x^a x^b \ ,
\end{aligned}
\tag{95}
$$

where $E^i{}_a$ satisfies (24) and (25). Substituting $y^i = E^i{}_a x^a$ in L_{rc}, one can now verify that one indeed obtains L_{bc} up to a total time-derivative. The way to see this without any calculation is to start from the complete geodesic Lagrangian in Rosen coordinates in the lightcone gauge $U = t$,

$$\mathcal{L} = \tfrac{1}{2}g_{\mu\nu}\dot{y}^\mu \dot{y}^\nu = \dot{V} + \tfrac{1}{2}g_{ij}(t)\dot{y}^i \dot{y}^j = \dot{V} + L_{rc}(y) \ . \tag{96}$$

This Lagrangian is still invariant under coordinate transformations of the remaining coordinates, and is hence equal, on the nose, to its Brinkmann coordinate counterpart (20),

$$L_{bc}(x) + \dot{v} = L_{rc}(y) + \dot{V} \ . \tag{97}$$

This implies that the two Lagrangians $L_{bc}(x)$ and $L_{rc}(y)$ differ only by a total time-derivative, namely the derivative of the shift of V in the coordinate transformation (95).

4.3. The explanation: from plane wave metrics to Yang-Mills actions

We can now come back to the two types of Yang-Mills actions S_{BC} (84) and S_{RC} (85), which are obviously in some sense non-Abelian counterparts of the classical mechanics Brinkmann and Rosen coordinate actions $S_{bc} = \int L_{bc}$ and $S_{rc} = \int L_{rc}$ discussed above. The claim is that these two actions are related (perhaps up to a total derivative term) by the linear transformation

$$Y^i = E^i_a X^a \tag{98}$$

of the scalar fields (matrix-valued coordinates) Y^i and X^a, where E^i_a is the vielbein that enters in the relation between Rosen and Brinkmann coordinates.

Even though in general non-Abelian coordinate transformations are a tricky issue, this particular transformation is easy to deal with since it is linear as well as diagonal in matrix (Lie algebra) space. Consider e.g. the quartic potential terms in (84) and (85). With the substitution (98), one obviously has

$$\begin{aligned} g_{ik}g_{jl}[Y^i, Y^j][Y^k, Y^l] &= g_{ik}g_{jl}E^i_a E^j_b E^k_c E^l_d [X^a, X^b][X^c, X^d] \\ &= \delta_{ac}\delta_{bd}[X^a, X^b][X^c, X^d] \ , \end{aligned} \tag{99}$$

so that the two quartic terms are indeed directly related by (98). Now consider the gauge covariant kinetic term for the scalars in (85). Since $E^i_a = E^i_a(t)$ depends only on (lightcone) time t, the spatial covariant derivatives transform as

$$\alpha \neq t : \quad D_\alpha Y^i = E^i_a(t) D_\alpha X^a \ , \tag{100}$$

so that the spatial derivative parts of the scalar kinetic terms are mapped into each other. It thus remains to discuss the term $\mathrm{Tr}\, g_{ij}(t) D_t Y^i D_t Y^j$ involving the covariant time-derivatives. For the ordinary partial derivatives, the argument is identical to that in section 4.2, and thus one finds

$$\tfrac{1}{2}\mathrm{Tr}\, g_{ij}(t)\dot{Y}^i \dot{Y}^j = \tfrac{1}{2}\mathrm{Tr}(\delta_{ab}\dot{X}^a \dot{X}^b + A_{ab}(t) X^a X^b) + \tfrac{d}{dt}(\ldots) \ . \tag{101}$$

The only remaining subtlety are terms involving the t-derivative $\dot{E}^i_a$ of E^i_a, arising from cross-terms like

$$\mathrm{Tr}\, g_{ij}(t)[A_t, Y^i]\partial_t Y^j = \mathrm{Tr}\, g_{ij}(t) E^i_a[A_t, X^a]\partial_t(E^j_b X^b) \ . \tag{102}$$

However, these terms do not contribute at all since

$$g_{ij}(t) E^i_a \dot{E}^j_b \,\mathrm{Tr}[A_t, X^a]X^b = g_{ij}(t) E^i_a \dot{E}^j_b \,\mathrm{Tr}\, A_t[X^a, X^b] = 0 \tag{103}$$

by the cyclic symmetry of the trace and the symmetry condition (25). It is pleasing to see that this symmetry condition, which already ensured several cancelations in the standard transformation from Rosen to Brinkmann coordinates (and thus also in establishing e.g. (101)), cooperatively also serves to eliminate some terms of genuinely non-Abelian origin.

Putting everything together, we have now established the claimed equivalence (86) between the two apparently quite different classes of Yang-Mills theories, namely standard Yang-Mills theories with (possibly time-dependent) mass-terms on the one hand, and Yang-Mills theories with non-standard time-dependent scalar couplings on the other.

I still owe you an explanation of where all of this comes from or what it is good for. The appropriate context for this is provided by a non-perturbative description of type IIA string theory in certain backgrounds known as *matrix string theory* [15]. In this context, the standard action (82), with $p = 1$ and $D = 10$, suitably supersymmetrized, and with gauge group $U(N)$, describes IIA string theory in a Minkowski background. The Yang-Mills coupling constant g_{YM} is inversely related to the string coupling constant g_s. At weak string (strong gauge) coupling, the quartic term forces the non-Abelian coordinates ϕ^a, with $a = 1, \ldots, 8$, to commute, so that they can be considered as ordinary coordinates. One can show that (oversimplifying things a bit, since this should really be thought of as a second quantized description) in this limit one reproduces the usual weak coupling lightcone quantization of the string.

However, the description of string theory based on the action (82) is equally well defined at strong string (weak gauge) coupling, where the full non-Abelian dynamics of the gauge theory becomes important. This is one indication that at strong coupling the target space geometry of a string may be described by a very specific kind of (matrix) non-commutative geometry.

The generalized actions (84,85) arise in the matrix string description of strings propagating in plane wave backgrounds, and the general covariance of this description leads one to a priori expect a relation of the kind (86). These kinds of models, generalizations of the Matrix Big Bang model of [16], become particularly interesting for singular plane waves with a singularity at strong string coupling (so that a perturbative string description is obviously inadequate), and one can investigate what the non-Abelian dynamics (non-commutative geometry) says about what happens at such a space-time singularity. Some of these issues will be explored in [3], from which also the entire discussion of this section 4 is taken.

References

[1] M. Blau, *Lecture Notes on Plane Waves and Penrose Limits* (2004, 87 p.), available from http://www.unine.ch/phys/string/Lecturenotes.html.

[2] M. Blau, M. O'Loughlin, *Homogeneous Plane Waves*, Nucl. Phys. B654 (2003) 135–176, arXiv:hep-th/0212135.

[3] M. Blau, D. Frank, G. Milanesi, M. O'Loughlin, S. Weiss, *in preparation*.

[4] J. Ehlers, W. Kundt, *Exact Solutions of the Gravitational Field Equations*, in *Gravitation: An Introduction to Current Research* (ed. L. Witten), Wiley, New York (1962) 49–101.

[5] D. Kramer, H. Stephani, E. Herlt, M. MacCallum, *Exact solutions of Einstein's field equations*, Cambridge University Press, Cambridge (1980).

[6] A.A. Tseytlin, *Exact solutions of closed string theory*, Class. Quant. Grav. 12 (1995) 2365–2410, arXiv:hep-th/9505052.

[7] M. Blau, J. Figueroa-O'Farrill, C. Hull, G. Papadopoulos, *A new maximally supersymmetric background of IIB superstring theory*, JHEP 0201 (2002) 047, arXiv:hep-th/0110242.

[8] R.R.Metsaev, *Type IIB Green-Schwarz superstring in plane wave Ramond-Ramond background*, Nucl. Phys. B625 (2002) 70–96, arXiv:hep-th/0112044.

[9] M. Blau, J. Figueroa-O'Farrill, C. Hull, G. Papadopoulos, *Penrose limits and maximal supersymmetry*, Class. Quant. Grav. 19 (2002) L87-L95, arXiv:hep-th/0201081.

[10] D. Berenstein, J. Maldacena, H. Nastase, *Strings in flat space and pp waves from $\mathcal{N} = 4$ Super Yang Mills*, JHEP 0204 (2002) 013, arXiv:hep-th/0202021.

[11] M. Blau, M. Borunda, M. O'Loughlin, G. Papadopoulos, *Penrose Limits and Space-time Singularities*, Class. Quant. Grav. 21 (2004) L43–L49, arXiv:hep-th/0312029.

[12] M. Blau, M. Borunda, M. O'Loughlin, G. Papadopoulos, *The universality of Penrose limits near space-time singularities*, JHEP 0407 (2004) 068, arXiv:hep-th/0403252

[13] M. Blau, D. Frank, S. Weiss, *Fermi Coordinates and Penrose Limits*, Class. Quantum Grav. 23 (2006) 3993–4010, arXiv:hep-th/0603109.

[14] H.R. Lewis, *Class of exact invariants for classical and quantum time-dependent harmonic oscillators*, J. Math. Phys. 9 (1968) 1976; H.R. Lewis, W.B. Riesenfeld, *An exact quantum theory of the time dependent harmonic oscillator and of a charged particle in a time dependent electromagnetic field*, J. Math. Phys. 10 (1969) 1458.

[15] R. Dijkgraaf, E. Verlinde, H. Verlinde, *Matrix String Theory*, Nucl. Phys. B500 (1997) 43-61, arXiv:hep-th/9703030v3; R. Dijkgraaf, E. Verlinde, H. Verlinde, *Notes on Matrix and Micro Strings*, Nucl. Phys. Proc. Suppl. 62 (1998) 348–362, arXiv:hep-th/9709107v2.

[16] B. Craps, S. Sethi, E. Verlinde, *A Matrix Big Bang*, JHEP 0510 (2005) 005, arXiv:hep-th/0506180v2; B. Craps, *Big Bang Models in String Theory*, Class. Quant. Grav. 23 (2006) 849–881, arXiv:hep-th/0605199v2.

Matthias Blau
Institut de Physique
Université de Neuchâtel
Rue Breguet 1
CH–2000 Neuchâtel
e-mail: matthias.blau@unine.ch

Quantum Field Theory
B. Fauser, J. Tolksdorf and E. Zeidler, Eds., 217–234
© 2009 Birkhäuser Verlag Basel/Switzerland

Canonical Quantum Gravity and Effective Theory

Martin Bojowald

Abstract. Effective actions provide a powerful way to understand low-energy, semiclassical or other regimes of quantum field theories. Although canonical quantum gravity theories, such as loop quantum gravity, are not directly accessible to the usual techniques of computing effective actions, equivalent methods to derive an effective Hamiltonian and effective equations have recently been developed. This provides means to study the semiclassical limit of loop quantum gravity, demonstrating the correct approach to classical behavior and including quantum correction terms. Salient features of loop quantum gravity, its effective equations and applications in cosmological regimes are discussed.

Mathematics Subject Classification (2000). Primary 81Q20; Secondary 81V17.

Keywords. Effective theory, canonical quantum gravity, quantum cosmology, loop quantum gravity, big bang versus big bounce.

1. Loop quantum gravity

Quantization proceeds by turning a chosen basic set of variables, which faithfully parameterize all phase space points and are chosen such that they form a closed set under taking Poisson brackets, and representing the resulting algebra on a Hilbert space. In a local field theory, one usually has a classical formulation of fields on a manifold, whose Poisson relations involve delta functions. The basic variables one chooses to base a quantum theory on, however, must form a well-defined algebra free of divergences. Only then does one have a well-defined algebra whose representations on a Hilbert space can be used as candidates for a quantization This can be achieved by not using local field values but rather integrated, so-called smeared, versions where spatial integrations over arbitrary 3-dimensional regions remove the delta functions. (Such integrations are also used in the definition of creation and annihilation operators via Fourier transformation, which then can be used to construct the Fock representation of local quantum field theory from the cyclic vacuum state.)

Gravity, like any other field theory, requires such smearings of its canonical fields before they are quantized. The difficulty here is that an integration measure or other structures are required to make the integrations well-defined, but in gravity the metric itself is dynamical and to be smeared before quantization. If one tries to use the metric for its own smearing, non-linear objects without any chance of forming a closed algebra under Poisson brackets result. The other option is to use a separate, non-physical background metric with the sole purpose of defining the integration used in the smearing of the physical metric. But then the quantization is likely to depend on this chosen background, and it is usually difficult to obtain background independent statements. Moreover, strong field effects as they are presented by space-time singularities of the gravitational field are difficult to control in a setting based on a fixed background.

It is thus advantageous to start the quantization, if possible, without introducing such a metric background in the first place. Loop quantum gravity [1, 2, 3] is the prime example for such an approach. It is based on the basic observation that connection variables [4, 5] are available for gravity and allow natural smearings to holonomies. Similarly, the momentum of the connection is a densitized vector field (the metric-independent dual of a 2-form) which can equally naturally be integrated to fluxes:

$$ h_e(A) = \mathcal{P} \exp \int_e A_a^i \tau_i \dot{e}^a \, \mathrm{d}t \quad , \quad F_S(E) = \int_S \mathrm{d}^2 y \, n_a E_i^a \tau_i \qquad (1.1) $$

with tangent vectors $\dot{e}^a$ to curves in space, co-normals n_a to surfaces, and Pauli matrices τ_i. For these variables, the Poisson bracket relations result in an algebra which has a unique cyclic and (spatial) diffeomorphism covariant representation [6, 7]. (Cyclicity of the representation is crucial for the uniqueness as demonstrated in [8].)

Curves and surfaces are to be taken as submanifolds of space, not space-time, because the formulation is canonical. Thus, also diffeomorphisms used for covariance of the representation are only spatial, while a further constraint will have to be imposed to recover space-time covariance. At this stage, an auxiliary structure becomes relevant which one can also view as a kind of background – the space-time foliation on which any canonical quantization is based. The classical canonical theory is independent of the foliation once the constraints of general relativity are imposed. After quantization, independence of this structure is not guaranteed automatically and has to be verified explicitly for the suggested quantum theory: The quantized constraints as operators have to form a first class algebra under commutation such that the number of gauge symmetries is preserved by the quantization. One thus has to test the anomaly issue for the quantum constraints, an important problem also for loop quantum gravity. Only an anomaly-free quantization – or a consistent deformation of the classical theory by quantum corrections – could be regarded background independent also from the space-time point of view.

The connection formulation is of first order, where at first the spatial metric q_{ab} is replaced by a co-triad e_a^i such that $q_{ab} = e_a^i e_b^i$. The momentum conjugate to the connection is directly related to the co-triad: it is the densitized triad $E_i^a = \left| \det(e_b^j) \right| e_i^a$. By itself it already defines a unique connection, the spin connection, by requiring that it is covariantly constant under this derivative. Explicitly, the spin connection is given in terms of the triad components by

$$\Gamma_a^i = -\epsilon^{ijk} e_j^b (\partial_{[a} e_{b]}^k + \tfrac{1}{2} e_k^c e_a^l \partial_{[c} e_{b]}^l) \,. \tag{1.2}$$

As a functional of e_a^i, it has vanishing Poisson brackets with E_i^a and cannot serve as the canonical connection. But from the spin connection we can define another connection, the Ashtekar connection $A_a^i = \Gamma_a^i + \gamma K_a^i$ [4] which, due to the presence of extrinsic curvature K_{ab} in $K_a^i := e_i^b K_{ab}$, is canonically conjugate to E_i^a. (Additional torsion components not present in (1.2) would occur when fermionic matter is coupled to gravity.) This connection is no longer uniquely defined; it rather contains a 1-parameter ambiguity labeled by the Barbero–Immirzi parameter $\gamma > 0$ [5, 9]. Classical dynamics does not depend on its value since it can be changed by a simple canonical transformation. Nevertheless, this parameter will have meaning in quantum gravity, where no unitary transformation is available to change the value. (This is related to a compactification which arises in the transition from the classical to the quantum configuration space.)

Loop quantum gravity is based on a canonical quantization of the phase space spanned by A_a^i and E_i^a using holonomies and fluxes as the separating and smeared set of functions. Even though holonomies are not linear in the basic fields, their algebra with fluxes under Poisson brackets is in fact linear and sufficiently simple to lead to direct representations. The representation space is most conveniently expressed as consisting of states in the connection representation, i.e. as functionals on the space of connections. After quantization, the quantum configuration space is an extension of the classical configuration space by distributional connections. This quantum configuration space turns out to be compact as a consequence of the loop quantization: Holonomies are used to represent the connection and, for a given edge, take values in a compact group, SU(2). The quantum configuration space remains compact even when all curves in space are allowed as necessary to capture all field theoretical degrees of freedom [10].

Spatial geometry is given through flux operators quantizing the densitized triad through invariant vector fields on SU(2). Since those operators have discrete spectra, spatial geometry reconstructed from flux operators is discrete [11, 12, 13]. In this non-perturbative quantization, states describe the excitations of all of spatial quantum geometry. As the classical spatial geometry is determined solely in terms of E_i^a, so is quantum geometry determined by expectation values of flux operators in a given state as a solution. By construction, there is no background metric which could contribute to the geometrical structure. Thus, what one can consider to be the "vacuum" of this theory has no fluxes excited, and no nondegenerate mean geometry at all. Every nearly smooth geometry must be a highly excited state, even Minkowski space or other backgrounds on which usual vacua of quantum field theories are based.

To make contact with quantum field theory on a curved spacetime, or even with (cosmological) observations, a "low energy description" is required. This should determine the behavior of excitations near local Minkowski or Friedmann–Robertson–Walker space, and can be used to verify the semiclassical limit dynamically. A low energy effective action as known from particle physics would be ideal for this purpose, but usual techniques of its derivation fail: A canonical quantization is used which cannot easily be formulated in terms of path integrations, and the lack of a useful vacuum implies that it is not even clear what state one could use to expand the theory around. Fortunately, a generalization is available which reproduces the 1-particle irreducible low energy effective action in regimes where the usual techniques apply, but which can also deal with canonical quantizations and with a different state structure [14, 15, 16].

2. Effective equations

The canonical formulation of effective descriptions starts with a straightforward extension of Ehrenfest's theorem. To highlight several of its physical and algebraic features, we first illustrate its application to the anharmonic oscillator with Hamiltonian

$$\hat{H} = \frac{1}{2m}\hat{p}^2 + V(\hat{q}) = \frac{1}{2m}\hat{p}^2 + \frac{1}{2}m\omega^2\hat{q}^2 + \frac{1}{3}\lambda\hat{q}^3$$

i.e. with anharmonic potential $V(q) = \frac{1}{2}m\omega^2 q^2 + \frac{1}{3}\lambda q^3$. In a semiclassical state, "classical" variables are given by expectation values of the basic operators $\hat{q}$ and $\hat{p}$. One can thus expect such expectation values to play a role in effective equations. However, due to the notorious difficulty of finding explicit dynamical semiclassical states for arbitrary interacting systems, it is useful to formulate equations of motion in general, without using any semiclassicality assumptions. Suitable conditions for semiclassical or coherent states to be used in an analysis of the resulting equations will be discussed self-consistently at a later stage.

Expectation values in any state satisfy equations of motion whose form is well-known from the Ehrenfest theorem:

$$\frac{\mathrm{d}}{\mathrm{d}t}\langle\hat{q}\rangle = \frac{1}{i\hbar}\langle[\hat{q},\hat{H}]\rangle = \frac{1}{m}\langle\hat{p}\rangle \tag{2.1}$$

$$\frac{\mathrm{d}}{\mathrm{d}t}\langle\hat{p}\rangle = \frac{1}{i\hbar}\langle[\hat{p},\hat{H}]\rangle = -m\omega^2\langle\hat{q}\rangle - \lambda\langle\hat{q}^2\rangle = -m\omega^2\langle\hat{q}\rangle - \lambda\langle\hat{q}\rangle^2 - \lambda(\Delta q)^2$$

$$= -V'(\langle\hat{q}\rangle) - \lambda(\Delta q)^2. \tag{2.2}$$

In general, as explicitly seen for this anharmonic system, quantum equations of motion couple expectation values to fluctuations such as $(\Delta q)^2 - \langle(\hat{q}-\langle\hat{q}\rangle)^2\rangle$, or higher moments of the state for higher monomials in the anharmonicity. The additional term in the equation of motion for $\langle\hat{p}\rangle$ requires a quantum correction to the classical dynamics. The effect of such terms is being captured in effective equations, which are to be computed from a more detailed analysis of the quantum evolution. Here, specific properties of semiclassical states will be derived and enter the final result.

2.1. Quantum back-reaction

Fluctuations as they appear in contributions to the equations of motion of expectation values are themselves dynamical: states in general spread and thus fluctuations depend on time. Their equations of motion can be derived as those of expectation values above, e.g.

$$
\begin{aligned}
\frac{\mathrm{d}}{\mathrm{d}t}(\Delta q)^2 &= \frac{\mathrm{d}}{\mathrm{d}t}(\langle \hat{q}^2 \rangle - \langle \hat{q} \rangle^2) = \frac{1}{i\hbar}\langle [\hat{q}^2, \hat{H}] \rangle - 2\langle \hat{q} \rangle \frac{\mathrm{d}}{\mathrm{d}t}\langle \hat{q} \rangle \\
&= \frac{1}{m}\langle \hat{q}\hat{p} + \hat{p}\hat{q} \rangle - \frac{2}{m}\langle \hat{q} \rangle \langle \hat{p} \rangle = \frac{2}{m}C_{qp}\,.
\end{aligned}
$$

This equation of motion, in turn, requires the covariance $C_{qp} = \frac{1}{2}\langle \hat{q}\hat{p} + \hat{p}\hat{q} \rangle - \langle \hat{q} \rangle \langle \hat{p} \rangle$ to be known as a function of time, which evolves according to

$$
\frac{\mathrm{d}}{\mathrm{d}t}C_{qp} = \frac{1}{m}C_{qp} + m\omega^2(\Delta q)^2 + 6\lambda\langle \hat{q} \rangle (\Delta q)^2 + 3\lambda G^{0,3}\,.
$$

Here, in addition to a non-linear term coupling the expectation value $\langle \hat{q} \rangle$ to the fluctuation of q we encounter the moment

$$
G^{0,3} := \langle (\hat{q} - \langle \hat{q} \rangle)^3 \rangle = \langle \hat{q}^3 \rangle - 3\langle \hat{q} \rangle (\Delta q)^2 - \langle \hat{q} \rangle^3
$$

of third order, which is related to deformations of the state away from a Gaussian, or its skewness. Iteration of this procedure shows that all infinitely many moments, or quantum variables

$$
G^{a,n} := \left\langle \left((\hat{q} - \langle \hat{q} \rangle)^{n-a}(\hat{p} - \langle \hat{p} \rangle)^a \right)_{\mathrm{symm}} \right\rangle \tag{2.3}
$$

of a state, defined using totally symmetric ordering, are coupled to each other and to expectation values.

This whole system of infinitely many ordinary differential equations is equivalent to the partial Schrödinger equation. Effective equations, by contrast, involve only finitely many local degrees of freedom. They thus require further knowledge to be gleaned from a partial solution of the equations of motion. If $(\Delta q)(\langle \hat{q} \rangle, \langle \hat{p} \rangle)$ can be determined by some means, for instance, inserting it into $\frac{\mathrm{d}}{\mathrm{d}t}\langle \hat{p} \rangle = -V'(\langle \hat{q} \rangle) - \lambda(\Delta q)^2$ results in effective equations for $q = \langle \hat{q} \rangle$ and $p = \langle \hat{p} \rangle$ in closed form. Since expectation values and fluctuations are independent of each other in general states, a truncation of the infinite system to a finite dimensional effective one can only be achieved for a specific class of states satisfying, for instance, certain semiclassicality conditions. The availability and form of any type of such conditions can be determined self-consistently from the fact that a truncation of the full system of equations or a partial solution to find, e.g., $\Delta q(q, p)$ independently of specific solutions for $q(t)$ and $p(t)$ must be possible.

For perturbative potentials around the harmonic oscillator, an adiabatic and semiclassical approximation decouples the equations and allows one to compute

$(\Delta q)(q, p)$ order by order. Semiclassicality implies that moments of higher order are suppressed by powers of $\hbar$, while adiabaticity is used for the quantum variables mirroring the fact that one is perturbing around the ground state of the harmonic oscillator which has quantum variables constant in time. Implicitly, corrections in effective equations take into account that quantum variables in the interacting ground state of the anharmonic system are no longer precisely constant but change only weakly in a perturbative setting.

Using the result to first order in $\hbar$ and second in the adiabatic approximation, and formulating the corresponding first order equations for q and p as a second order equation for q yields [14]

$$\left(m + \frac{\hbar U'''(q)^2}{32 m^2 \omega^5 \left(1 + \frac{U''(q)}{m\omega^2}\right)^{\frac{5}{2}}} \right) \ddot{q} + m\omega^2 q + U'(q) + \frac{\hbar U'''(q)}{4 m \omega \left(1 + \frac{U''(q)}{m\omega^2}\right)^{\frac{1}{2}}}$$

$$+ \frac{\hbar \left(4 m \omega^2 U'''(q) U''''(q) \left(1 + \frac{U''(q)}{m\omega^2}\right) - 5 U'''(q)^3\right)}{128 m^3 \omega^7 \left(1 + \frac{U''(q)}{m\omega^2}\right)^{\frac{7}{2}}} \dot{q}^2 = 0$$

with general anharmonic potential $U(q)$. This is indeed in agreement with the 1-particle irreducible low energy effective action [17]

$$\Gamma_{\text{eff}}[q] = \int dt \left(\left(m + \frac{\hbar(U''')^2}{32 m^2 \left(\omega^2 + \frac{U''}{m}\right)^{\frac{5}{2}}} \right) \frac{\dot{q}^2}{2} \right.$$

$$\left. - \frac{1}{2} m \omega^2 q^2 - U - \frac{\hbar\omega}{2} \left(1 + \frac{U''}{m\omega^2}\right)^{\frac{1}{2}} \right).$$

2.2. General procedure

These basic principles of the quantum back-reaction of a spreading and deforming state are more widely applicable than the 1-particle irreducible effective action. In particular, the derivation was purely canonical, and so analogous ones can be performed for canonical quantizations such as loop quantum gravity. There is, however, one requirement for its feasibility: there must be a relation to a free system where quantum variables decouple. In such a system one needs to solve only finitely many coupled equations, without any approximations or quantum corrections from quantum coupling terms. This is realized only in very special cases, but often one can use perturbations around those systems and derive effective equations as approximations even in the presence of couplings.

Decoupling is realized if the Hamiltonian is quadratic in canonical variables, since $[\cdot, \hat{H}]$ is then linear in basic operators and equations of motion for their expectation values depend only on the expectation values themselves. The prime example is the harmonic oscillator as used in the derivation above, but decoupling can happen more generally. It occurs for any linear system defined by basic variables J_i which (i) form a linear commutator algebra and (ii) have a linear combination which equals the Hamiltonian $\hat{H}$. This will be used later in quantum cosmology.

Given a linear system, one can explicitly solve its equations of motion for quantum variables, determining the moments of its dynamical coherent states. For perturbations around such a system one obtains coupled equations and has to analyze their form to find suitable conditions and approximations in which they can be truncated. While one often makes use of a semiclassical approximation to justify ignoring most of the quantum variables, other steps such as the adiabaticity assumption used for the low energy effective action may not always be valid. This assumption is based on more specific properties of the solvable system, namely the fact that the ground state of the harmonic oscillator has constant quantum variables. They can thus remain adiabatic even under perturbations of the system. But not all linear systems have a ground (or other) state with constant quantum variables and there may be no other way to realize adiabaticity. Then, the $G^{a,n}$ may not be solvable explicitly as functions of J_i which one could then insert into the equations of motion for expectation values. Still, the semiclassical approximation allows a decoupling of almost all quantum variables; one will, to a given order, only have to keep a finite number of them as independent quantum degrees of freedom in addition to the classical-type expectation values. In this way, one obtains higher dimensional effective systems which are dynamical systems formulated in terms of more than just the classical variables. In addition to quantum correction terms such as those in an effective potential, such systems have additional variables and equations for true quantum degrees of freedom.

This emphasis on observables, i.e. expectation values, fluctuations and other moments, rather than states provides a more algebraic viewpoint which has additional advantages beyond the fact that one obtains a good approximation to quantum behavior. By directly solving equations of motion for moments, rather than first solving for a state and then integrating out to obtain moments, one can bypass a calculational step which is often analytically difficult or numerically time consuming. Especially in semiclassical regimes states are often highly oscillatory, which would require numerical integrations in expectation values with very fine resolutions. This can be avoided by directly using numerical solutions to the coupled equations for expectation values and some moments. Also analytically, after one successfully truncated the infinite set of equations by semiclassical or other approximations, one has only one step of solving coupled ordinary differential equations instead of solving a partial differential equation for the state and then still having to integrate it to find observables of real interest.

The procedure has an underlying geometrical picture [18, 19, 20]: The dynamics is formulated on the quantum phase space which is given by the (projective) Hilbert space. Points in the phase space are thus states, and so are tangent vectors since the Hilbert space is linear. The symplectic form as a 2-form is then an antisymmetric bilinear form which one can define as $\Omega(\cdot, \cdot) = 2\hbar \text{Im}\langle \cdot | \cdot \rangle$ in terms of the imaginary part of the inner product. (The real part defines a metric, which together with the symplectic form and the usual complex structure of quantum mechanics makes the quantum phase space Kähler. This extra structure is, however, not necessary for discussing dynamics.)

The Hamiltonian operator, like any operator, defines a function, the quantum Hamiltonian $H_Q(\cdot) = \langle \cdot | \hat{H} | \cdot \rangle$, on the quantum phase space. It is the Hamiltonian which on the phase space generates the flow defined by the Schrödinger equation for states. For an efficient evaluation, one can expand the quantum Hamiltonian as a formal series in quantum variables, which we do in dimensionless variables $\tilde{G}^{a,n} = \hbar^{-n/2}(m\omega)^{n/2-a}G^{a,n}$. In a formal Taylor expansion, this yields

$$
H_Q = \langle H(\hat{q}, \hat{p}) \rangle = \langle H(q + (\hat{q} - q), p + (\hat{p} - p)) \rangle = \frac{1}{2m}p^2 + \frac{1}{2}m\omega^2 q^2 \quad (2.4)
$$

$$
+ U(q) + \frac{\hbar\omega}{2}(\tilde{G}^{0,2} + \tilde{G}^{2,2}) + \sum_{n>2}\frac{1}{n!}\left(\frac{\hbar}{m\omega}\right)^{n/2} U^{(n)}(q)\tilde{G}^{0,n}
$$

showing explicitly the coupling terms between expectation values and quantum variables for a potential of higher than quadratic order. (For a quadratic Hamiltonian the quantum Hamiltonian does depend on quantum variables, but only in its zero-point energy which does not provide any coupling terms.) Using the symplectic structure and the Poisson relations it implies for quantum variables one can see that this quantum Hamiltonian produces the same equations of motion as derived before from expectation values of commutators.

We end this general discussion with a list of several additional advantages to this canonical treatment of effective equations:

- Given a solvable model, in which quantum variables decouple from expectation values, properties of dynamical coherent states, i.e. all moments, can be computed order by order in a semiclassical expansion around the free theory. In quantum field theory language, one computes the interacting vacuum perturbatively. This is possible even if no explicit form for such states is known, which represents the typical case faced for interacting systems. Here, coherent states are computed by first solving equations of motion for fluctuations, as we saw them in earlier examples, and then saturating the uncertainty relations

$$
G^{AA}G^{BB} - (G^{AB})^2 \geq \frac{1}{4}(i\langle[\hat{A}, \hat{B}]\rangle)^2 \quad (2.5)
$$

which hold for any self-adjoint $\hat{A}$, $\hat{B}$ (with a straightforward change of notation in the quantum variables compared to (2.3)).

- Other advantages appear for constrained systems such as gravity: In analogy to the quantum Hamiltonian, for constrained systems a major role is played by the effective constraint $\langle \cdot | \hat{C} | \cdot \rangle$ defined through expectation values of constraint operators. (This case is more involved, however, since one needs more constraints even if there is only one classical constraint. Just as there are now infinitely many quantum variables, we require infinitely many effective constraints through higher powers, e.g., $\langle \cdot | \hat{C}^n | \cdot \rangle$ to remove all unwanted quantum variables.) The advantage of the effective treatment now is that one can rather straightforwardly address the physical inner product, which is simply implemented by reality conditions for the moments. No integral representation with an explicit measure factor is required, which can be found systematically only in a small number of examples. Reality conditions for expectation values and quantum variables can, on the other hand, be directly imposed, immediately achieving the correct adjointness conditions for basic operators which the physical inner product is intended to ensure. Also here, foregoing specific representations of states helps enormously in analyzing the quantum system in an economic way.

• One other issue of importance for constrained systems can be analyzed with significantly more ease: that of anomalies. We can compute effective constraints as expectation values without worrying about potential anomalies in the constraint operators. Problems will only show up as inconsistencies if we try to solve the effective constraints of an anomalous quantization. Instead of determining a complete, anomaly-free quantization which is often difficult, one can first compute effective constraints and then make sure that they are anomaly-free. This may not tell one the complete anomaly-free quantum theory, or even demonstrate its existence. But one can already test whether obstructions to an anomaly-free quantization exist, and if not derive possible physical consequences of quantum effects.

3. A solvable model for cosmology

Gravitational or cosmological systems are not closely related to a harmonic oscillator and thus we need to look for other linear systems if our aim is to make effective equations available for cosmology. Fortunately, such systems exist: The Friedmann equation (the Hamiltonian constraint for isotropic gravity) for a spatially flat model sourced by a free, massless scalar ϕ with momentum p_ϕ is

$$c^2 \sqrt{p} = \frac{4\pi G}{3} p^{-3/2} p_\phi^2$$

written here in variables $c = \dot{a}$ and p with $p^{3/2} = a^3$, which are closer to Ashtekar variables. Solving this for p_ϕ yields $|p_\phi| \propto |cp| =: H$, which can be interpreted as the Hamiltonian generating the flow in the variable ϕ (rather than a coordinate time whose flow is generated by the Hamiltonian constraint). We thus choose ϕ as our internal time variable, describing evolution not with respect to a coordinate time but relationally between matter and geometry. This does not only easily eliminate the problem of time in this system – in the absence of a potential, ϕ is monotonic and thus provides a global internal time – it also gives us the desired solvable system: the ϕ-Hamiltonian is quadratic in canonical variables (c, p) and we have the same decoupling of classical and quantum variables as realized for the harmonic oscillator. (In fact, a simple canonical transformation shows that the Hamiltonian cp is an upside-down harmonic oscillator.)

The easy availability of a global internal time and the structure of the constraint equation were exploited in [21, 22] to determine the physical Hilbert space of this system by techniques known from the Klein–Gordon particle, and to analyze some solutions given by Gaussian initial states numerically. The solvable nature of this system was realized in [23], which explained several features of the numerical solutions which were unexpected from general experience of quantum systems.

Quantum equations of motion for the solvable system with Hamiltonian[1] $H = cp$ are

$$\dot{c} = c \quad , \quad \dot{p} = p \quad , \quad \dot{G}^{0,2} = 2G^{0,2} \quad , \quad \dot{G}^{1,2} = 0 \quad , \quad \dot{G}^{2,2} = -2G^{2,2}$$

and so on for higher moments, which are easily solved by $c(t) = c_1 e^t$, $p(t) = c_2 e^{-t}$, $G^{0,2}(t) = c_3 e^{-2t}$, $G^{1,2}(t) = c_4$ and $G^{2,2}(t) = c_5 e^{2t}$. The solution for p directly shows the classical singularity since $p = 0$ is reached for $\phi \to \infty$ which, as one can check by gauging the system, corresponds to a finite proper time interval. One clearly sees here that the classical singularity is realized in the same way when the system is quantized in the Schrödinger representation, as done in Wheeler–DeWitt quantum cosmology. In fact, due to the solvability there are no quantum correction terms whatsoever, and the quantum equations of motion for expectation values are exactly the same as the classical equations of motion.

For semiclassical issues, one can analyze the solutions for fluctuations. They are subject to the uncertainty relation

$$G^{0,2} G^{2,2} - (G^{1,2})^2 \geq \frac{1}{4}\hbar^2 \tag{3.1}$$

which for the constants of integration in our general solutions implies $c_3 c_5 \geq \hbar^2/4 + c_4^2$. If a state is initially chosen to be semiclassical, this property is preserved during evolution: we have constant relative spreads $(\Delta p)/p = \sqrt{G^{2,2}}/p$ and $(\Delta c)/c = \sqrt{G^{0,2}}/c$. This also illustrates why the classical singularity cannot be resolved: There are simply no significant quantum effects in this system, no matter how close we get to the classical singularity.

This is different in a loop formulation. Here, in loop quantum cosmology [25, 26], we do not have an operator for a connection component like c, but only for holonomies as basic operators. While the precise form of a loop quantized Hamiltonian constraint can be complicated, the main effect is that now a periodic function such as $\sin c$ replaces c in the Friedmann equation. This provides a type of quantum geometry effect through the addition of higher powers of extrinsic curvature. We now expect deviations from classical behavior when c becomes large, just when we approach the classical big bang singularity.

But it also means that the loop Hamiltonian $H = p \sin c$ now is non-quadratic in canonical variables and thus, in this form, not solvable. The corresponding

---

[1]Note that we are dropping the absolute value to have a truly quadratic Hamiltonian. This can be shown to be a safe procedure when properties of semiclassical states are of interest [24].

quantum back-reaction would be too complicated for solving the equations directly. While a perturbative treatment around the free Wheeler–DeWitt model with Hamiltonian cp would be possible for small c, this would not give us reliable information about what happens to the classical singularity. Fortunately, the system is, indeed, solvable if we only choose a specific factor ordering and formulate it in new basic variables whose algebra under Poisson brackets is not of the canonical form.

To some degree, the use of new variables is anyway required because we have to refer to "holonomies"[2] e^{ic} instead of c itself, so that we have basic Poisson brackets $\{p, e^{ic}\} = -ie^{ic}$. But also in these variables the system is not solvable. We rather introduce $\hat{J} = \hat{p}\widehat{e^{ic}}$, already defined as an operator because its factor ordering is important for solvability properties, and obtain the linear Hamiltonian $\hat{H} = -\frac{1}{2}i(\hat{J} - \hat{J}^\dagger)$ which defines its factor ordering in the form used below. (Due to the use of $\widehat{e^{ic}}$ this is a finite shift operator: the dynamical equation of loop quantum gravity is a difference equation for states [27]. Here, however, we will not use this difference equation directly but rather describe its solutions by more powerful means directly providing effective equations.)

A linear Hamiltonian in canonical variables would directly imply solvability, but we use non-canonical ones $(\hat{p}, \hat{J})$. As one can easily check, these operators obey a (trivially) centrally extended $sl(2, \mathbb{R})$ algebra

$$[\hat{p}, \hat{J}] = \hbar\hat{J} \quad , \quad [\hat{p}, \hat{J}^\dagger] = -\hbar\hat{J}^\dagger \quad , \quad [\hat{J}, \hat{J}^\dagger] = -2\hbar\hat{p} - \hbar^2 \, . \tag{3.2}$$

Since the Hamiltonian is a linear combination of two of the basic operators, the dynamical system is linear and thus indeed solvable. We also mention here already that the new variables force us to work with non-symmetric basic operators. Thus, we will have to impose reality conditions after finding solutions, which corresponds to implementing the correct physical inner product to normalize wave functions in which expectation values and moments are computed.

The equations of motion, derived as before, are

$$\dot{p} = -\tfrac{1}{2}(J + \bar{J}) \quad , \quad \dot{J} = -\tfrac{1}{2}(p + \hbar) = \dot{\bar{J}} \tag{3.3}$$

for expectation values (which variables without hat now refer to) with general solution

$$p(\phi) = \tfrac{1}{2}(c_1 e^{-\phi} + c_2 e^{\phi}) - \tfrac{1}{2}\hbar \tag{3.4}$$

$$J(\phi) = \tfrac{1}{2}(c_1 e^{-\phi} - c_2 e^{\phi}) + iH \, . \tag{3.5}$$

---

[2] The use of these exponentials derives from the use of holonomies as basic variables in the full theory. This would lead one to consider all exponentials $e^{i\mu c}$ with real μ rather than only functions on a circle of fixed periodicity [26]. The configuration space of isotropic connections is thus not periodically identified to a circle, but instead compactified to the Bohr compactification of the real line on which almost periodic functions, i.e. countable linear combinations $\sum_I f_I e^{i\mu_I c}$ for $\mu_I \in \mathbb{R}$, form the C^*-algebra of all continuous functions. Dynamically, however, it turns out to be sufficient to consider only integer powers of a periodic e^{ic} due to superselection.

In contrast to the Wheeler–DeWitt model, we indeed see a resolution of the classical singularity: There is a "bounce" in the sense that $|p| \to \infty$ for $\phi \to \pm\infty$. However, without further analysis one could not be sure that the state does not enter the quantum regime too deeply, where other quantum geometry corrections not included in the solvable model would set in (e.g. those of [28], based on [43] in the full theory).

The model itself could thus break down, invalidating the naive conclusion about the bounce. In particular, the solutions show that, for $c_1 c_2 < 0$, p can become zero where the model could no longer be trusted. (Nevertheless, the underlying difference equation is non-singular since it extends the wave function uniquely through the classical singularity [29, 30]. But this behavior of the wave function itself does not easily provide an effective or geometrical picture of how the classical singularity is resolved.) The precise form in which a deep quantum regime could be entered depends, by its very nature, on details of the quantum model, in particular the factor ordering of the constraint. There may be choices for which the expectation value of $\hat{p}$ never becomes zero, but this does not rule out that a deep quantum regime can be entered, where the model would break down. In any case one has to provide a detailed analysis if one tries to justify the bounce under general conditions.

The situation is simpler when one assumes that a state starts out semiclassically at large volume, and then follows its evolution to smaller scales and possibly back to larger ones after a bounce. As in the Wheeler–DeWitt model, the state would not change its semiclassicality properties strongly while it is still at rather large volume. Semiclassicality properties can, however, change in the bounce transition between two branches.

For a correct analysis of the corresponding solutions we now have to impose the reality conditions. Classically we have $J\bar{J} = p^2$ for $J = p\exp(ic)$. When quantized, $\exp(ic)$ must become an unitary operator on the physical Hilbert space, which implies reality conditions for our basic operators. Although we use $\hat{J}$ instead of $\widehat{e^{ic}}$ and cannot directly implement unitarity, there is a similar condition related to unitarity for the basic operators: $\hat{J}\hat{J}^\dagger = \hat{p}^2$. This is not linear, and so taking an expectation values of this relation provides a condition involving expectation values as well as fluctuations,

$$|J|^2 - (p + \tfrac{1}{2}\hbar)^2 = \tfrac{1}{4}\hbar^2 - G^{J\bar{J}} + G^{pp}. \tag{3.6}$$

(We are now denoting the second order quantum variables by subscripts of the operators used. This is more convenient than the notation $G^{a,n}$ because we have to distinguish between $\hat{J}$ and $\hat{J}^\dagger$ in addition to $\hat{p}$.)

Inserting solutions, this implies $c_1 c_2 = H^2 + O(\hbar)$ for the expectation values where we assumed an initial semiclassical state where fluctuations contribute only terms of the order $\hbar$. Since this relation only depends on constants of motion (including the $O(\hbar)$-terms since one can see easily that $G^{J\bar{J}} - G^{pp}$ is preserved), it is valid at all times and even at the bounce. Moreover, for semiclassical states the H^2-term dominates since it represents the large matter contribution of the universe, implying that $c_1 c_2$ is positive and $p = 0$ is never reached by solutions which are semiclassical at one time. This is directly seen from the bouncing solutions

$$p(\phi) = H\cosh(\phi - \delta) - \hbar \quad , \quad J(\phi) = -H(\sinh(\phi - \delta) - i) \qquad (3.7)$$

(with $e^{2\delta} = c_2/c_1$) which follow from (3.4) and (3.5) taking into account the reality conditions.

Equations of motion for fluctuations are

$$
\begin{aligned}
\dot{G}^{pp} &= -G^{pJ} - G^{p\bar{J}} \\
\dot{G}^{JJ} &= -2G^{pJ} \quad , \quad \dot{G}^{\bar{J}\bar{J}} = -2G^{p\bar{J}} \\
\dot{G}^{pJ} &= -\frac{1}{2}G^{JJ} - \frac{1}{2}G^{J\bar{J}} - G^{pp} \quad , \quad \dot{G}^{p\bar{J}} = -\frac{1}{2}G^{\bar{J}\bar{J}} - \frac{1}{2}G^{J\bar{J}} - G^{pp} \\
\dot{G}^{J\bar{J}} &= -G^{pJ} - G^{p\bar{J}}
\end{aligned}
$$

whose solutions determine properties of dynamical coherent states of this system [24]. For $H \gg \hbar$, volume fluctuations are of the form

$$(\Delta p)^2 = G^{pp} \approx \hbar H \cosh(2(\phi - \delta_2)) \, .$$

As one can see, the minimum of fluctuations is determined by a parameter δ_2 independent of δ_1 [24]. The ratio of the two parameters describes the squeezing of the state, and is thus a quantum property which cannot be seen in the expectation values alone. Fluctuations thus do not need to be symmetric around the bounce: they could have been much larger, or smaller, before the big bang than afterwards; see Fig. 1. For known fluctuations, the uncertainty relation (3.1) provides an upper bound on correlations and thus on squeezing. In our model, this implies a bound on the asymmetry of fluctuations once their size can be bounded. However, realistic observations one could make after the big bang cannot provide sharp enough bounds on the squeezing and thus on pre-big bang fluctuations. The infinitely many higher moments are even less restricted, and so it is impossible to know the precise state of the universe before the big bang. Thus, even though the state equations are deterministic, information on the state before the big bang is practically lost during cosmic evolution [31].

3.1. Interactions

The free bounce model may replace the harmonic oscillator as the "free" gravitational theory even in a loop quantization. Realistic ingredients such as a matter potential, anisotropy or inhomogeneities can be included perturbatively to derive effective equations for those models. For a non-zero potential, for instance, we

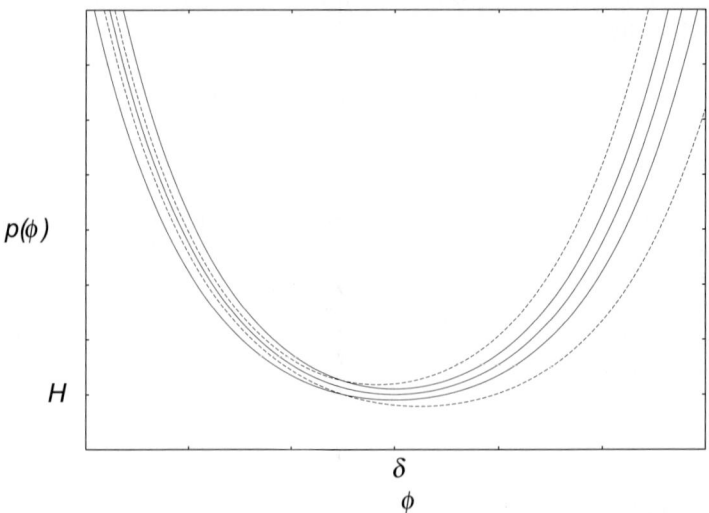

FIGURE 1. Two different solutions of expectation values and fluctuations of the solvable cosmological model, illustrating the bounce and the possible difference in fluctuations before and after the bounce. The solid lines represent the mean and spread of an unsqueezed state, while the state corresponding to the dashed lines is squeezed.

obtain effective equations [32]

$$
\begin{aligned}
\dot{p} ={}& -\frac{J+\bar{J}}{2} + \frac{J+\bar{J}}{(J-\bar{J})^2}p^3 V(\phi) \\
&+3\frac{p^3(J+\bar{J})}{(J-\bar{J})^4}(G^{JJ}+G^{\bar{J}\bar{J}}-2G^{J\bar{J}})V(\phi) \\
&-6\frac{p^2(J+\bar{J})}{(J-\bar{J})^3}(G^{pJ}-G^{p\bar{J}})V(\phi)+3\frac{p(J+\bar{J})}{(J-\bar{J})^2}G^{pp}\,V(\phi) \\
&-\frac{2p^3}{(J-\bar{J})^3}(G^{JJ}-G^{\bar{J}\bar{J}})V(\phi)+\frac{3p^2}{(J-\bar{J})^2}(G^{pJ}+G^{p\bar{J}})V(\phi)
\end{aligned}
$$

for p, accompanied by similar equations for $\dot{J}$ as well as the moments $\dot{G}^{a,2}$ which here appear as independent variables. This is an example for a higher dimensional effective system of nine real variables (the expectation values p, $\mathrm{Re}J$, $\mathrm{Im}J$ and fluctuations and covariances G^{pp}, $G^{J\bar{J}}$, $\mathrm{Re}G^{pJ}$, $\mathrm{Im}G^{pJ}$, $\mathrm{Re}G^{JJ}$ and $\mathrm{Im}G^{JJ}$). Unlike the case of perturbations around the harmonic oscillator, quantum variables do not behave adiabatically and assuming so would force one to violate uncertainty relations; see [32] for details. Thus, we cannot solve for the $G^{a,2}$ in terms of expectation values directly, and rather have to keep them as independent variables in a higher dimensional effective system.

4. Effective quantum gravity

Many field theoretical issues present in this approach to effective equations remain to be worked out, which would be required for an application to quantum gravity beyond cosmological models. In this context, it is encouraging that variables similar to the $G^{a,n}$ were used in early (non-canonical) developments of effective field theories in the context of symmetry breaking [33, 34]. Some difficulties may be overcome by using a mode decomposition of fields, made possible by the choice of a background geometry which enters the effective description as part of the semi-classical states used. This strategy is applied to inflationary structure formation in [35].

For loop quantum gravity, moreover, one may be in a manageable situation because the discreteness of graphs implies essentially a finite number of degrees of freedom in any finite volume, in contrast to infinitely many ones in ordinary quantum field theory. One can thus use moments between all independent graph-associated degrees of freedom, and then perform a continuum approximation for the resulting effective expressions. Only after performing this limit would the infinitely many degrees of freedom of the classical theory arise. This procedure, without explicitly computing all quantum correction terms, is outlined in [36].

This allows one to apply existing techniques available for the derivation of effective systems to a derivation of cosmological perturbation equations [37, 38, 39, 40, 41]. As a result one obtains quantum corrections arising from the loop quantization, which can then be directly analyzed in cosmological phenomenology, or for fundamental questions such as the issue of local Lorentz invariance at low energies. Moreover, from a combination of the perturbation equations one can recover the Newton potential to classical order and also here derive quantum corrections.

As already mentioned, one can also study some part of the anomaly issue of constrained systems at the effective level. Effective constraints can first be computed from operators which have not yet been checked for possible anomalies. For the effective constraints anomalies can then be analyzed much more straightforwardly by computing their Poisson algebra. Even before complete effective constraints have been computed, which is so far not available in inhomogeneous systems, one may include only one type of quantum corrections into an incomplete effective constraint. In addition to quantum back-reaction effects, whose computation is typically most involved, there are several effects arising from the discreteness of quantum geometry. The latter include higher order terms due to the use of holonomies in the basic representation, and corrections to inverse densitized triad components [42, 43]. Any of these terms can be introduced as part of an effective constraint, and their effect on anomalies studied in separation.

There have recently been several results in particular for inverse triad corrections: Such non-trivial quantum corrections are possible even for anomaly-free constraints. This is rather surprising as it has often been expected that quantum corrections due to the underlying discreteness necessarily break covariance. As the calculations show, this may happen for some form of the corrections, and possible quantization ambiguities which may exist at the kinematical level are thus restricted. But anomaly-freedom does not require such quantum corrections to be completely absent and allows non-trivial quantum effects in a covariant quantum space-time. Specific cases which have been dealt with in this way include the linearized cosmological mode equations of [39, 40, 44] and the fully non-linear spherically symmetric vacuum sector [45].

A different issue is the performance of the continuum limit for a sector of the quantum theory defined by introducing a background via a specific class of states (such as those of semiclassical geometries peaked at the given background). Although one would not usually need to perform such a limit, given that the full theory is already a continuum quantum theory, it is of interest for making contact with local quantum field theory on a background. At this point, one expects that renormalization would have to be performed, whose role in discrete quantum gravity could thus be studied.

Acknowledgement

Work described here was supported in part by NSF grants PHY-0554771 and PHY-0653127.

References

[1] C. Rovelli, *Quantum Gravity*, Cambridge University Press, Cambridge, UK, 2004.

[2] A. Ashtekar and J. Lewandowski, *Background independent quantum gravity: A status report* Class. Quantum Grav. **21** (2004), R53–R152, [gr-qc/0404018].

[3] T. Thiemann, *Introduction to Modern Canonical Quantum General Relativity*, [gr-qc/0110034].

[4] A. Ashtekar, *New Hamiltonian Formulation of General Relativity* Phys. Rev. D **36** (1987), 1587–1602.

[5] J. F. Barbero G., *Real Ashtekar Variables for Lorentzian Signature Space-Times* Phys. Rev. D **51** (1995), 5507–5510, [gr-qc/9410014].

[6] J. Lewandowski, A. Okołów, H. Sahlmann, and T. Thiemann, *Uniqueness of diffeomorphism invariant states on holonomy-flux algebras* Commun. Math. Phys. **267** (2006), 703–733, [gr-qc/0504147].

[7] C. Fleischhack, *Representations of the Weyl Algebra in Quantum Geometry*, [math-ph/0407006].

[8] M. Varadarajan, *Towards new background independent representations for Loop Quantum Gravity*, [arXiv:0709.1680].

[9] G. Immirzi, *Real and Complex Connections for Canonical Gravity* Class. Quantum Grav. **14** (1997), L177–L181.

[10] A. Ashtekar, J. Lewandowski, D. Marolf, J. Mourão, and T. Thiemann, *Quantization of Diffeomorphism Invariant Theories of Connections with Local Degrees of Freedom* J. Math. Phys. **36** (1995), 6456–6493, [gr-qc/9504018].

[11] C. Rovelli and L. Smolin, *Discreteness of Area and Volume in Quantum Gravity* Nucl. Phys. B **442** (1995), 593–619, [gr-qc/9411005], Erratum: *Nucl. Phys. B* 456 (1995) 753.

[12] A. Ashtekar and J. Lewandowski, *Quantum Theory of Geometry I: Area Operators* Class. Quantum Grav. **14** (1997), A55–A82, [gr-qc/9602046].

[13] A. Ashtekar and J. Lewandowski, *Quantum Theory of Geometry II: Volume Operators* Adv. Theor. Math. Phys. **1** (1997), 388–429, [gr-qc/9711031].

[14] M. Bojowald and A. Skirzewski, *Effective Equations of Motion for Quantum Systems* Rev. Math. Phys. **18** (2006), 713–745, [math-ph/0511043].

[15] A. Skirzewski, *Effective Equations of Motion for Quantum Systems*, PhD thesis, Humboldt-Universität Berlin, 2006.

[16] M. Bojowald and A. Skirzewski, *Quantum Gravity and Higher Curvature Actions* Int. J. Geom. Math. Mod. Phys. **4** (2007), 25–52, [hep-th/0606232].

[17] F. Cametti, G. Jona-Lasinio, C. Presilla, and F. Toninelli, *Comparison between quantum and classical dynamics in the effective action formalism* Proceedings of the International School of Physics "Enrico Fermi", Course CXLIII, pages 431–448 (IOS Press, Amsterdam, 2000), [quant-ph/9910065].

[18] T. W. B. Kibble, *Geometrization of quantum mechanics* Commun. Math. Phys. **65** (1979), 189–201.

[19] A. Heslot, *Quantum mechanics as a classical theory* Phys. Rev. D **31** (1985), 1341–1348.

[20] A. Ashtekar and T. A. Schilling, Geometrical Formulation of Quantum Mechanics, In A. Harvey, editor, *On Einstein's Path: Essays in Honor of Engelbert Schücking*, pages 23–65, Springer-Verlag, 1999, [gr-qc/9706069].

[21] A. Ashtekar, T. Pawlowski, and P. Singh, *Quantum Nature of the Big Bang* Phys. Rev. Lett. **96** (2006), 141301, [gr-qc/0602086].

[22] A. Ashtekar, T. Pawlowski, and P. Singh, *Quantum Nature of the Big Bang: An Analytical and Numerical Investigation* Phys. Rev. D **73** (2006), 124038, [gr-qc/0604013].

[23] M. Bojowald, *Large scale effective theory for cosmological bounces* Phys. Rev. D **75** (2007), 081301(R), [gr-qc/0608100].

[24] M. Bojowald, *Dynamical coherent states and physical solutions of quantum cosmological bounces* Phys. Rev. D **75** (2007), 123512, [gr-qc/0703144].

[25] M. Bojowald, *Isotropic Loop Quantum Cosmology* Class. Quantum Grav. **19** (2002), 2717–2741, [gr-qc/0202077].

[26] A. Ashtekar, M. Bojowald, and J. Lewandowski, *Mathematical structure of loop quantum cosmology* Adv. Theor. Math. Phys. **7** (2003), 233–268, [gr-qc/0304074].

[27] M. Bojowald, *Loop Quantum Cosmology IV: Discrete Time Evolution* Class. Quantum Grav. **18** (2001), 1071–1088, [gr-qc/0008053].

[28] M. Bojowald, *Inverse Scale Factor in Isotropic Quantum Geometry* Phys. Rev. D **64** (2001), 084018, [gr-qc/0105067].

[29] M. Bojowald, *Absence of a Singularity in Loop Quantum Cosmology* Phys. Rev. Lett. **86** (2001), 5227–5230, [gr-qc/0102069].

[30] M. Bojowald, *Singularities and Quantum Gravity* AIP Conf. Proc. **910** (2007), 294–333, [gr-qc/0702144].

[31] M. Bojowald, *What happened before the big bang?* Nature Physics **3** (2007), 523–525.

[32] M. Bojowald, H. Hernández, and A. Skirzewski, *Effective equations for isotropic quantum cosmology including matter* Phys. Rev. D **76** (2007), 063511, [arXiv:0706.1057].

[33] K. Symanzik, *Renormalizable models with simple symmetry breaking I. Symmetry breaking by a source term* Comm. Math. Phys. **16** (1970), 48–80.

[34] L. Dolan and R. Jackiw, *Gauge-invariant signal for gauge-symmetry breaking* Phys. Rev. D **9** (1974), 2904–2912.

[35] M. Bojowald and A. Skirzewski, *Effective theory for the cosmological generation of structure* Adv. Sci. Lett., to appear.

[36] M. Bojowald, H. Hernández, M. Kagan, and A. Skirzewski, *Effective constraints of loop quantum gravity* Phys. Rev. D **75** (2007), 064022, [gr-qc/0611112].

[37] M. Bojowald, H. Hernández, M. Kagan, P. Singh, and A. Skirzewski, *Formation and evolution of structure in loop cosmology* Phys. Rev. Lett. **98** (2007), 031301, [astro-ph/0611685].

[38] M. Bojowald, H. Hernández, M. Kagan, P. Singh, and A. Skirzewski, *Hamiltonian cosmological perturbation theory with loop quantum gravity corrections* Phys. Rev. D **74** (2006), 123512, [gr-qc/0609057].

[39] M. Bojowald and G. Hossain, *Cosmological vector modes and quantum gravity effects* Class. Quantum Grav. **24** (2007), 4801–4816, [arXiv:0709.0872].

[40] M. Bojowald and G. Hossain, *Quantum gravity corrections to gravitational wave dispersion* Phys. Rev. D **77** (2008), 023508, [arXiv:0709.2365].

[41] M. Bojowald, G. Hossain, M. Kagan, D. Mulryne, N. Nunes, and S. Shankaranarayanan, in preparation.

[42] T. Thiemann, *Quantum Spin Dynamics (QSD)* Class. Quantum Grav. **15** (1998), 839–873, [gr-qc/9606089].

[43] T. Thiemann, *QSD V: Quantum Gravity as the Natural Regulator of Matter Quantum Field Theories* Class. Quantum Grav. **15** (1998), 1281–1314, [gr-qc/9705019].

[44] M. Bojowald and F. Hinterleitner, *Isotropic loop quantum cosmology with matter* Phys. Rev. D **66** (2002), 104003, [gr-qc/0207038].

[45] M. Bojowald and J. D. Reyes, in preparation.

Martin Bojowald
Institute for Gravitation and the Cosmos
The Pennsylvania State University
104 Davey Lab, University Park, PA 16802, USA
e-mail: bojowald@gravity.psu.edu

Quantum Field Theory
B. Fauser, J. Tolksdorf and E. Zeidler, Eds., 235–259
© 2009 Birkhäuser Verlag Basel/Switzerland

From Discrete Space-Time to Minkowski Space: Basic Mechanisms, Methods and Perspectives

Felix Finster

Abstract. This survey article reviews recent results on fermion systems in discrete space-time and corresponding systems in Minkowski space. After a basic introduction to the discrete setting, we explain a mechanism of spontaneous symmetry breaking which leads to the emergence of a discrete causal structure. As methods to study the transition between discrete space-time and Minkowski space, we describe a lattice model for a static and isotropic space-time, outline the analysis of regularization tails of vacuum Dirac sea configurations, and introduce a Lorentz invariant action for the masses of the Dirac seas. We mention the method of the continuum limit, which allows to analyze interacting systems. Open problems are discussed.

Mathematics Subject Classification (2000). Primary 83C99; Secondary 70G75.

Keywords. discrete space-time, emergent space-time, special relativity, causality, fermionic projector, Dirac sea.

1. Introduction

It is generally believed that the concept of a space-time continuum (like Minkowski space or a Lorentzian manifold) should be modified for distances as small as the Planck length. The principle of the fermionic projector [4] proposes a new model of space-time, which should be valid down to the Planck scale. This model is introduced as a system of quantum mechanical wave functions defined on a finite number of space-time points and is referred to as a *fermion system in discrete space-time*. The interaction is described via a variational principle where we minimize an action defined for the ensemble of wave functions. A-priori, there are no relations between the space-time points; in particular, there is no nearest-neighbor relation and no notion of causality. The idea is that these additional structures

I would like to thank the organizers of the workshop "Recent Developments in Quantum Field Theory" (Leipzig, July 2007) for helpful discussions and encouragement. I thank Joel Smoller and the referee for helpful comments on the manuscript.

should be generated spontaneously. More precisely, in order to minimize the action, the wave functions form specific configurations; this can be visualized as a "self-organization" of the particles. As a consequence of this self-organization, the wave functions induce non-trivial relations between the space-time points. We thus obtain additional structures in space-time, and it is conjectured that, in a suitable limit where the number of particles and space-time points tends to infinity, these structures should give rise to the local and causal structure of Minkowski space. In this limit, the configuration of the wave functions should go over to a Dirac sea structure.

This conjecture has not yet been proved, but recent results give a detailed picture of the connection between discrete space-time and Minkowski space. Also, mathematical methods were developed to shed light on particular aspects of the problem. In this survey article we report on the present status, explain basic mechanisms and outline the analytical methods used so far. The presentation is self-contained and non-technical. The paper concludes with a discussion of open problems.

2. Fermion systems in discrete space-time

We begin with the basic definitions in the discrete setting (for more details see [5]). Let $(H, <.|.>)$ be a finite-dimensional complex inner product space. Thus $<.|.>$ is linear in its second and anti-linear in its first argument, and it is symmetric,

$$\overline{<\Psi \,|\, \Phi>} \;=\; <\Phi \,|\, \Psi> \qquad \text{for all } \Psi, \Phi \in H \,,$$

and non-degenerate,

$$<\Psi \,|\, \Phi> \;=\; 0 \;\text{ for all } \Phi \in H \quad \Longrightarrow \quad \Psi = 0 \,.$$

In contrast to a scalar product, $<.|.>$ need *not* be positive.

A *projector* A in H is defined just as in Hilbert spaces as a linear operator which is idempotent and self-adjoint,

$$A^2 = A \qquad \text{and} \qquad <A\Psi \,|\, \Phi> = <\Psi \,|\, A\Phi> \quad \text{for all } \Psi, \Phi \in H \,.$$

Let M be a finite set. To every point $x \in M$ we associate a projector E_x. We assume that these projectors are orthogonal and complete in the sense that

$$E_x \, E_y \;=\; \delta_{xy} \, E_x \qquad \text{and} \qquad \sum_{x \in M} E_x \;=\; \mathbb{1} \,. \tag{1}$$

Furthermore, we assume that the images $E_x(H) \subset H$ of these projectors are non-degenerate subspaces of H, which all have the same signature (n, n). We refer to n as the *spin dimension*. The points $x \in M$ are called *discrete space-time points*, and the corresponding projectors E_x are the *space-time projectors*. The structure $(H, <.|.>, (E_x)_{x \in M})$ is called *discrete space-time*.

We next introduce the so-called *fermionic projector* P as a projector in H whose image $P(H) \subset H$ is *negative definite*. The vectors in the image of P have the interpretation as the quantum states of the particles of our system. Thus the rank of P gives the *number of particles* $f := \dim P(H)$. The name "fermionic projector" is motivated from the correspondence to Minkowski space, where our particles should go over to Dirac particles, being fermions (see Section 6 below). We call the obtained structure $(H, <.|.>, (E_x)_{x \in M}, P)$ a *fermion system in discrete space-time*. Note that our definitions involve only three integer parameters: the spin dimension n, the number of space-time points m, and the number of particles f.

The above definitions can be understood as a mathematical reduction to some of the structures present in relativistic quantum mechanics, in such a way that the *Pauli Exclusion Principle*, a *local gauge principle* and the *equivalence principle* are respected (for details see [4, Chapter 3]). More precisely, describing the many-particle system by a projector P, every vector $\Psi \in H$ either lies in the image of P or it does not. In this way, the fermionic projector encodes for every state the occupation numbers 1 and 0, respectively, but it is impossible to describe higher occupation numbers. More technically, choosing a basis $\Psi_1, \ldots \Psi_f$ of $P(H)$, we can form the anti-symmetric many-particle wave function

$$\Psi = \Psi_1 \wedge \cdots \wedge \Psi_f.$$

Due to the anti-symmetrization, this definition of Ψ is (up to a phase) independent of the choice of the basis $\Psi_1, \ldots, \Psi_f$. In this way, we can associate to every fermionic projector a fermionic many-particle wave function, which clearly respects the Pauli exclusion principle. To reveal the local gauge principle, we consider unitary operators U (i.e. operators which for all $\Psi, \Phi \in H$ satisfy the relation $<U\Psi|U\Phi> = <\Psi|\Phi>$) which do not change the space-time projectors,

$$E_x = U E_x U^{-1} \qquad \text{for all } x \in M. \tag{2}$$

We transform the fermionic projector according to

$$P \rightarrow U P U^{-1}. \tag{3}$$

Such transformations lead to physically equivalent fermion systems. The conditions (2) mean that U maps every subspace $E_x(H)$ onto itself. In other words, U acts "locally" on the subspaces associated to the individual space-time points. The transformations (2, 3) can be identified with local gauge transformations in physics (for details see [4, §3.1]). The equivalence principle is built into our framework in a very general form by the fact that our definitions do not distinguish an ordering between the space-time points. Thus our definitions are symmetric under permutations of the space-time points, generalizing the diffeomorphism invariance in general relativity.

Obviously, important physical principles are missing in our framework. In particular, our definitions involve *no locality* and *no causality*, and not even relations like the nearest-neighbor relations on a lattice. The idea is that these additional structures, which are of course essential for the formulation of physics, should emerge as a consequence of a spontaneous symmetry breaking and a self-organization of the particles as described by a variational principle. Before explaining in more detail how this is supposed to work (Section 6), we first introduce the variational principle (Section 3), explain the mechanism of spontaneous symmetry breaking (Section 4), and discuss the emergence of a discrete causal structure (Section 5).

3. A variational principle

In order to introduce an interaction of the particles, we now set up a variational principle. For any $u \in H$, we refer to the projection $E_x u \in E_x(H)$ as the *localization* of u at x. We also use the short notation $u(x) = E_x u$ and sometimes call $u(x)$ the *wave function* corresponding to the vector u. Furthermore, we introduce the short notation

$$P(x,y) \ = \ E_x \, P \, E_y \,, \qquad x, y \in M \,. \tag{4}$$

This operator product maps $E_y(H) \subset H$ to $E_x(H)$, and it is often useful to regard it as a mapping only between these subspaces,

$$P(x,y) \ : \ E_y(H) \ \to \ E_x(H) \,.$$

Using the properties of the space-time projectors (1), we find

$$(Pu)(x) \ = \ E_x \, Pu \ = \ \sum_{y \in M} E_x \, P \, E_y \, u \ = \ \sum_{y \in M} (E_x \, P \, E_y) \, (E_y \, u) \,,$$

and thus

$$(Pu)(x) \ = \ \sum_{y \in M} P(x,y) \, u(y) \,. \tag{5}$$

This relation resembles the representation of an operator with an integral kernel, and thus we refer to $P(x,y)$ as the *discrete kernel* of the fermionic projector. Next we introduce the *closed chain* A_{xy} as the product

$$A_{xy} \ := \ P(x,y) \, P(y,x) \ = \ E_x \, P \, E_y \, P \, E_x \,; \tag{6}$$

it maps $E_x(H)$ to itself. Let $\lambda_1, \ldots, \lambda_{2n}$ be the roots of the characteristic polynomial of A_{xy}, counted with multiplicities. We define the *spectral weight* $|A_{xy}|$ by

$$|A_{xy}| \ = \ \sum_{j=1}^{2n} |\lambda_j| \,.$$

Similarly, one can take the spectral weight of powers of A_{xy}, and by summing over the space-time points we get positive numbers depending only on the form of the fermionic projector relative to the space-time projectors. Our variational principle is to

$$\text{minimize} \quad \sum_{x,y \in M} |A_{xy}^2| \tag{7}$$

by considering variations of the fermionic projector which satisfy for a given real parameter κ the constraint

$$\sum_{x,y \in M} |A_{xy}|^2 = \kappa \,. \tag{8}$$

In the variation we also keep the number of particles f as well as discrete space-time fixed. Clearly, we need to choose κ such that there is at least one fermionic projector which satisfies (8). It is easy to verify that (7) and (8) are invariant under the transformations (2, 3), and thus our variational principle is gauge invariant.

The above variational principle was first introduced in [4]. In [5] it is analyzed mathematically, and it is shown in particular that minimizers exist:

Theorem 3.1. *The minimum of the variational principle (7, 8) is attained.*

Using the method of Lagrange multipliers, for every minimizer P there is a real parameter μ such that P is a stationary point of the *action*

$$\mathcal{S}_\mu[P] \;=\; \sum_{x,y \in M} \mathcal{L}_\mu[A_{xy}] \tag{9}$$

with the *Lagrangian*

$$\mathcal{L}_\mu[A] \;=\; |A^2| - \mu\,|A|^2 \,. \tag{10}$$

A useful method for constructing stationary points for a given value of the Lagrange multiplier μ is to minimize the action $\mathcal{S}_\mu$ without the constraint (8). This so-called *auxiliary variational principle* behaves differently depending on the value of μ. If $\mu < \frac{1}{2n}$, the action is bounded from below, and it is proved in [5] that minimizers exist. In the case $\mu > \frac{1}{2n}$, on the other hand, the action is not bounded from below, and thus there are clearly no minimizers. In the remaining so-called *critical case* $\mu = \frac{1}{2n}$, partial existence results are given in [5], but the general existence problem is still open. The critical case is important for the physical applications. For simplicity, we omit the subscript $\mu = \frac{1}{2n}$ and also refer to the auxiliary variational principle in the critical case as the *critical variational principle*. Writing the critical Lagrangian as

$$\mathcal{L}[A] \;=\; \frac{1}{4n} \sum_{i,j=1}^{2n} \left(|\lambda_i| - |\lambda_j|\right)^2 , \tag{11}$$

we get a good intuitive understanding of the critical variational principle: it tries to achieve that for every $x, y \in M$, all the roots of the characteristic polynomial of the closed chain A_{xy} have the same absolute value.

We next derive the corresponding *Euler-Lagrange equations* (for details see [4, §3.5 and §5.2]). Suppose that P is a critical point of the action (9). We consider a variation $P(\tau)$ of projectors with $P(0) = P$. Denoting the gradient of the Lagrangian by $\mathcal{M}$,

$$\mathcal{M}_\mu[A]_\beta^\alpha \;:=\; \frac{\partial \mathcal{L}_\mu[A]}{\partial A_\alpha^\beta}\,, \qquad \text{with } \alpha, \beta \in \{1, \ldots, 2n\}\,, \tag{12}$$

we can write the variation of the Lagrangian as a trace on $E_x(H)$,

$$\delta\mathcal{L}_\mu[A_{xy}] \;=\; \frac{d}{d\tau}\mathcal{L}_\mu[A_{xy}(\tau)]\Big|_{\tau=0} \;=\; \mathrm{Tr}\left(E_x\,\mathcal{M}_\mu[A_{xy}]\,\delta A_{xy}\right).$$

Using the Leibniz rule

$$\delta A_{xy} \;=\; \delta P(x,y)\, P(y,x) \;+\; P(x,y)\, \delta P(y,x)$$

together with the fact that the trace is cyclic, after summing over the space-time points we find

$$\sum_{x,y\in M} \delta\mathcal{L}_\mu[A_{xy}] \;=\; \sum_{x,y\in M} 4\,\mathrm{Tr}\left(E_x\, Q_\mu(x,y)\, \delta P(y,x)\right),$$

where we set

$$Q_\mu(x,y) \;=\; \frac{1}{4}\left(\mathcal{M}_\mu[A_{xy}]\, P(x,y) \;+\; P(x,y)\,\mathcal{M}_\mu[A_{yx}]\right). \tag{13}$$

Thus the first variation of the action can be written as

$$\delta\mathcal{S}_\mu[P] \;=\; 4\,\mathrm{Tr}\left(Q_\mu\,\delta P\right), \tag{14}$$

where Q_μ is the operator in H with kernel (13). This equation can be simplified using that the operators $P(\tau)$ are all projectors of fixed rank. Namely, there is a family of unitary operators $U(\tau)$ with $U(\tau) = \mathbb{1}$ and

$$P(\tau) \;=\; U(\tau)\, P\, U(\tau)^{-1}.$$

Hence $\delta P = i[B,P]$, where we set $B = -iU'(0)$. Using this relation in (14) and again using that the trace is cyclic, we find $\delta\mathcal{S}_\mu[P] = 4i\,\mathrm{Tr}\left([P,Q_\mu]\,B\right)$. Since B is an arbitrary self-adjoint operator, we conclude that

$$[P,Q_\mu] \;=\; 0. \tag{15}$$

This commutator equation with Q_μ given by (13) are the Euler-Lagrange equations corresponding to our variational principle.

4. A mechanism of spontaneous symmetry breaking

In the definition of fermion systems in discrete space-time, we did not distinguish an ordering of the space-time points; all our definitions are symmetric under permutations of the points of M. However, this does not necessarily mean that a given fermion system will have this permutation symmetry. The simplest counterexample is to take a fermionic projector consisting of one particle which is localized at the first space-time point, i.e. in bra/ket-notation

$$P \;=\; -|u\!><\!u| \quad\text{with}\quad <\!u\,|\,u\!> = -1 \quad\text{and}$$
$$E_1 u \;=\; u, \quad E_x u \;=\; 0 \quad\text{for all } x = 2,\dots,m.$$

Then the fermionic projector distinguishes the first space-time point and thus breaks the permutation symmetry. In [6] it is shown under general assumptions on the number of particles and space-time points that, no matter how we choose the fermionic wave functions, it is impossible to arrange that the fermionic projector respects the permutation symmetry. In other words, the fermionic projector necessarily breaks the permutation symmetry of discrete space-time. We first specify the result and explain it afterwards. The group of all permutations of the space-time points is the symmetric group, denoted by S_m.

Definition 4.1. *A subgroup $\mathcal{O} \subset S_m$ is called* **outer symmetry group** *of the fermion system in discrete space-time if for every $\sigma \in \mathcal{O}$ there is a unitary transformation U such that*

$$U P U^{-1} = P \quad \text{and} \quad U E_x U^{-1} = E_{\sigma(x)} \quad \text{for all } x \in M. \tag{16}$$

Theorem 4.2. (**spontaneous breaking of the permutation symmetry**) *Suppose that $(H, <.|.>, (E_x)_{x \in M}, P)$ is a fermion system in discrete space-time of spin dimension n. Assume that the number of space-time points m is sufficiently large,*

$$m > \begin{cases} 3 & \text{if } n = 1 \\ \max\left(2n + 1,\ 4\left[\log_2 n\right] + 6\right) & \text{if } n > 1 \end{cases} \tag{17}$$

(where $[x]$ is the Gauß bracket), and that the number of particles f lies in the range

$$n < f < m - 1. \tag{18}$$

Then the fermion system cannot have the outer symmetry group $\mathcal{O} = S_m$.

For clarity we note that this theorem does not refer to the variational principle of Section 3. To explain the result, we now give an alternative proof in the simplest situation where the theorem applies: the case $n = 1$, $f = 2$ and $m = 4$. For systems of two particles, the following construction from [2] is very useful for visualizing the fermion system. The image of P is a two-dimensional, negative definite subspace of H. Choosing an orthonormal basis (u_1, u_2) (i.e. $<u_i|u_j> = -\delta_{ij}$), the fermionic projector can be written in bra/ket-notation as

$$P = -|u_1><u_1| - |u_2><u_2|. \tag{19}$$

For any space-time point $x \in M$ we introduce the so-called *local correlation matrix* F_x by

$$(F_x)^i_j = -<u_i \,|\, E_x u_j>. \tag{20}$$

The matrix F_x is Hermitian on the standard Euclidean $\mathbb{C}^2$. Thus we can decompose it in the form

$$F_x = \frac{1}{2}\left(\rho_x \mathbb{1} + \vec{v}_x \vec{\sigma}\right), \tag{21}$$

where $\vec{\sigma} = (\sigma^1, \sigma^2, \sigma^3)$ are the Pauli matrices. We refer to the $\vec{v}_x$ as the *Pauli vectors*. The local correlation matrices are obviously invariant under unitary transformations in H. But they do depend on the arbitrariness in choosing the orthonormal basis (u_1, u_2) of $P(H)$. More precisely, the choice of the orthonormal basis involves a $U(2)$-freedom and, according to the transformation of Pauli spinors in non-relativistic quantum mechanics, this gives rise to orientation preserving rotations of all Pauli vectors. Hence the local correlation matrices are unique up to the transformations

$$\vec{v}_x \longrightarrow R \vec{v}_x \quad \text{with} \quad R \in SO(3). \tag{22}$$

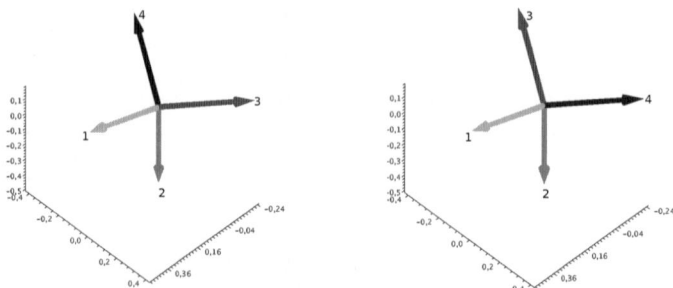

FIGURE 1. Tetrahedron configurations of the Pauli vectors

Let us collect a few properties of the local correlation matrices. Summing over x and using the completeness relation (1), we find that $\sum_{x \in M} F_x = \mathbb{1}$ or, equivalently,

$$\sum_{x \in M} \rho_x = 2 \quad \text{and} \quad \sum_{x \in M} \vec{v}_x = \vec{0} . \tag{23}$$

Furthermore, as the inner product in (20) has signature $(1, 1)$, the matrix F_x can have at most one positive and at most one negative eigenvalue. Expressed in terms of the decomposition (21), this means that

$$|\vec{v}_x| \geq \rho_x \quad \text{for all } x \in M . \tag{24}$$

Now assume that a fermion system with $m = 4$ space-time points is permutation symmetric. Then the scalars ρ_x must all be equal. Using the left equation in (23), we conclude that $\rho_x = 1/2$. Furthermore, the Pauli vectors must all have the same length. In view of (24), this means that

$$|\vec{v}_x| = v \geq \frac{1}{2} \quad \text{for all } x \in M .$$

Moreover, the angles between any two vectors $\vec{v}_x, \vec{v}_y$ with $x \neq y$ must coincide. The only configuration with these properties is that the vectors $\vec{v}_x$ form the vertices of a tetrahedron, see Figure 1. Labeling the vertices by the corresponding space-time points distinguishes an orientation of the tetrahedron; in particular, the two tetrahedra in Figure 1 cannot be mapped onto each other by an orientation-preserving rotation (22). This also implies that with the transformation (22) we cannot realize odd permutations of the space-time points. Hence the fermion system cannot be permutation symmetric, a contradiction.

Theorem 4.2 makes the effect of spontaneous symmetry breaking rigorous and shows that the fermionic projector induces non-trivial relations between the space-time points. But unfortunately, the theorem gives no information on what the resulting smaller outer symmetry group is, nor how the induced relations on the space-time points look like. For answering these questions, the setting of Theorem 4.2 is too general, because the particular form of our variational principle

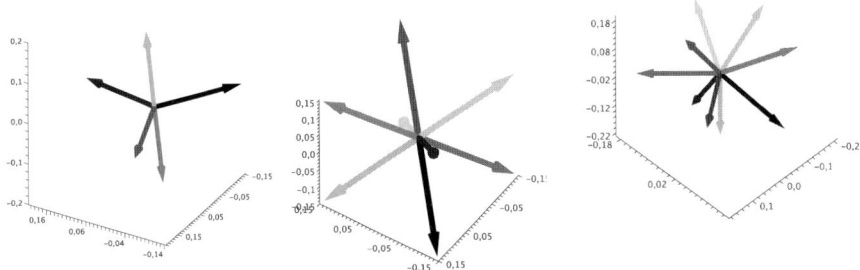

FIGURE 2. Pauli vectors of the minimizers for five, eight and nine space-time points

becomes important. The basic question is which symmetries the minimizers have. In [2] the minimizers of the critical action are constructed numerically for two particles and up to nine space-time points. For four space-time points, the Pauli vectors of the minimizers indeed form a tetrahedron. In Figure 2, the Pauli vectors of minimizers are shown in a few examples. Qualitatively, one sees that for many space-time points, the vectors $\vec{v}_x$ all have approximately the same length $2/m$ and can thus be identified with points on a two-dimensional sphere of radius $2/m$. The critical variational principle aims at distributing these points uniformly on the sphere. The resulting structure is similar to a lattice on the sphere. Thus we can say that for the critical action in the case $f = 2$ and in the limit $m \to \infty$, there is numerical evidence that the spontaneous symmetry breaking leads to the emergence of the structure of a two-dimensional lattice.

The above two-particle systems exemplify the spontaneous generation of additional structures in discrete space-time. However, one should keep in mind that for the transition to Minkowski space one needs to consider systems which involve many particles and are thus much more complicated. Before explaining how this transition is supposed to work, we need to consider how causality arises in the discrete framework.

5. Emergence of a discrete causal structure

In an indefinite inner product space, the eigenvalues of a self-adjoint operator A need not be real, but alternatively they can form complex conjugate pairs (see [10] or [5, Section 3]). This simple fact can be used to introduce a notion of causality.

Definition 5.1. (discrete causal structure) *Two discrete space-time points $x, y \in M$ are called* **timelike** *separated if the roots λ_j of the characteristic polynomial of A_{xy}*

*are all real. They are said to be **spacelike** separated if all the λ_j form complex conjugate pairs and all have the same absolute value.*

As we shall see in Section 6 below, for Dirac spinors in Minkowski space this definition is consistent with the usual notion of causality. Moreover, the definition can be understood within discrete space-time in that it reflects the structure of the critical action. Namely, suppose that two space-time points x and y are spacelike separated. Then the critical Lagrangian (11) vanishes. A short calculation shows that the first variation $\mathcal{M}[A_{xy}]$, (12), also vanishes, and thus A_{xy} does not enter the Euler-Lagrange equations. This can be seen in analogy to the usual notion of causality that points with spacelike separation cannot influence each other.

In [2, Section 5.2] an explicit example is given where the spontaneous symmetry breaking gives rise to a non-trivial discrete causal structure. We now outline this example, omitting a few technical details. We consider minimizers of the variational principle with constraint (7, 8) in the case $n = 1$, $f = 2$ and $m = 3$. We found numerically that in the range of κ under consideration here, the minimizers are permutation symmetric. Thus in view of (23, 24), the local correlation matrices are of the form (21) with

$$|\vec{v}_x| =: v \geq \frac{2}{3} = \rho_x \, ,$$

and the three Pauli vectors form an equilateral triangle. In [2, Lemma 4.4] it is shown that any such choice of local correlation matrices can indeed be realized by a fermionic projector. Furthermore, it is shown that all fermionic projectors corresponding to the same value of v are gauge equivalent. Thus we have, up to gauge transformations, a one-parameter family of fermionic projectors, parameterized by $v \geq 2/3$.

We again represent the fermionic projector as in (19). Then the closed chain can be written as

$$A_{xy} = \sum_{i,j=1}^{2} |E_x \, u_i><u_i|E_y \, u_j><u_j|E_x \, .$$

Using the identity $\det(BC - \lambda) = \det(CB - \lambda)$, cyclically commuting the factors does not change the spectrum, and thus A_{xy} is isospectral to the matrix

$$\sum_{i=1}^{2} <u_j|E_x \, u_i><u_i|E_y \, u_k> = (F_x F_y)_{jk} \, .$$

This makes it possible to express the roots $\lambda_\pm$ of the characteristic polynomial of A_{xy} in terms of the local correlation matrices. A direct computation gives (see [2, Proposition 4.1])

$$\lambda_\pm = \frac{1}{4} \left(\rho_x \rho_y + \vec{v}_x \vec{v}_y \pm \sqrt{|\rho_x \vec{v}_y + \rho_y \vec{v}_x|^2 - |\vec{v}_x \times \vec{v}_y|^2} \right) .$$

If $x = y$, the cross product vanishes, and thus the $\lambda_\pm$ are real. Hence each space-time point has timelike separation from itself (we remark that this is valid in

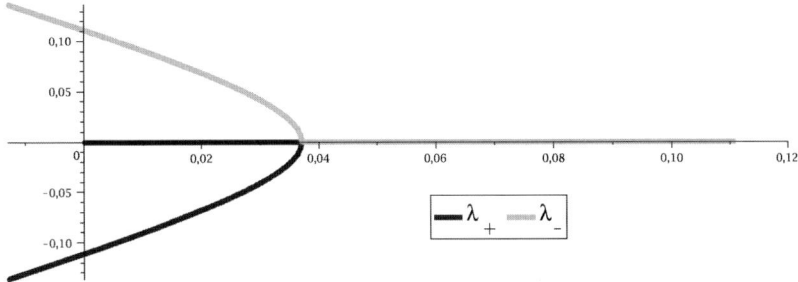

FIGURE 3. Plots of λ_+ and λ_- in the complex plane for varying v.

general, see [2, Proposition 2.7] and [5, Lemma 4.2]). In the case $x \neq y$, the eigenvalues λ_+ and λ_- are shown in Figure 3 for different values of v. If $v = 2/3$, the eigenvalue λ_- vanishes, whereas $\lambda_+ = 1/9$. If $v = \frac{4\sqrt{3}}{9}$, the values of λ_- and λ_+ coincide. If v is further increased, the $\lambda_\pm$ become complex and form a complex conjugate pair. Hence different space-time points have timelike separation if $v \leq \frac{4\sqrt{3}}{9}$, whereas they have spacelike separation if $v > \frac{4\sqrt{3}}{9}$. In the latter case the discrete causal structure is non-trivial, because some pairs of points have spacelike and other pairs timelike separation. Finally, a direct computation of the constraint (8) gives a relation between v and κ. One finds that $v > \frac{4\sqrt{3}}{9}$ if and only if $\kappa > \frac{68}{81}$. We conclude that in the case $\kappa > \frac{68}{81}$, the spontaneous symmetry breaking leads to the emergence of non-trivial discrete causal structure.

We point out that the discrete causal structure of Definition 5.1 differs from the definition of a causal set (see [1]) in that it does not distinguish between future and past directed separations. In the above example with three space-time points, the resulting discrete causal structure is also a causal set, albeit in a rather trivial way where each point has timelike separation only from itself.

6. A first connection to Minkowski space

In this section we describe how to get a simple connection between discrete space-time and Minkowski space. In the last Sections 7–9 we will proceed by explaining the first steps towards making this intuitive picture precise. The simplest method for getting a correspondence to relativistic quantum mechanics in Minkowski space is to replace the discrete space-time points M by the space-time continuum $\mathbb{R}^4$ and the sums over M by space-time integrals. For a vector $\Psi \in H$, the corresponding localization $E_x\Psi$ should be a 4-component Dirac wave function, and the scalar product $<\Psi(x) \mid \Phi(x)>$ on $E_x(H)$ should correspond to the usual Lorentz invariant scalar product on Dirac spinors $\overline{\Psi}\Phi$ with $\overline{\Psi} = \Psi^\dagger\gamma^0$ the adjoint spinor. Since this last scalar product is indefinite of signature $(2,2)$, we are led to choosing $n = 2$. In view of (5), the discrete kernel should go over to the integral kernel of an operator P on the Dirac wave functions,

$$(P\Psi)(x) = \int_M P(x,y)\,\Psi(y)\,d^4y\,.$$

The image of P should be spanned by the occupied fermionic states. We take Dirac's concept literally that in the vacuum all negative-energy states are occupied by fermions forming the so-called *Dirac sea*. Thus we are led to describe the vacuum by the integral over the lower mass shell

$$P(x,y) = \int \frac{d^4k}{(2\pi)^4}\,(k\!\!\!/ + m)\,\delta(k^2 - m^2)\,\Theta(-k^0)\,e^{-ik(x-y)}$$

(here Θ is the Heaviside function). Likewise, if we consider several generations of particles, we take a sum of such Fourier integrals,

$$P(x,y) = \sum_{\beta=1}^{g} \rho_\beta \int \frac{d^4k}{(2\pi)^4}\,(k\!\!\!/ + m_\beta)\,\delta(k^2 - m_\beta^2)\,\Theta(-k^0)\,e^{-ik(x-y)}\,, \qquad (25)$$

where g denotes the number of generations, and the $\rho_\beta > 0$ are weight factors for the individual Dirac seas (for a discussion of the weight factors see [7, Appendix A]). Computing the Fourier integrals, one sees that $P(x,y)$ is a smooth function, except on the light cone $\{(y-x)^2 = 0\}$, where it has poles and singular contributions (for more details see (41) below).

Let us find the connection between Definition 5.1 and the usual notion of causality in Minkowski space. Even without computing the Fourier integral (25), it is clear from the *Lorentz symmetry* that for every x and y for which the Fourier integral exists, $P(x,y)$ can be written as

$$P(x,y) = \alpha\,(y-x)_j\gamma^j + \beta\,\mathbb{1} \qquad (26)$$

with two complex coefficients α and β. Taking the complex conjugate of (25), we see that

$$P(y,x) = \overline{\alpha}\,(y-x)_j\gamma^j + \overline{\beta}\,\mathbb{1}\,.$$

As a consequence,

$$A_{xy} = P(x,y)\,P(y,x) = a\,(y-x)_j\gamma^j + b\,\mathbb{1} \qquad (27)$$

with real parameters a and b given by

$$a = \alpha\overline{\beta} + \beta\overline{\alpha}\,, \qquad b = |\alpha|^2\,(y-x)^2 + |\beta|^2\,, \qquad (28)$$

where $(y-x)^2 = (y-x)_j(y-x)^j$, and for the signature of the Minkowski metric we use the convention $(+---)$. Applying the formula $(A_{xy} - b\mathbb{1})^2 = a^2\,(y-x)^2$, one can easily compute the roots of the characteristic polynomial of A_{xy},

$$\lambda_1 = \lambda_2 = b + \sqrt{a^2\,(y-x)^2}\,, \qquad \lambda_3 = \lambda_4 = b - \sqrt{a^2\,(y-x)^2}\,. \qquad (29)$$

If the vector $(y-x)$ is timelike, we see from the inequality $(y-x)^2 > 0$ that the λ_j are all real. Conversely, if the vector $(y-x)$ is spacelike, the term $(y-x)^2 < 0$

is negative. As a consequence, the λ_j form complex conjugate pairs and all have the same absolute value. This shows that for Dirac spinors in Minkowski space, Definition 5.1 is consistent with the usual notion of causality.

We next consider the Euler-Lagrange equations corresponding to the critical Lagrangian (11). If $(y - x)$ is spacelike, the λ_j all have the same absolute value, and thus the Lagrangian vanishes. If on the other hand $(y - x)$ is timelike, the λ_j as given by (29) are all real, and a simple computation using (28) yields that $\lambda_1 \lambda_2 \geq 0$, so that all the λ_j have the same sign (we remark that this is true in more generality, see [7, Lemma 2.1]). Hence the Lagrangian (11) simplifies to

$$\mathcal{L}[A_{xy}] = \begin{cases} \text{Tr}(A_{xy}^2) - \dfrac{1}{4} \text{Tr}(A_{xy})^2 & \text{if } (y - x) \text{ is timelike} \\[2mm] 0 & \text{if } (y - x) \text{ is spacelike} \,. \end{cases} \tag{30}$$

Now we can compute the gradient (12) to obtain (for details see [7, Section 2.2])

$$\mathcal{M}[A_{xy}] = \begin{cases} 2A_{xy} - \dfrac{1}{2} \text{Tr}(A_{xy})\, \mathbb{1} & \text{if } (y - x) \text{ is timelike} \\[2mm] 0 & \text{if } (y - x) \text{ is spacelike} \,. \end{cases} \tag{31}$$

Using (27), we can also write this for timelike $(y - x)$ as

$$\mathcal{M}[A_{xy}] = 2a(x, y)\, (y - x)^j \gamma_j \,. \tag{32}$$

Furthermore, using (28) we obtain that

$$\mathcal{M}[A_{xy}] = \mathcal{M}[A_{yx}] \,. \tag{33}$$

Combining the relations (33, 32, 26), we find that the two summands in (13) coincide, and thus

$$Q(x, y) = \frac{1}{2} \mathcal{M}[A_{xy}]\, P(x, y) \tag{34}$$

(we remark that the last identity holds in full generality, see [4, Lemma 5.2.1]).

We point out that this calculation does not determine $\mathcal{M}$ on the light cone, and due to the singularities of $P(x, y)$, the Lagrangian is indeed ill-defined if $(y - x)^2 = 0$. However, as an important special feature of the critical Lagrangian, we can make sense of the Euler-Lagrange equations (15), if we only assume that $\mathcal{M}$ is well-defined as a distribution. We now explain this argument, which will be crucial for the considerations in Sections 8 and 9. More precisely, we assume that the gradient of the critical Lagrangian is a Lorentz invariant distribution, which away from the light cone coincides with (31), has a vector structure (32) and is symmetric (33). Then this distribution, which we denote for clarity by $\tilde{\mathcal{M}}$, can be written as

$$\tilde{\mathcal{M}}(\xi) = 2\slashed{\xi}\, a(\xi^2)\, \Theta(\xi^2)\, \epsilon(\xi^0) \,, \tag{35}$$

where we set $\xi \equiv y - x$ and $\slashed{\xi} \equiv \xi^j \gamma_j$, and ϵ is the step function (defined by $\epsilon(x) = 1$ if $x \geq 0$ and $\epsilon(x) = -1$ otherwise). We now consider the Fourier transform of the

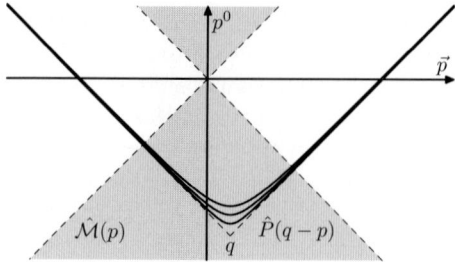

FIGURE 4. The convolution $\hat{\mathcal{M}} * \hat{P}$.

distribution $\tilde{\mathcal{M}}(\xi)$, denoted by $\hat{\mathcal{M}}(k)$. The factor $\not\xi$ corresponds to the differential operator $i\partial_k$ in momentum space, and thus

$$\hat{\mathcal{M}}(k) = 2i\,\partial_k \int d^4\xi\, a(\xi^2)\,\Theta(\xi^2)\,\epsilon(\xi^0)\, e^{-ik\xi}\,. \tag{36}$$

This Fourier integral vanishes if $k^2 < 0$. Namely, due to Lorentz symmetry, in this case we may assume that k is purely spatial, $k = (0, \vec{k})$. But then the integrand of the time integral in (36) is odd because of the step function, and thus the whole integral vanishes. As in [4], we denote the *mass cone* as well as the upper and lower mass cone by

$$\mathcal{C} = \{k \mid k^2 > 0\}\,, \quad \mathcal{C}^\vee = \{k \in \mathcal{C} \mid k^0 > 0\}\,, \quad \mathcal{C}^\wedge = \{k \in \mathcal{C} \mid k^0 < 0\}\,, \tag{37}$$

respectively. Then the above argument shows that the distribution $\hat{\mathcal{M}}$ is *supported in the closed mass cone*, $\operatorname{supp}\hat{\mathcal{M}} \subset \overline{\mathcal{C}}$. Next we rewrite the pointwise product in (34) as a convolution in momentum space,

$$\hat{Q}(q) = \frac{1}{2}\,(\hat{\mathcal{M}} * \hat{P})(q) = \frac{1}{2}\int \frac{d^4p}{(2\pi)^4}\,\hat{\mathcal{M}}(p)\,\hat{P}(q - p)\,. \tag{38}$$

If q is in the lower mass cone $\mathcal{C}^\wedge$, the integrand of the convolution has compact support (see Figure 4), and the integral is finite (if however $q \notin \overline{\mathcal{C}^\wedge}$, the convolution integral extends over an unbounded region and is indeed ill-defined). We conclude that $\hat{Q}(q)$ is well-defined inside the lower mass cone. Since the fermionic projector (25) is also supported in the lower mass cone, this is precisely what we need in order to make sense of the operator products $\hat{P}(k)\,\hat{\mathcal{M}}(k)$ and $\tilde{\mathcal{M}}(k)\,\hat{P}(k)$ which appear in the commutator (15). In this way we have given the Euler-Lagrange equations a mathematical meaning.

In the above consideration we only considered the critical Lagrangian. To avoid misunderstandings, we now briefly mention the physical significance of the variational principle with constraint (7, 8) and explain the connection to the above arguments. In order to describe a realistic physical system involving different types of fermions including left-handed neutrinos, for the fermionic projector of the vacuum one takes a direct sum of fermionic projectors of the form (25) (for details see [4, §5.1]). On the direct summands involving the neutrinos, the closed chain A_{xy} vanishes identically, and also the Euler-Lagrange equations are trivially satisfied. On all the other direct summands, we want the operator $\mathcal{M}$ to be of the form (35), so that the above considerations apply again. In order to arrange this, the value of the Lagrange multiplier μ must be larger than the critical value $\frac{1}{2n}$. Thus we are in the case $\mu > \frac{1}{2n}$ where the auxiliary variational principle has no minimizers. This is why we need to consider the variational principle with constraint (7, 8). Hence the fermionic projector of fundamental physics should be a minimizer of the variational principle with constraint (7, 8) corresponding to a value $\mu > \frac{1}{2n}$ of the Lagrange multiplier (such minimizers with $\mu > \frac{1}{2n}$ indeed exist, see [2, Proposition 5.2] for a simple example). The physical significance of the critical variational principle lies in the fact that restricting attention to one direct summand of the form (25) (or more generally to a subsystem which does not involve chiral particles), the Euler-Lagrange equations corresponding to (7, 8) coincide with those for the critical Lagrangian as discussed above. For more details we refer to [4, Chapter 5].

7. A static and isotropic lattice model

Our concept is that for many particles and many space-time points, the mechanism explained in Sections 4 and 5 should lead to the spontaneous emergence of the structure of Minkowski space or a Lorentzian manifold. The transition between discrete space-time and the space-time continuum could be made precise by proving conjectures of the following type.

Conjecture 7.1. *In spin dimension* $(2,2)$, *there is a series of fermion systems in discrete space-time* $(H^{(l)}, <.|.>, (E_x^{(l)})_{x \in M^{(l)}}, P^{(l)})$ *with the following properties:*

(1) *The fermionic projectors* $P^{(l)}$ *are minimizers of the auxiliary variational principle (9) in the critical case* $\mu = \frac{1}{4}$.

(2) *The number of space-time points* $m^{(l)}$ *and the number of particles* $f^{(l)}$ *scale in* l *as follows,*

$$m^{(l)} \sim l^4, \qquad f^{(l)} \sim l^3.$$

(3) *There are positive constants* $c^{(l)}$, *embeddings* $\iota^{(l)} : M^{(l)} \hookrightarrow \mathbb{R}^4$ *and isomorphisms* $\alpha^{(l)} : H^{(l)} \to L^2(\Phi^{(l)}(M^{(l)}), <.|.>)$ *(where* $<.|.>$ *is the standard inner product on Dirac spinors* $<\Phi|\Psi> = \Phi^\dagger \gamma^0 \Psi$*), such that for any test wave functions* $\Psi, \Phi \in C_0^\infty(\mathbb{R}^4)^4$,

$$c^{(l)} \sum_{x,y \in M^{(l)}} \Phi(\iota x)^\dagger \alpha^{(l)} E_x^{(l)} P^{(l)} E_y^{(l)} (\alpha^{(l)})^{-1} \Psi(\iota y)$$

$$\xrightarrow{l \to \infty} \int d^4x \int d^4y \, \Phi(x)^\dagger P(x,y) \, \Psi(y),$$

where $P(x,y)$ *is the distribution (25).*

(4) *As* $l \to \infty$, *the operators* $\mathcal{M}[A^{(l)}_{xy}]$ *converge likewise to the distribution* $\tilde{\mathcal{M}}(\xi)$, *(35).*

Similarly, one can formulate corresponding conjectures for systems involving several Dirac seas, where the variational principle (7) with constraint (8) should be used if chiral particles are involved. Moreover, it would be desirable to specify that minimizers of the above form are in some sense generic. Ultimately, one would like to prove that under suitable generic conditions, every sequence of minimizing fermion systems has a subsequence which converges in the above weak sense to an interacting physical system defined on a Lorentzian manifold.

Proving such conjectures is certainly difficult. In preparation, it seems a good idea to analyze particular aspects of the problem. One important task is to understand why discrete versions of Dirac sea configurations (25) minimize the critical action. A possible approach is to analyze discrete fermion systems numerically. In order to compare the results in a reasonable way to the continuum, one clearly needs systems involving many space-time points and many particles. Unfortunately, large discrete systems are difficult to analyze numerically. Therefore, it seems a good idea to begin the numerical analysis with simplified systems, which capture essential properties of the original system but are easier to handle. In [9] such a simplified system is proposed, where we employ a spherically symmetric and static Ansatz for the fermionic projector. We now briefly outline the derivation of this model and discuss a few results.

For the derivation we begin in Minkowski space with a *static* and *isotropic* system, which means that the fermionic projector $P(x,y)$ depends only on the difference $\xi = y - x$ and is spherically symmetric. We take the Fourier transform,

$$P(\xi) = \int \frac{d^4p}{(2\pi)^4} \, \hat{P}(p) \, e^{ip\xi} , \tag{39}$$

and take for $\hat{P}$ an Ansatz involving a *vector-scalar structure*, i.e.

$$\hat{P}(p) = v_j(p) \, \gamma^j + \phi(p) \, \mathbb{1} \tag{40}$$

with real functions v_j and ϕ. Using the spherical symmetry, we can choose polar coordinates and carry out the angular integrals in (39). This leaves us with a two-dimensional Fourier integral from the momentum variables ($\omega = p^0, k = |\vec{p}|$) to the position variables ($t = \xi^0, r = |\vec{\xi}|$). In order to discretize the system, we restrict the position variables to a finite lattice $\mathfrak{L}$,

$$(t,r) \in \mathfrak{L} := \Big\{ 0, 2\pi, \ldots, 2\pi(N_t - 1) \Big\} \times \Big\{ 0, 2\pi, \ldots, 2\pi(N_r - 1) \Big\} .$$

Here the integer parameters N_t and N_r describe the size of the lattice, and by scaling we arranged that the lattice spacing is equal to 2π. Then the momentum variables are on the corresponding dual lattice $\hat{\mathfrak{L}}$,

$$(\omega, k) \in \hat{\mathfrak{L}} := \Big\{ -(N_t - 1), \ldots, -1, 0 \Big\} \times \Big\{ 1, \ldots, N_r \Big\} .$$

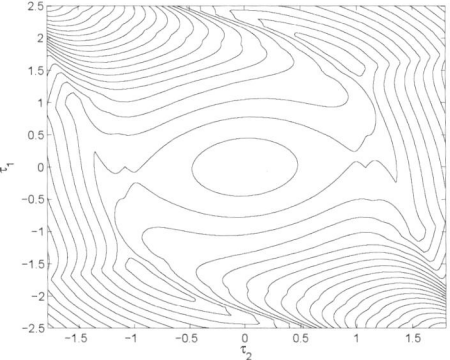

FIGURE 5. Action for a lattice system with two occupied states

Defining the closed chain by $A(t,r) = P(t,r)P(t,r)^*$, we can again introduce the critical Lagrangian (10) with $\mu = \frac{1}{4}$. For the action, we modify (9) to

$$\mathcal{S}[P] = \sum_{(t,r)\in\mathfrak{L}} \rho_t(t)\,\rho_r(r)\,\mathcal{L}(t,r)\,,$$

where the weight factors ρ_t and ρ_r take into account that we only consider positive t and that a point (t,r) corresponds to many states on a sphere of radius r. When varying the action we need to take into account two constraints, called the *trace condition* and the *normalization condition*, which take into account that the total number of particles is fixed and that the fermionic projector should be idempotent.

In [9, Proposition 6.1] the existence of minimizers is proved, and we also present first numerical results for a small lattice system. More precisely, we consider an 8×6-lattice and occupy one state with $k = 1$ and one with $k = 2$. The absolute minimum is attained when occupying the lattice points $(\omega_1 = -1, k = 1)$ and $(\omega_2 = -2, k = 2)$. Introducing a parameter τ by the requirement that the spatial component of the vector v in (40) should satisfy the relation $|\vec{v}| = \phi \sinh \tau$, the trace and normalization conditions fix our system up to the free parameters τ_1 and τ_2 at the two occupied space-time points. In Figure 5 the action is shown as a function of these two free parameters. The minimum at the origin corresponds to the trivial configuration where the two vectors v_i are both parallel to the ω-axis. However, this is only a local minimum, whereas the absolute minimum of the action is attained at the two non-trivial points $(\tau_1 \approx 1.5, \tau_2 \approx 1)$ and $(\tau_1 \approx -1.5, \tau_2 \approx -1)$.

Obviously, a system of two occupied states on an 8×6-lattice is much too small for modeling a Dirac sea structure. But at least, our example shows that our variational principle generates a non-trivial structure on the lattice where the occupied points distinguish specific lattice points, and the corresponding vectors v are not all parallel.

8. Analysis of regularization tails

Another important task in making the connection to Minkowski space rigorous is to justify the distribution $\tilde{\mathcal{M}}$ in (35). To explain the difficulty, let us assume that we have a family of fermion systems $(H^{(l)}, <.|.>, (E_x^{(l)})_{x \in M^{(l)}}, P^{(l)})$ having the properties (1)-(3) of Conjecture 7.1. We can then regard the operators $\alpha^{(l)} P^{(l)} (\alpha^{(l)})^{-1}$ as regularizations of the continuum fermionic projector (25). It is easier to consider more generally a family of regularizations $(P^\varepsilon)_{\varepsilon > 0}$ in Minkowski space with

$$P^\varepsilon(x, y) \xrightarrow{\varepsilon \searrow 0} P(x, y) \quad \text{in the distributional sense.}$$

The parameter ε should be the length scale of the regularization. In order to justify (35) as well as the convolution integral (38), our regularization should have the following properties:

Definition 8.1. *The fermionic projector satisfies the assumption of a **distributional** $\mathcal{M}P$-**product** if the following conditions are satisfied:*

   (i) *There is a distribution $\tilde{\mathcal{M}}(\xi)$ of the form (35) such that $\lim_{\varepsilon \searrow 0} \mathcal{M}[A_{xy}^\varepsilon] = \tilde{\mathcal{M}}(\xi)$ in the distributional sense.*
   (ii) *For every k for which $\lim_{\varepsilon \searrow 0} \hat{Q}^\varepsilon(k)$ exists, the convolution integral (38) is well-defined and $\lim_{\varepsilon \searrow 0} \hat{Q}^\varepsilon(k) = \hat{Q}(k)$.*

This notion was introduced in [4, §5.6] and used as an ad-hoc assumption on the regularization. Justifying this assumption is not just a technicality, but seems essential for getting a detailed understanding of how the connection between discrete space-time and Minkowski space is supposed to work. Namely, if one takes a simple ultraviolet regularization (for example a cutoff in momentum space), then, due to the distributional singularity of $P(x, y)$ on the light cone, the product $A_{xy}^\varepsilon = P^\varepsilon(x, y) P^\varepsilon(y, x)$ will in the limit $\varepsilon \searrow 0$ develop singularities on the light cone, which are ill-defined even in the distributional sense. Thus, in order to satisfy the conditions of Definition 8.1, we need to construct *special regularizations*, such that the divergences on the light cone cancel. In [7] it is shown that this can indeed be accomplished. The method is to consider a class of spherically symmetric regularizations involving many free parameters, and to adjust these parameters such that all the divergences on the light cone and near the origin compensate each other. It seems miraculous that it is possible to cancel all the divergencies; this can be regarded as a confirmation for our approach. If one believes that the regularized fermionic projector describes nature, we get concrete hints on how the vacuum should look like on the Planck scale. More specifically, the admissible regularizations give rise to a *multi-layer structure* near the light cone involving *several length scales*.

In this survey article we cannot enter into the constructions of [7]. Instead, we describe a particular property of Dirac sea configurations which is crucial for making the constructions work. Near the light cone, the distribution $P(x, y)$ has an expansion of the following form

$$
\begin{aligned}
P(x, y) \;=\; & +iC_0 \, \xi\!\!\!/ \, \frac{\mathrm{PP}}{\xi^4} + C_1 \, \frac{\mathrm{PP}}{\xi^2} + iC_2 \, \xi\!\!\!/ \, \frac{\mathrm{PP}}{\xi^2} + C_3 \, \log(\xi^2) + \cdots \\
& + \epsilon(\xi^0)\Big(D_0 \, \xi\!\!\!/ \, \delta'(\xi^2) + iD_1 \, \delta(\xi^2) + D_2 \, \xi\!\!\!/ \delta(\xi^2) + iD_3 \, \Theta(\xi^2) + \cdots \Big) \quad (41)
\end{aligned}
$$

with real constants C_j and D_j (PP denotes the principal part; for more details see [7, Section 3]). Let us consider the expression $\mathcal{M}[A_{xy}]$, (31), for timelike ξ. Computing the closed chain by $A_{xy} = P(x, y) \, P(x, y)^*$, from (41) we obtain away from the light cone the expansion

$$
A_{xy} \;=\; \frac{C_0^2}{\xi^6} + \frac{C_1^2 + 2C_0 C_2}{\xi^4} + 2C_0 D_3 \, \frac{\xi\!\!\!/ \, \epsilon(\xi^0)}{\xi^4} + \cdots \qquad (\xi^2 > 0). \qquad (42)
$$

It is remarkable that there is no contribution proportional to $\xi\!\!\!//\xi^6$. This is because the term $\sim C_0 C_1$ is imaginary, and because the contributions corresponding to D_0, D_1 and D_2 are supported on the light cone. Taking the trace-free part, we find

$$
\mathcal{M}[A_{xy}] \;=\; 4C_0 D_3 \, \frac{\xi\!\!\!/ \, \epsilon(\xi^0)}{\xi^4} + \cdots \qquad (\xi^2 > 0). \qquad (43)
$$

The important point is that, due to the specific form of the Dirac sea configuration, the leading pole of $\mathcal{M}[A_{xy}]$ on the light cone is of lower order than expected from a naive scaling. This fact is extremely useful in the constructions of [7]. Namely, if we consider regularizations of the distribution (41), the terms corresponding to C_0, C_1 and C_2 will be "smeared out" and will thus no longer be supported on the light cone. In particular, the contribution $\sim C_0 D_1$ no longer vanishes, and this *vector contribution* can be used to modify (43). In simple terms, this effect means that the contributions by the regularization are amplified, making it possible to modify $\mathcal{M}[A_{xy}^\varepsilon]$ drastically by small regularization terms. In [7] we work with *regularization tails*, which are very small but spread out on a large scale ε^γ with $\gamma < 1$. Taking many tails with different scales gives rise to the above-mentioned multi-layer structure. Another important effect is that the regularization yields *bilinear contributions* to $\mathcal{M}[A_{xy}^\varepsilon]$ of the form $\sim iC_0 D_0 \gamma^t \gamma^r$ (with $\gamma^r = \vec{\xi}\vec{\gamma}/|\vec{\xi}|$), which are even more singular on the light cone than the vector contributions. The bilinear contributions tend to make the roots λ_j complex (as can be understood already from the fact that $(i\gamma^t \gamma^r)^2 = -\mathbb{1}$). This can be used to make a neighborhood of the light cone space-like; more precisely,

$$
\mathcal{M}[A_{xy}^\varepsilon] \equiv 0 \qquad \text{if } |\xi^0| < |\vec{\xi}| + \epsilon^\gamma \, |\vec{\xi}|^{-\frac{1}{\alpha}} \text{ with } \gamma < 1 \text{ and } \alpha > 1. \qquad (44)
$$

The analysis in [7] also specifies the singularities of the distribution $\tilde{\mathcal{M}}$ on the light cone (recall that by (31), $\tilde{\mathcal{M}}$ is determined only away from the light cone).

We find that $\tilde{\mathcal{M}}$ is unique up to the contributions

$$\tilde{\mathcal{M}}(x,y) \asymp c_0 \, \slashed{\xi} \, \delta'(\xi^2) \, \epsilon(\xi^0) + c_1 \, \slashed{\xi} \, \delta(\xi^2) \, \epsilon(\xi^0) \tag{45}$$

with two free parameters $c_0, c_1 \in \mathbb{R}$. Moreover, the regularization tails give us additional freedom to modify $\mathcal{M}[A^\varepsilon_{xy}]$ near the origin $\xi = 0$. This makes it possible to go beyond the distributional $\mathcal{M}P$-product by arranging extra contributions supported at the origin. Expressed in momentum space, we may modify $\hat{Q}(q)$ by a polynomial in Q; namely (see [7, Theorem 2.4])

$$\hat{Q}(q) := \lim_{\varepsilon \searrow 0} \hat{Q}^\varepsilon(q) = \frac{1}{2} \, (\hat{\mathcal{M}} * \hat{P})(q) + c_2 + c_3 \, \slashed{q} + c_4 \, q^2 \qquad (q \in \mathcal{C}^\wedge) \tag{46}$$

with additional free parameters $c_2, c_3, c_4 \in \mathbb{R}$.

9. A variational principle for the masses of the Dirac seas

With the analysis of the regularization tails in the previous section we have given the Euler-Lagrange equations for a vacuum Dirac sea configuration a rigorous mathematical meaning. All the formulas are well-defined in Minkowski space without any regularization. The freedom to choose the regularization of the fermionic projector is reflected by the free real parameters $c_0, \ldots, c_4$ in (45) and (46). This result is the basis for a more detailed analysis of the Euler-Lagrange equations for vacuum Dirac sea configurations as carried out recently in [8]. We now outline the methods and results of this paper.

We first recall the notion of *state stability* as introduced in [4, §5.6]. We want to analyze whether the vacuum Dirac sea configuration is a stable local minimum of the critical variational principle within the class of static and homogeneous fermionic projectors in Minkowski space. Thus we consider variations where we take an occupied state of one of the Dirac seas and bring the corresponding particle to any other unoccupied state $q \in \mathcal{Q}^\wedge$. Taking into account the vector-scalar structure in the Ansatz (25) and the negative definite signature of the fermionic states, we are led to the variations (for details see [4, §5.6])

$$\delta P = -c(\slashed{k} + m) \, e^{-ik(x-y)} + c(\slashed{l} + m) \, e^{-iq(x-y)} \tag{47}$$

with $m \in \{m_1, \ldots, m_g\} \not\ni \sqrt{q^2}$, $k^2 = l^2 = m^2$ and $k^0, l^0 < 0$. We demand that such variations should not decrease the action,

$$S[P + \delta P] \geq S[P] \quad \text{for all variations (47)}.$$

For the proper normalization of the fermionic states, we need to consider the system in finite 3-volume. Since the normalization constant c in (47) tends to zero in the infinite volume limit, we may treat δS as a first order perturbation. Hence computing the variation of the action by (14), we obtain the condition stated in the next definition. Note that, according to (45), (46) and (38), we already know that $\hat{Q}$ is well-defined inside the lower mass cone and has a vector scalar structure, i.e.

$$\hat{Q}(k) = a \frac{\not{k}}{|k|} + b, \qquad k \in \mathcal{C}^{\wedge}, \tag{48}$$

where we set $|k| = \sqrt{k^2}$, and $a = a(k^2)$, $b = b(k^2)$ are two real-valued, Lorentz invariant functions.

Definition 9.1. *The fermionic projector of the vacuum is called* **state stable** *if the functions a and b in the representation (48) of $\hat{Q}(k)$ have the following properties:*

(1) *a is non-negative.*
(2) *The function $a + b$ is minimal on the mass shells,*

$$(a + b)(m_\alpha^2) = \inf_{q \in \mathcal{C}^{\wedge}} (a + b)(q^2) \qquad \text{for all } \alpha \in \{1, \ldots, g\}. \tag{49}$$

It is very helpful for the understanding and the analysis of state stability that the condition (49) can be related to the Euler-Lagrange equations of a corresponding variational principle. This variational principle was introduced in [8] for unregularized Dirac sea configurations of the form (25) and can be regarded as a Lorentz invariant analog of the critical variational principle. To define this variational principle, we expand the trace-free part of the closed chain inside the light cone similar to (41, 42) as follows,

$$A_0(\xi) := A_{xy} - \frac{1}{4} \text{Tr}(A_{xy}) = \not{\xi} \, \epsilon(\xi^0) \left(\frac{\mathfrak{m}_3}{\xi^4} + \frac{\mathfrak{m}_5}{\xi^2} + \mathcal{O}(\log \xi^2) \right),$$

where the coefficients $\mathfrak{m}_3$ and $\mathfrak{m}_5$ are functions of the parameters ρ_β and m_β in (25). Using the simplified form (30) of the critical Lagrangian, we thus obtain for ξ in the interior of the light cone the expansion

$$\mathcal{L} = \text{Tr}(A_0(\xi)^2) = \frac{\mathfrak{m}_3}{\xi^6} + \frac{2\mathfrak{m}_3\mathfrak{m}_5}{\xi^4} + \mathcal{O}(\xi^{-2} \log \xi^2) \qquad (\xi^2 > 0).$$

The naive adaptation of the critical action (9) would be to integrate $\mathcal{L}$ over the set $\{\xi^2 > 0\}$ (for details see [8, Section 2]). However, this integral diverges because the hyperbolas $\{\xi^2 = \text{const}\}$, where $\mathcal{L}$ is constant, have infinite measure. To avoid this problem, we introduce the variable $z = \xi^2$ and consider instead the one-dimensional integral $\int_0^\infty \mathcal{L}(z) \, z \, dz$, which has the same dimension of length as the integral $\int \mathcal{L} \, d^4\xi$. Since this new integral is still divergent near $z = 0$, we subtract suitable counter terms and set

$$S = \lim_{\varepsilon \searrow 0} \left(\int_\varepsilon^\infty \mathcal{L}(z) \, z \, dz - \frac{\mathfrak{m}_3^2}{\varepsilon} + 2 \, \mathfrak{m}_3 \, \mathfrak{m}_5 \, \log \varepsilon \right). \tag{50}$$

In order to build in the free parameters c_0, c_1 in (45) and c_2, c_3 in (46), we introduce the *extended action* by adding extra terms,

$$S_{\text{ext}} = S + F(\mathfrak{m}_3, \mathfrak{m}_5) + c_3 \sum_{\beta=1}^{g} \rho_\beta \, m_\beta^4 + c_4 \sum_{\beta=1}^{g} \rho_\beta \, m_\beta^5, \tag{51}$$

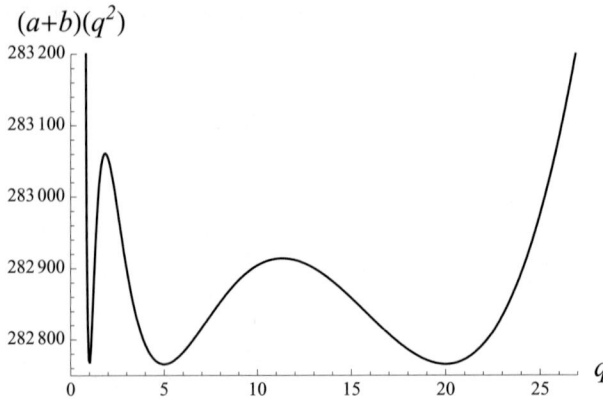

FIGURE 6. A state stable Dirac sea structure with three generations

where F is an arbitrary real function (note that the parameter c_2 in (46) is irrelevant for state stability because it merely changes the function b in (48) by a constant). In our **Lorentz invariant variational principle** we minimize (51), varying the parameters ρ_β and m_β under the constraint

$$\sum_{\beta=1}^{g} m_\beta \, \rho_\beta^3 = \text{const} .$$

This constraint is needed to rule out trivial minimizers; it can be understood as replacing the condition in discrete space-time that the number of particles is fixed.

In [8] it is shown that, allowing for an additional "test Dirac sea" of mass m_{g+1} and weight ρ_{g+1} (with $\rho_{g+1} = 0$ but $\delta\rho_{g+1} \neq 0$), the corresponding Euler-Lagrange equations coincide with (49). The difficult point in the derivation is to take the Fourier transform of the Lorentz invariant action and to reformulate the ε-regularization in (50) in momentum space. In [8] we proceed by constructing numerical solutions of the Euler-Lagrange equations which in addition satisfy the condition (1) in Definition 9.1. We thus obtain state stable Dirac sea configurations. Figure 6 shows an example with three generations and corresponding values of the parameters $m_1 = 1$, $m_2 = 5$, $m_3 = 20$ and $\rho_1 = 1$, $\rho_2 = 10^{-4}$, $\rho_3 = 9.696 \times 10^{-6}$.

10. The continuum limit

The continuum limit provides a method for analyzing the Euler-Lagrange equations (15) for interacting systems in Minkowski space. For details and results we refer to [4, Chapters 6-8]; here we merely put the procedure of the continuum limit in the context of the methods outlined in Sections 6–9. As explained in Section 8, the regularization yields bilinear contributions to A_{xy}^ε, which make a neighborhood of the light cone spacelike (44). Hence near the light cone, the roots λ_j of the characteristic polynomial form complex conjugate pairs and all have the same absolute value,

$$|\lambda_i| = |\lambda_j| \qquad \text{for all } i, j, \tag{52}$$

so that the critical Lagrangian (11) vanishes identically. If we introduce an interaction (for example an additional Dirac wave function or a classical gauge field), the corresponding perturbation of the fermionic projector will violate (52). We thus obtain corresponding contributions to $\mathcal{M}[A^\varepsilon_{xy}]$ in a strip of size $\sim \varepsilon$ around the light cone. These contributions diverge if the regularization is removed. For small ε, they are much larger than the contributions by the regularization tails as discussed in Section 8; this can be understood from the fact that they are much closer to the light cone. The formalism of the continuum limit is obtained by an expansion of these divergent contributions in powers of the regularization length ε. The dependence of the expansion coefficients on the regularization is analyzed using the *method of variable regularization*; we find that this dependence can be described by a small number of free parameters, which take into account the unknown structure of space-time on the Planck scale. The dependence on the gauge fields can be analyzed explicitly using the method of integration along characteristics or, more systematically, by performing a *light-cone expansion* of the fermionic projector. In this way, one can relate the Euler-Lagrange equations to an effective interaction in the framework of second quantized Dirac fields and classical bosonic fields.

11. Outlook and open problems

In this paper we gave a detailed picture of the transition from discrete space-time to the usual space-time continuum. Certain aspects have already been worked out rigorously. But clearly many questions are still open. Generally speaking, the main task for making the connection between discrete space-time and Minkowski space rigorous is to clarify the symmetries and the discrete causal structure of the minimizers for discrete systems involving many particles and many space-time points. More specifically, we see the following directions for future work:

1. *Numerics for large lattice models:* The most direct method to clarify the connection between discrete and continuous models is to the static and isotropic lattice model [9] for systems which are so large that they can be compared in a reasonable way to the continuum. Important questions are whether the minimizers correspond to Dirac sea configurations and what the resulting discrete causal structure is. The next step will be to analyze the connection to the regularization effects described in [7]. In particular, does the lattice model give rise to a multi-layer structure near the light cone? What are the resulting values of the constants $c_0, \ldots, c_4$ in (45, 46)?

2. *Numerics for fermion systems in discrete space-time:* For more than two particles, almost nothing is known about the minimizers of our variational principles. A systematic numerical study could answer the question whether for many particles and many space-time points, the minimizers have outer

symmetries which can be associated to an underlying lattice structure. A numerical analysis of fermion systems in discrete space-time could also justify the spherically symmetric and static Ansatz in [9].

3. *Analysis and estimates for discrete systems:* In the critical case, the general existence problem for minimizers is still open. Furthermore, using methods of [6], one can study fermion systems with prescribed outer symmetry analytically. One question of interest is whether for minimizers the discrete causal structure is compatible with the structure of a corresponding causal set. It would be extremely useful to have a method for analyzing the minimizers asymptotically for a large number of space-time points and many particles. As a first step, a good approximation technique (maybe using methods from quantum statistics?) would be very helpful.

4. *Analysis of the Lorentz invariant variational principle:* In [8] the variational principle for the masses of the vacuum Dirac seas is introduced and analyzed. However, the existence theory has not yet been developed. Furthermore, the structure of the minimizers still needs to be worked out systematically.

5. *Analysis of the continuum limit:* It is a major task to analyze the continuum limit in more detail. The next steps are the derivation of the field equations and the analysis of the spontaneous symmetry breaking of the chiral gauge group. Furthermore, except for [3, Appendix B], no calculations for gravitational fields have been made so far. The analysis of the continuum limit should also give constraints for the weight factors ρ_β in (25) (see [8, Appendix A]).

6. *Field quantization:* As explained in [4, §3.6], the field quantization effects should be a consequence of a "discreteness" of the interaction described by our variational principle. This effect could be studied and made precise for small discrete systems.

Apart from being a challenge for mathematics, these problems have the physical perspective of clarifying the microscopic structure of our universe and explaining the emergence of space and time.

References

[1] L. Bombelli, J. Lee, D. Meyer, R. Sorkin, "Space-time as a causal set," *Phys. Rev. Lett.* **59** (1987) 521-524

[2] A. Diethert, F. Finster, D. Schiefeneder, "Fermion systems in discrete space-time exemplifying the spontaneous generation of a causal structure," arXiv:0710.4420 [math-ph] (2007)

[3] F. Finster, "Light-cone expansion of the Dirac sea to first order in the external potential," hep-th/9707128, *Michigan Math. J.* **46** (1999) 377-408

[4] F. Finster, "The Principle of the Fermionic Projector," *AMS/IP Studies in Advanced Mathematics* **35** (2006)

[5] F. Finster, "A variational principle in discrete space-time – existence of minimizers," math-ph/0503069, *Calc. Var. and Partial Diff. Eq.* **29** (2007) 431-453

[6] F. Finster, "Fermion systems in discrete space-time – outer symmetries and spontaneous symmetry breaking," math-ph/0601039, *Adv. Theor. Math. Phys.* **11** (2007) 91-146

[7] F. Finster, "On the regularized fermionic projector of the vacuum," math-ph/0612003v2, to appear in *J. Math. Phys.* (2008)

[8] F. Finster, S. Hoch, "An action principle for the masses of Dirac particles," arXiv:0712.0678 [math-ph] (2007)

[9] F. Finster, W. Plaum, "A lattice model for the fermionic projector in a static and isotropic space-time," arXiv:0712.0676 [math-ph], to appear in *Math. Nachr.* (2008)

[10] I. Gohberg, P. Lancaster, L. Rodman, "Matrices and Indefinite Scalar Products," *Birkhäuser Verlag* (1983)

Felix Finster
NWF I – Mathematik
Universität Regensburg
D–93040 Regensburg
e-mail: `Felix.Finster@mathematik.uni-regensburg.de`

Quantum Field Theory
B. Fauser, J. Tolksdorf and E. Zeidler, Eds., 261–283
© 2009 Birkhäuser Verlag Basel/Switzerland

Towards a q-Deformed Quantum Field Theory

Hartmut Wachter

In memory of Julius Wess.

Abstract. The formalism of q-deformation seems to enable the formulation of physical theories on discrete spacetime structures that do not suffer from the absence of well-known spacetime symmetries. In this manner q-deformation can lead to a new regularization mechanism in quantum field theory and requires a substantial revision of the mathematical formalism quantum theory is based on. The article gives a review of the motivation for q-deformation in physics, describes the main ideas of the new formalism and outlines the current status of its application to quantum theory.

Mathematics Subject Classification (2000). Primary 81R50; Secondary 16W30; 17B37.

Keywords. q-deformation, quantum groups, quantum spaces, q-deformed free Schrödinger particle, q-calculus.

1. Introduction

I would like to start with some historical notes. From the great mathematician Bernhard Riemann we know that the answer to the question for the right geometry of spacetime is essential to physics. The founders of modern quantum theory, i.e. people like Bohr, Born, Dirac, or Heisenberg were of the opinion that the deeper reason for the infinities in quantum field theory lies in an inadequate description of the geometry of spacetime at very small distances. They believed that spacetime itself had to be incorporated into a quantization scheme. On these grounds it is not very astonishing that the idea of quantizing spacetime is nearly as old as modern quantum theory itself [1, 2]. Up to the present day many researchers took up this idea over and over again [3–8] (for a survey see, for example, Ref. [9]) but their approaches often suffer from conceptual difficulties such as the breakdown of Lorentz symmetry or the prediction of new, rather unusual phenomena.

Feynman, Schwinger, and Tomonaga suggested a different method to over-come the difficulties with infinities in quantum field theory. Their approach is known as renormalization theory [10–12]. With the great experimental success of renormalization theory after the Second World War most efforts towards quantiz-ing spacetime came to an end. Although theoreticians at first felt rather uncom-fortable with renormalization theory they began to accept it since it gave them a tool at hand by which they could make numerical predictions that were in aston-ishing agreement with experiment. Nowadays the ideas of renormalization theory are well understood and much progress in quantum field theory was due to the assumption that any sensible physical theory should be renormalizable.

Nevertheless, theoreticians are convinced that renormalizable theories are only a low-energy approximation to a more fundamental, possibly non-renormaliza-ble theory. This belief is nourished by the following observation. Renormalizability allows us to neglect certain quantum processes yielding divergent diagrams in a perturbative treatment. The reason for this lies in the fact that such quantum processes should be highly suppressed at accessible energies. They only have an influence on the masses and charges that are actually measured. However, when quantum corrections to the gravitational field become important these assumptions seem no longer acceptable. This claim is underpinned by the continued failure to find a renormalizable theory of gravitation.

As renormalizability seems to be not a fundamental physical requirement, the old idea of quantizing spacetime enjoys a revival. In this respect, quantum groups and quantum spaces that arise from q-deformation provide a suitable framework for formulating physical theories on quantized spaces [13–23]. They imply dis-cretizations of well-known spacetime structures [24–29], which, in turn, lead to a very powerful regularization mechanism, as is demonstrated in Sec. 2 of this article.

The attractiveness of q-deformed quantum groups and quantum spaces stems from the fact that they describe discrete spaces that do not suffer from the ab-sence of the well-known spacetime symmetries. However, there is a high price we have to pay for this feature, since the usage of quantum groups and quantum spaces requires a complete revision of the mathematics important physical ideas are formulated by. In Sec. 3 we show how the elements of analysis like partial derivatives, integrals, and so on change under q-deformation. On the other hand the mathematics of q-deformed quantum groups and quantum spaces is sophisti-cated enough to ensure that well-established physical laws pertain. The treatment of the free non-relativistic q-deformed particle underlines this statement as will be seen in section 4.

2. q-Regularization

In this section we would like to illustrate the regularization mechanism induced by q-deformation. First of all, let us recall that quantizing spacetime can single out a lattice which, in turn, implies the existence of a smallest length given by the lattice spacing. On such a discrete spacetime plane-waves of wave-length smaller than twice the lattice spacing could not propagate. Since wave-length and momentum of a plane-wave are inversely proportional to each other it then follows that momentum space would be bounded. As a consequence all integrals over loop momenta should take on finite values.

As we will see below, q-deformation realizes these general ideas in a more subtle way. To this end, let us consider the following q-analogs of trigonometrical functions [30]:

$$\cos_q(x) \equiv \sum_{k=0}^{\infty} \frac{(-1)^k x^{2k}}{[2k]_q!}, \qquad \sin_q(x) \equiv q^{1/2} \sum_{k=0}^{\infty} \frac{(-1)^k x^{2k+1}}{[2k+1]_q!}. \qquad (2.1)$$

The symmetrical q-numbers and their factorials are respectively given by [31]

$$[n]_q \equiv \frac{q^n - q^{-n}}{q - q^{-1}}, \qquad q > 1, \qquad (2.2)$$

and

$$[n]_q! \equiv [n]_q [n-1]_q \dots [1]_q, \quad [0]_q! \equiv 1. \qquad (2.3)$$

From Fig. 1 we see that the q-trigonometrical functions are more or less periodic, but their maxima and minima increase in height exponentially.

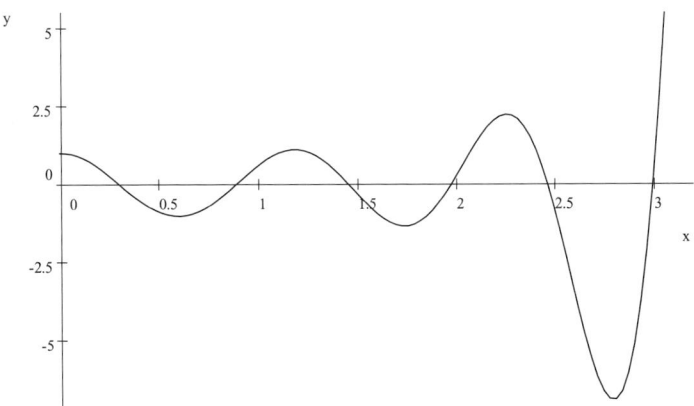

FIGURE 1. Fig. 1: Plot of the q-trigonometrical function $\cos_q(x/(q-q^{-1}))$, section at $q = 1.1$

The q-regularization mechanism is based on the following very remarkable property of q-trigonometrical functions [32,33]: There are real constants a and b such that

$$\lim_{n \to \infty} \cos_q(aq^{2n}) = \lim_{n \to \infty} \sin_q(bq^{2n+1}) = 0. \qquad (2.4)$$

If the q-trigonometrical functions are given by the expressions in (2.1) these constants take on the values

$$a = \frac{q^{-1/2}}{q - q^{-1}}, \qquad b = \frac{q^{1/2}}{q - q^{-1}}. \tag{2.5}$$

The deeper reason for this behavior of q-trigonometrical functions lies in the fact that their sequence of roots approach the values aq^{2n} or bq^{2n+1} if n tends to infinity. Table 1 contains the absolute values of $\cos_q(aq^{2n})$ one obtains when n runs through all integers from 1 to 20. A short glance at the table tells us that they are decreasing very rapidly with increasing n (notice that the smallest values appear in the top line and the biggest ones in the bottom line).

Tab. 1: The absolute values of $\cos_q(aq^{2n})$ for $n = 1, \ldots, 20$, $q = 1.1$

1.0954837827,	0.3806453606,	1.3002562658,	0.7466139278,
0.2073439674,	0.0332735098,	0.0033145196,	0.0002126316,
8.97256×10^{-6},	2.52305×10^{-7},	4.76707×10^{-9},	6.08464×10^{-11},
5.26538×10^{-13},	3.09651×10^{-15},	1.23955×10^{-17},	3.38126×10^{-20},
6.28977×10^{-23},	7.98276×10^{-26},	6.91482×10^{-29},	4.08903×10^{-32}.

In complete analogy to the undeformed case the q-trigonometrical functions in (2.1) represent the real and imaginary part of q-deformed plane-waves. Transition amplitudes in a q-deformed quantum field theory are then given by Fourier transforms that are formulated by means of q-trigonometrical functions and so-called *Jackson integrals*. The latter are given by [34]

$$\int_0^\infty d_q x \, f(x) \equiv \sum_{k=-\infty}^{\infty} (q-1)(c\,q^k) f(c\,q^k), \tag{2.6}$$

where c stands for an undetermined real constant. In this manner, the following expressions are viewed as q-analogs of one-dimensional Fourier transforms:

$$\int_0^\infty d_{q^2} x \, f(x) \cdot \cos_q(x\,p) \equiv \sum_{k=-\infty}^{\infty} (q^2-1) \, (aq^{2k}) \, f(aq^{2k}) \cdot \cos_q(aq^{2k}p), \tag{2.7}$$

$$\int_0^\infty d_{q^2} x \, f(x) \cdot \sin_q(x\,p) \equiv \sum_{k=-\infty}^{\infty} (q^2-1) \, (bq^{2k+1}) \, f(bq^{2k+1}) \cdot \sin_q(bq^{2k+1}p). \tag{2.8}$$

Notice that in (2.7) and (2.8) we set c equal to a and b, respectively [cf. Eq. (2.5)].

Essential for us is the following observation. If we require that the value of the momentum variable p takes on even powers of q, the sums on the right-hand side of (2.7) and (2.8) even converge for functions $f(x)$ that have no classical Fourier transform. To illustrate this point let us turn our attention to the expression

$$\int_a^\infty d_{q^2} x \, x^{10} \cdot \cos_q(x) = \sum_{k=0}^{\infty} (q^2-1) \, (a^{11} q^{22k}) \cdot \cos_q(aq^{2k}). \tag{2.9}$$

It is part of the calculation of a q-deformed Fourier transform of the function $f(x) = x^{10}$. Table 2 contains the numerical values of all partial sums of the Jackson integral in (2.9) that contain no more than twenty summands. Once again, we see that their sequence converges so fast that we are allowed to truncate the sum in (2.9) after twenty summations. It should finally be mentioned that this observation is a consequence of the property in (2.4).

Tab. 2: Partial sums with no more than twenty summands , $q = 1.1$

8.57483×10^7,	2.96844×10^8,	-5.03939×10^9,	1.76356×10^{10},
-2.89647×10^{10},	2.63758×10^{10},	-1.44196×10^{10},	4.94746×10^9,
-1.10037×10^9,	1.58134×10^8,	-1.78318×10^7,	-1.21080×10^6,
-2.27518×10^6,	-2.22886×10^6,	-2.23023×10^6,	-2.23021×10^6,
-2.23021×10^6,	-2.23021×10^6,	-2.23021×10^6,	-2.23021×10^6.

The considerations so far should now enable us to understand the regularization mechanism in a q-deformed field theory. It can be described as follows: The Jackson integral is determined by the values its integrand takes on for the points $c\,q^{2n}$ or $c\,q^{2n+1}$, where $n \in \mathbb{Z}$. These points establish a so-called q-lattice with offset c. A suitable choice for c guarantees that the values of the q-deformed trigonometrical functions rapidly diminish on the q-lattice points with increasing distance from the origin. This feature carries over to the integrands of q-deformed Fourier integrals and so insures their convergency.

3. Basic ideas of the mathematical formalism

This section is devoted to mathematical ideas which are indispensable for understanding q-deformed versions of physical theories. The presentation gives a brief outline of the subject, only. In this respect, it shall serve as a kind of 'appetizer'. For details, we refer the reader to Refs. [35–42].

3.1. What are quantum groups and quantum spaces?

Let us first recall that spacetime symmetries are normally described by Lie groups. Roughly speaking, quantum groups are nothing other than natural generalizations of such groups. To understand this assertion in more detail let us also recall that a group G is a set on which an associative product is defined such that a unit element e exists and such that every element u has an inverse u^{-1}. Instead of working with the group elements itself, it is possible to restrict attention to linear functions on the group. The latter constitute an algebra which is often denoted by $\mathcal{F}(G)$. One finds that the group multiplication induces on $\mathcal{F}(G)$ an algebraic structure making $\mathcal{F}(G)$ into a so-called *Hopf algebra*. To each group we can construct a Hopf algebra but the converse does not hold in general, since the variety of Hopf algebras is larger than that of groups. Quantum groups, for example, are Hopf algebras, but they show a non-commutative algebraic structure that prevents us from identifying them with an algebra of linear functions on an ordinary group.

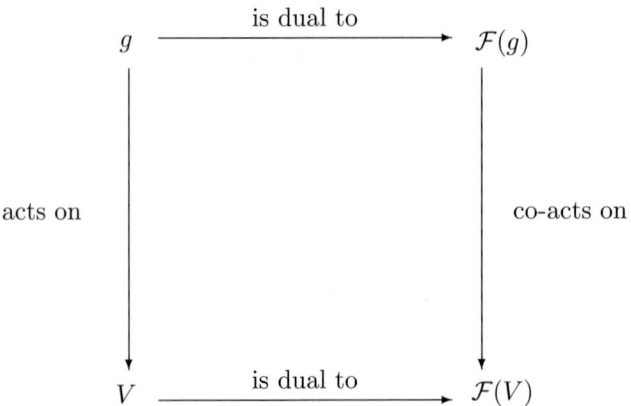

FIGURE 2. Relation between a group g and its function algebra $\mathcal{F}(g)$

Nevertheless, quantum groups show features that are reminiscent of function algebras on a group. In this sense, it should become clear that quantum groups can be viewed as generalizations of groups.

Perhaps, the reader may have realized that the notion of a quantum group arises from a process of dualization in which the group structure is replaced by a Hopf structure. There is a similar relationship between ordinary spaces and quantum spaces. To this end let us recall that a physical space such as a coordinate space is often described as a vector space V some group G acts upon. In other words, physical spaces are representations of their symmetry groups and are often referred to as modules. Interestingly, the vector space $\mathcal{F}(V)$ being dual to a given module V contains the same information as the module itself. In mathematical terms, this observation amounts to the assertion that $\mathcal{F}(V)$ becomes a so-called *co-representation* or *co-module* of $\mathcal{F}(G)$ (see Fig. 2).

Quantum spaces are generalizations of ordinary spaces insofar as they are defined as co-modules of quantum groups. In this respect, they are non-commutative spaces whose symmetry is governed by quantum groups. To sum up, dualizing the notion of a group and that of its representations we arrive at generalizations of ordinary physical spaces and their symmetries.

The simplest example of a quantum space is given by the famous Manin plane, which is spanned by the coordinates X^1 and X^2 subject to

$$X^1 X^2 = q X^2 X^1, \qquad q > 1. \tag{3.1}$$

The Manin plane is a co-representation of the quantum group $SL_q(2)$. The latter is a Hopf algebra generated by four non-commuting elements which we can arrange in matrix form:

$$M^i{}_j = \begin{pmatrix} a & b \\ c & d \end{pmatrix}. \tag{3.2}$$

If we do so, the coactions of $SL_q(2)$ on quantum plane coordinates become

$$\beta(X^i) = M^i{}_j \otimes X^j. \tag{3.3}$$

The requirement of covariance, i.e.

$$\beta(X^1)\beta(X^2) = q\beta(X^2)\beta(X^1), \tag{3.4}$$

implies the relations

$$ab = qba, \quad ac = qca, \quad bc = cb, \quad bd = qdb,$$
$$cd = qdc, \quad ad - da = (q - q^{-1})bc. \tag{3.5}$$

3.2. How do we multiply on quantum spaces?

From the example in the last subsection we saw that quantum spaces are spanned by non-commuting elements. The essential feature of our approach is that we realize the non-commutative structure of our quantum space algebras on commutative coordinate spaces. To achieve this, we make use of the fact that there are vector space isomorphisms that map normal ordered monomials of the non-commuting quantum space generators to the corresponding monomials of ordinary coordinates, i.e.

$$\mathcal{W}^{-1} : V_q \longrightarrow V,$$
$$\mathcal{W}^{-1}((X^1)^{i_1} \ldots (X^n)^{i_n}) \equiv (x^1)^{i_1} \ldots (x^n)^{i_n}. \tag{3.6}$$

Notice that we denote quantum space generators by capital letters. It should also be clear that the vector space isomorphism $\mathcal{W}$ is completely determined by (3.6), since each quantum space element can be written as a formal power series of normal ordered monomials.

Interestingly, one can extend the above vector space isomorphism to an algebra isomorphism by introducing a non-commutative product in V, the so-called *star product* [43–45]. This product is defined via the relation

$$\mathcal{W}(f \circledast g) = \mathcal{W}(f) \cdot \mathcal{W}(g), \tag{3.7}$$

being tantamount to

$$f \circledast g \equiv \mathcal{W}^{-1}(\mathcal{W}(f) \cdot \mathcal{W}(g)), \tag{3.8}$$

where f and g are formal power series in the commutative coordinate algebra V.

In the case of the Manin plane, for example, the star product can be calculated from the operator expression

$$f(x^i) \circledast g(x^j) = [q^{-\hat{n}_{x^2}\hat{n}_{y^1}} f(x^i) g(y^j)]_{y \to x}$$
$$= f(x^i) g(x^j) + O(h), \quad \text{with} \quad h = \ln q, \tag{3.9}$$

where we introduced the operators

$$\hat{n}_{x^i} \equiv x^i \frac{\partial}{\partial x^i}, \quad i = 1, 2. \tag{3.10}$$

The second equality in (3.9) tells us that the star product modifies the commutative product by terms depending on $h = \ln q$. Evidently, this modifications vanish in the classical limit $q \to 1$.

3.3. What are q-deformed translations?

Translations in spacetime play a very important role in physics as the fundamental physical laws have to be invariant under translations. From a mathematical point of view, translations are described by a mapping (We use alternatively $X \cong X \otimes 1$ and $Y \cong 1 \otimes X$)

$$\Delta : V_q \to V_q \otimes V_q, \tag{3.11}$$

with

$$\Delta(X^i) \equiv [X^i]_{(1)} \otimes [X^i]_{(2)} = X^i \otimes 1 + 1 \otimes X^i = X^i + Y^i. \tag{3.12}$$

The expressions $\Delta(X^i)$, $i = 1, \ldots, n$, are the components of a vector obtained from two other vectors with components X^i and Y^i via vector addition. For this reason the vector components $\Delta(X^i)$ should obey the same algebraic properties as the quantum space coordinates X^i. Concretely, we require

$$\Delta(X^i Z^j) = \Delta(X^i)\Delta(Z^j), \qquad \Delta(\beta(X^i)) = \beta(\Delta(X^i)), \tag{3.13}$$

i.e. the mapping in (3.11) has to be a *co-module algebra* homomorphism.

The first relation in (3.13) implies that the components $\Delta(X^i)$ again fulfil the commutation relations between quantum space coordinates. The second relation in (3.13) tells us that the components $\Delta(X^i)$ behave under symmetry transformations in very much the same way as quantum space coordinates X^i.

However, to satisfy the requirements in (3.13) we have to introduce a non-commutative tensor product by

$$(1 \otimes X^i) \cdot (X^j \otimes 1) = k\hat{R}^{ij}_{kl} X^k \otimes X^l, \quad k \in \mathbb{R}, \tag{3.14}$$

or, alternatively, by

$$(1 \otimes X^i) \cdot (X^j \otimes 1) = k^{-1}(\hat{R}^{-1})^{ij}_{kl} X^k \otimes X^l, \tag{3.15}$$

where $\hat{R}$ denotes the so-called *R-matrix* of the quantum space under consideration. It should be clear that for ordinary spaces the R-matrix and its inverse degenerate to the twist matrix $\delta^i_k \delta^j_l$.

In principle, we have everything together to calculate translations of monomials of quantum space generators. This can be achieved in the following manner:

$$\begin{aligned}
\Delta((X^1)^{i_1} \ldots (X^n)^{i_n}) &= ((\Delta(X^1))^{i_1} \ldots (\Delta(X^n))^{i_n}) \\
&= (X^1 \otimes 1 + 1 \otimes X^1)^{i_1} \ldots (X^n \otimes 1 + 1 \otimes X^n)^{i_n} \\
&= [(X^1)^{i_1} \ldots (X^n)^{i_n}]_{(1)} \otimes [(X^1)^{i_1} \ldots (X^n)^{i_n}]_{(2)}. \tag{3.16}
\end{aligned}$$

The first step is the homomorphism property of Δ and the second step makes use of (3.12). For the last step we have to apply (3.14) and end up with a kind of binomial theorem. Instead of (3.14) we could also work with (3.15). Doing so would lead us to a second q-analog of the binomial theorem.

In the case of Manin plane these considerations yield, for example,

$$\Delta((X^1)^{n_1}(X^2)^{n_2}) = \sum_{k_1=0}^{n_1} \sum_{k_2=0}^{n_2} q^{-(n_1-k_1)k_2} \begin{bmatrix} n_1 \\ k_1 \end{bmatrix}_{q^{-2}} \begin{bmatrix} n_2 \\ k_2 \end{bmatrix}_{q^{-2}}$$
$$\times (X^1)^{k_1}(X^2)^{k_2} \otimes (X^1)^{n_1-k_1}(X^2)^{n_2-k_2}, \qquad (3.17)$$

where the q-deformed binomial coefficients are defined in analogy to their undeformed counterparts:

$$\begin{bmatrix} \alpha \\ m \end{bmatrix}_{q^b} \equiv \frac{[[\alpha]]_{q^b} [[\alpha-1]]_{q^b} \cdots [[\alpha-m+1]]_{q^b}}{[[m]]_{q^b}!}, \quad b \in \mathbb{C}, m \in \mathbb{N}. \qquad (3.18)$$

If we want to deal with inverse translations we need a further mapping given by

$$S : V_q \to V_q, \quad S(X^i) \equiv -X^i. \qquad (3.19)$$

This mapping is an anti-algebra homomorphism. The values of S on monomials are determined in such a way that it holds

$$m \circ (1 \otimes S) \circ \Delta((X^1)^{i_1} \dots (X^n)^{i_n}) = m \circ (S \otimes 1) \circ \Delta((X^1)^{i_1} \dots (X^n)^{i_n}) = 0. \ (3.20)$$

Without going into the details, it should be mentioned that the operations in (3.11), (3.14), (3.15), and (3.19) again have realizations on commutative coordinate spaces. For these realizations we introduce the following notation:

$$\Delta \to f(x^i \oplus_A y^j), \quad S \to f(\ominus_A x^i),$$

$$(1 \otimes a) \cdot (b \otimes 1) \to f(x^i) \overset{x|y}{\odot}_A g(y^j). \qquad (3.21)$$

Notice that we have two different versions of each of the above operations according to the two possible choices for the non-commutative tensor product [cf. Eqs. (3.14) and (3.15)]. To distinguish them we introduced the label $A \in \{L, \bar{L}\}$.

It should also be mentioned that q-translations fulfil some interesting calculational rules. In terms of the operations on the commutative spaces they take a form that is strongly reminiscent of the group axioms satisfied by ordinary translations. We wish to illustrate this statement by the identities

$$f((x^i \oplus_A y^j) \oplus (\ominus_A y^k)) = f(x^i \oplus_A (y^j \oplus (\ominus_A y^k)))$$
$$= f(x^i \oplus_A 0) = f(x^i). \qquad (3.22)$$

The first equality in (3.22) means associativity of q-deformed translations and the second equality concerns the existence of inverse elements.

Finally, we would like to point out for the mathematically oriented reader that q-deformed translations are connected with braided Hopf algebras [46, 47]. In this respect, the equalities in (3.22) follow from the axioms of a braided Hopf algebra.

3.4. How to differentiate and integrate on quantum spaces

Partial derivatives on quantum spaces should generate infinitesimal q-translations. In this sense, we can introduce q-deformed partial derivatives by

$$\partial_i \triangleright f(x^j) \equiv \lim_{y^k \to 0} \frac{f(y^k \oplus_{\bar{L}} x^j) - f(x^j)}{y^i},$$

$$\hat{\partial}_i \bar{\triangleright} f(x^j) \equiv \lim_{y^k \to 0} \frac{f(y^k \oplus_L x^j) - f(x^j)}{y^i}, \qquad (3.23)$$

$$f(x^j) \bar{\triangleleft} \partial^i \equiv \lim_{y^k \to 0} \frac{f(x^j) - f(x^j \oplus_{\bar{L}} y^k)}{y_i},$$

$$f(x^j) \triangleleft \hat{\partial}^i \equiv \lim_{y^k \to 0} \frac{f(x^j) - f(x^j \oplus_L y^k)}{y_i}. \qquad (3.24)$$

The symbols $\triangleright$, $\bar{\triangleright}$, $\triangleleft$, and $\bar{\triangleleft}$ denote left and right actions. Objects with lower and upper indices transform into each other via a quantum metric (for the details, see Ref. [48]).

There are operator representations for the action of partial derivatives. They can be read off from the operator expressions for q-deformed translations, since we have

$$f(x^i \oplus_L y^j) = 1 \otimes f(y^j) + x^k \otimes \hat{\partial}_k \bar{\triangleright} f(y^j) + O(x^2),$$

$$f(x^i \oplus_{\bar{L}} y^j) = 1 \otimes f(y^j) + x^k \otimes \partial_k \triangleright f(y^j) + O(x^2), \qquad (3.25)$$

$$f(y^i \oplus_L x^j) = f(y^i) \otimes 1 - f(y^i) \triangleleft \hat{\partial}^k \otimes x_k + O(x^2),$$

$$f(y^i \oplus_{\bar{L}} x^j) = f(y^i) \otimes 1 - f(y^i) \bar{\triangleleft} \partial^k \otimes x_k + O(x^2), \qquad (3.26)$$

where repeated indices are to be summed over.

As representations of partial derivatives on the Manin plane we obtain, for example,

$$\partial_1 \triangleright f(x^1, x^2) = D^1_{q^2} f(x^1, q^2 x^2),$$

$$\partial_2 \triangleright f(x^1, x^2) = D^2_{q^2} f(qx^1, x^2). \qquad (3.27)$$

In the above formulae we introduced the so-called *Jackson derivatives* [49], which are given by

$$D^i_{q^a} f(x^i) \equiv \frac{f(q^a x^i) - f(x^i)}{(q^a - 1)x^i}. \qquad (3.28)$$

Being difference operators the Jackson derivatives represent discretized versions of ordinary derivatives.

Next, we come to integrals on quantum spaces. They can be seen as operations being inverse to partial derivatives. More concretely, an integral gives a solution to the equation

$$\partial^i \triangleright F = f, \qquad (3.29)$$

where f is a given function. The following reasonings about calculating integrals apply to all types of partial derivatives introduced in (3.23) and (3.24).

To find solutions to Eq. (3.29) it is helpful to realize that the operator expressions for partial derivatives on quantum spaces split into a classical part ∂^i_{cl} and corrections ∂^i_{cor} vanishing in the undeformed limit $q \to 1$, i.e.

$$\partial^i \triangleright F = \left(\partial^i_{cl} + \partial^i_{cor} \right) \triangleright F. \tag{3.30}$$

The classical part is always invertible, so we have

$$F = (\partial^i)^{-1} \triangleright f = \frac{1}{\partial^i_{cl} + \partial^i_{cor}} \triangleright f$$

$$= \frac{1}{\partial^i_{cl} \left(1 + (\partial^i_{cl})^{-1} \partial^i_{cor} \right)} \triangleright f = \frac{1}{1 + (\partial^i_{cl})^{-1} \partial^i_{cor}} \cdot \frac{1}{\partial^i_{cl}} \triangleright f$$

$$= \sum_{k=0}^{\infty} (-1)^k \left[(\partial^i_{cl})^{-1} \partial^i_{cor} \right]^k (\partial^i_{cl})^{-1} \triangleright f. \tag{3.31}$$

Once again, we wish to demonstrate our considerations by a simple example. Comparing the expressions in (3.27) with the right-hand side of Eq. (3.30) we find that

$$(\partial_1)_{cl} \triangleright f = D^1_{q^2} f(q^2 x^2), \qquad (\partial_1)_{cor} \triangleright f = 0,$$
$$(\partial_2)_{cl} \triangleright f = D^2_{q^2} f(q x^1), \qquad (\partial_2)_{cor} \triangleright f = 0, \tag{3.32}$$

which, in turn, leads to

$$(\partial_1)^{-1} \triangleright f = (\partial_1)^{-1}_{cl} \triangleright f = (D^1_{q^2})^{-1} f(q^{-2} x^2),$$
$$(\partial_2)^{-1} \triangleright f = (\partial_2)^{-1}_{cl} \triangleright f = (D^2_{q^2})^{-1} f(q^{-1} x^1), \tag{3.33}$$

where $(D^i_{q^a})^{-1}$ denotes the *Jackson integral* operator[cf. Eq. (2.6)].

Partial derivatives and integrals on quantum spaces lead us into the realm of multi-dimensional q-analysis. This subject was completely worked out in Ref. [41], where we derived various calculational rules such as Leibniz rules for partial derivatives, rules for integration by parts or q-analogs of the fundamental theorem of calculus.

3.5. Fourier transformations on quantum spaces

Fourier transformations play a very important role in quantum physics. Thus, it arises the question what are Fourier transformations on quantum spaces. To answer this question we first introduce q-analogs of plane-waves. In complete analogy to the undeformed case they are defined to be eigenfunctions of q-deformed partial derivatives [38, 50, 51]. Notice that q-deformed partial derivatives can act in different ways on a function and for this reason there are different versions of q-deformed plane-waves, i.e.

$$\partial^i \overset{x}{\triangleright} \exp(x^k|\partial^j)_{\bar{R},L} = \exp(x^k|\partial^j)_{\bar{R},L} \overset{\partial}{\circledast} \partial^i,$$

$$\hat{\partial}^i \overset{x}{\triangleright} \exp(x^k|\hat{\partial}^j)_{R,\bar{L}} = \exp(x^k|\hat{\partial}^j)_{R,\bar{L}} \overset{\partial}{\circledast} \hat{\partial}^i, \qquad (3.34)$$

$$\exp(\partial^j|x^k)_{\bar{R},L} \overset{x}{\triangleleft} \partial^i = \partial^i \overset{\partial}{\circledast} \exp(\partial^j|x^k)_{\bar{R},L},$$

$$\exp(\hat{\partial}^j|x^k)_{R,\bar{L}} \overset{x}{\triangleleft} \hat{\partial}^i = \hat{\partial}^i \overset{\partial}{\circledast} \exp(\hat{\partial}^j|x^k)_{R,\bar{L}}. \qquad (3.35)$$

However, q-deformed plane-waves are not completely determined by the relations in (3.34) and (3.35). Thus, we additionally require that our q-deformed plane-waves are normalized according to

$$\exp(x^i|\partial^j)_{\bar{R},L}|_{x^i=0} = \exp(x^i|\partial^j)_{\bar{R},L}|_{\partial^j=0} = 1,$$

$$\exp(x^i|\hat{\partial}^j)_{R,\bar{L}}|_{x^i=0} = \exp(x^i|\hat{\partial}^j)_{R,\bar{L}}|_{\partial^j=0} = 1. \qquad (3.36)$$

These normalization conditions together with (3.34) and (3.35) allow us to calculate expressions for q-deformed plane-waves. In the case of the Manin plane, for example, one finds

$$\exp(x^i|\hat{\partial}^j)_{R,\bar{L}} = \sum_{n_1,n_2=0}^{\infty} \frac{(x^1)^{n_1}(x^2)^{n_2} \otimes (\hat{\partial}_2)^{n_2}(\hat{\partial}_1)^{n_1}}{[[n_1]]_{q^{-2}}! \, [[n_2]]_{q^{-2}}!}. \qquad (3.37)$$

Interestingly, q-deformed plane-waves fulfil addition laws which can be written in the form

$$\exp(x^i \oplus_L y^j|\hat{\partial}^k)_{R,\bar{L}} = \exp(x^i| \exp(y^j|\hat{\partial}^k)_{R,\bar{L}} \overset{\partial}{\circledast} \hat{\partial}^l)_{R,\bar{L}},$$

$$\exp(x^i \oplus_{\bar{L}} y^j|\partial^k)_{\bar{R},L} = \exp(x^i| \exp(y^j|\partial^k)_{\bar{R},L} \overset{\partial}{\circledast} \partial^l)_{\bar{R},L}, \qquad (3.38)$$

$$\exp(x^k|\hat{\partial}^j \oplus_L \hat{\partial}'^i)_{R,\bar{L}} = \exp(x^l \overset{x}{\circledast} \exp(x^k|\hat{\partial}^j)_{R,\bar{L}}|\hat{\partial}'^i)_{R,\bar{L}},$$

$$\exp(x^k|\partial^j \oplus_{\bar{L}} \partial'^i)_{\bar{R},L} = \exp(x^l \overset{x}{\circledast} \exp(x^k|\partial^j)_{\bar{R},L}|\partial'^i)_{\bar{R},L}. \qquad (3.39)$$

Furthermore, our framework enables us to introduce 'inverse' q-deformed plane-waves:

$$\exp(\ominus_{\bar{L}} x^i|\partial^j)_{\bar{R},L} = \exp(x^i|\ominus_{\bar{L}} \partial^j)_{\bar{R},L},$$

$$\exp(\ominus_L x^i|\hat{\partial}^j)_{R,\bar{L}} = \exp(x^i|\ominus_L \hat{\partial}^j)_{R,\bar{L}}, \qquad (3.40)$$

$$\exp(\ominus_L x^i|\partial^j)_{\bar{R},L} = \exp(x^i|\ominus_L \partial^j)_{\bar{R},L},$$

$$\exp(\ominus_{\bar{L}} x^i|\hat{\partial}^j)_{R,\bar{L}} = \exp(x^i|\ominus_{\bar{L}} \hat{\partial}^j)_{R,\bar{L}}. \qquad (3.41)$$

In principle, q-deformed plane-waves can be manipulated in a similar fashion as their undeformed counterparts. Without going into the details, the following calculation shall serve as an example:

$$\exp(x^i| \overset{x}{\circledast} \exp(\ominus_{\bar{L}} x^j|\partial^k)_{\bar{R},L} \overset{\partial}{\circledast} \partial^l)_{\bar{R},L} = \exp(x^i \oplus_{\bar{L}} (\ominus_{\bar{L}} x^j)|\partial^k)_{\bar{R},L}$$

$$= \exp(x^i|\partial^l)_{\bar{R},L}|_{x^i=0} = 1. \qquad (3.42)$$

It should be obvious that the above result justifies to identify the expressions in (3.40) and (3.41) with inverse q-deformed plane-waves.

Now, we have everything together to introduce Fourier transformations on quantum spaces. Due to the different versions of q-deformed plane-waves, there are the following possibilities for defining Fourier transformations on quantum spaces [52, 53]:

$$\mathcal{F}_L(f)(p^k) \equiv \int_{-\infty}^{+\infty} d_1^n x \, f(x^i) \overset{x}{\circledast} \exp(x^j|i^{-1}p^k)_{\bar{R},L},$$

$$\mathcal{F}_{\bar{L}}(f)(p^k) \equiv \int_{-\infty}^{+\infty} d_2^n x \, f(x^i) \overset{x}{\circledast} \exp(x^j|i^{-1}p^k)_{R,\bar{L}}, \qquad (3.43)$$

$$\mathcal{F}_{\bar{R}}(f)(p^k) \equiv \int_{-\infty}^{+\infty} d_1^n x \, \exp(i^{-1}p^k|x^j)_{\bar{R},L} \overset{x}{\circledast} f(x^i),$$

$$\mathcal{F}_R(f)(p^k) \equiv \int_{-\infty}^{+\infty} d_2^n x \, \exp(i^{-1}p^k|x^j)_{R,\bar{L}} \overset{x}{\circledast} f(x^i). \qquad (3.44)$$

Let us make some comments on these definitions. First of all, we applied the substitution $\partial^k = i^{-1}p^k$ to our q-deformed plane-waves. The integrals over the whole space are obtained by applying one-dimensional integrals in succession. The volume elements of the integrals carry an index which is a consequence of the fact that we have two q-deformed differential calculi.

There is a second set of q-deformed Fourier transformations. They are given by

$$\mathcal{F}_L^*(f)(x^k) \equiv \frac{1}{\text{vol}_1} \int_{-\infty}^{+\infty} d_1^n p \, \exp(i^{-1}p^l|\ominus_L x^k)_{\bar{R},L} \overset{x|p}{\odot}_{\bar{L}} f(p^j),$$

$$\mathcal{F}_{\bar{L}}^*(f)(x^k) \equiv \frac{1}{\text{vol}_2} \int_{-\infty}^{+\infty} d_2^n p \, \exp(i^{-1}p^l|\ominus_{\bar{L}} x^k)_{R,\bar{L}} \overset{x|p}{\odot}_L f(\hat{p}^j), \qquad (3.45)$$

$$\mathcal{F}_R^*(f)(x^k) \equiv \frac{1}{\text{vol}_2} \int_{-\infty}^{+\infty} d_2^n p \, f(p^j) \overset{p|x}{\odot}_{\bar{R}} \exp(\ominus_R x^k|i^{-1}p^l)_{R,\bar{L}},$$

$$\mathcal{F}_{\bar{R}}^*(f)(x^k) \equiv \frac{1}{\text{vol}_1} \int_{-\infty}^{+\infty} d_1^n p \, f(p^j) \overset{p|x}{\odot}_R \exp(\ominus_{\bar{R}} x^k|i^{-1}p^l)_{\bar{R},L}. \qquad (3.46)$$

Notice that we introduced some sort of normalization constants $(\text{vol}_\alpha)^{-1}$, $\alpha \in \{1,2\}$. They are determined in such a way that the Fourier transforms in (3.45) and (3.46) are inverse to those in (3.43) and (3.44). To be more specific, we have

$$(\mathcal{F}_{\bar{R}}^* \circ \mathcal{F}_L)(f)(x^k) = f(\kappa x^k), \qquad (\mathcal{F}_L \circ \mathcal{F}_{\bar{R}}^*)(f)(x^k) = \kappa^{-n} f(\kappa^{-1} x^k),$$

$$(\mathcal{F}_R^* \circ \mathcal{F}_{\bar{L}})(f)(x^k) = f(\kappa^{-1} x^k), \qquad (\mathcal{F}_{\bar{L}} \circ \mathcal{F}_R^*)(f)(x^k) = \kappa^n f(\kappa x^k), \qquad (3.47)$$

$$(\mathcal{F}_{\bar{L}}^* \circ \mathcal{F}_R)(f)(x^k) = f(\kappa^{-1} x^k), \qquad (\mathcal{F}_R \circ \mathcal{F}_{\bar{L}}^*)(f)(x^k) = \kappa^n f(\kappa x^k),$$

$$(\mathcal{F}_L^* \circ \mathcal{F}_{\bar{R}})(f)(x^k) = f(\kappa x^k), \qquad (\mathcal{F}_{\bar{R}} \circ \mathcal{F}_L^*)(f)(x^k) = \kappa^{-n} f(\kappa^{-1} x^k), \qquad (3.48)$$

where κ denotes a certain constant.

From the undeformed case we know that taking the Fourier transform of unity yields a delta function. In this manner, q-analogs of the delta function should be given by

$$\delta_1^n(p^k) \equiv \mathcal{F}_L(1)(p^k) = \int_{-\infty}^{+\infty} d_1^n x \, \exp(x^j | \mathrm{i}^{-1} p^k)_{\bar{R},L},$$

$$\delta_2^n(p^k) \equiv \mathcal{F}_{\bar{L}}(1)(p^k) = \int_{-\infty}^{+\infty} d_2^n x \, \exp(x^j | \mathrm{i}^{-1} p^k)_{R,\bar{L}}. \qquad (3.49)$$

In analogy to their undeformed counterparts q-deformed delta functions fulfil as characteristic properties

$$\int_{-\infty}^{+\infty} d_\alpha^m y \, f(y^i) \overset{y}{\circledast} \delta_\beta^n(y^j \oplus_C (\ominus_C x^k)) = \mathrm{vol}_{\alpha,\beta} \, f(\kappa_C^{-1} x^k), \qquad (3.50)$$

and

$$\int_{-\infty}^{+\infty} d_\alpha^m y \, \delta_\beta^n((\ominus_C x^k) \oplus_C y^j) \overset{y}{\circledast} f(y^i) = \mathrm{vol}_{\alpha,\beta} \, f(\kappa_C^{-1} x^k), \qquad (3.51)$$

where

$$\mathrm{vol}_{\alpha,\beta} \equiv \int_{-\infty}^{+\infty} d_\alpha^m x \, \delta_\beta^n(x^k), \qquad \alpha, \beta \in \{1,2\}, \qquad (3.52)$$

and

$$\kappa_L = \kappa_{\bar{R}} = \kappa, \qquad \kappa_{\bar{L}} = \kappa_R = \kappa^{-1}. \qquad (3.53)$$

Moreover, one can show that the Fourier transform of a q-deformed delta function yields a q-deformed plane-wave and vice versa.

Finally, let us mention that q-deformed Fourier transformations interchange the star-product with q-deformed differentiation. The following identities shall serve as example:

$$\mathcal{F}_L(f \overset{x}{\triangleleft} \partial^j)(p^k) = \mathcal{F}_L(f)(p^k) \overset{p}{\circledast} (\mathrm{i}^{-1} p^j),$$

$$\mathcal{F}_L(f \overset{x}{\circledast} x^j)(p^k) = \mathcal{F}_L(f)(p^k) \overset{p}{\triangleleft} (\mathrm{i}\partial^j). \qquad (3.54)$$

4. Applications to physics

In the last section we gave a short introduction into analysis on q-deformed quantum spaces. It is now our aim to show how the new formalism applies to the non-relativistic one-particle problem. Once again, we restrict attention to some fundamental aspects, only. A rather complete treatment of the subject can be found in Refs. [54–56].

The three-dimensional q-deformed Euclidean space provides a suitable framework for dealing with non-relativistic physics in a q-deformed setting. It is spanned by three coordinates X^A, $A \in \{+, 3, -\}$, subject to the relations

$$X^3 X^+ = q^2 X^+ X^3, \quad X^- X^3 = q^2 X^3 X^-,$$
$$X^- X^+ = X^+ X^- + \lambda X^3 X^3, \tag{4.1}$$

where $\lambda = q - q^{-1}$.

This algebra can be extended by a central time element t. In doing so, we obtain a spacetime structure in which time behaves like a commutative and continuous variable, while space coordinates establish a q-deformed quantum space. Since time is completely decoupled from space, the time evolution operator is the usual one. For the same reason, the Schrödinger and Heisenberg equations of motion remain unchanged compared to the undeformed case.

4.1. Plane-wave solutions to the free-particle Schrödinger equation

We first seek a suitable Hamiltonian operator describing a free non-relativistic particle on three-dimensional q-deformed Euclidean space. An obvious choice is given by

$$H_0 \equiv g^{AB} P_B P_A (2m)^{-1}, \tag{4.2}$$

where g^{AB} denotes the quantum metric of the three-dimensional q-deformed Euclidean space and m stands for a mass parameter. The momentum operators are represented by partial derivatives, i.e. $P_A = i\partial_A$.

The Schrödinger equation for the free q-deformed particle reads

$$i\partial_t \overset{t}{\triangleright} \phi(x^A, t) = H_0 \overset{x}{\triangleright} \phi(x^A, t). \tag{4.3}$$

As solutions to this equation one finds

$$(u_{\bar{R},L})_{p,m}(x^A, t) =$$

$$= \sum_{n=0}^{\infty} \sum_{k=0}^{n_0-1} \frac{(i^{-1})^{n_0}(-\lambda_+)^{n_0-k} q^{-2k+2n_3(n_0-k)}}{n_0! \, [[n_+]]_{q^4}! \, [[n_3]]_{q^2}! \, [[n_-]]_{q^4}!} \begin{bmatrix} n_0 \\ k \end{bmatrix}_{q^4}$$

$$\times t^{n_0} (x^+)^{n_+} (x^3)^{n_3} (x^-)^{n_-}$$

$$\otimes (i^{-1} p_-)^{n_- + n_0 - k} (i^{-1} p_3)^{n_3 + 2k} (i^{-1} p_+)^{n_+ + n_0 - k} (2m)^{-n_0}, \tag{4.4}$$

where $\lambda_+ = q + q^{-1}$. The above result can also be written as product of a q-deformed plane-wave and some sort of time-dependent phase factor, i.e.

$$(u_{\bar{R},L})_{p,m}(x^A, t) = \exp(x^A | i^{-1} p_B)_{\bar{R},L} \overset{p}{\circledast} \exp(-itp^2(2m)^{-1})_{\bar{R},L}, \tag{4.5}$$

with

$$\exp(-itp^2(2m)^{-1})_{\bar{R},L} \equiv \sum_{n=0}^{\infty} \frac{(-it)^n}{n!} \sum_{k=0}^{n} q^{-2k}(-\lambda_+)^{n-k} \begin{bmatrix} n \\ k \end{bmatrix}_{q^4}$$
$$\times (p_-)^{n-k}(p_3)^{2k}(p_+)^{n-k}(2m)^{-n}. \qquad (4.6)$$

Our solutions to the Schrödinger equation of the free q-deformed particle can be regarded as q-analogs of momentum eigenfunctions of definite energy. This observation follows from the identities

$$P_A \overset{x}{\triangleright} (u_{\bar{R},L})_{p,m}(x^A,t) = i\partial_A \overset{x}{\triangleright} (u_{\bar{R},L})_{p,m}(x^A,t)$$
$$= (u_{\bar{R},L})_{p,m}(x^A,t) \overset{p}{\circledast} p_A, \qquad (4.7)$$

$$H_0 \overset{x}{\triangleright} (u_{\bar{R},L})_{p,m}(x^A,t) = (2m)^{-1}P^2 \overset{x}{\triangleright} (u_{\bar{R},L})_{p,m}(x^A,t)$$
$$= (u_{\bar{R},L})_{p,m}(x^A,t) \overset{p}{\circledast} p^2(2m)^{-1}. \qquad (4.8)$$

The above relations tell us that the wave function $(u_{\bar{R},L})_{p,m}$ corresponds to a particle with momentum components p_A, $A \in \{+,3,-\}$, and energy

$$E = p^2(2m)^{-1} = g^{AB}p_B \overset{p}{\circledast} p_A(2m)^{-1}. \qquad (4.9)$$

In complete analogy to the undeformed case, the time-evolution of our q-deformed momentum eigenfunctions is given by

$$(u_{\bar{R},L})_{p,m}(x^A,t) = \exp(-itH_0) \overset{H_0|x}{\triangleright} \exp(x^A|i^{-1}p_B)_{\bar{R},L}, \qquad (4.10)$$

i.e. we simply have to apply the usual time-evolution operator to a q-deformed plane-wave of three-dimensional q-deformed Euclidean space. It is not very difficult to check that (4.10) indeed gives a solution to the Schrödinger equation of the free q-deformed particle. To this end, one inserts the expression in (4.10) into Eq. (4.3) and then applies the time derivative to the time evolution operator.

Our q-analogs of momentum eigenfunctions are normalized to q-deformed delta functions. In this manner, they establish an orthonormal set of functions. Concretely, we have

$$\int_{-\infty}^{+\infty} d_1^3x \, (u_{\bar{R},L})_{\ominus_L \tilde{p},m}(x^A, -q^{-\zeta}t) \overset{\tilde{p}|x}{\odot_{\bar{L}}} (u_{\bar{R},L})_{p,m}(x^B,t) =$$
$$= (\text{vol}_1)^{-1} \delta_1^3((\ominus_L \tilde{p}_C) \oplus_L p_D), \qquad (4.11)$$

with

$$\text{vol}_1 \equiv \int_{-\infty}^{+\infty} d_1^3x \int_{-\infty}^{+\infty} d_1^3p \, \exp(x^A|i^{-1}p_B)_{\bar{R},L}. \qquad (4.12)$$

It should be mentioned that the tensor multiplication $\odot_{\bar{L}}$ on the left-hand side of Eq. (4.11) is due to the fact that q-deformed momentum eigenfunctions live in the tensor product of a position space with a momentum space.

Our q-deformed momentum eigenfunctions also fulfil a completeness relation. It reads

$$\int_{-\infty}^{+\infty} d_1^3 p \, (u_{\bar{R},L})_{p,m}(x^A, t) \overset{p|y}{\odot_{\bar{L}}} (u_{\bar{R},L})_{\ominus L p,m}(y^B, -q^{-\varsigma}t) =$$

$$= (\text{vol}_1)^{-1} \delta_1^3 (x^A \oplus_L (\ominus_L y^B)). \tag{4.13}$$

Notice that the time variable of the momentum eigenfunction with negative momentum is affected by a scaling. This scaling ensures that the completeness relation for q-deformed momentum eigenfunctions is time-independent. For the same reason the orthonormality relation in Eq. (4.11) does not vary with time.

Now, we are in a position to write down expressions for q-deformed wave packets. As usual, they arise from superposition of q-deformed momentum eigenfunctions. In this manner, we have

$$(\phi_1^*)_m(x^i) = (\text{vol}_1)^{1/2} \int_{-\infty}^{+\infty} d_1^3 p \, (u_{\bar{R},L})_{p,m}(x^i) \overset{p}{\circledast} (c_1^*)_{\kappa^{-1}p}. \tag{4.14}$$

The reader should not be confused about the different labels like the asterisk, since their meaning is not important in this context. The volume integral in Eq. (4.14) is a q-deformed one and for the sake of simplicity we took the convention $x^i \equiv (x^A, t)$.

Conversely, the completeness relations for q-deformed momentum eigenfunctions together with the properties of q-deformed delta functions allow us to regain the expansion coefficients $(c_1^*)_p$ from the wave packet $(\phi_1^*)_m$. This can be achieved by means of the formula

$$(c_1^*)_p = \frac{1}{(\text{vol}_1)^{1/2}} \int_{-\infty}^{+\infty} d_1^3 x \, (u_{\bar{R},L})_{\ominus L p,m}(x^A, -q^{-\varsigma}t) \overset{p|x}{\odot_{\bar{L}}} (\phi_1^*)_m(x^i). \tag{4.15}$$

The q-deformed wave-packets are again solutions to the free-particle Schrödinger equation in (4.3), i.e.

$$i\partial_t \overset{t}{\triangleright} (\phi_1^*)_m(x^i) = H_0 \overset{x}{\triangleright} (\phi_1^*)_m(x^i). \tag{4.16}$$

This observation is a direct consequence of the fact that a q-deformed wave-packet is a superposition of q-deformed momentum eigenfunctions. For the same reason Eq. (4.10) implies that

$$(\phi_1^*)_m(x^A, t) = \exp(-it H_0) \overset{H_0|x}{\triangleright} (\phi_1^*)_m(x^A, t=0). \tag{4.17}$$

4.2. The propagator of the free q-deformed particle

Once the wave function of a quantum system is known at a certain time the time-evolution operator enables us to find the wave function at any later time. However, there is another way to solve the time-evolution problem. It requires to know the so-called propagator. It is now our aim to discuss the propagator of the free non-relativistic q-deformed particle.

To this end, we insert the expression for the expansion coefficients given by Eq. (4.15) into the expansion in Eq. (4.14). In doing so, we obtain an integral operator that acts on an initial wave function to yield the final wave function:

$$(\phi_1^*)_m(x^i) = \int_{-\infty}^{+\infty} d_1^3 y \, (K_1^*)_m(x^i; y^j) \overset{y}{\circledast} (\phi_1^*)_m(\kappa y^A, t_y). \qquad (4.18)$$

The kernel of this integral operator is the wanted propagator and takes the form

$$(K_1^*)_m(x^i; y^j) =$$

$$= \kappa^3 \int_{-\infty}^{+\infty} d_1^3 p \, (u_{\bar{R},L})_{p,m}(x^A, t_x) \overset{p|x}{\odot_{\bar{L}}} (u_{\bar{R},L})_{\ominus L p,m}(y^B, -q^{-\zeta} t_y). \qquad (4.19)$$

We have to impose the causality requirement on the propagator, since the wave function at time t should not be influenced by wave functions at times $t' > t$. This leads us to the retarded q-deformed Green's function

$$(K_1^*)_{m+}(x^i; y^j) \equiv \theta(t_x - t_y) \, (K_1^*)_m(x^A, t_x; y^B, t_y), \qquad (4.20)$$

where $\theta(t)$ stands for the Heaviside function

$$\theta(t) = \begin{cases} 1 & \text{if } t \geq 0, \\ 0 & \text{otherwise.} \end{cases} \qquad (4.21)$$

In analogy to the undeformed case the propagator in Eq. (4.19) satisfies the boundary condition

$$\lim_{t_x \to t_y} (K_1^*)_m(y^i; x^j) = \kappa^3 (\text{vol}_1)^{-1} \delta_1^3(x^A \oplus_L (\ominus_L y^B)). \qquad (4.22)$$

Due to this condition the retarded q-deformed Green's function is a solution to a Schrödinger equation with a q-deformed delta function as inhomogeneous part, i.e.

$$i\partial_t \overset{t_x}{\triangleright} (K_1^*)_{m+}(x^i; y^j) - H_0 \overset{y}{\triangleright} (K_1^*)_{m+}(x^i; y^j) =$$

$$= i\kappa^3 (\text{vol}_1)^{-1} \delta(t_x - t_y) \delta_1^3(x^A \oplus_L (\ominus_L y^B)). \qquad (4.23)$$

Notice that the last result follows from reasonings being very similar to those for the undeformed case.

Using Eq. (4.23) it is not very difficult to show that the q-deformed Green's function in Eq. (4.20) generates solutions to the inhomogeneous Schrödinger equation given by

$$i\partial_t \overset{t_x}{\triangleright} (\psi_1^*)_\varrho(x^i) - H_0 \overset{x}{\triangleright} (\psi_1^*)_\varrho(x^i) = \varrho(x^i). \qquad (4.24)$$

These solutions can be written in the form

$$(\psi_1^*)_\varrho(x^i) = -i \int_{-\infty}^{+\infty} dt_y \int_{-\infty}^{+\infty} d_1^3 y \, (K_1^*)_{m+}(x^i; y^j) \overset{y}{\circledast} \varrho(\kappa y^A, t_y), \qquad (4.25)$$

as one readily checks by insertion.

The considerations so far reveal a remarkable analogy to the undeformed case. It seems that there is a q-analog to each identity playing an important role in physics. This assertion is confirmed by the observation that the retarded q-deformed Green's function again fulfils a composition property:

$$(K_1^*)_{m+}(x^i; y^j) = \int_{-\infty}^{+\infty} d_1^3 z \, (K_1^*)_{m+}(x^i; z^A, t_z) \overset{z}{\circledast} (K_1^*)_{m+}(\kappa z^B, t_z; y^j). \quad (4.26)$$

4.3. Scattering of q-deformed particles

In scattering theory one typically considers the situation that the Hamilton operator H can be divided into the Hamiltonian H_0 of a free-particle and an interaction V, i.e.

$$H = H_0 + V. \quad (4.27)$$

It is now our aim to seek solutions to the Schrödinger equation

$$i\partial_t \overset{t}{\triangleright} (\psi_1^*)_{m+}(x^i) = H \overset{x}{\triangleright} (\psi_1^*)_{m+}(x^i). \quad (4.28)$$

Moreover, we require that the solutions describe q-deformed particles being free in the remote past. For this reason, we impose the following condition on $(\psi_1^*)_{m+}$:

$$\lim_{t_x \to -\infty} ((\psi_1^*)_m(x^A, t_x) - (\phi_1^*)_m(x^A, t_x)) = 0, \quad (4.29)$$

where $(\phi_1^*)_m$ denotes a solution to the free-particle Schrödinger equation in (4.3).

The point now is that Eq. (4.28) is identical to Eq. (4.24) if we make the identification

$$\varrho(x^A, t_x) = V(x^A, t_x) \overset{x}{\circledast} (\psi_1^*)_{m+}(x^B, t_x). \quad (4.30)$$

Plugging this into the right-hand side of Eq. (4.25) we obtain a q-analog of the Lippmann-Schwinger equation:

$$(\psi_1^*)_{m+}(x^i) = (\phi_1^*)_m(x^i)$$

$$- i \int_{-\infty}^{+\infty} dt_y \int_{-\infty}^{+\infty} d_1^3 y \, (K_1^*)_{m+}(x^i; y^j) \overset{y}{\circledast} V(\kappa y^A, t_y)$$

$$\overset{y}{\circledast} (\psi_1^*)_{m+}(\kappa y^B, t_y). \quad (4.31)$$

Notice that each solution to this integral equation shows the correct boundary condition, since the q-deformed Green's function on the right-hand side of Eq. (4.31) vanishes as $t_x \to -\infty$.

To solve the q-deformed Lippmann-Schwinger equation it is sometimes convenient to introduce a new Green's function $(G_1^*)_{m+}$. It is defined via the identity

$$(\psi_1^*)_{m+}(x^i) = \lim_{t_y \to -\infty} \int_{-\infty}^{+\infty} d_1^3 y \, (G_1^*)_{m+}(x^i; y^A, t_y) \overset{y}{\circledast} (\phi_1^*)_m(\kappa y^B, t_y). \quad (4.32)$$

From this equation one can prove that $(G_1^*)_{m+}$ satisfies the differential equation

$$i\partial_t \overset{t_y}{\triangleright} (G_1^*)_{m+}(x^i; y^j) - H \overset{y}{\triangleright} (G_1^*)_{m+}(x^i; y^j) =$$

$$= i\kappa^3 (\text{vol}_1)^{-1} \delta(t_x - t_y) \delta_1^3(x^A \oplus_L (\ominus_L y^B)), \quad (4.33)$$

One can also derive an integral equation for $(G_1^*)_{m+}$. To this end, we replace $(\psi_1^*)_{m+}$ in the q-deformed Lippmann-Schwinger equation by the right-hand side of Eq. (4.32). Proceeding in this manner should enable us to read off the following equation:

$$(G_1^*)_{m+}(x^i; y^j) = (K_1^*)_{m+}(x^i; y^j)$$

$$- i \int_{-\infty}^{+\infty} dt_z \int_{-\infty}^{+\infty} d_1^3 z \, (K_1^*)_{m+}(x^i; z^k) \overset{z}{\circledast} V(\kappa z^A, t_z)$$

$$\overset{z}{\circledast} (G_1^*)_{m+}(\kappa z^B, t_z; y^j). \tag{4.34}$$

Finally, it should be noted that the Green's function $(G_1^*)_{m+}$ obeys the composition rule

$$(G_1^*)_{m+}(x^A, t_x; y^B, t_y) =$$

$$= \int_{-\infty}^{+\infty} d_1^3 z \, (G_1^*)_{m+}(x^A, t_x; z^C, t_z) \overset{z}{\circledast} (G_1^*)_{m+}(\kappa z^D, t_z; y^B, t_y), \tag{4.35}$$

where $t_x > t_z > t_y$.

Let us return to the q-deformed Lippmann-Schwinger equation as it is given in (4.31). We can solve it iteratively if the interaction V is rather small. This way, we obtain a q-analog of the famous Born series:

$$(\psi_1^*)_{m+}(x^i) = (\phi_1^*)_m(x^i)$$

$$+ i^{-1} \int_{-\infty}^{+\infty} dt_1 \int_{-\infty}^{+\infty} d_1^3 y_1 \, (K_1^*)_{m+}(x^i; y_1^j) \overset{y_1}{\circledast} V(\kappa y_1^A, t_1) \overset{y_1}{\circledast} (\phi_1^*)_m(\kappa y_1^B, t_1)$$

$$+ i^{-2} \int_{-\infty}^{+\infty} dt_1 \int_{-\infty}^{+\infty} d_1^3 y_1 \int_{-\infty}^{+\infty} dt_2 \int_{-\infty}^{+\infty} d_1^3 y_2 \, (K_1^*)_{m+}(x^i; y_1^j)$$

$$\overset{y_1}{\circledast} V(\kappa y_1^A, t_1) \overset{y_1}{\circledast} (K_1^*)_{m+}(\kappa y_1^B, t_1; y_2^k) \overset{y_2}{\circledast} V(\kappa y_2^C, t_2)$$

$$\overset{y_2}{\circledast} (\phi_1^*)_m(\kappa y_2^D, t_2) + \ldots \tag{4.36}$$

It should be obvious that the last result corresponds to the expansion

$$(G_1^*)_{m+}(x^i, z^j) = (K_1^*)_{m+}(x^i; z^j)$$

$$+ i^{-1} \int_{-\infty}^{+\infty} dt_1 \int_{-\infty}^{+\infty} d_1^3 y_1 \, (K_1^*)_{m+}(x^i; y_1^k) \overset{y_1}{\circledast} V(\kappa y_1^A, t_1)$$

$$\overset{y_1}{\circledast} (K_1^*)_{m+}(\kappa y_1^B, t_1; z^j)$$

$$+ i^{-2} \int_{-\infty}^{+\infty} dt_1 \int_{-\infty}^{+\infty} d_1^3 y_1 \int_{-\infty}^{+\infty} dt_2 \int_{-\infty}^{+\infty} d_1^3 y_2 \, (K_1^*)_{m+}(x^i; y_1^k)$$

$$\overset{y_1}{\circledast} V(\kappa y_1^A, t_1) \overset{y_1}{\circledast} (K_1^*)_{m+}(\kappa y_1^B, t_1; y_2^l) \overset{y_2}{\circledast} V(\kappa y_2^C, t_2)$$

$$\overset{y_2}{\circledast} (K_1^*)_{m+}(\kappa y_2^D, t_2; z^j) + \ldots \tag{4.37}$$

The expressions in (4.36) and (4.37) show a striking similarity to their undeformed counterparts. For this reason they are interpreted along the same line of reasonings as in the undeformed case.

Last but not least, it should be mentioned that the matrix elements of the Green's function $(G_1^*)_{m^+}$ are related to q-deformed scattering amplitudes. A correct treatment of this subject, however, is a little bit more involved. Thus, we are not discussing that issue here.

5. Conclusion

Let us once more summarize the main objectives of this paper. We first demonstrated that q-deformation implies a regularization mechanism. This mechanism has its origin in a discrete description of spacetime, but avoids difficulties of other approaches for discretizing spacetime, since it does not suffer from the absence of spacetime symmetries.

However, rewriting physical theories in a q-deformed setting requires to muse on mathematical ideas with the aim to become aware of their essence. More concretely, one has to think about a q-deformed version of multi-dimensional analysis. Their constituents are mainly determined by the requirement of being compatible with q-deformed spacetime symmetries.

The mathematical framework we obtain in this manner is so powerful that it enables a consistent and complete revision of well-established theories. This can be done along the same line of reasonings as in the undeformed case. In this manner, our approach has the advantage that it helps to generalize and extend physical theories in a way that the relationship to their undeformed limit is rather clear. So to speak, the undeformed theories can be seen as approximations or simplifications of a more detailed description provided by q-deformation. The non-relativistic Schrödinger theory of the free q-deformed particle serves as an example to underline this assertion.

Acknowledgement

I am very grateful to Prof. Eberhard Zeidler for financial support. His interest in my ideas and his encouragement helped me to finish this work. I am also very grateful to Prof. Julius Wess who brought me into contact with the subject of q-deformation during my PhD. Furthermore, I would like to thank Alexander Schmidt for useful discussions and his steady support. Finally, I thank Prof. Dieter Lüst for kind hospitality.

References

[1] L. Landau and R. Peierls, Z. Phys. **69** (1931) 56.

[2] W. Heisenberg, Ann. Phys. **32** (1938) 20.

[3] H. S. Snyder, Phys. Rev. **71** (1947) 38.

[4] H.T. Flint, Phys. Rev. **74** (1948) 209.

[5] E.L. Hill, Phys. Rev. **100** (1955) 1780.

[6] Yu. A. Gol'fand, Sov. Phys. JETP **16** (1963) 184.

[7] D. Finkelstein, Int. J. Theor. Phys. **28** (1989) 441.

[8] C.J. Isham, in H. Lee, ed., Proceedings of the Advanced Summer Institute on Physics, Geometry and Topology, Plenum, New York (1990).

[9] E. Prugovečki, *Principles of Quantum General Relativity*, World Scientific, Singapore, 1995.

[10] S. Tomonaga, *Prog. Theor. Phys.* **1** (1946) 27; *Phys. Rev.* **74** (1948) 224.

[11] J. Schwinger, *Phys. Rev.* **75** (1949) 651; *Phys. Rev.* **76** (1949) 790.

[12] R.P. Feynman, *Phys. Rev.* **76** (1949) 749.

[13] P.P. Kulish and N. Yu. Reshetikin, J. Sov. Math. **23** (1983) 2345.

[14] V.G. Drinfeld, Sov. Math. Dokl. **32** (1985) 254.

[15] M. Jimbo, Lett. Math. Phys. **10** (1985) 63.

[16] V.G. Drinfeld, in A.M. Gleason, ed., Proceedings of the International Congress of Mathematicians, Amer. Math. Soc. (1986).

[17] Yu. I. Manin, *Quantum Groups and Non-Commutative Geometry*, Centre de Recherche Mathématiques, Montreal (1988).

[18] S.L. Woronowicz, Commun. Math. Phys. **111** (1987) 613.

[19] M. Takeuchi, Israel J. Math. **72** (1990) 232.

[20] N. Yu. Reshetikhin, L.A. Takhtadzhyan and L.D. Faddeev, Leningrad Math. J. **1** (1990) 193.

[21] J. Wess and B. Zumino, Nucl. Phys. B. Suppl. **18** (1991) 302.

[22] S. Majid, *Foundations of Quantum Group Theory,* University Press, Cambridge (1995).

[23] M. Chaichian and A.P. Demichev, *Introduction to Quantum Groups*, World Scientific, Singapore, 1996.

[24] M. Fichtenmüller, A. Lorek and J. Wess, Z. Phys. C **71** (1996) 533, [hep-th/9511106].

[25] J. Wess, Int. J. Mod. Phys. A **12** (1997) 4997.

[26] B.L. Cerchiai and J. Wess, Eur. Phys. J. C **5** (1998) 553, [math.qa/9801104].

[27] H. Grosse, C. Klimčik and P. Prešnajder, Int. J. Theor. Phys. **35** (1996) 231, [hep-th/9505175].

[28] S. Majid, Int. J. Mod. Phys. A **5** (1990) 4689.

[29] R. Oeckl, Commun. Math. Phys. **217** (2001) 451.

[30] T.H. Koornwinder, R.F. Swarttouw, Trans. AMS **333**, 445 (1992).

[31] V. Kac and P. Cheung, *Quantum Calculus,* Springer-Verlag, Berlin (2000).

[32] J. Wess, *q-deformed Heisenberg Algebras*, in H. Gausterer, H. Grosse and L. Pittner, eds., Proceedings of the 38. Internationale Universitätswochen für Kern- und Teilchenphysik, no. 543 in Lect. Notes in Phys., Springer-Verlag, Schladming (2000), [math-ph/9910013].

[33] J. Seifert, *q-Deformierte Einteilchen-Quantenmechanik*, Ph.D. thesis, Ludwig-Maximilians-Universität München, Fakultät für Physik, München (1996).

[34] F.H. Jackson, Proc. Durham Phil. Soc. **7** (1927) 182.

[35] H. Wachter and M. Wohlgenannt, Eur. Phys. J. C **23** (2002) 761, [hep-th/0103120].

[36] C. Bauer and H. Wachter, Eur. Phys. J. C **31** (2003) 261, [math-ph/0201023].

[37] H. Wachter, Eur. Phys. J. C **32** (2004) 281, [hep-th/0206083].

[38] H. Wachter, Eur. Phys. J. C **37** (2004) 379, [hep-th/0401113].

[39] H. Wachter, *q-Translations on quantum spaces,* preprint, [hep-th/0410205].

[40] H. Wachter, *Braided products for quantum spaces,* preprint, [math-ph/0509018].

[41] H. Wachter, Int. J. Mod. Phys. A **22** (2007) 95, [math-ph/0604028].

[42] H. Wachter, *Quantum kinematics on q-deformed quantum spaces I,* preprint, [quant-ph/0612173].

[43] F. Bayen, M. Flato, C. Frønsdal, A. Lichnerowicz and D. Sternheimer, Ann. Phys. **111** (1978) 61.

[44] J.E. Moyal, Proc. Camb. Phil. Soc. **45** (1949) 99.

[45] J. Madore, S. Schraml, P. Schupp and J. Wess, Eur. Phys. J. C **16** (2000) 161, [hep-th/0001203].

[46] S. Majid, *Introduction to Braided Geometry and q-Minkowski Space,* preprint (1994), [hep-th/9410241].

[47] S. Majid, *Beyond Supersymmetry and Quantum Symmetry (an introduction to braided groups and braided matrices),* in M.L. Ge and H.J. de Vega, eds., Quantum Groups, Integrable Statistical Models and Knot Theory, World Scientific, 231 (1993).

[48] A. Lorek, W. Weich and J. Wess, *Non-commutative Euclidean and Minkowski Structures,* Z. Phys. C **76** (1997) 375, [q-alg/9702025].

[49] F.H. Jackson, trans. Roy. Edin. **46** (1908) 253.

[50] S. Majid, J. Math. Phys. **34** (1993) 4843.

[51] A. Schirrmacher, J. Math. Phys. **36** (3) (1995) 1531.

[52] A. Kempf and S. Majid, J. Math. Phys. **35** (1994) 6802.

[53] H. Wachter, *Quantum kinematics on q-deformed quantum spaces II,* preprint, [quant-ph/0612174].

[54] H. Wachter, *Non-relativistic Schrödinger theory on q-deformed quantum spaces I,* preprint, quant-ph/0703070.

[55] H. Wachter, *Non-relativistic Schrödinger theory on q-deformed quantum spaces II,* preprint, quant-ph/0703072.

[56] H. Wachter, *Non-relativistic Schrödinger theory on q-deformed quantum spaces III,* preprint, quant-ph/0703073.

Hartmut Wachter
Max Planck Institute for Mathematics in the Sciences
Inselstrasse 22
DE–04103 Leipzig
e-mail: Hartmut.Wachter@physik.uni-muenchen.de

Quantum Field Theory
B. Fauser, J. Tolksdorf and E. Zeidler, Eds., 285–302
© 2009 Birkhäuser Verlag Basel/Switzerland

Towards a q-Deformed Supersymmetric Field Theory

Alexander Schmidt

In memory of Julius Wess.

Abstract. This article intends to show the possibility of building noncommutative, q-deformed structures in a way that is analog to the commutative case. Especially, some aspects concerning q-deformed superspaces and q-deformed superalgebras are considered.

Mathematics Subject Classification (2000). Primary 81R50; Secondary 16W30; 17B37.

Keywords. noncommutative geometry, quantum groups, supersymmetry, q-deformation, quantum field theory.

1. Introduction

The idea that one should change the continuous nature of space-time at extremely small length-scales, like the Planck scale, to a discrete structure has a long and prominent history in physics [1, 2]. Let me give two nice quotes to confirm the statement about the long and prominent history of the idea to change geometry. Bernhard Riemann already thought about this possibility in his inaugural lecture from 1854:

> Now it seems that the empirical notions on which the metric determinations of Space are based, the concept of a solid body and light ray, lose their validity in the infinitely small; it is therefore quite definitely conceivable that the metric relations of Space in the infinitely small do not conform the hypothesis of geometry; and in fact, one ought to assume this as soon as it permits a simpler way of explaining phenomena

Albert Einstein made the observation [3]

> ...that the introduction of a space-time continuum may be considered as contrary to nature in view of the molecular structure of everything

which happens on a small scale. It is maintained that perhaps the success of the Heisenberg method points to a purely algebraical method of description of nature, that is to the elimination of continuous functions from physics. Then, however, we must also give up, by principle, the space-time continuum. It is not unimaginable that human ingenuity will some day find methods which will make it possible to proceed along such a path.

One way to achieve this is the use of noncommutative geometry. In this article we want to concentrate only on a special aspect of noncommutative geometry, namely the q-deformation of spaces. One hope associated with it is to find a new method to regularize quantum field theories [4–8]. It was also shown, that q-deformation can lead to discrete versions of space-time via the discretizations of the spectra of space-time observables [9, 10].

The aim of this article is to show that one can build supersymmetric objects and structures in a quantum group covariant manner on q-deformed algebras or spaces. The work presented extends some well-known structures used in supersymmetry, i.e. superalgebras as well as their representation on superspaces including supercovariant derivatives, to q-deformed spaces. Most of this work was done in Refs. [11–14].

Let us describe the article's content more detailed. In Sec. 2 of the present article we review some well-known fundamental algebraic concepts. We also define the notion of a q-commutator and review the fermionic quantum plane. With q-commutators at hand we are able to introduce quantum symmetry algebras in a way that mirrors the classical, i.e. undeformed limit. (In the remainder of this article the terms 'classical' and 'undeformed' are always synonymous. In this respect, one should notice that our quantum symmetry algebras arise from classical space-time symmetries via q-deformation.)

After this reviews the next sections introduce some new results. Section 3 presents the main part of the article: the construction of q-deformed superalgebras for some special types of quantum groups like the q-Poincaré algebra. We demonstrate how to merge the concept of supersymmetric algebras with that of quantum symmetry algebras. As example we treat the case of the q-deformed Lorentz algebra and extend it to the q-deformed Super-Poincaré algebra.

In the last section we restrict attention to the three-dimensional q-deformed Super-Euclidean algebra and outline how to construct its operator representations on a corresponding quantum superspace. In addition to this, we give expressions for q-deformed analogs of supersymmetric covariant derivatives and mention some of the arising peculiarities.

The work presented in the article can be seen as a first step to the extension of supersymmetric field theories to q-deformed spaces. It is based on the foundations of q-deformed algebras laid in [15]. Thus, the article should be readable with a basic knowledge of Hopf-algebras and quantum algebras.

Let us close with some remarks concerning our notation. Throughout the article it is understood that $\lambda \equiv q - q^{-1}$ and $\lambda_+ \equiv q + q^{-1}$. The symbol $\varepsilon^{\alpha\beta}$ denotes the two-dimensional q-deformed spinor metric with nonzero entries $\varepsilon^{12} = -q^{-1/2}$ and $\varepsilon^{21} = q^{1/2}$. It holds $\varepsilon^{\alpha\beta} = -\varepsilon_{\alpha\beta}$.

Similarly, g_{AB} is the metric of three-dimensional q-deformed Euclidean space with $g^{AB} = g_{AB}$. Its nonzero entries are $g_{+-} = -q$, $g_{33} = 1$, and $g_{-+} = -q^{-1}$.

2. Fundamental Algebraic Concepts

In this section we want to review some of the fundamental notions we need for our further explanations. In Sec. 3.1 of Ref. [16] the general idea behind matrix quantum groups was shortly explained [18,19]. The following considerations, however, are based on so-called quantum enveloping algebras as they were proposed by Drinfeld and Jimbo [20–22]. It should be mentioned that both approaches are equivalent in a similar manner as one can deal either with Lie groups or Lie algebras.

The most relevant mathematical structure all of our reasonings are based on is that of a q-deformed symmetry algebra. Let us take as example the symmetry algebra for three-dimensional q-deformed Euclidean space [23, 25–27]. For the sake of completeness and to introduce our conventions we write down all relevant formulas, though this is a well-known example. It is given by the quantum algebra $U_q(su(2))$ [23–27], which can be viewed as q-analog of the algebra of three-dimensional angular momentum. The algebra is generated by the elements L^+, L^-, L^3, and τ^3 subject to the relations

$$L^{\pm}\tau^3 = q^{\pm 4}\tau^3 L^{\pm}, \qquad L^3 \tau^3 = \tau^3 L^3,$$
$$L^- L^+ - L^+ L^- = (\tau^3)^{-1/2} L^3,$$
$$L^{\pm} L^3 - L^3 L^{\pm} = q^{\pm 1} L^{\pm}(\tau^3)^{-1/2}. \tag{2.1}$$

If we recognize that τ^3 tends to 1 for $q \to 1$, we regain the common algebra of three-dimensional angular momentum in the classical limit.

Finally, let us mention that the quantum algebra $U_q(su(2))$ is a Hopf algebra. On its generators the corresponding Hopf structure is given by

$$\Delta(L^{\pm}) = L^{\pm} \otimes (\tau^3)^{-1/2} + 1 \otimes L^{\pm},$$
$$\Delta(L^3) = L^3 \otimes (\tau^3)^{-1/2} + (\tau^3)^{1/2} \otimes L^3,$$
$$+ \lambda(\tau^3)^{1/2}\left(q^{-1}L^- \otimes L^+ + qL^+ \otimes L^-\right), \tag{2.2}$$

$$S(L^{\pm}) = -L^{\pm}(\tau^3)^{1/2},$$
$$S(L^3) = (\tau^3)^{1/2}(q^2 L^+ L^- - q^{-2} L^- L^+), \tag{2.3}$$

$$\epsilon(L^A) = 0, \qquad A \in \{+, 3, -\}. \tag{2.4}$$

On these grounds, the Hopf structure also allows us to generalize the notion of classical commutators. These leads us to so-called q-commutators (see, for example, Ref. [13]).

With the help of q-commutators we can write the algebra in a form which resembles the classical limit in a nice way. The q-commutator is given by the adjoint action of a Hopf algebra onto itself. Recalling the definition of the adjoint action the q-commutators become

$$[L^A, V]_q \equiv L^A \triangleright V = L^A_{(1)} V S(L^A_{(2)}),$$
$$[V, L^A]_q \equiv V \triangleleft L^A = S^{-1}(L^A_{(2)}) V L^A_{(1)}, \qquad (2.5)$$

where we wrote the coproduct in Sweedler notation. From their very definition it follows that q-commutators obey the q-deformed Jacobi identities

$$[L^A, [L^B, V]_q]_q = [[L^A_{(1)}, L^B]_q, [L^A_{(2)}, V]_q]_q,$$
$$[[L^A, L^B]_q, V]_q = [L^A_{(1)}, [L^B, [S(L^A_{(2)}), V]_q]_q]_q. \qquad (2.6)$$

Now, we are able to introduce the notion of a quantum Lie algebra as it was given in Refs. [28–30]. A quantum Lie algebra can be regarded as a subspace of a q-deformed enveloping algebra $U_q(g)$ being invariant under the adjoint action of $U_q(g)$. The point now is that the L^A are the components of a tensor operator and this is the reason why their adjoint action on each other equals a linear combination of the L^A [31]. In this sense, the L^A span a quantum Lie algebra with

$$[L^A, L^B]_q (= L^A \triangleright L^B = L^A \triangleleft L^B) = (C^A)^B_C L^C, \qquad (2.7)$$

where the C^A are the so-called quantum structure constants and summation over repeated indices is understood.

For the quantum Lie algebra of $U_q(su(2))$, for example, we get as q-commutators

$$[L^A, L^A]_q = 0, \qquad A \in \{+, -\},$$
$$[L^3, L^3]_q = -\lambda L^3,$$
$$[L^\pm, L^3]_q = \mp q^{\pm 1} L^\pm,$$
$$[L^\pm, L^\mp]_q = \mp L^3. \qquad (2.8)$$

More compactly, this can be written as

$$[L^A, L^B]_q = q^2 \varepsilon^{AB}{}_C L^C, \qquad (2.9)$$

where $\varepsilon^{AB}{}_C$ denotes a q-deformed analog to the antisymmetric ε-tensor [23]. Notice that the last result is very similar to that in the undeformed case. For details and a treatment concerning q-deformed four-dimensional Euclidean space as well as q-deformed Minkowski space we refer to Ref. [13].

As a second step we want to review the notion of a quantum space using the example of the Manin plane. A quantum space is defined as module algebra of a quantum algebra. For the definition of a module algebra see, for example, Ref. [27].

A physical relevant example for such a module algebra is the quantum plane [32]. Its algebra is generated by the elements x^1 and x^2 subject to the condition

$$x^1 x^2 = q x^2 x^1. \tag{2.10}$$

This relation determines what is called the two-dimensional bosonic quantum plane.

In addition to this, there is an antisymmetric variant of the two-dimensional quantum plane, the so-called fermionic quantum plane. It is generated by the spinor coordinates θ^1 and θ^2 that satisfy the relations

$$(\theta^1)^2 = (\theta^2)^2 = 0, \qquad \theta^1 \theta^2 = -q^{-1} \theta^2 \theta^1, \tag{2.11}$$

showing the correct classical limit for $q \to 1$. Alternatively, these relations can be written with the $\hat{R}$-matrix of $U_q(su(2))$ [15]:

$$\theta^\alpha \theta^\beta = -q^{-1} \hat{R}^{\alpha\beta}{}_{\gamma\delta}\, \theta^\gamma \theta^\delta, \tag{2.12}$$

where

$$\hat{R}^{\alpha\beta}{}_{\gamma\delta} = \begin{pmatrix} q & 0 & 0 & 0 \\ 0 & \lambda & 1 & 0 \\ 0 & 1 & 0 & 0 \\ 0 & 0 & 0 & q \end{pmatrix}. \tag{2.13}$$

For later purposes we introduce conjugate spinor coordinates $\bar{\theta}^{\dot\alpha}$. They again obey the commutation relations in (2.12). There is a braiding between the unconjugate and the conjugate quantum plane given by

$$\bar{\theta}^{\dot\alpha} \theta^\beta = -q^{-1} \hat{R}^{\dot\alpha\beta}{}_{\gamma\dot\delta}\, \theta^\gamma \bar{\theta}^{\dot\delta}. \tag{2.14}$$

We can merge the quantum plane and its symmetry algebra resulting in a crossed product. For this, we have to fix relations between the spinor coordinates and the $U_q(su(2))$-generators. These are

$$L^+ \theta^1 = \theta^1 L^+ - q^{1/2} \lambda_+^{-1/2} \theta^2 (\tau^3)^{-1/2},$$
$$L^+ \theta^2 = \theta^2 L^+, \tag{2.15}$$

$$L^3 \theta^1 = q\theta^1 L^3 - q^{-1/2} \lambda \lambda_+^{-1/2} \theta^2 L^- - q\lambda_+^{-1} \theta^1 (\tau^3)^{-1/2},$$
$$L^3 \theta^2 = q^{-1} \theta^2 L^3 + q^{1/2} \lambda \lambda_+^{-1/2} \theta^1 L^+ + q^{-1} \lambda_+^{-1} \theta^2 (\tau^3)^{-1/2}, \tag{2.16}$$

$$L^- \theta^1 = \theta^1 L^-,$$
$$L^- \theta^2 = \theta^2 L^- + q^{-1/2} \lambda_+^{-1/2} \theta^1 (\tau^3)^{-1/2}. \tag{2.17}$$

We have to take into account the scaling operators Λ_s which have to be added as additional generators to the algebra generated by the L^A's. They have trivial commutation relations with the L^A's, with spinor coordinates they commute according to

$$\Lambda_s \theta^\alpha = -q^{1/2} \theta^\alpha \Lambda_s, \qquad \Lambda_s \bar{\theta}^\alpha = -q^{1/2} \bar{\theta}^\alpha \Lambda_s. \tag{2.18}$$

The spinor coordinates carry a Hopf structure, too. Their coproduct reads as

$$\Delta(\theta^1) = \theta^1 \otimes 1 + \Lambda_s(\tau^3)^{-1/4} \otimes \theta^1,$$

$$\Delta(\theta^2) = \theta^2 \otimes 1 + \Lambda_s(\tau^3)^{1/4} \otimes \theta^2 + q^{1/2}\lambda\lambda_+^{1/2}\Lambda_s(\tau^3)^{1/4}L^+ \otimes \theta^1, \qquad (2.19)$$

$$\Delta(\bar{\theta}^1) = \bar{\theta}^1 \otimes 1 + \Lambda_s^{-1}(\tau^3)^{1/4} \otimes \bar{\theta}^1 + q^{-1/2}\lambda\lambda_+^{1/2}\Lambda_s^{-1}(\tau^3)^{1/4}L^- \otimes \bar{\theta}^2,$$

$$\Delta(\bar{\theta}^2) = \bar{\theta}^2 \otimes 1 + \Lambda_s^{-1}(\tau^3)^{-1/4} \otimes \bar{\theta}^2. \qquad (2.20)$$

With the help of these coproducts we are in a position to generate the braiding relations in (2.14), since we have

$$\theta^\alpha \bar{\theta}^{\dot{\beta}} = (\theta^\alpha_{(1)} \triangleright \bar{\theta}^{\dot{\beta}})\theta^\alpha_{(2)} = \bar{\theta}^{\dot{\beta}}_{(2)}(\theta^\alpha \triangleleft \bar{\theta}^{\dot{\beta}}_{(1)}),$$

$$\bar{\theta}^{\dot{\alpha}}\theta^\beta = (\bar{\theta}^{\dot{\alpha}}_{(1)} \triangleright \theta^\beta)\bar{\theta}^{\dot{\alpha}}_{(2)} = \theta^\beta_{(2)}(\bar{\theta}^{\dot{\alpha}} \triangleleft \theta^\beta_{(1)}). \qquad (2.21)$$

The symbol $\triangleright$ and $\triangleleft$ denote the right- and left adjoint action, respectively. The left- and right-adjoint actions are defined as given in (2.5).

3. q-Deformed Superalgebras

In this section we show that it is even possible to extend the symmetry algebras of q-deformed quantum spaces to q-deformed analogs of superalgebras. If not stated otherwise, in this section we restrict attention to q-deformed Minkowski space and its symmetry algebras, since they are most interesting from a physical point of view. We build the superalgebra in several steps. At first we extend the q-Lorentz algebra adding the momentum generators in a consistent way. Doing so we get the well-known q-Poincaré algebra. In a second step we add supergenerators $Q^\alpha, \bar{Q}^{\dot{\alpha}}$ to get a q-deformed Super-Poincaré algebra. The resulting algebra will be an example of what we call a q-deformed superalgebra. In the classical limit it turns over into the well-known Super-Poincaré algebra.

Now let us first introduce the subalgebra generated by the components of four-momentum. The momentum algebra of q-deformed Minkowski space is spanned by the momentum components P^μ, $\mu \in \{0, +, -, 3\}$ subject to

$$P^\mu P^0 = P^0 P^\mu, \quad \mu \in \{0, +, -, 3\},$$

$$P^3 P^\pm - q^{\pm 2}P^\pm P^3 = -q\lambda P^0 P^\pm,$$

$$P^- P^+ - P^+ P^- = \lambda(P^3 P^3 - P^0 P^3). \qquad (3.1)$$

Let us recall that the P^μ, $\mu \in \{0, +, -, 3\}$, behave as a four-vector operator under q-deformed Lorentz transformations. This property is a consequence of the commutation relations between generators of the q-deformed Lorentz algebra and the corresponding momentum components. They take the form (see also Ref. [13])

$$[V^{\mu\nu}, P^\rho]_q = -q^{-1}(P_A)^{\mu\nu}{}_{\nu'\rho'}\eta^{\rho'\rho}P^{\nu'}, \qquad (3.2)$$

where P_A is a q-analog of an antisymmetrizer for q-Minkowski space. The $V^{\mu\nu}$ are the generators of the q-Lorentz algebra written as tensor generators. A detailed treatment of this algebra can be found in [13].

As a matter of principle the procedure to construct a deformed superalgebra is quite simple. The non-supersymmetric part of the algebra is already known. It is just the quantum symmetry algebra for the space under consideration combined with the corresponding momentum algebra.

What remains is to include spinorial generators in a consistent way. We restrict ourselves to the case of $N = 1$ supersymmetry. In complete analogy to the undeformed case we introduce the new generators

$$Q^1, Q^2, \bar{Q}^1, \bar{Q}^2. \tag{3.3}$$

How the spinorial generators commute with the Lorentz algebra generators $V^{\mu\nu}$ is described in Ref. [13]. The result written down in terms of q-commutators is

$$[V^{\mu\nu}, Q^\alpha]_q = q^{-1}\lambda_+^{-1}(\sigma^{\mu\nu})_\beta{}^\alpha Q^\beta,$$

$$[V^{\mu\nu}, \bar{Q}^{\dot\alpha}]_q = q^{-1}\lambda_+^{-1}(\bar\sigma^{\mu\nu})_{\dot\beta}{}^{\dot\alpha}\bar{Q}^{\dot\beta}. \tag{3.4}$$

For the definition of the Pauli matrices see appendix A. Note that the above relations are equivalent to the commutation relations in (2.15)-(2.17).

Next we have to fix some sets of relations. These are the relations the supercharges fulfil among themselves, the relations between the supercharges and the generators of the symmetry algebra, and finally the relations between momenta and supercharges. This has to be done in a way that the complete algebra generated from $V^{\mu\nu}, P^\mu, Q^\alpha$, and $\bar{Q}^{\dot\alpha}$ is consistent with all relations among them, i.e., every relation remains unchanged after acting with any generator on it.

Since the two types of supercharges Q^α and $\bar{Q}^{\dot\alpha}$ transform as spinors, we require for them to span quantum planes. In this manner, the supercharges have to fulfil

$$Q^\alpha Q^\alpha = \bar{Q}^{\dot\alpha}\bar{Q}^{\dot\alpha} = 0, \qquad \alpha, \dot\alpha = 1, 2, \tag{3.5}$$

$$Q^1 Q^2 = -q^{-1}Q^2 Q^1, \qquad \bar{Q}^1\bar{Q}^2 = -q^{-1}\bar{Q}^2\bar{Q}^1. \tag{3.6}$$

Introducing the q-anticommutator

$$\{\theta^\alpha, \tilde{\theta}^\beta\}_k \equiv \theta^\alpha\tilde{\theta}^\beta + k\hat{R}^{\alpha\beta}{}_{\alpha'\beta'}\tilde{\theta}^{\alpha'}\theta^{\beta'}, \tag{3.7}$$

the relations in (3.6) become

$$\{Q^\alpha, Q^\beta\}_q = \{\bar{Q}^\alpha, \bar{Q}^\beta\}_q = 0. \tag{3.8}$$

What remains is to determine the sets of relations $Q^\alpha\bar{Q}^{\dot\beta}$, $P^\mu Q^\alpha$ and $P^\mu\bar{Q}^{\dot\beta}$ for $\mu \in \{0, +, 3, -\}$ as well as $\alpha, \dot\beta \in \{1, 2\}$. For these relations we first make general Ansätze. Then we multiply each relation from the right with a generator, commute the generator to the left, and determine the unknown coefficients from the requirement that the relation remains unchanged.

Proceeding this way, we get for the $Q\bar{Q}$-relations of the q-deformed Super-Poincaré algebra

$$\bar{Q}^1 Q^1 + Q^1 \bar{Q}^1 = c\,q^{-1} P^-,$$
$$\bar{Q}^1 Q^2 + q^{-1} Q^2 \bar{Q}^1 = -q^{-1}\lambda Q^1 \bar{Q}^2 + c\,q^{-1/2}\lambda_+^{-1/2}(P^3 + q^{-2}P^0),$$
$$\bar{Q}^2 Q^1 + q^{-1} Q^1 \bar{Q}^2 = -c\,q^{-3/2}\lambda_+^{-1/2}(P^0 - P^3),$$
$$\bar{Q}^2 Q^2 + Q^2 \bar{Q}^2 = c\,q^{-1} P^+. \tag{3.9}$$

For the PQ- and $P\bar{Q}$-relations we likewise have

$$P^+ Q^1 = q^{-2} Q^1 P^+, \quad P^+ Q^2 = Q^2 P^+,$$
$$P^0 Q^1 = q^{-1} Q^1 P^0, \quad P^0 Q^2 = q^{-1} Q^2 P^0,$$
$$P^3 Q^1 = q^{-1} Q^1 P^3,$$
$$P^3 Q^2 = q^{-1} Q^2 P^3 + q^{-3/2}\lambda\lambda_+^{1/2} Q^1 P^+,$$
$$P^- Q^1 = Q^1 P^-,$$
$$P^- Q^2 = q^{-2} Q^2 P^- + q^{-3/2}\lambda\lambda_+^{1/2} Q^1 P^3, \tag{3.10}$$

and

$$P^0 \bar{Q}^1 = q\bar{Q}^1 P^0, \quad P^0 \bar{Q}^2 = q\bar{Q}^2 P^0,$$
$$P^- \bar{Q}^1 = \bar{Q}^1 P^-, \quad P^- \bar{Q}^2 = q^2 \bar{Q}^1 P^-,$$
$$P^3 \bar{Q}^1 = q\bar{Q}^1 P^3 - q^{3/2}\lambda\lambda_+^{1/2} \bar{Q}^2 P^-,$$
$$P^3 \bar{Q}^2 = q\bar{Q}^2 P^3,$$
$$P^+ \bar{Q}^1 = q^2 \bar{Q}^1 P^+ - q^{3/2}\lambda\lambda_+^{1/2} \bar{Q}^2 P^3,$$
$$P^+ \bar{Q}^2 = \bar{Q}^2 P^+. \tag{3.11}$$

Notice that the relations in (3.9) are determined up to a constant c, only. One can check that all of the above relations in (3.6), (3.9), (3.10), and (3.11) are compatible with the relations concerning the generators $V^{\mu\nu}$ of the q-Lorentz algebra.

At last, we list the complete q-deformed supersymmetric Poincaré algebra. To this end, we apply our q-commutators introduced earlier since they enable us in a very nice way to show the close relationship with the classical limit [14]:

$$[V^{\mu\nu}, V^{\rho\sigma}]_q = -q^{-1}\lambda_+ (P_A)^{\mu\nu}{}_{\nu'\rho''} (P_A)^{\rho\sigma}{}_{\rho'\sigma'} \eta^{\rho''\rho'} V^{\nu'\sigma'},$$

$$[V^{\mu\nu}, P^{\rho}]_q = -q^{-1}(P_A)^{\mu\nu}{}_{\nu'\rho'} \eta^{\rho'\rho} P^{\nu'},$$

$$[V^{\mu\nu}, Q^{\alpha}]_q = q^{-1}\lambda_+^{-1} (\sigma^{\mu\nu})_{\beta}{}^{\alpha} Q^{\beta},$$

$$[V^{\mu\nu}, \bar{Q}^{\dot\alpha}]_q = q^{-1}\lambda_+^{-1} (\bar{\sigma}^{\mu\nu})_{\dot\beta}{}^{\dot\alpha} \bar{Q}^{\dot\beta},$$

$$(P_A)^{\mu\nu}{}_{\mu'\nu'} P^{\mu'} P^{\nu'} = 0,$$

$$[P^{\mu}, Q^{\alpha}]_q = 0, \qquad [P^{\mu}, \bar{Q}^{\dot\alpha}]_{\bar q} = 0,$$

$$\{Q^{\alpha}, Q^{\beta}\}_q = 0, \qquad \{\bar{Q}^{\dot\alpha}, \bar{Q}^{\dot\beta}\}_q = 0,$$

$$\{\bar{Q}^{\dot\alpha}, Q^{\beta}\}_{q^{-1}} = c\,(\bar{\sigma}_{\mu}^{-1})^{\dot\alpha\beta} P^{\mu}. \tag{3.12}$$

Note that $(P_A)^{\mu\nu}{}_{\mu'\nu'}$ denotes a q-analog of an antisymmetrizer of q-deformed Minkowski space. We used q-deformed Pauli matrices too. They are defined in the appendix A.

4. q-Deformed Superspaces and Operator Representations

In this concluding section we show that it is possible to build q-deformed superspaces and operator representations along the same line of reasonings as in the undeformed case. As an example we treat the three-dimensional q-deformed Euclidean superspace. Its bosonic sector is spanned by three coordinates named X^+, X^3, and X^-. These coordinates are subject to the relations

$$X^3 X^{\pm} = q^{\pm 2} X^{\pm} X^3,$$

$$X^- X^+ - X^+ X^- = \lambda X^3 X^3, \tag{4.1}$$

and have the coproducts

$$\begin{aligned}
\Delta(X^-) &= X^- \otimes 1 + \Lambda^{-1/2}(\tau^3)^{-\frac{1}{2}} \otimes X^-, \\
\Delta(X^3) &= X^3 \otimes 1 + \Lambda^{-1/2} \otimes X^3 + \lambda\lambda_+ \Lambda^{-1/2} L^+ \otimes X^-, \\
\Delta(X^+) &= X^+ \otimes 1 + \Lambda^{-1/2}(\tau^3)^{\frac{1}{2}} \otimes X^+ + q\lambda\lambda_+ \Lambda^{-1/2}(\tau^3)^{\frac{1}{2}} L^+ \otimes X^3 \\
&+ q^2\lambda^2\lambda_+ \Lambda^{-1/2}(\tau^3)^{\frac{1}{2}} (L^+)^2 \otimes X^-.
\end{aligned} \tag{4.2}$$

Notice that the above expressions for the coproduct include the scaling operator Λ.

To build a superspace, we have to merge the bosonic space with the two quantum planes spanned by the Weyl spinors θ^{α} as well as $\bar{\theta}^{\dot\alpha}$. This has to be done in a covariant way, i.e., the action of angular momentum generators on the relations should be zero. Using reasonable Ansätze, we end up with

$$X^+\bar{\theta}^1 = q\bar{\theta}^1 X^+ - q^{1/2}\lambda\lambda_+^{1/2}\bar{\theta}^2 X^3,$$
$$X^+\bar{\theta}^2 = q^{-1}\bar{\theta}^2 X^+,$$
$$X^3\bar{\theta}^1 = \bar{\theta}^1 X^3 - q^{1/2}\lambda\lambda_+^{1/2}\bar{\theta}^2 X^-,$$
$$X^3\bar{\theta}^2 = \bar{\theta}^2 X^3,$$
$$X^-\bar{\theta}^1 = q^{-1}\bar{\theta}^1 X^-, \quad X^-\bar{\theta}^2 = q\bar{\theta}^2 X^-. \tag{4.3}$$

These relations can also be generated from the coproducts in (4.2) as well as from the coproducts in (2.20) via

$$X^A\bar{\theta}^{\dot\alpha} = (X^A_{(1)} \rhd \bar{\theta}^{\dot\alpha})X^A_{(2)} = \bar{\theta}^{\dot\alpha}_{(2)}(X^A \lhd \bar{\theta}^{\dot\alpha}_{(1)}). \tag{4.4}$$

Applying these identities requires to take account of the fact that Λ acts trivially on spinor coordinates:

$$\Lambda \rhd \bar{\theta}^{\dot\alpha} = \bar{\theta}^{\dot\alpha}, \tag{4.5}$$

as well as Λ_s on bosonic coordinates

$$\Lambda_s \rhd X^A = X^A. \tag{4.6}$$

In Ref. [35] a second construction for the X^A was presented, which gives rise to a second set of new, consistent commutation relations between the X^A and the unconjugate spinor coordinates θ^α. Alternatively, this second set follows from conjugating the relations in (4.3) if we take into account the conjugation assignment [15]

$$\overline{\theta^\alpha} = -\varepsilon_{\alpha\beta}\bar{\theta}^{\dot\beta}, \quad \overline{\bar{\theta}^{\dot\alpha}} = \varepsilon_{\alpha\beta}\theta^\beta,$$
$$\overline{X^A} = g_{AB}X^B. \tag{4.7}$$

Finally, it should be noted that the coproducts in (2.20) and (4.2) respect the commutation relations in (4.3), i.e. they are consistent with the homomorphism property

$$\Delta(X^A\bar{\theta}^{\dot\alpha}) = \Delta(X^A)\Delta(\bar{\theta}^{\dot\alpha}). \tag{4.8}$$

The coproducts in (4.2) can be extended to q-analogs of supercoproducts. To achieve this we have to add fermionic parts $\Delta_f(X^A)$ to the bosonic coproducts $\Delta(X^A)$. These fermionic parts are completely determined by making some reasonable claims. First, the new supercoproducts $\Delta_s(X^A) = \Delta(X^A) + \Delta_f(X^A)$ should behave under symmetry transformations in the same way as the vector components X^A. In formulas this means $\Delta_s(L^A \rhd X^B) = \Delta(L^A) \rhd \Delta_s(X^B)$. Further, the supercoproducts should respect the XX-relations in (4.1). Finally, we demand that the relations in (4.3) are consistent with supercoproducts and the coproducts in (2.20).

To fulfil these requirements we are enforced to introduce a new scaling operator Λ_k, which helps to adapt the fermionic parts of our supercoproducts to their bosonic parts. Then the fermionic parts corresponding to the coproducts in (4.2) are given by

$$\Delta_f(X^+) = \bar{\theta}^2 \Lambda_s^{-1} \Lambda_k^{-1} (\tau^3)^{-1/4} \otimes \theta^2 \Lambda_k,$$

$$\Delta_f(X^3) = q^{1/2} \lambda_+^{-1/2} \bar{\theta}^1 \Lambda_s^{-1} \Lambda_k^{-1} (\tau^3)^{-1/4} \otimes \theta^2 \Lambda_k$$
$$+ q^{-1/2} \lambda_+^{-1/2} \bar{\theta}^2 \Lambda_s^{-1} \Lambda_k^{-1} (\tau^3)^{1/4} \otimes \theta^1 \Lambda_k$$
$$+ q^{-1} \lambda \bar{\theta}^2 \Lambda_s^{-1} \Lambda_k^{-1} (\tau^3)^{1/4} L^- \otimes \theta^2 \Lambda_k,$$

$$\Delta_f(X^-) = \bar{\theta}^1 \Lambda_s^{-1} \Lambda_k^{-1} (\tau^3)^{1/4} \otimes \theta^1 \Lambda_k$$
$$+ q^{-1/2} \lambda_+^{1/2} \bar{\theta}^1 \Lambda_s^{-1} \Lambda_k^{-1} (\tau^3)^{1/4} L^- \otimes \theta^2 \Lambda_k. \qquad (4.9)$$

The complete supercoproducts $\Delta_s(X^A) = \Delta(X^A) + \Delta_f(X^A)$ indeed fulfil

$$\Delta_s(X^\mu X^\nu) = \Delta_s(X^\mu)\Delta_s(X^\nu), \quad \Delta_s(X^\mu \bar{\theta}^\alpha) = \Delta_s(X^\mu)\Delta_s(\bar{\theta}^\alpha), \qquad (4.10)$$

if the following holds:

$$\Lambda_k X^\mu = q^{-2} X^\mu \Lambda_k,$$
$$\Lambda_k \theta^\alpha = q^{-1} \theta^\alpha \Lambda_k, \qquad \Lambda_k \bar{\theta}^\alpha = q^{-1} \bar{\theta}^\alpha \Lambda_k. \qquad (4.11)$$

Now we want to describe operator representations for three-dimensional q-deformed superspace. They should represent the q-deformed superalgebra for three-dimensional Euclidean space. We regain it from (3.1) and (3.8)-(3.11) if we set P^0 equal to zero. This reflects the fact, that the q-deformed three-dimensional algebra is a real subalgebra for the Super-Poincaré algebra.

To construct a superspace with operator representations, we are forced to represent the super charges Q^α, $\bar{Q}^{\dot\alpha}$ as well as to give consistent, covariant relations between the bosonic derivatives ∂_X^A and the fermionic objects f^α, with f^α standing for $\theta^\alpha, \bar{\theta}^\alpha, \partial_\theta^\alpha$ or $\hat{\partial}_\theta^{\dot\alpha}$, respectively.

However, in the quantum space case there occur some differences to the classical limit concerning operator representations. First, it is not possible to construct one unique superspace. Instead, one finds two distinct operator representations. In each representation either the unconjugate or the conjugate supercharges are represented in a nontrivial way, but not both. Second, if we try to incorporate super covariant derivatives in a consistent way, we have to extend the algebra by new scaling operators Λ_n and Λ_m. At last, each operator representation leads to its own set of covariant commutation relations between the generators ∂_X^A, θ^α, $\bar{\theta}^\alpha$, ∂_θ^α, $\hat{\partial}_\theta^{\dot\alpha}$. Thus, we must carefully specify the set of relations we have to use for a certain operation. The differential calculus we use is

$$\partial_\theta^\alpha \theta^\beta = -(\hat{R}^{-1})^{\alpha\beta}{}_{\gamma\delta} \theta^\gamma \partial_\theta^\delta, \quad \partial_\theta^\alpha \bar{\theta}^{\dot\beta} = \varepsilon^{\alpha\dot\beta} - (\hat{R}^{-1})^{\alpha\dot\beta}{}_{\dot\gamma\delta} \bar{\theta}^{\dot\gamma} \partial_\theta^\delta,$$

$$\hat{\partial}_\theta^{\dot\alpha} \theta^\beta = \varepsilon^{\dot\alpha\beta} - \hat{R}^{\dot\alpha\beta}{}_{\gamma\dot\delta} \theta^\gamma \hat{\partial}_\theta^{\dot\delta}, \quad \hat{\partial}_\theta^{\dot\alpha} \bar{\theta}^{\dot\beta} = -\hat{R}^{\dot\alpha\dot\beta}{}_{\dot\gamma\dot\delta} \bar{\theta}^{\dot\gamma} \hat{\partial}_\theta^{\dot\delta}. \qquad (4.12)$$

Now we turn to the explicit form of the operator representation with conjugate supercharges being represented trivially:

$$Q^1 = q^{-1}\partial_\theta^1 \Lambda_m^{-1} + q^{3/2}\theta^1 \partial_X^3 \Lambda_m - q^3 \lambda_+ \theta^2 \partial_X^- \Lambda_m,$$
$$Q^2 = q^{-1}\partial_\theta^2 \Lambda_m^{-1} - q^{7/2}\theta^2 \partial_X^3 \Lambda_m + q^2 \lambda_+ \theta^1 \partial_X^+ \Lambda_m, \tag{4.13}$$

$$\bar{Q}^1 = q^{-1}\hat{\partial}_\theta^1 \Lambda_n, \quad \bar{Q}^2 = q^{-1}\hat{\partial}_\theta^2 \Lambda_n. \tag{4.14}$$

Using the q-deformed Pauli matrices (for their explicit form see App. A) we are able to rewrite the above result as

$$Q^\alpha = q^{-1}\partial_\theta^\alpha \Lambda_m^{-1} - q^2 (\sigma_A)^\alpha{}_\beta \theta^\beta \partial_X^A \Lambda_m,$$
$$\bar{Q}^{\dot\alpha} = q^{-1}\hat{\partial}_\theta^{\dot\alpha} \Lambda_n. \tag{4.15}$$

The relations of the q-deformed superalgebra given in Eq. (3.12) are satisfied by the expressions in (4.15) together with $P^A = \partial_X^A \Lambda_n \Lambda_m$ if we assume as commutation relations

$$\partial_X^+ \theta^1 = \theta^1 \partial_X^+ - q^{-1/2}\lambda\lambda_+^{1/2}\theta^2 \partial_X^3,$$
$$\partial_X^+ \theta^2 = q^{-2}\theta^2 \partial_X^+,$$
$$\partial_X^3 \theta^1 = q^{-1}\theta^1 \partial_X^3 - q^{-1/2}\lambda\lambda_+^{1/2}\theta^2 \partial_X^-,$$
$$\partial_X^3 \theta^2 = q^{-1}\theta^2 \partial_X^3,$$
$$\partial_X^- \theta^1 = q^{-2}\theta^1 \partial_X^-, \quad \partial_X^- \theta^2 = \theta^2 \partial_X^-, \tag{4.16}$$

and

$$\partial_X^+ \bar{\theta}^1 = q^2 \bar{\theta}^1 \partial_X^+ - q^{3/2}\lambda\lambda_+^{1/2}\bar{\theta}^2 \partial_X^3,$$
$$\partial_X^+ \bar{\theta}^2 = \bar{\theta}^2 \partial_X^+,$$
$$\partial_X^3 \bar{\theta}^1 = q\bar{\theta}^1 \partial_X^3 - q^{3/2}\lambda\lambda_+^{1/2}\bar{\theta}^2 \partial_X^-,$$
$$\partial_X^3 \bar{\theta}^2 = q\bar{\theta}^2 \partial_X^3,$$
$$\partial_X^- \bar{\theta}^1 = \bar{\theta}^1 \partial_X^-, \quad \partial_X^- \bar{\theta}^2 = q^2 \bar{\theta}^2 \partial_X^-, \tag{4.17}$$

as well as

$$\partial_X^+ \partial_\theta^1 = q^{-2}\partial_\theta^1 \partial_X^+, \quad \partial_X^+ \partial_\theta^2 = \partial_\theta^2 \partial_X^+,$$
$$\partial_X^3 \partial_\theta^1 = q^{-1}\partial_\theta^1 \partial_X^3,$$
$$\partial_X^3 \partial_\theta^2 = q^{-1}\partial_\theta^2 \partial_X^3 + q^{-3/2}\lambda\lambda_+ \partial_\theta^1 \partial_X^+,$$
$$\partial_X^- \partial_\theta^1 = \partial_\theta^1 \partial_X^-,$$
$$\partial_X^- \partial_\theta^2 = q^{-2}\partial_\theta^2 \partial_X^- + q^{-3/2}\lambda\lambda_+ \partial_\theta^1 \partial_X^-. \tag{4.18}$$

and

$$\partial_X^+ \hat{\partial}_\theta^1 = \hat{\partial}_\theta^1 \partial_X^+ - q^{-1/2} \lambda \lambda_+ \hat{\partial}_\theta^2 \partial_X^3,$$

$$\partial_X^+ \hat{\partial}_\theta^2 = q^{-2} \hat{\partial}_\theta^2 \partial_X^+,$$

$$\partial_X^3 \hat{\partial}_\theta^1 = q \hat{\partial}_\theta^1 \partial_X^3 - q^{-1/2} \lambda \lambda_+ \hat{\partial}_\theta^2 \partial_X^-,$$

$$\partial_X^3 \hat{\partial}_\theta^2 = q^{-2} \hat{\partial}_\theta^2 \partial_X^3,$$

$$\partial_X^- \hat{\partial}_\theta^1 = q^{-2} \hat{\partial}_\theta^1 \partial_X^-, \quad \partial_X^- \hat{\partial}_\theta^2 = \hat{\partial}_\theta^2 \partial_X^-. \tag{4.19}$$

As already mentioned in the introduction, there are q-analogs of super-covariant derivatives. The super-covariant derivatives, corresponding to the representations in (4.15), take the form

$$\bar{D}^{\dot{\alpha}} = -i \hat{\partial}_\theta^{\dot{\alpha}} \Lambda_m^{-1} - i q^{-1} (\sigma_A)^\alpha{}_{\dot{\beta}} \bar{\theta}^{\dot{\beta}} \partial_X^A \Lambda_m,$$

$$D^\alpha = i q^2 \partial_\theta^\alpha \Lambda_n^{-1}. \tag{4.20}$$

They again fulfil the relations of the superalgebra in (3.12) if we replace the supercharges Q^α and $\bar{Q}^{\dot{\alpha}}$ with D^α and $\bar{D}^{\dot{\alpha}}$, respectively. However, there are some subtleties we have to take into account. First, we now have to use commutation relations that follow from those in (4.16)-(4.19) by applying the conjugation properties

$$\overline{\theta^\alpha} = -\varepsilon_{\alpha\dot{\beta}} \bar{\theta}^{\dot{\beta}}, \quad \overline{\bar{\theta}^{\dot{\alpha}}} = \varepsilon_{\alpha\beta} \theta^\beta,$$

$$\overline{\partial_\theta^\alpha} = -\varepsilon_{\alpha\beta} \bar{\partial}_\theta^{\dot{\beta}}, \quad \overline{\bar{\partial}_\theta^{\dot{\alpha}}} = \varepsilon_{\alpha\beta} \partial_\theta^\beta,$$

$$\overline{\partial_X^A} = g_{AB} \partial_X^B. \tag{4.21}$$

Notice, that $\hat{\partial}^{\dot{\alpha}} = -q \bar{\partial}^{\dot{\alpha}}$. Further, we have to replace the momenta P^A in (3.9) with $\partial_X^A \Lambda_n^{-1} \Lambda_m$.

In addition to this, let us say a few words about the scaling operators Λ_m and Λ_n. They obey the conditions

$$\Lambda_m \theta^\alpha = q \, \theta^\alpha \Lambda_m, \quad \Lambda_m \bar{\theta}^{\dot{\alpha}} = q^{-1} \bar{\theta}^{\dot{\alpha}} \Lambda_m, \quad \Lambda_m X^A = X^A \Lambda_m, \tag{4.22}$$

and

$$\Lambda_n \theta^\alpha = q \, \theta^\alpha \Lambda_n, \quad \Lambda_n \bar{\theta}^{\dot{\alpha}} = q \, \bar{\theta}^{\dot{\alpha}} \Lambda_n, \quad \Lambda_n X^A = q^2 \, X^A \Lambda_n. \tag{4.23}$$

Comparing the last line with (4.11), we see that we can make the identification $\Lambda_n = \Lambda_k^{-1}$, which enables us to reduce the number of necessary scaling operators to two, namely Λ_k and Λ_m. The corresponding derivatives are scaled with inverse factors compared to coordinates.

Super-covariant derivatives commute with supercharges as follows:

$$D^\alpha Q^\beta = -q^{-1} (\hat{R}^{-1})^{\alpha\beta}{}_{\gamma\delta} Q^\gamma D^\delta, \quad D^\alpha \bar{Q}^{\dot{\beta}} = -q^3 (\hat{R}^{-1})^{\alpha\dot{\beta}}{}_{\dot{\gamma}\delta} \bar{Q}^{\dot{\gamma}} D^\delta,$$

$$\bar{D}^{\dot{\alpha}} Q^\beta = -q \hat{R}^{\dot{\alpha}\beta}{}_{\gamma\dot{\delta}} Q^\gamma \bar{D}^{\dot{\delta}}, \quad \bar{D}^{\dot{\alpha}} \bar{Q}^{\dot{\beta}} = -q^{-1} (\hat{R}^{-1})^{\dot{\alpha}\dot{\beta}}{}_{\dot{\gamma}\dot{\delta}} \bar{Q}^{\dot{\gamma}} \bar{D}^{\dot{\delta}}. \tag{4.24}$$

Moreover, we have to specify the commutation relations between supercharges and fermionic coordinates as well as the commutation relations between covariant derivatives and fermionic coordinates. The former are

$$Q^\alpha \theta^\beta = -q^{-1}(\hat{R}^{-1})^{\alpha\beta}{}_{\gamma\delta}\theta^\gamma Q^\delta,$$

$$Q^\alpha \bar{\theta}^{\dot{\beta}} = \varepsilon^{\alpha\dot{\beta}}\Lambda_m^{-1} - q(\hat{R}^{-1})^{\alpha\dot{\beta}}{}_{\dot{\gamma}\delta}\bar{\theta}^{\dot{\gamma}}Q^\delta,$$

$$\bar{Q}^{\dot{\alpha}} \theta^\beta = \varepsilon^{\dot{\alpha}\beta}\Lambda_n - q\hat{R}^{\dot{\alpha}\beta}{}_{\gamma\dot{\delta}}\theta^\gamma \bar{Q}^{\dot{\delta}},$$

$$\bar{Q}^{\dot{\alpha}} \bar{\theta}^{\dot{\beta}} = -q\hat{R}^{\dot{\alpha}\dot{\beta}}{}_{\dot{\gamma}\dot{\delta}}\bar{\theta}^{\dot{\gamma}}\bar{Q}^{\dot{\delta}}, \tag{4.25}$$

and, likewise, for the latter

$$D^\alpha \theta^\beta = -q^{-1}(\hat{R}^{-1})^{\alpha\beta}{}_{\gamma\delta}\theta^\gamma D^\delta,$$

$$D^\alpha \bar{\theta}^{\dot{\beta}} = iq^{-1}\varepsilon^{\alpha\dot{\beta}}\Lambda_n^{-1} - q^{-1}(\hat{R}^{-1})^{\alpha\dot{\beta}}{}_{\dot{\gamma}\delta}\bar{\theta}^{\dot{\gamma}}D^\delta,$$

$$\bar{D}^{\dot{\alpha}} \theta^\beta = iq^{-1}\varepsilon^{\dot{\alpha}\beta}\Lambda_m^{-1} - q^{-1}\hat{R}^{\dot{\alpha}\beta}{}_{\gamma\dot{\delta}}\theta^\gamma \bar{D}^{\dot{\delta}},$$

$$\bar{D}^{\dot{\alpha}} \bar{\theta}^{\dot{\beta}} = -q\hat{R}^{\dot{\alpha}\dot{\beta}}{}_{\dot{\gamma}\dot{\delta}}\bar{\theta}^{\dot{\gamma}}\bar{D}^{\dot{\delta}}. \tag{4.26}$$

The relations in (4.16)-(4.19) guarantee that the operator representations of supercharges fulfil (4.25). However, the operator representations for super-covariant derivatives are only consistent with (4.26) if we use relations that result from (4.16)-(4.19) via the conjugation (4.7). In both cases, the actions of the scaling operators are given by (4.22) and (4.23). Finally, it should be mentioned that we can write the relations in (4.25) and (4.26) with the help of q-anticommutators [cf. Eq. (3.7)] as

$$\{Q^\alpha, D^\beta\}_q = \{\bar{Q}^{\dot{\alpha}}, D^\beta\}_{q^{-3}} = \{\bar{D}^{\dot{\alpha}}, Q^\beta\}_q = \{\bar{Q}^{\dot{\beta}}, \bar{D}^{\dot{\alpha}}\}_q = 0. \tag{4.27}$$

$$\{\theta^\alpha, Q^\beta\}_q = \{\bar{Q}^{\dot{\alpha}}, \bar{\theta}^{\dot{\beta}}\}_q = 0,$$

$$\{\bar{\theta}^{\dot{\beta}}, Q^\alpha\}_{q^{-1}} = \varepsilon^{\alpha\dot{\beta}}\Lambda_m^{-1}, \quad \{\bar{Q}^{\dot{\alpha}}, \theta^\beta\}_q = \varepsilon^{\dot{\alpha}\beta}\Lambda_n. \tag{4.28}$$

$$\{\theta^\alpha, D^\beta\}_q = \{\bar{D}^{\dot{\alpha}}, \bar{\theta}^{\dot{\beta}}\}_q = 0,$$

$$\{\bar{\theta}^{\dot{\alpha}}, D^\beta\}_q = iq^{-1}\varepsilon^{\dot{\alpha}\beta}\Lambda_m^{-1}, \quad \{\bar{D}^{\dot{\beta}}, \theta^\alpha\}_q = iq^{-1}\varepsilon^{\alpha\dot{\beta}}\Lambda_n^{-1}. \tag{4.29}$$

Appendix A. q-Analogs of Pauli matrices and spin matrices

In this appendix we collect some essential ideas from Ref. [36] concerning q-deformed spinor calculus. We used q-deformed Pauli matrices in Sects. 3 and 4 and showed that they enable us to write identities from q-deformed supersymmetry in a way that reveals a remarkable analogy with the classical case.

Let us recall that the Pauli matrices tell us how to combine two spinors x^α, $\bar{x}^{\dot\beta}$ to form a four-vector X^μ:

$$X^\mu = \sum_{\alpha=1,\,\dot\beta=1}^{2} x^\alpha (\sigma^\mu)_{\alpha\dot\beta} \bar{x}^{\dot\beta}, \qquad X^\mu = \sum_{\dot\alpha=1,\,\beta=1}^{2} \bar{x}^{\dot\alpha} (\bar{\sigma}^\mu)_{\dot\alpha\beta} x^\beta. \qquad (A.1)$$

In order to express tensor products of two spinor components by vector components we use so-called *inverse* Pauli matrices:

$$X^{\alpha\dot\beta} \equiv x^\alpha \bar{x}^{\dot\beta} = \sum_{\mu=1}^{4} X^\mu (\sigma_\mu^{-1})^{\alpha\dot\beta},$$

$$X^{\dot\alpha\beta} \equiv \bar{x}^{\dot\alpha} x^\beta = \sum_{\mu=1}^{4} X^\mu (\bar{\sigma}_\mu^{-1})^{\dot\alpha\beta}. \qquad (A.2)$$

Explicitly, the q-deformed Pauli matrices are given by

a) (three-dimensional Euclidean space)

$$(\sigma^+)_{\alpha\dot\beta} = q^{1/2}\lambda_+^{1/2} \begin{pmatrix} 0 & 0 \\ 0 & 1 \end{pmatrix}, \qquad (\sigma^-)_{\alpha\dot\beta} = q^{1/2}\lambda_+^{1/2} \begin{pmatrix} 1 & 0 \\ 0 & 0 \end{pmatrix},$$

$$(\sigma^3)_{\alpha\dot\beta} = \begin{pmatrix} 0 & q \\ 1 & 0 \end{pmatrix}, \qquad (A.3)$$

b) (four-dimensional Euclidean space)

$$(\sigma^1)_{\alpha\dot\beta} = \begin{pmatrix} 1 & 0 \\ 0 & 0 \end{pmatrix}, \qquad (\sigma^2)_{\alpha\dot\beta} = \begin{pmatrix} 0 & 0 \\ 1 & 0 \end{pmatrix},$$

$$(\sigma^3)_{\alpha\dot\beta} = \begin{pmatrix} 0 & 1 \\ 0 & 0 \end{pmatrix}, \qquad (\sigma^4)_{\alpha\dot\beta} = \begin{pmatrix} 0 & 0 \\ 0 & 1 \end{pmatrix}, \qquad (A.4)$$

c) (Minkowski space)

$$(\sigma^+)_{\alpha\dot\beta} = \begin{pmatrix} 0 & 0 \\ 0 & q \end{pmatrix}, \qquad (\sigma^3)_{\alpha\dot\beta} = q\lambda_+^{-1/2} \begin{pmatrix} 0 & q^{1/2} \\ q^{-1/2} & 0 \end{pmatrix},$$

$$(\sigma^-)_{\alpha\dot\beta} = \begin{pmatrix} q & 0 \\ 0 & 0 \end{pmatrix}, \qquad (\sigma^0)_{\alpha\dot\beta} = \lambda_+^{-1/2} \begin{pmatrix} 0 & -q^{-1/2} \\ q^{1/2} & 0 \end{pmatrix}. \qquad (A.5)$$

In Sec. 4 we use the index convention

$$(\sigma_A)^\alpha{}_\beta = g_{A'A} (\sigma^{A'})_{\alpha'\beta} \varepsilon^{\alpha'\alpha}. \qquad (A.6)$$

In the case of three-dimensional q-deformed Euclidean space and q-deformed Minkowski space we have

$$(\bar{\sigma}^\mu)_{\dot\gamma\delta} = q^{-1}(\hat{R}^{-1})^{\alpha\dot\beta}{}_{\dot\gamma\delta}(\sigma^\mu)_{\alpha\dot\beta}, \qquad (A.7)$$

where $(\hat{R}^{-1})^{\alpha\dot\beta}{}_{\dot\gamma\delta}$ denotes the inverse of the $\hat{R}$-matrix for $U_q(su(2))$ [cf. Eq. (2.13)]. For four-dimensional q-deformed Euclidean space, however, σ^μ does not differ from

$\bar{\sigma}^{\mu}$. Finally, the entries of the *inverse* Pauli matrices are determined by the orthogonality relations

$$(\sigma^{\mu})_{\alpha\dot{\beta}}(\sigma_{\nu}^{-1})^{\alpha\dot{\beta}} = \delta^{\mu}{}_{\nu}, \qquad (\bar{\sigma}^{\mu})_{\dot{\alpha}\beta}(\bar{\sigma}_{\nu}^{-1})^{\dot{\alpha}\beta} = \delta^{\mu}{}_{\nu}. \tag{A.8}$$

In analogy to the undeformed case the q-deformed two-dimensional spin matrices are defined by

$$(\sigma^{\mu\nu})_{\alpha}{}^{\beta} \equiv (P_A)^{\mu\nu}{}_{\kappa\lambda}(\sigma^{\kappa})_{\alpha\dot{\alpha}}\,\varepsilon^{\dot{\alpha}\dot{\alpha}'}(\bar{\sigma}^{\lambda})_{\dot{\alpha}'\beta'}\,\varepsilon^{\beta'\beta},$$

$$(\bar{\sigma}^{\mu\nu})_{\dot{\alpha}}{}^{\dot{\beta}} \equiv (P_A)^{\mu\nu}{}_{\kappa\lambda}(\bar{\sigma}^{\kappa})_{\dot{\alpha}\alpha}\,\varepsilon^{\alpha\alpha'}(\sigma^{\lambda})_{\alpha'\dot{\beta}'}\,\varepsilon^{\dot{\beta}'\dot{\beta}}, \tag{A.9}$$

$$(\sigma_{\mu\nu}^{-1})_{\alpha}{}^{\beta} \equiv (P_A)^{\kappa\lambda}{}_{\mu\nu}\,\varepsilon_{\alpha\alpha'}(\sigma_{\kappa}^{-1})^{\alpha'\dot{\beta}'}\varepsilon_{\dot{\beta}'\dot{\beta}}(\bar{\sigma}_{\lambda}^{-1})^{\dot{\beta}\beta},$$

$$(\bar{\sigma}_{\mu\nu}^{-1})_{\dot{\alpha}}{}^{\dot{\beta}} \equiv (P_A)^{\kappa\lambda}{}_{\mu\nu}\,\varepsilon_{\dot{\alpha}\dot{\alpha}'}(\bar{\sigma}_{\kappa}^{-1})^{\dot{\alpha}'\beta'}\varepsilon_{\beta'\beta}(\sigma_{\lambda}^{-1})^{\beta\dot{\beta}}. \tag{A.10}$$

Note that P_A stands for q-analogs of antisymmetrizers.

Acknowledgement

I am especially grateful to Prof. Eberhard Zeidler. All the results described in the present article could only be achieved because he gave me the ability to apply myself, boundlessly, to research activities for the last years. Nowadays, this is absolutely not self-evident and was only possible because of his special interest in my work, his confidence and his financial support.

Most of the work presented here was done with my colleague Hartmut Wachter. I am very grateful to him for his steady support and years of interesting collaboration. Finally I thank Prof. Dieter Lüst for kind hospitality.

References

[1] W. Heisenberg, 1930 *Z. Phys.* **64** 4 & 1938 *Ann. Phys.* **32** 20.

[2] H.S. Snyder, *Quantized space-time*, Phys. Rev. **71** (1947) 38-41.

[3] A. Einstein *Out of My Later Years*

[4] H. Grosse, C. Klimčik, and P. Prešnajder, Towards finite quantum field theory in non-commutative geometry, 1996 *Int. J. Theor. Phys.* **35** 231 [hep-th/9505175]

[5] S. Majid, On the q-regularisation, 1990 *Int. J. Mod. Phys. A* **5** 4689

[6] R. Oeckl, Braided Quantum Field Theory, 2001 *Commun. Math. Phys.* **217** 451

[7] M. Chaichian, A.P. Demichev, *Phys. Lett. B* **320** 273 (1994) [hep-th/9310001]

[8] C. Blohmann, Free q-deformed relativistic wave equations by representation theory, 2003 *Eur. Phys. J. C* **30** 435 [hep-th/0111172]

[9] M. Fichtmüller, A. Lorek, and J. Wess, q-deformed Phase Space and its Lattice Structure, 1996 *Z. Phys. C* **71** 533 [hep-th/9511106]; J. Wess, q-deformed phase space and its lattice structure, 1997 *Int. J. Mod. Phys. A* **12** 4997

[10] B.L. Cerchiai and J. Wess, q-Deformed Minkowski Space based on a q-Lorentz Algebra, 1998 *Eur. Phys. J. C* **5** 553 [math.QA/9801104]

[11] D. Mikulovic, A. Schmidt and H. Wachter, *Grassmann variables on quantum spaces*, Eur. Phys. J. C **45** (2006) 529, [hep-th/0407273].

[12] A. Schmidt and H. Wachter, *Superanalysis on quantum spaces*, JHEP **0601** (2006) 84, [hep-th/0411180].

[13] A. Schmidt and H. Wachter, *q-Deformed quantum Lie algebras*, J. Geom. Phys. **56** (2006), [mat-ph/0500932].

[14] A. Schmidt, H. Wachter *q-Deformed Superalgebras*, preprint [hep-th/0705.1683].

[15] J. Wess, *q-deformed Heisenberg Algebras*, in H. Gausterer, H. Grosse and L.Pittner, Eds., Proceedings of the 38. Internationale Universitätswochen für Kern- und Teilchen physik, no. 543 in Lect. Notes in Phys., Springer-Verlag, Schladming (2000), math-phy/9910013.

[16] H. Wachter *Towards a q-Deformed Quantum Field Theory*, Printed in this volume.

[17] H. Wachter, *Analysis on q-deformed quantum spaces*, Int. J. Mod. Phys. A **22** (2007) 95.

[18] S. L. Woronowicz, *Compact matrix pseudogroups* , Commun. Math. Phys. **111** (1987) 613.

[19] N. Yu. Reshetikhin, L. A. Takhtadzhyan and L. D. Faddeev, *Quantization of Lie Groups and Lie Algebras*, Leningrad Math. J. **1** (1990) 193.

[20] V. G. Drinfeld, *Hopf algebras and the quantum Yang-Baxter equation*, Sov. Math. Dokl. **32** (1985) 254.

[21] M. Jimbo, *A q-analogue of U(g) and the Yang-Baxter equation*, Lett. Math. Phys. **10** (1985) 63.

[22] V. G. Drinfeld, *Quantum groups*, in A. M. Gleason, ed., Proceedings of the International Congress of Mathematicians, Amer. Math. Soc., 798 (1986).

[23] A. Lorek, W. Weich and J. Wess, *Non-commutative Euclidean and Minkowski Structures*, Z. Phys. C **76** (1997) 375, [q-alg/9702025].

[24] P. P. Kulish and N. Yu. Reshetikin, *Quantum linear problem for the Sine-Gordon equation and higher representations*, J. Sov. Math. **23** (1983) 2345.

[25] S. Majid, *Foundations of Quantum Group Theory*, University Press, Cambridge (1995).

[26] M. Chaichian and A. P. Demichev, *Introduction to Quantum Groups*, World Scientific, Singapore (1996).

[27] A. Klimyk, K. Schmüdgen, *Quantum Groups and their Representations,* Springer-Verlag, 1997.

[28] P. Schupp, P. Watts and B. Zumino, *Bicovariant Quantum Algebras and Quantum Lie algebras*, Commun. Math. Phys. **157** (1993) 305.

[29] S. Majid, *Quantum and Braided Lie algebras*, J. Geom. Phys. **13** (1994) 307.

[30] A. Sudbery, *Quantum differential calculus and lie algebras*, Int. J. Mod. Phys. **A** (1993) 228.

[31] L.C. Biedenharn, M. Tartini, *On q-Tensor Operators for Quantum Groups,* Lett. Math. Phys. **20** (1990) 271-278.

[32] Yu. I. Manin, *Quantum Groups and Non-Commutative Geometry*, Centre de Recherche Mathématiques, Montreal (1988).

[33] S. Majid, *Free braided differential calculus, braided binomial theorem and the braided exponential map*, 1993 *J. Mat. Phys.* **34** 4843.

[34] S. Majid, Introduction to Braided Geometry and q-Minkowski Space, *Preprint*, 1994 [hep-th/9410241]

[35] C. Bauer and H. Wachter, Operator representations on quantum spaces, 2003 *Eur. Phys. J. C* **31** 261 [math-ph/0201023]

[36] A. Schmidt and H. Wachter, *Spinor calculus for q-deformed quantum spaces I*, preprint, [hep-th/0705.1640].

Alexander Schmidt
Arnold-Sommerfeld-Center
Theresienstr. 37
D–80333 München
e-mail: schmidt@theorie.physik.uni-muenchen.de

Quantum Field Theory
B. Fauser, J. Tolksdorf and E. Zeidler, Eds., 303–424
© 2009 Birkhäuser Verlag Basel/Switzerland

L_∞-Algebra Connections and Applications to String- and Chern-Simons n-Transport

Hisham Sati, Urs Schreiber and Jim Stasheff

Abstract. We give a generalization of the notion of a Cartan-Ehresmann connection from Lie algebras to L_∞-algebras and use it to study the obstruction theory of lifts through higher String-like extensions of Lie algebras. We find (generalized) Chern-Simons and BF-theory functionals this way and describe aspects of their parallel transport and quantization.

It is known that over a D-brane the Kalb-Ramond background field of the string restricts to a 2-bundle with connection (a gerbe) which can be seen as the obstruction to lifting the $PU(H)$-bundle on the D-brane to a $U(H)$-bundle. We discuss how this phenomenon generalizes from the ordinary central extension $U(1) \to U(H) \to PU(H)$ to higher categorical central extensions, like the String-extension $\mathbf{B}U(1) \to \mathrm{String}(G) \to G$. Here the obstruction to the lift is a 3-bundle with connection (a 2-gerbe): the Chern-Simons 3-bundle classified by the first Pontrjagin class. For $G = \mathrm{Spin}(n)$ this obstructs the existence of a String-structure. We discuss how to describe this obstruction problem in terms of Lie n-algebras and their corresponding categorified Cartan-Ehresmann connections. Generalizations even beyond String-extensions are then straightforward. For $G = \mathrm{Spin}(n)$ the next step is "Fivebrane structures" whose existence is obstructed by certain generalized Chern-Simons 7-bundles classified by the second Pontrjagin class.

Mathematics Subject Classification (2000). Primary 83E30; Secondary 55P20; 81T30; 55R45.

Keywords. Cartan-Ehresman connection, L_∞-algebra, Chern-Simons theory, BF-theory, 2-bundles, Eilenberg-MacLane spaces, differential greded algebras, branes, strings.

1. Introduction

The study of extended n-dimensional relativistic objects which arise in string theory has shown that these couple to background fields which can naturally be thought of as n-fold categorified generalizations of fiber bundles with connection.

These structures, or various incarnations of certain special cases of them, are probably most commonly known as (bundle-)$(n-1)$-gerbes with connection. These are known to be equivalently described by Deligne cohomology, by abelian gerbes with connection ("and curving") and by Cheeger-Simons differential characters. Following [6, 9] we address them as n-bundles with connection.

fundamental object	background field
n-particle	n-bundle
$(n-1)$-brane	$(n-1)$-gerbe

TABLE 1. **The two schools of counting** higher dimensional structures. Here n is in $\mathbb{N} = \{0, 1, 2, \cdots\}$.

In string theory, the first departure from bundles with connections to higher bundles with connection occurred with the fundamental (super)string coupling to the Neveu-Schwarz (NS) B-field. Locally, the B-field is just an $\mathbb{R}$-valued two-form. However, the study of the path integral, which amounts to 'exponentiation', reveals that the B-field can be thought of as an abelian gerbe with connection whose curving corresponds to the H-field H_3 or as a Cheeger-Simons differential character, whose holonomy [34] can be described [20] in the language of bundle gerbes [62].

The next step up occurs with the M-theory (super)membrane which couples to the C-field [11]. In supergravity, this is viewed locally as an $\mathbb{R}$-valued differential three-form. However, the study of the path integral has shown that this field is quantized in a rather nontrivial way [79]. This makes the C-field not precisely a 2-gerbe or degree 3 Cheeger-Simons differential character but rather a shifted version [28] that can also be modeled using the Hopkins-Singer description of differential characters [46]. Some aspects of the description in terms of Deligne cohomology is given in [26].

From a purely formal point of view, the need of higher connections for the description of higher dimensional branes is not a surprise: n-fold categorified bundles with connection should be precisely those objects that allow us to define a consistent assignment of "phases" to n-dimensional paths in their base space. We address such an assignment as **parallel n-transport**. This is in fact essentially the definition of Cheeger-Simons differential characters [25] as these are consistent assignments of phases to chains. However, abelian bundle gerbes, Deligne cohomology and Cheeger-Simons differential characters all have one major restriction: they only know about assignments of elements in $U(1)$.

While the group of phases that enter the path integral is usually abelian, more general n-transport is important nevertheless. For instance, the latter plays a role at intermediate stages. This is well understood for $n = 2$: over a D-brane the abelian bundle gerbe corresponding to the NS field has the special property that it measures the obstruction to lifting a $PU(H)$-bundle to a $U(H)$-bundle, i.e. lifting a bundle with structure group the infinite projective unitary group on a Hilbert space H to the corresponding unitary group [15], [16]. Hence, while itself an abelian 2-structure, it is crucially related to a nonabelian 1-structure.

That this phenomenon deserves special attention becomes clear when we move up the dimensional ladder: The Green-Schwarz anomaly cancelation [40] in the heterotic string leads to a 3-structure with the special property that, over the target space, it measures the obstruction to lifting an $E_8 \times \mathrm{Spin}(n)$-bundle to a certain nonabelian principal 2-bundle, called a *String 2-bundle*. Such a 3-structure is also known as a Chern-Simons 2-gerbe [21]. By itself this is abelian, but its structure is constrained by certain nonabelian data. Namely this string 2-bundle with connection, from which the Chern-Simons 3-bundle arises, is itself an instance of a structure that yields parallel 2-transport. It can be described neither by abelian bundle gerbes, nor by Cheeger-Simons differential characters, nor by Deligne cohomology.

In anticipation of such situations, previous works have considered nonabelian gerbes and nonabelian bundle gerbes with connection. However, it turns out that care is needed in order to find the right setup. For instance, the kinds of nonabelian gerbes with connection studied in [17], [3], although very interesting, are not sufficiently general to capture String 2-bundles. Moreover, it is not easy to see how to obtain the parallel 2-transport assignment from these structures. For the application to string physics, it would be much more suitable to have a nonabelian generalization of the notion of a Cheeger-Simons differential character, and thus a structure which, by definition, knows how to assign generalized phases to n-dimensional paths.

The obvious generalization that is needed is that of a parallel transport n-functor. Such a notion was described in [9], [72]: a structure defined by the very fact that it labels n-paths by algebraic objects that allow composition in n different directions, such that this composition is compatible with the gluing of n-paths. One can show that such transport n-functors encompass abelian and nonabelian gerbes with connection as special cases [72]. However, these n-functors are more general. For instance, String 2-bundles with connection are given by parallel transport 2-functors. Ironically, the strength of the latter – namely their knowledge about general phase assignments to higher dimensional paths – is to some degree also a drawback: for many computations, a description *entirely* in terms of differential form data would be more tractable. However, the passage from parallel n-transport to the corresponding differential structure is more or less straightforward: a parallel transport n-functor is essentially a morphism of Lie n-groupoids. As such, it can be sent, by a procedure generalizing the passage from Lie groups to Lie algebras, to a morphism of Lie n-algebroids.

The aim of this paper is to describe two topics: First, to set up a formalism for higher bundles with connections entirely in terms of L_∞-algebras, which may be thought of as a categorification of the theory of Cartan-Ehresmann connections.

This is supposed to be the differential version of the theory of parallel transport n-functors, but an exhaustive discussion of the differentiation procedure is not given here. Instead we discuss a couple of examples and then show how the lifting problem has a nice description in this language. To do so, we present a family of L_∞-algebras that govern the gauge structure of p-branes, as above, and discuss the lifting problem for them. By doing so, we characterize Chern-Simons 3-forms as local connection data on 3-bundles with connection which arise as the obstruction to lifts of ordinary bundles to the corresponding String 2-bundles, governed by the String Lie 2-algebra.

The formalism immediately allows the generalization of this situation to higher degrees. Indeed we indicate how certain 7-dimensional generalizations of Chern-Simons 3-bundles obstruct the lift of ordinary bundles to certain 6-bundles governed by the Fivebrane Lie 6-algebra. The latter correspond to what we define as the fivebrane structure, for which the degree seven NS field H_7 plays the role that the degree three dual NS field H_3 plays for the $n = 2$ case.

The paper is organized in such a way that section 2 serves more or less as a self-contained description of the basic ideas and construction, with the rest of the document having all the details and all the proofs.

In this paper we make use of the homotopy algebras usually referred to as L_∞-algebras. These algebras also go by other names such as sh-Lie algebras [57]. In our context we may also call such algebras Lie ∞-algebras which we think of as the abstract concept of an ∞-vector space with an antisymmetric bracket ∞-functor on it, which satisfies a Jacobi identity up to coherent equivalence, whereas "L_∞-algebra" is concretely a codifferential coalgebra of sorts. In this paper we will nevertheless follow the standard notation of L_∞-algebra.

2. The setting and plan

We set up a useful framework for describing higher order bundles with connection entirely in terms of Lie n-algebras, which can be thought of as arising from a categorification of the concept of an Ehresmann connection on a principal bundle. Then we apply this to the study of Chern-Simons n-bundles with connection as obstructions to lifts of principal G-bundles through higher String-like extensions of their structure Lie algebra.

2.1. L_∞-algebras and their String-like central extensions

A Lie group has all the right properties to locally describe the phase change of a charged particle as it traces out a worldline. A Lie n-group is a higher structure with precisely all the right properties to describe locally the phase change of a charged $(n - 1)$-brane as it traces out an n-dimensional worldvolume.

2.1.1. L_∞-algebras. Just as ordinary Lie groups have Lie algebras, Lie n-groups have Lie n-algebras. If the Lie n-algebra is what is called *semistrict*, these are [5]

precisely L_∞-algebras [57] which have come to play a significant role in cohomo-logical physics. A ("semistrict" and finite dimensional) Lie n-algebra is any of the following three equivalent structures:

- an L_∞-algebra structure on a graded vector space $\mathfrak{g}$ concentrated in the first n degrees $(0, ..., n-1)$;
- a quasi-free differentially graded-commutative algebra ("qDGCA": free as a graded-commutative) algebra on the dual of that vector space: this is the Chevalley-Eilenberg algebra $CE(\mathfrak{g})$ of $\mathfrak{g}$;
- an n-category internal to the category of *graded* vector spaces and equipped with a skew-symmetric linear bracket functor which satisfies a Jacobi identity up to higher coherent equivalence.

For every L_∞-algebra $\mathfrak{g}$, we have the following three qDGCAs:

- the **Chevalley-Eilenberg algebra** $CE(\mathfrak{g})$
- the **Weil algebra** $W(\mathfrak{g})$
- the algebra of **invariant polynomials** or **basic forms** $inv(\mathfrak{g})$.

These sit in a sequence

$$CE(\mathfrak{g}) \longleftarrow\!\!\!\longleftarrow W(\mathfrak{g}) \longleftarrow\!\!\!\longrightarrow inv(\mathfrak{g}) \ , \qquad (2.1)$$

where all morphisms are morphisms of dg-algebras. This sequence plays the role of the sequence of differential forms on the "universal $\mathfrak{g}$-bundle".

FIGURE 1. (see next page) **The universal G-bundle** in its various incarnations. That the ordinary universal G bundle is the realization of the nerve of the groupoid which we denote here by $\mathrm{INN}(G)$ is an old result by Segal (see [65] for a review and a discussion of the situation for 2-bundles). This groupoid $\mathrm{INN}(G)$ is in fact a 2-group. The corresponding Lie 2-algebra (2-term L_∞-algebra) we denote by $inn(\mathfrak{g})$. Regarding this as a codifferential coalgebra and then dualizing that to a differential algebra yields the Weil algebra of the Lie algebra $\mathfrak{g}$. This plays the role of differential forms on the universal G-bundle, as already known to Cartan. The entire table is expected to admit an ∞-ization. Here we concentrate on discussing ∞-bundles with connection in terms just of L_∞-algebras and their dual dg-algebras. An integration of this back to the world of ∞-groupoids should proceed along the lines of [35, 43], but is not considered here.

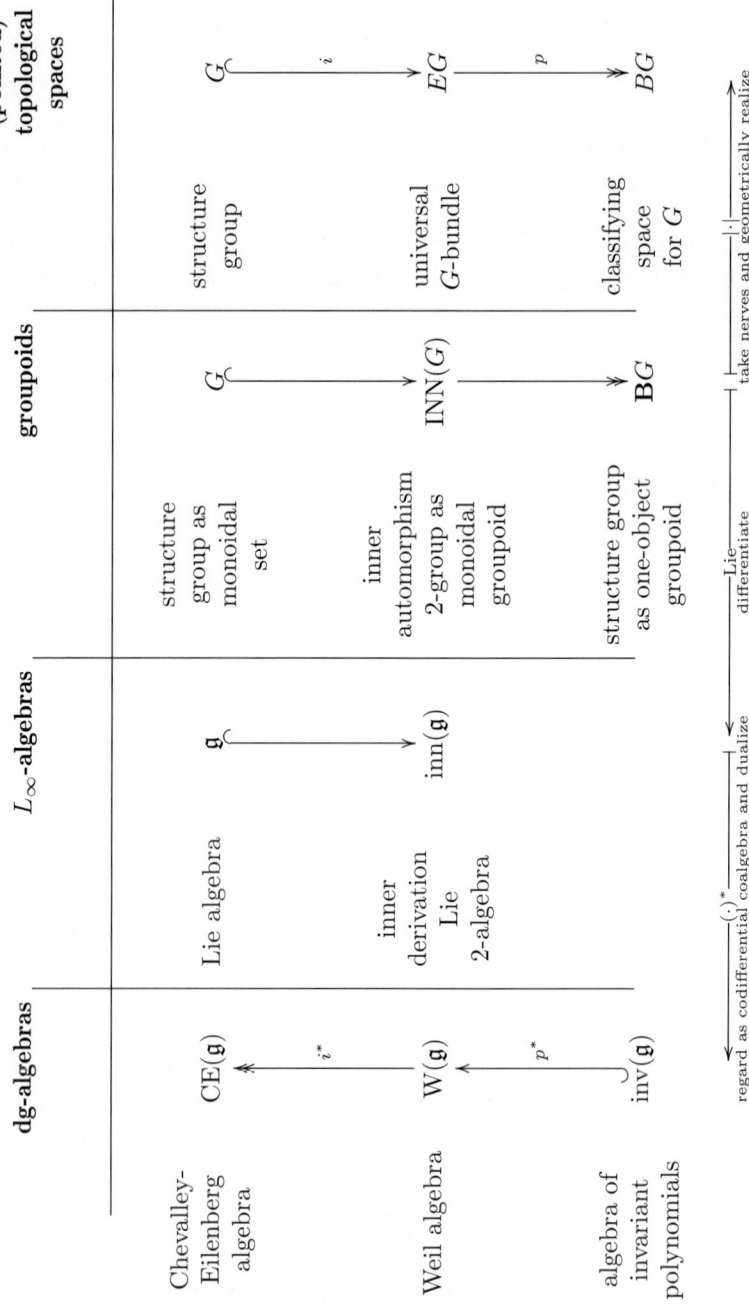

2.1.2. L_∞-algebras from cocycles: String-like extensions. A simple but important source of examples for higher Lie n-algebras comes from the abelian Lie algebra $\mathfrak{u}(1)$ which may be shifted into higher categorical degrees. We write $b^{n-1}\mathfrak{u}(1)$ for the Lie n-algebra which is entirely trivial except in its nth degree, where it looks like $\mathfrak{u}(1)$. Just as $\mathfrak{u}(1)$ corresponds to the Lie group $U(1)$, so $b^{n-1}\mathfrak{u}(1)$ corresponds to the iterated classifying space $B^{n-1}U(1)$, realizable as the topological group given by the Eilenberg-MacLane space $K(\mathbb{Z}, n)$. Thus an important source for interesting Lie n-algebras comes from extensions

$$0 \to b^{n-1}\mathfrak{u}(1) \to \hat{\mathfrak{g}} \to \mathfrak{g} \to 0 \tag{2.2}$$

of an ordinary Lie algebra $\mathfrak{g}$ by such a shifted abelian Lie n-algebra $b^{n-1}\mathfrak{u}(1)$. We find that, for each $(n+1)$-cocycle μ in the Lie algebra cohomology of $\mathfrak{g}$, we do obtain such a central extension, which we describe by

$$0 \to b^{n-1}\mathfrak{u}(1) \to \mathfrak{g}_\mu \to \mathfrak{g} \to 0\,. \tag{2.3}$$

Since, for the case when $\mu = \langle \cdot, [\cdot, \cdot] \rangle$ is the canonical 3-cocycle on a semisimple Lie algebra $\mathfrak{g}$, this $\mathfrak{g}_\mu$ is known ([7] and [43]) to be the Lie 2-algebra of the String 2-group, we call these central extensions *String-like* central extensions. (We also refer to these as Lie n-algebras "of Baez-Crans type" [5].) Moreover, whenever the cocycle μ is related by transgression to an invariant polynomial P on the Lie algebra, we find that $\mathfrak{g}_\mu$ fits into a short *homotopy* exact sequence of Lie $(n+1)$-algebras

$$0 \to \mathfrak{g}_\mu \to \mathrm{cs}_P(\mu) \to \mathrm{ch}_P(\mu) \to 0\,. \tag{2.4}$$

Here $\mathrm{cs}_P(\mathfrak{g})$ is a Lie $(n+1)$-algebra governed by the Chern-Simons term corresponding to the transgression element interpolating between μ and P. In a similar fashion $\mathrm{ch}_P(\mathfrak{g})$ knows about the characteristic (Chern) class associated with P.

In summary, from elements of the cohomology of $\mathrm{CE}(\mathfrak{g})$ together with related elements in $\mathrm{W}(\mathfrak{g})$ we obtain the String-like extensions of Lie algebras to Lie $2n$-algebras and the associated Chern- and Chern-Simons Lie $(2n-1)$-algebras:

Lie algebra cocycle	μ	Baez-Crans Lie n-algebra	$\mathfrak{g}_\mu$
invariant polynomial	P	Chern Lie n-algebra	$\mathrm{ch}_P(\mathfrak{g})$
transgression element	cs	Chern-Simons Lie n-algebra	$\mathrm{cs}_P(\mathfrak{g})$

2.1.3. L_∞-algebra differential forms. For $\mathfrak{g}$ an ordinary Lie algebra and Y some manifold, one finds that dg-algebra morphisms $\mathrm{CE}(\mathfrak{g}) \to \Omega^\bullet(Y)$ from the Chevally-Eilenberg algebra of $\mathfrak{g}$ to the DGCA of differential forms on Y are in bijection with $\mathfrak{g}$-valued 1-forms $A \in \Omega^1(Y, \mathfrak{g})$ whose ordinary curvature 2-form

$$F_A = dA + [A \wedge A] \tag{2.5}$$

vanishes. Without the flatness, the correspondence is with algebra morphisms *not* respecting the differentials. But dg-algebra morphisms $A : \mathrm{W}(\mathfrak{g}) \to \Omega^\bullet(Y)$ are in bijection with arbitrary $\mathfrak{g}$-valued 1-forms. These are flat precisely if A factors through $\mathrm{CE}(\mathfrak{g})$. This situation is depicted in the following diagram:

$$
\begin{array}{ccc}
\mathrm{CE}(\mathfrak{g}) & \longleftarrow & \mathrm{W}(\mathfrak{g}) \\
{\scriptstyle (A,F_A=0)}\Big\downarrow & & \Big\downarrow{\scriptstyle (A,F_A)} \\
\Omega^\bullet(Y) & \underset{=}{\longrightarrow} & \Omega^\bullet(Y)
\end{array}
\tag{2.6}
$$

This has an obvious generalization for $\mathfrak{g}$ an arbitrary L_∞-algebra. For $\mathfrak{g}$ any L_∞-algebra, we write

$$
\Omega^\bullet(Y,\mathfrak{g}) = \mathrm{Hom}_{\mathrm{dg-Alg}}(\mathrm{W}(\mathfrak{g}),\Omega^\bullet(Y))
\tag{2.7}
$$

for the collection of $\mathfrak{g}$-**valued differential forms** and

$$
\Omega^\bullet_{\mathrm{flat}}(Y,\mathfrak{g}) = \mathrm{Hom}_{\mathrm{dg-Alg}}(\mathrm{CE}(\mathfrak{g}),\Omega^\bullet(X))
\tag{2.8}
$$

for the collection of **flat $\mathfrak{g}$-valued differential forms**.

2.2. L_∞-algebra Cartan-Ehresmann connections

2.2.1. $\mathfrak{g}$-bundle descent data.

A *descent object* for an ordinary principal G-bundle on X is a surjective submersion $\pi : Y \to X$ together with a functor $g : Y \times_X Y \to \mathbf{B}G$ from the groupoid whose morphisms are pairs of points in the same fiber of Y, to the groupoid $\mathbf{B}G$ which is the one-object groupoid corresponding to the group G. Notice that the groupoid $\mathbf{B}G$ is not itself the classifying space BG of G, but the geometric realization of its nerve, $|\mathbf{B}G|$, is: $|\mathbf{B}G| = BG$.

We may take Y to be the disjoint union of some open subsets $\{U_i\}$ of X that form a good open cover of X. Then g is the familiar concept of a transition function describing a bundle that has been locally trivialized over the U_i. But one can also use more general surjective submersions. For instance, for $P \to X$ any principal G-bundle, it is sometimes useful to take $Y = P$. In this case one obtains a canonical choice for the cocycle

$$
g : Y \times_X Y = P \times_X P \to \mathbf{B}G
\tag{2.9}
$$

since P being principal means that

$$
P \times_X P \simeq_{\mathrm{diffeo}} P \times G .
\tag{2.10}
$$

This reflects the fact that every principal bundle canonically trivializes when pulled back to its own total space. The choice $Y = P$ differs from that of a good cover crucially in the following aspect: if the group G is connected, then also the fibers of $Y = P$ are connected. Cocycles over surjective submersions with connected fibers have special properties, which we will utilize: *When the fibers of Y are connected, we may think of the assignment of group elements to pairs of points in one fiber as arising from the parallel transport with respect to a flat vertical 1-form $A_{\mathrm{vert}} \in \Omega^1_{\mathrm{vert}}(Y,\mathfrak{g})$, flat along the fibers.* As we shall see, this can be thought of as the vertical part of a Cartan-Ehresmann connection 1-form. This provides a morphism

$$\Omega_{\mathrm{vert}}^{\bullet}(Y) \xleftarrow{\ A_{\mathrm{vert}}\ } \mathrm{CE}(\mathfrak{g}) \tag{2.11}$$

of differential graded algebras from the Chevalley-Eilenberg algebra of $\mathfrak{g}$ to the vertical differential forms on Y.

Unless otherwise specified, *morphism* will always mean *homomorphism of differential graded algebra*. A_{vert} has an obvious generalization: for $\mathfrak{g}$ any Lie n-algebra, we say that a $\mathfrak{g}$-bundle descent object for a $\mathfrak{g}$-n-bundle on X is a surjective submersion $\pi : Y \to X$ together with a morphism $\Omega_{\mathrm{vert}}^{\bullet}(Y) \xleftarrow{\ A_{\mathrm{vert}}\ } \mathrm{CE}(\mathfrak{g})$. Now $A_{\mathrm{vert}} \in \Omega_{\mathrm{vert}}^{\bullet}(Y, \mathfrak{g})$ encodes a collection of vertical p-forms on Y, each taking values in the degree p-part of $\mathfrak{g}$ and all together satisfying a certain flatness condition, controlled by the nature of the differential on $\mathrm{CE}(\mathfrak{g})$.

2.2.2. Connections on n-bundles: the extension problem.
Given a descent object $\Omega_{\mathrm{vert}}^{\bullet}(Y) \xleftarrow{\ A_{\mathrm{vert}}\ } \mathrm{CE}(\mathfrak{g})$ as above, a **flat connection** on it is an extension of the morphism A_{vert} to a morphism A_{flat} that factors through differential forms on Y

$$
\begin{array}{ccc}
\Omega_{\mathrm{vert}}^{\bullet}(Y) & \xleftarrow{\ A_{\mathrm{vert}}\ } & \mathrm{CE}(\mathfrak{g}) \ . \\
\uparrow{\scriptstyle i^*} & \nearrow{\scriptstyle A_{\mathrm{flat}}} & \\
\Omega^{\bullet}(Y) & &
\end{array}
\tag{2.12}
$$

In general, such an extension does not exist. A general **connection** on a $\mathfrak{g}$-descent object A_{vert} is a morphism

$$\Omega^{\bullet}(Y) \xleftarrow{\ (A, F_A)\ } \mathrm{W}(\mathfrak{g}) \tag{2.13}$$

from the Weil algebra of $\mathfrak{g}$ to the differential forms on Y together with a morphism

$$\Omega^{\bullet}(Y) \xleftarrow{\ \{K_i\}\ } \mathrm{inv}(\mathfrak{g}) \tag{2.14}$$

from the invariant polynomials on $\mathfrak{g}$, as in 2.1.1, to the differential forms on X, such that the following two squares commute:

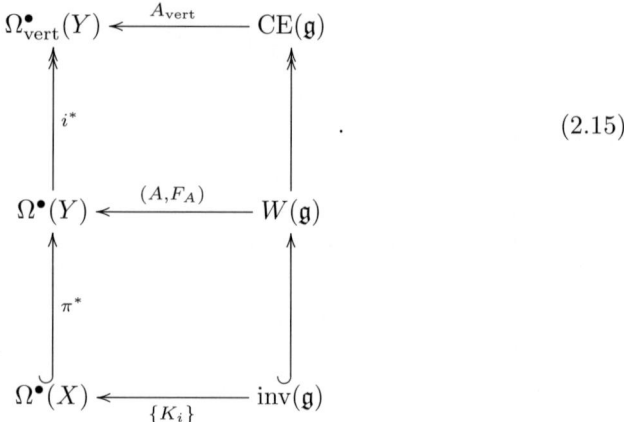

$$(2.15)$$

Whenever we have such two commuting squares, we say

- $A_{\mathrm{vert}} \in \Omega^\bullet_{\mathrm{vert}}(Y, \mathfrak{g})$ is a $\mathfrak{g}$-bundle **descent object** (playing the role of a **transition function**);
- $A \in \Omega^\bullet(Y, \mathfrak{g})$ is a (Cartan-Ehresmann) **connection** with values in the L_∞-algebra $\mathfrak{g}$ on the total space of the surjective submersion;
- $F_A \in \Omega^{\bullet+1}(Y, \mathfrak{g})$ are the corresponding **curvature** forms;
- and the set $\{K_i \in \Omega^\bullet(X)\}$ are the corresponding **characteristic forms**, whose classes $\{[K_i]\}$ in de Rham cohomology

$$\Omega^\bullet(X) \xleftarrow{\quad \{K_i\} \quad} \mathrm{inv}(\mathfrak{g}) \qquad (2.16)$$

$$H^\bullet_{\mathrm{deRham}}(X) \xleftarrow{\quad \{[K_i]\} \quad} H^\bullet(\mathrm{inv}(\mathfrak{g}))$$

are the corresponding **characteristic classes** of the given descent object A_{vert}.

Hence we realize the curvature of a $\mathfrak{g}$-connection as the *obstruction* to extending a $\mathfrak{g}$-descent object to a *flat* $\mathfrak{g}$-connection.

2.3. Higher string and Chern-Simons n-transport: the lifting problem

Given a $\mathfrak{g}$-descent object

$$\mathrm{CE}(\mathfrak{g}) \ , \qquad (2.17)$$

$$\swarrow A_{\mathrm{vert}}$$

$$\Omega^\bullet_{\mathrm{vert}}(Y)$$

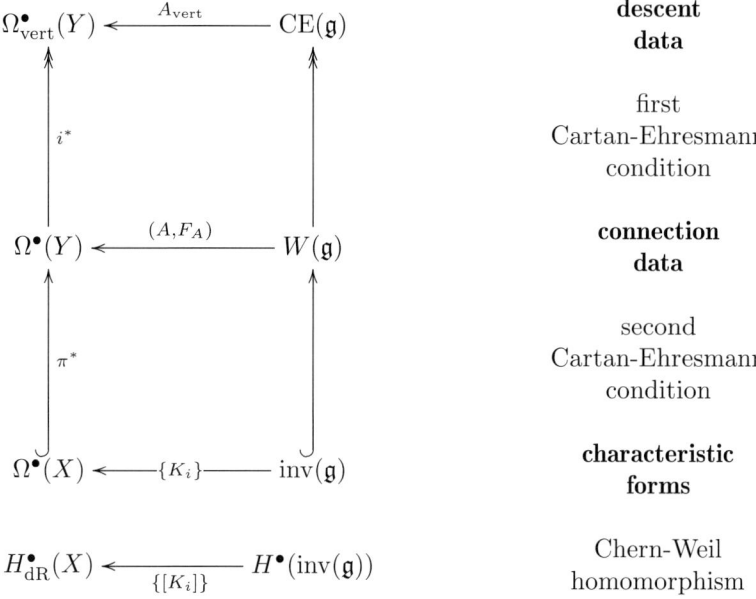

FIGURE 2. **A $\mathfrak{g}$-connection descent object and its interpretation.**
For $\mathfrak{g}$-any L_∞-algebra and X a smooth space, a $\mathfrak{g}$-connection on
X is an equivalence class of pairs $(Y, (A, F_A))$ consisting of a surjective submersion $\pi : Y \to X$ and dg-algebra morphisms forming
the above commuting diagram. The equivalence relation is concordance of such diagrams. The situation for ordinary Cartan-
Ehresmann (1-)connections is described in 6.2.1.

and given an extension of $\mathfrak{g}$ by a String-like L_∞-algebra

$$\mathrm{CE}(b^{n-1}\mathfrak{u}(1)) \xleftarrow{\quad i \quad} \mathrm{CE}(\mathfrak{g}_\mu) \xleftarrow{\qquad\qquad} \mathrm{CE}(\mathfrak{g}) \ , \qquad (2.18)$$

we ask if it is possible to *lift the descent object* through this extension, i.e. to find
a dotted arrow in

$$\mathrm{CE}(b^{n-1}\mathfrak{u}(1)) \xleftarrow{\quad} \mathrm{CE}(\mathfrak{g}_\mu) \xleftarrow{\qquad\qquad} \mathrm{CE}(\mathfrak{g}) \ . \qquad (2.19)$$
$$\Omega^\bullet_{\mathrm{vert}}(Y)$$

In general this is not possible. We seek a straightforward way to compute the
obstruction to the existence of the lift. The strategy is to form the *weak* (homotopy)

kernel of

$$CE(b^{n-1}\mathfrak{u}(1)) \xleftarrow{\quad i \quad} CE(\mathfrak{g}_\mu) \qquad (2.20)$$

which we denote by $CE(b^{n-1}\mathfrak{u}(1) \hookrightarrow \mathfrak{g}_\mu)$ and realize as a mapping cone of qDGCAs. This is governed by the diagram

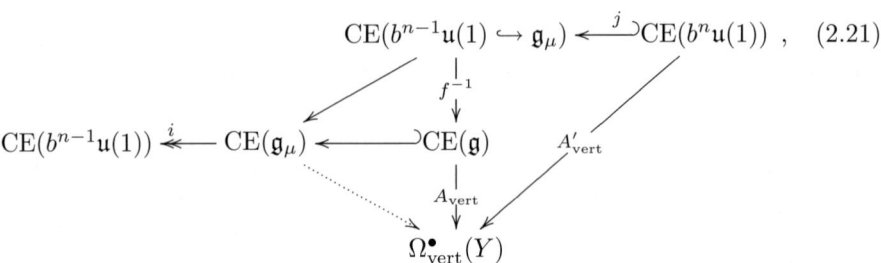

which we now describe.

We have canonically a morphism f from $CE(\mathfrak{g})$ to $CE(b^{n-1}\mathfrak{u}(1) \hookrightarrow \mathfrak{g}_\mu)$ which happens to have a *weak* inverse f^{-1}. While the lift to a $\mathfrak{g}_\mu$-cocycle may not always exist, the lift to a $(b^{n-1}\mathfrak{u}(1) \hookrightarrow \mathfrak{g}_\mu)$-cocycle does always exist, $A_{\text{vert}} \circ f^{-1}$. The failure of this lift to be a true lift to $\mathfrak{g}_\mu$ is measured by the component of $A_{\text{vert}} \circ f^{-1}$ on $b^{n-1}\mathfrak{u}(1)[1] \simeq b^n\mathfrak{u}(1)$. Formally this is the composite $A'_{\text{vert}} := A_{\text{vert}} \circ f^{-1} \circ j$. The nontriviality of the $b^n\mathfrak{u}(1)$-descent object A'_{vert} is the obstruction to constructing the desired lift.

We thus find the following results, for any $\mathfrak{g}$-cocycle μ which is in transgression with the invariant polynomial P on $\mathfrak{g}$,

- The characteristic classes (in de Rham cohomology) of $\mathfrak{g}_\mu$-bundles are those of the corresponding $\mathfrak{g}$-bundles modulo those coming from the invariant polynomial P.
- The lift of a $\mathfrak{g}$-valued connection to a $\mathfrak{g}_\mu$-valued connection is obstructed by a $b^n\mathfrak{u}(1)$-valued $(n+1)$-connection whose $(n+1)$-form curvature is $P(F_A)$, i.e. the image under the Chern-Weil homomorphism of the invariant polynomial corresponding to μ.
- Accordingly, the $(n+1)$-form connection of the obstructing $b^n\mathfrak{u}(1)$ $(n+1)$-bundle is a Chern-Simons form for this characteristic class.

We call the obstructing $b^n\mathfrak{u}(1)$ $(n+1)$-descent object the corresponding Chern-Simons $(n+1)$-bundle. For the case when $\mu = \langle \cdot, [\cdot, \cdot] \rangle$ is the canonical 3-cocycle on a semisimple Lie algebra $\mathfrak{g}$, this structure (corresponding to a 2-gerbe) has a 3-connection given by the ordinary Chern-Simons 3-form and has a curvature 4-form given by (the image in de Rham cohomology of) the first Pontrjagin class of the underlying $\mathfrak{g}$-bundle.

3. Statement of the main results

We define, for any L_∞-algebra $\mathfrak{g}$ and any smooth space X, a notion of

- $\mathfrak{g}$-descent objects over X;

and an extension of these to

- $\mathfrak{g}$-connection descent objects over X.

These descent objects are to be thought of as the data obtained from locally trivializing an n-bundle (with connection) whose structure n-group has the Lie n-algebra $\mathfrak{g}$. Being differential versions of n-functorial descent data of such n-bundles, they consist of morphisms of quasi free differential graded-commutative algebras (qDGCAs).

We define for each L_∞-algebra $\mathfrak{g}$ a dg-algebra $\mathrm{inv}(\mathfrak{g})$ of *invariant polynomials* on $\mathfrak{g}$. We show that every $\mathfrak{g}$-connection descent object gives rise to a collection of de Rham classes on X: its *characteristic classes*. These are images of the elements of $\mathrm{inv}(\mathfrak{g})$. Two descent objects are taken to be equivalent if they are *concordant* in a natural sense.

Our first main result is

Theorem 3.1 (Characteristic classes). *Characteristic classes are indeed characteristic of $\mathfrak{g}$-descent objects (but do not necessarily fully characterize them) in the following sense:*

- *Concordant $\mathfrak{g}$-connection descent objects have the same characteristic classes.*
- *If the $\mathfrak{g}$-connection descent objects differ just by a gauge transformation, they even have the same characteristic forms.*

This is our proposition 6.9 and corollary 6.10.

Remark. We expect that this result can be strengthened. Currently our characteristic classes are just in de Rham cohomology. One would expect that these are images of classes in integral cohomology. While we do not attempt here to discuss integral characteristic classes in general, we discuss some aspects of this for the case of abelian Lie n-algebras $\mathfrak{g} = b^{n-1}\mathfrak{u}(1)$ in 6.1.1 by relating $\mathfrak{g}$-descent objects to Deligne cohomology.

The reader should also note that in our main examples to be discussed in section 7 we start with an L_∞-connection which happens to be an ordinary Cartan-Ehresmann connection on an ordinary bundle and is hence known to have integral classes. It follows from our results then that also the corresponding Chern-Simons 3-connections in particular have an integral class.

We define String-like extensions $\mathfrak{g}_\mu$ of L_∞-algebras coming from any L_∞-algebra cocycle μ: a closed element in the Chevalley-Eilenberg dg-algebra $\mathrm{CE}(\mathfrak{g})$ corresponding to $\mathfrak{g}$: $\mu \in \mathrm{CE}(\mathfrak{g})$. These generalize the String Lie 2-algebra which governs the dynamics of (heterotic) superstrings.

Our second main results is

Theorem 3.2 (String-like extensions and their properties). *For every degree $(n+1)$-cocycle μ on an L_∞-algebra $\mathfrak{g}$ we obtain the String-like extension $\mathfrak{g}_\mu$ which sits in an exact sequence*

$$0 \to b^{n-1}\mathfrak{u}(1) \to \mathfrak{g}_\mu \to \mathfrak{g} \to 0.$$

When μ is in transgression with an invariant polynomial P we furthermore obtain a weakly exact sequence

$$0 \to \mathfrak{g}_\mu \to \mathrm{cs}_P(\mathfrak{g}) \to \mathrm{ch}_P(\mu) \to 0$$

of L_∞-algebras, where $\mathrm{cs}_P(\mathfrak{g}) \simeq \mathrm{inn}(\mathfrak{g}_\mu)$ is trivializable (equivalent to the trivial L_∞-algebra). There is an algebra of invariant polynomials on $\mathfrak{g}$ associated with $\mathrm{cs}_P(\mathfrak{g})$ and we show that it is the algebra of invariant polynomials of $\mathfrak{g}$ modulo the ideal generaled by P.

This is proposition 5.29, proposition 5.30 and proposition 5.33.

Our third main result is

Theorem 3.3 (Obstructions to lifts through String-like extensions). *For $\mu \in \mathrm{CE}(\mathfrak{g})$ any degree $n+1$ $\mathfrak{g}$-cocycle that transgresses to an invariant polynomial $P \in \mathrm{inv}(\mathfrak{g})$, the obstruction to lifting a $\mathfrak{g}$-descent object to a $\mathfrak{g}_\mu$-descent object is a $(b^n\mathfrak{u}(1))$-descent object whose single characteristic class is the class corresponding to P of the original $\mathfrak{g}$-descent object.*

This is reflected by the fact that the cohomology of the basic forms on the Chevalley-Eilenberg algebra of the corresponding Chern-Simons L_∞-algebra $\mathrm{cs}_P(\mathfrak{g})$ is that of the algebra of basic forms on $\mathrm{inv}(\mathfrak{g})$ modulo the ideal generated by P.

This is our proposition 5.33 and proposition 7.12.

While we do not discuss it here, our L_∞-connections may be integrated to full nonabelian differential cocycles along the lines of [70].

We discuss the following **applications**.

- For $\mathfrak{g}$ an ordinary semisimple Lie algebra and μ its canonical 3-cocycle, the obstruction to lifting a $\mathfrak{g}$-bundle to a String 2-bundle is a Chern-Simons 3-bundle with characteristic class the Pontrjagin class of the original bundle. This is a special case of our proposition 7.12 which is spelled out in detail in in 7.3.1.

  The vanishing of this obstruction is known as a String structure [52, 56, 63]. In categorical language, this issue was first discussed in [76].

  By passing from our Lie ∞-algebraic description to smooth spaces along the lines of section 4.1 and then forming fundamental n-groupoids of these spaces, one can see that our construction of obstructing n-bundles to lifts through String-like extensions reproduces the construction [18, 19] of Čech cocycles representing characteristic classes. This, however, will not be discussed here.

- This result generalizes to all String-like extensions. Using the 7-cocycle on $\mathfrak{so}(n)$ we obtain lifts through extensions by a Lie 6-algebra, which we call the Fivebrane Lie 6-algebra. Accordingly, fivebrane structures on string structures are obstructed by the second Pontrjagin class.

  This pattern continues and one would expect our obstruction theory for lifts through String-like extensions with respect to the 11-cocycle on $\mathfrak{so}(n)$ to correspond to *Ninebrane* structure.

The issue of p-brane structures for higher p was discussed before in [60]. In contrast to the discussion there, we here see p-brane structures only for $p = 4n+1$, corresponding to the list of invariant polynomials and cocycles for $\mathfrak{so}(n)$. While our entire obstruction theory applies to all cocycles on all Lie ∞-algebras, it is only for those on $\mathfrak{so}(n)$ and maybe $\mathfrak{e}_8$ for which the physical interpretation in the sense of p-brane structures is understood.

- We discuss how the action functional of the topological field theory known as BF-theory arises from an invariant polynomial on a strict Lie 2-algebra, in a generalization of the integrated Pontrjagin 4-form of the topological term in Yang-Mills theory. See proposition 5.26 and the example in 5.6.1.

  This is similar to but different from the Lie 2-algebraic interpretation of BF theory indicated in [38, 39], where the "cosmological" bilinear in the connection 2-form is not considered and a constraint on the admissible strict Lie 2-algebras is imposed.

- We indicate in 8.1 the notion of parallel transport induced by a $\mathfrak{g}$-connection, relate it to the n-functorial parallel transport of [9, 72, 73, 74] and point out how this leads to σ-model actions in terms of dg-algebra morphisms. See section 8.

- We indicate in 8.3.1 how by forming configuration spaces by sending DGCAs to smooth spaces and then using the internal hom of smooth space, we obtain for every $\mathfrak{g}$-connection descent object configuration spaces of maps equipped with an action functional induced by the transgressed $\mathfrak{g}$-connection. We show that the algebra of differential forms on these configuration spaces naturally supports the structure of the corresponding BRST-BV complex, with the iterated ghost-of-ghost structure inherited from the higher degree symmetries induced by $\mathfrak{g}$.

  This construction is similar in spirit to the one given in [1], reviewed in [66], but also, at least superficially, a bit different.

- We indicate also in 8.3.1 how this construction of configuration spaces induces the notion of transgression of n-bundles on X to $(n-k)$-bundles on spaces of maps from k-dimensional spaces into X. An analogous integrated description of transgression in terms of inner homs is in [73]. We show in 8.3.1 in particular that this transgression process relates the concept of String-structures in terms of 4-classes down on X with the corresponding 3-classes on LX, as discussed for instance in [56]. Our construction immediately generalizes to fivebrane and higher classes.

All of our discussion here pertains to *principal* L_∞-connections. One can also discuss *associated* $\mathfrak{g}$-connections induced by $(\infty-)$representations of $\mathfrak{g}$ (for instance as in [58]) and then study the collections of "sections" or "modules" of such associated $\mathfrak{g}$-connections.

The extended quantum field theory of a $(n-1)$-brane charged under an n-connection ("a charged n-particle", definition 8.5) should (see for instance [32, 33, 76, 45]) assign to each d-dimensional part Σ of the brane's parameter space ("worldvolume") the collection (an $(n-d-1)$-category, really) of sections/modules of the transgression of the n-bundle to the configuration space of maps from Σ.

For instance, the space of sections of a Chern-Simons 3-connection transgressed to maps from the circle should yield the representation category of the Kac-Moody extension of the corresponding loop group.

Our last proposition 8.6 points in this direction. But a more detailed discussion will not be given here.

4. Differential graded-commutative algebra

Differential $\mathbb{N}$-graded commutative algebras (DGCAs) play a prominent role in our discussion. One way to understand what is special about DGCAs is to realize that every DGCA can be regarded, essentially, as the algebra of *differential forms on some generalized smooth space*.

We explain what this means precisely in section 4.1. The underlying phenomenon is essentially the familiar governing principle of Sullivan models in rational homotopy theory [44, 77], but instead of working with simplicial spaces, we here consider presheaf categories. This will not become relevant, though, until the discussion of configuration spaces, parallel transport and action functionals in section 8.

4.1. Differential forms on smooth spaces

We can think of every differential graded commutative algebra essentially as being the algebra of differential forms on *some* space, possibly a generalized space.

Definition 4.1. *Let S be the category whose objects are the open subsets of $\mathbb{R} \cup \mathbb{R}^2 \cup \mathbb{R}^3 \cup \cdots$ and whose morphisms are smooth maps between these. We write*

$$S^\infty := \mathrm{Set}^{S^{\mathrm{op}}} \tag{4.1}$$

for the category of set-valued presheaves on S.

So an object X in S^∞ is an assignment of sets $U \mapsto X(U)$ to each open subset U, together with an assignment

$$(U \xrightarrow{\phi} V) \mapsto (X(U) \xleftarrow{\phi_X^*} X(V)) \tag{4.2}$$

of maps of sets to maps of smooth subsets which respects composition. A morphism

$$f : X \to Y \tag{4.3}$$

of smooth spaces is an assignment $U \mapsto (X(U) \xrightarrow{f_U} Y(U))$ of maps of sets to open subsets, such that for all smooth maps of subsets $U \xrightarrow{\phi} V$ we have that the square

$$X(V) \xrightarrow{\ f_V\ } Y(V) \qquad\qquad (4.4)$$

$$\downarrow \phi_X^* \qquad\qquad \downarrow \phi_Y^*$$

$$X(U) \xrightarrow{\ f_U\ } Y(U)$$

commutes. We think of the objects of S^∞ as smooth spaces. The set $X(U)$ that such a smooth space X assigns to an open subset U is to be thought of as the set of smooth maps from U into X. As opposed to manifolds which are *locally isomorphic* to an object in S, smooth spaces can hence be thought of as being objects which are just required to have the property that they may be *probed* by objects of S. Every open subset V becomes a smooth space by setting

$$V : U \mapsto \mathrm{Hom}_{S^\infty}(U, V) \,. \qquad\qquad (4.5)$$

This are the *representable* presheaves. Similarly, every ordinary manifold X becomes a smooth space by setting

$$X : U \mapsto \mathrm{Hom}_{\mathrm{manifolds}}(U, X) \,. \qquad\qquad (4.6)$$

The special property of smooth spaces which we need here is that they form a (cartesian) *closed* category:

- for any two smooth spaces X and Y there is a cartesian product $X \times Y$, which is again a smooth space, given by the assignment

$$X \times Y : U \mapsto X(U) \times Y(U) \,; \qquad\qquad (4.7)$$

  where the cartesian product on the right is that of sets;
- the collection $\mathrm{hom}(X, Y)$ of morphisms from one smooth space X to another smooth space Y is again a smooth space, given by the assignment

$$\mathrm{hom}_{S^\infty}(X, Y) : U \mapsto \mathrm{Hom}_{S^\infty}(X \times U, Y) \,. \qquad\qquad (4.8)$$

A very special smooth space is the smooth space of differential forms.

Definition 4.2. *We write $\Omega^\bullet$ for the smooth space which assigns to each open subset the set of differential forms on it*

$$\Omega^\bullet : U \mapsto \Omega^\bullet(U) \,. \qquad\qquad (4.9)$$

Using this object we define the DGCA of differential forms on any smooth space X to be the set

$$\Omega^\bullet(X) := \mathrm{Hom}_{S^\infty}(X, \Omega^\bullet) \qquad\qquad (4.10)$$

equipped with the obvious DGCA structure induced by the local DGCA structure of each $\Omega^\bullet(U)$.

Therefore the object $\Omega^\bullet$ is in a way both a smooth space as well as a differential graded commutative algebra: it is a DGCA-valued presheaf. Such objects are known as *schizophrenic* [48] or better *ambimorphic* [81] objects: they relate two different worlds by duality. In fact, the process of mapping into these objects provides an adjunction between the dual categories:

Definition 4.3. *There are contravariant functors from smooth spaces to DGCAs given by*

$$\Omega^\bullet : S^\infty \to \text{DGCAs}$$
$$X \mapsto \Omega^\bullet(X) \tag{4.11}$$

and

$$\text{Hom}(-, \Omega^\bullet(-)) : \text{DGCAs} \to S^\infty$$
$$A \mapsto X_A \tag{4.12}$$

These form an adjunction of categories. The unit

$$\text{DGCAs} \xrightarrow{\text{Hom}(-,\Omega^\bullet(-))} S^\infty \xrightarrow{\Omega^\bullet} \text{DGCAs} \tag{4.13}$$

of this adjunction is a natural transformation whose component map embeds each DGCA A into the algebra of differential forms on the smooth space it defines

$$A \hookrightarrow \Omega^\bullet(X_A) \tag{4.14}$$

by sending every $a \in A$ to the map of presheaves

$$(f \in \text{Hom}_{\text{DGCAs}}(A, \Omega^\bullet(U))) \mapsto (f(a) \in \Omega^\bullet(U)). \tag{4.15}$$

This way of obtaining forms on X_A from elements of A will be crucial for our construction of differential forms on spaces of maps, $\text{hom}(X,Y)$, used in section 8.3.

Using this adjunction, we can "pull back" the internal hom of S^∞ to DGCAs. Since the result is not literally the internal hom in DGCAs (which does not exist since DGCAs are not *profinite* as opposed to codifferential coalgebras [35]) we call it "maps" instead of "hom".

Definition 4.4 (Forms on spaces of maps). *Given any two DGCAs A and B, we define the DGCA of "maps" from B to A*

$$\text{maps}(B, A) := \Omega^\bullet(\text{hom}_{S^\infty}(X_A, X_B)). \tag{4.16}$$

This is a functor

$$\text{maps} : \text{DGCAs}^{\text{op}} \times \text{DGCAs} \to \text{DGCAs}. \tag{4.17}$$

Notice the fact (for instance corollary 35.10 in [53] and theorem 2.8 in [61]) that for any two smooth spaces X and Y, algebra homomorphisms $C^\infty(X) \xleftarrow{\phi^*} C^\infty(Y)$ and hence DGCA morphisms $\Omega^\bullet(X) \xleftarrow{\phi^*} \Omega^\bullet(Y)$ are in bijection with smooth maps $\phi : X \to Y$.

It follows that an element of $\mathrm{hom}(X_A, X_B)$ is, over test domains U and V a natural map of sets

$$\mathrm{Hom}_{\mathrm{DGCAs}}(A, \Omega^\bullet(V)) \times \mathrm{Hom}_{\mathrm{DGCAs}}(\Omega^\bullet(U), \Omega^\bullet(V)) \to \mathrm{Hom}_{\mathrm{DGCAs}}(B, \Omega^\bullet(V)).$$
$$(4.18)$$

One way to obtain such maps is from pullback along algebra homomorphisms

$$B \to A \otimes \Omega^\bullet(U).$$

This will be an important source of DGCAs of maps for the case that A is the Chevalley-Eilenberg algebra of an L_∞-algebra, as described in section 4.1.1.

4.1.1. Examples.

Diffeological spaces. Particularly useful are smooth spaces X which, while not quite manifolds, have the property that there is a set X_s such that

$$X : U \mapsto X(U) \subset \mathrm{Hom}_{\mathrm{Set}}(U, X_s) \tag{4.19}$$

for all $U \in S$. These are the Chen-smooth or diffeological spaces used in [9, 73, 74]. In particular, all spaces of maps $\mathrm{hom}_{S^\infty}(X, Y)$ for X and Y manifolds are of this form. This includes in particular loop spaces.

Forms on spaces of maps. When we discuss parallel transport and its transgression to configuration spaces in section 8.3, we need the following construction of differential forms on spaces of maps.

Definition 4.5 (Currents). *For A any DGCA, we say that a current on A is a smooth linear map*

$$c : A \to \mathbb{R}. \tag{4.20}$$

For $A = \Omega^\bullet(X)$ this reduces to the ordinary notion of currents.

Proposition 4.6. *Let A be a quasi free DGCAs in positive degree (meaning that the underlying graded commutative algebras are freely generated from some graded vector space in positive degree). For each element $b \in B$ and current c on A, we get an element in $\Omega^\bullet(\mathrm{Hom}_{\mathrm{DGCAs}}(B, A \otimes \Omega^\bullet(-)))$ by mapping, for each $U \in S$*

$$\mathrm{Hom}_{\mathrm{DGCAs}}(B, A \otimes \Omega^\bullet(U)) \quad \to \quad \Omega^\bullet(U)$$
$$f^* \quad \mapsto \quad c(f^*(b)). \tag{4.21}$$

If b is in degree n and c in degree $m \leq n$, then this differential form is in degree $n - m$.

The superpoint. Most of the DGCAs we shall consider here are non-negatively graded or even positively graded. These can be thought of as Chevalley-Eilenberg algebras of Lie n-algebroids and Lie n-algebras, respectively, as discussed in more detail in section 5. However, DGCAs of arbitrary degree do play an important role, too. Notice that a DGCA of non-positive degree is in particular a cochain complex of non-positive degree. But that is the same as a chain complex of non-negative degree.

The following is a very simple but important example of a DGCA in non-positive degree.

Definition 4.7 (Superpoint). *The "algebra of functions on the superpoint" is the DGCA*

$$C(\mathbf{pt}) := (\mathbb{R} \oplus \mathbb{R}[-1], d_{\mathbf{pt}}) \tag{4.22}$$

where the product on $\mathbb{R} \oplus \mathbb{R}[-1]$ is the tensor product over $\mathbb{R}$, and where the differential $d_{\mathbf{pt}} : \mathbb{R}[-1] \to \mathbb{R}$ is the canonical isomorphism.

The smooth space associated to this algebra according to definition 4.3 is just the ordinary point, because for any test domain U the set

$$\mathrm{Hom}_{\mathrm{DGCAs}}(C(\mathbf{pt}), \Omega^\bullet(U)) \tag{4.23}$$

contains only the morphism which sends $1 \in \mathbb{R}$ to the constant unit function on U, and which sends $\mathbb{R}[-1]$ to 0. However, as is well known from the theory of supermanifolds, the algebra $C(\mathbf{pt})$ is important in that morphisms from any other DGCA A into it compute the (shifted) *tangent space* corresponding to A. From our point of view here this manifests itself in particular by the fact that for X any manifold, we have a canonical injection

$$\Omega^\bullet(TX) \hookrightarrow \Omega^\bullet(\mathrm{maps}(C^\infty(X), C(\mathbf{pt}))) \tag{4.24}$$

of the differential forms on the tangent bundle of X into the differential forms on the smooth space of algebra homomorphisms of $C^\infty(X)$ to $C(\mathbf{pt})$:

for every test domain U an element in $\mathrm{Hom}_{\mathrm{DGCAs}}(C^\infty(X), C(\mathbf{pt} \otimes \Omega^\bullet(U)))$ comes from a pair consisting of a smooth map $f : U \to X$ and a vector field $v \in \Gamma(TX)$. Together this constitutes a smooth map $\hat{f} : U \to TX$ and hence for every form $\omega \in \Omega^\bullet(TX)$ we obtain a form on $\mathrm{maps}(C^\infty(X), C(\mathbf{pt}))$ by the assignment

$$((f, v) \in \mathrm{Hom}_{\mathrm{DGCAs}}(C^\infty(X), C(\mathbf{pt} \otimes \Omega^\bullet(U)))) \mapsto (\hat{f}^*\omega \in \Omega^\bullet(U)) \tag{4.25}$$

over each test domain U.

In section 5.1.1 we discuss how in the analogous fashion we obtain the Weil algebra $\mathrm{W}(\mathfrak{g})$ of any L_∞-algebra $\mathfrak{g}$ from its Chevalley-Eilenberg algebra $\mathrm{CE}(\mathfrak{g})$ by mapping that to $C(\mathbf{pt})$. This says that the Weil algebra is like the space of functions on the shifted tangent bundle of the "space" that the Chevalley-Eilenberg algebra is the space of functions on. See also figure 3.

4.2. Homotopies and inner derivations

When we forget the algebra structure of DGCAs, they are simply cochain complexes. As such they naturally live in a 2-category $\mathrm{Ch}^\bullet$ whose objects are cochain complexes $(V^\bullet, d_V)$, whose morphisms

$$(V^\bullet, d_V) \xleftarrow{\;\;f^*\;\;} (W^\bullet, d_W) \tag{4.26}$$

are degree preserving linear maps $V^\bullet \xleftarrow{\;f^*\;} W^\bullet$ that do respect the differentials,

$$[d, f^*] := d_V \circ f^* - f^* \circ d_W = 0, \tag{4.27}$$

and whose 2-morphisms

$$
(V^\bullet, d_V) \quad \overset{f^*}{\underset{g^*}{\rightrightarrows}} \Big\Downarrow \rho \quad (W^\bullet, d_W) \tag{4.28}
$$

are cochain homotopies, namely linear degree -1 maps $\rho : W^\bullet \to V^\bullet$ with the property that

$$
g^* = f^* + [d, \rho] = f^* + d_V \circ \rho + \rho \circ d_W . \tag{4.29}
$$

Later in section 5.2 we will also look at morphisms that do preserve the algebra structure, and homotopies of these. Notice that we can compose a 2-morphism from left and right with 1-morphisms, to obtain another 2-morphism

$$
(U^\bullet, d_U) \xleftarrow{\ h^*\ } (V^\bullet, d_V) \quad \overset{f^*}{\underset{g^*}{\rightrightarrows}} \Big\Downarrow \rho \quad (W^\bullet, d_W) \xleftarrow{\ j^*\ } (X^\bullet, d_X)
$$

$$\tag{4.30}$$

whose component map now is

$$
h^* \circ \rho \circ j^* : \quad X^\bullet \xrightarrow{\ j^*\ } W^\bullet \xrightarrow{\ \rho\ } V^\bullet \xrightarrow{\ h^*\ } U^\bullet . \tag{4.31}
$$

This will be important for the interpretation of the diagrams we discuss, of the type (4.40) and (4.46) below.

Of special importance are linear endomorphisms $V^\bullet \xleftarrow{\ \rho\ } V^\bullet$ of DGCAs which are algebra derivations. Among them, the *inner* derivations in turn play a special role:

Definition 4.8 (Inner derivations). *On any DGCA* $(V^\bullet, d_V)$, *a degree 0 endomorphism*

$$
(V^\bullet, d_V) \xleftarrow{\ L\ } (V^\bullet, d_V) \tag{4.32}
$$

is called an inner derivation *if*

- *it is an algebra derivation of degree 0;*
- *it is connected to the 0-derivation, i.e. there is a 2-morphims*

$$
(V^\bullet, d_V) \quad \overset{0}{\underset{L=[d_V,\rho]}{\rightrightarrows}} \Big\Downarrow \rho \quad (V^\bullet, d_V) , \tag{4.33}
$$

where ρ comes from an algebra derivation of degree -1.

Remark. Inner derivations generalize the notion of a Lie derivative on differential forms, and hence they encode the notion of vector fields in the context of DGCAs.

4.2.1. Examples.
Lie derivatives on ordinary differential forms. The formula sometimes known as "*Cartan's magic formula*", which says that on a smooth space Y the Lie derivative $L_v \omega$ of a differential form $\omega \in \Omega^\bullet(Y)$ along a vector field $v \in \Gamma(TY)$ is given by

$$L_v \omega = [d, \iota_v]\omega \,, \tag{4.34}$$

where $\iota_v : \Omega^\bullet(Y) \to \Omega^\bullet(Y)$, says that Lie derivatives on differential forms are inner derivations, in our sense. When Y is equipped with a smooth projection $\pi : Y \to X$, it is of importance to distinguish the vector fields vertical with respect to π. The abstract formulation of this, applicable to arbitrary DGCAs, is given in (4.3) below.

4.3. Vertical flows and basic forms
We will prominently be dealing with surjections

$$A \xleftarrow{\;\;i^*\;\;} B \tag{4.35}$$

of differential graded commutative algebras that play the role of the dual of an injection

$$F \xhookrightarrow{\;\;i\;\;} P \tag{4.36}$$

of a fiber into a bundle. We need a way to identify in the context of DGCAs which inner derivations of P are *vertical* with respect to i. Then we can find the algebra corresponding to the *basis* of P as those elements of B which are annihilated by all vertical derivations.

Definition 4.9 (Vertical derivations). *Given any surjection of differential graded algebras*

$$F \xleftarrow{\;\;i^*\;\;} P \tag{4.37}$$

we say that the vertical inner derivations

$$\tag{4.38}$$

(this diagram is in the category of cochain complexes, compare the beginning of section 5.2) on P with respect to i^* are those inner derivations

- *for which there exists an inner derivation of F*

$$(4.39)$$

such that

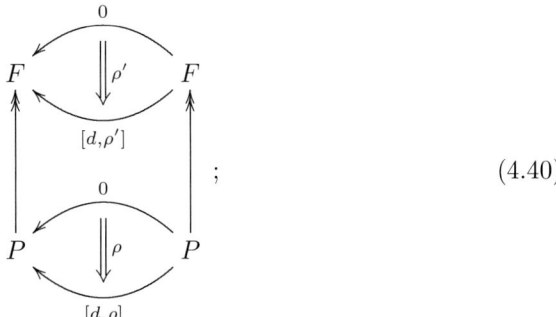

$$; (4.40)$$

- *and where ρ' is a* contraction, *$\rho' = \iota_x$, i.e. a derivation which sends indecomposables to degree 0.*

Definition 4.10 (Basic elements). *Given any surjection of differential graded algebras*

$$F \xleftarrow{\;i^*\;} P (4.41)$$

we say that the algebra

$$P_{\text{basic}} = \bigcap_{\rho \text{ vertical}} \ker(\rho) \cap \ker(\rho \circ d_p) (4.42)$$

of basic elements of P (with respect to the surjection i^*) is the subalgebra of P of all those elements $a \in P$ which are annihilated by all i^*-vertical derivations ρ, in that

$$\begin{aligned} \rho(a) &= 0 (4.43)\\ \rho(d_p a) &= 0. (4.44) \end{aligned}$$

We have a canonical inclusion

$$P \xleftarrow{\;p^*\;} P_{\text{basic}} \;. (4.45)$$

Diagrammatically the above condition says that

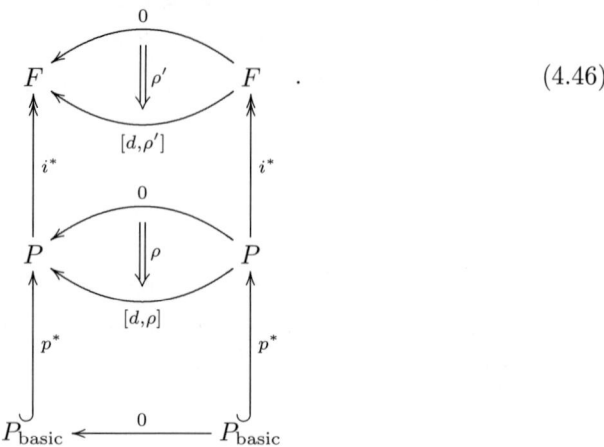

(4.46)

4.3.1. Examples.

Basic forms on a bundle. As a special case of the above general definition, we reobtain the standard notion of basic differential forms on a smooth surjective submersion $\pi : Y \to X$ with connected fibers.

Definition 4.11. *Let $\pi : Y \to X$ be a smooth map. The **vertical de Rham complex** , $\Omega^\bullet_{\mathrm{vert}}(Y)$, with respect to Y is the de Rham complex of Y modulo those forms that vanish when restricted in all arguments to vector fields in the kernel of $\pi_* : \Gamma(TY) \to \Gamma(TX)$, namely to vertical vector fields.*

The induced differential on $\Omega^\bullet_{\mathrm{vert}}(Y)$ sends $\omega_{\mathrm{vert}} = i^*\omega$ to

$$d_{\mathrm{vert}} : i^*\omega \mapsto i^*d\omega . \qquad (4.47)$$

Proposition 4.12. *This is well defined. The quotient $\Omega^\bullet_{\mathrm{vert}}(Y)$ with the differential induced from $\Omega^\bullet(Y)$ is indeed a dg-algebra, and the projection*

$$\Omega^\bullet_{\mathrm{vert}}(Y) \xleftarrow{\quad i^* \quad} \Omega^\bullet(Y) \qquad (4.48)$$

is a homomorphism of dg-algebras (in that it does respect the differential).

Proof. Notice that if $\omega \in \Omega^\bullet(Y)$ vanishes when evaluated on vertical vector fields then obviously so does $\alpha \wedge \omega$, for any $\alpha \in \Omega^\bullet(Y)$. Moreover, due to the formula

$$d\omega(v_1, \cdots , v_{n+1}) \;=\; \sum_{\sigma \in \mathrm{Sh}(1,n+1)} \pm v_{\sigma_1}\omega(v_{\sigma_2}, \cdots , v_{\sigma_{n+1}})$$

$$+ \sum_{\sigma \in \mathrm{Sh}(2,n+1)} \pm \omega([v_{\sigma_1}, v_{\sigma_2}], v_{\sigma_3}, \cdots , v_{\sigma_{n+1}}) \quad (4.49)$$

and the fact that for v, w vertical so is $[v, w]$ and hence $d\omega$ is also vertical. This gives that vertical differential forms on Y form a dg-subalgebra of the algebra of all forms on Y. Therefore if $i^*\omega = i^*\omega'$ then

$$di^*\omega' = i^*d\omega' = i^*d(\omega + (\omega' - \omega)) = i^*d\omega + 0 = di^*\omega. \qquad (4.50)$$

Hence the differential is well defined and i^* is then, by construction, a morphism of dg-algebras.

□

Recall the following standard definition of basic differential forms.

Definition 4.13 (Basic forms). *Given a surjective submersion $\pi : Y \to X$, the basic forms on Y are those with the property that they and their differentials are annihilated by all vertical vector fields*

$$\omega \in \Omega^\bullet(Y)_{\text{basic}} \quad \Leftrightarrow \quad \forall v \in \ker(\pi) : \iota_v\omega = \iota_v d\omega = 0. \qquad (4.51)$$

It is a standard result that

Proposition 4.14. *If $\pi : Y \to X$ is locally trivial and has connected fibers, then the basic forms are precisely those coming from pullback along π*

$$\Omega^\bullet(Y)_{\text{basic}} \simeq \Omega^\bullet(X). \qquad (4.52)$$

Remark. The reader should compare this situation with the definition of invariant polynomials in section 5.3.

The next proposition asserts that these statements about ordinary basic differential forms are indeed a special case of the general definition of basic elements with respect to a surjection of DGCAs, definition 4.10.

Proposition 4.15. *Given a surjective submersion $\pi : Y \to X$ with connected fibers, then*

- *the inner derivations of $\Omega^\bullet(Y)$ which are vertical with respect to*

  $\Omega^\bullet_{\text{vert}}(Y) \xleftarrow{\ i^*\ } \Omega^\bullet(Y)$ *according to the general definition 4.9, come precisely from contractions ι_v with vertical vector fields $v \in \ker(\pi_*) \subset \Gamma(TY)$;*
- *the basic differential forms on Y according to definition 4.13 coincide with the basic elements of $\Omega^\bullet(Y)$ relative to the above surjection*

  $$\Omega^\bullet_{\text{vert}}(Y) \xleftarrow{\ i^*\ } \Omega^\bullet(Y) \qquad (4.53)$$

  according to the general definition 4.10.

Proof. We first show that if $\Omega^\bullet(Y) \xleftarrow{\ \rho\ } \Omega^\bullet(Y)$ is a vertical algebra derivation, then ρ has to annihilate all forms in the image of π_*. Let $\alpha \in \Omega^\bullet(Y)$ be any

1-form and $\omega = \pi^*\beta$ for $\beta \in \Omega^1(X)$. Then the wedge product $\alpha \wedge \omega$ is annihilated by the projection to $\Omega^\bullet_{\mathrm{vert}}(Y)$ and we find

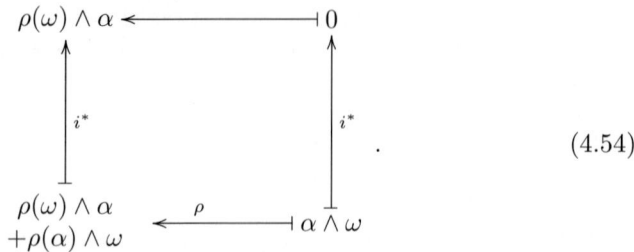

$$(4.54)$$

We see that $\rho(\omega) \wedge \alpha$ has to vanish for all α. Therefore $\rho(\omega)$ has to vanish for all ω pulled back along π^*. Hence ρ must be contraction with a vertical vector field. It then follows from the condition (4.40) that a basic form is one annihilated by all such ρ and all such $\rho \circ d$.   □

Possibly the most familiar kinds of surjective submersions are

- Fiber bundles.

  Indeed, the standard Cartan-Ehresmann theory of connections of principal bundles is obtained in our context by fixing a Lie group G and a principal G-bundle $p : P \to X$ and then using $Y = P$ itself as the surjective submersion. The definition of a connection on P in terms of a $\mathfrak{g}$-valued 1-form on P can be understood as the descent data for a connection on P obtained with respect to canonical trivialization of the pullback of P to $Y = P$. Using for the surjective submersion Y a principal G-bundle $P \to X$ is also most convenient for studying all kinds of higher n-bundles obstructing lifts of the given G-bundle. This is why we will often make use of this choice in the following.

- Covers by open subsets.

  The disjoint union of all sets in a cover of X by open subsets of X forms a surjective submersion $\pi : Y \to X$. In large parts of the literature on descent (locally trivialized bundles), these are the only kinds of surjective submersions that are considered. We will find here, that in order to characterize principal n-bundles entirely in terms of L_∞-algebraic data, open covers are too restrictive and the full generality of surjective submersions is needed. The reason is that, for $\pi : Y \to X$ a cover by open subsets, there are no nontrivial vertical vector fields

$$\ker(\pi) = 0, \qquad\qquad (4.55)$$

hence

$$\Omega^\bullet_{\mathrm{vert}}(Y) = 0 . \qquad\qquad (4.56)$$

With the definition of $\mathfrak{g}$-descent objects in section 6.1 this implies that all $\mathfrak{g}$-descent objects over a cover by open subsets are trivial.

There are two important subclasses of surjective submersions $\pi : Y \to X$:

- those for which Y is (smoothly) contractible;
- those for which the fibers of Y are connected.

We say Y is (smoothly) contractible if the identity map $\mathrm{Id} : Y \to Y$ is (smoothly) homotopic to a map $Y \to Y$ which is constant on each connected component. Hence Y is a disjoint union of spaces that are each (smoothly) contractible to a point. In this case the Poincaré lemma says that the dg-algebra $\Omega^\bullet(Y)$ of differential forms on Y is contractible; each closed form is exact:

$$
\Omega^\bullet(Y) \underset{[d,\tau]}{\overset{0}{\underset{\tau}{\rightleftarrows}}} \Omega^\bullet(Y) \ . \tag{4.57}
$$

Here τ is the familiar homotopy operator that appears in the proof of the Poincaré lemma. In practice, we often make use of the best of both worlds: surjective submersions that are (smoothly) contractible to a discrete set but still have a sufficiently rich collection of vertical vector fields. The way to obtain these is by **refinement**: starting with any surjective submersion $\pi : Y \to X$ which has good vertical vector fields but might not be contractible, we can cover Y itself with open balls, whose disjoint union, Y', then forms a surjective submersion $Y' \to Y$ over Y. The composite π'

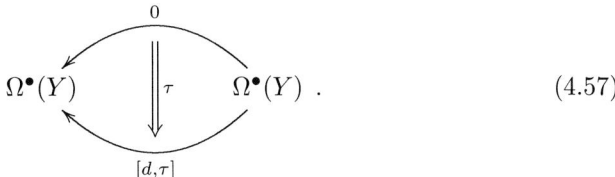

$$\tag{4.58}$$

is then a contractible surjective submersion of X. We will see that all our descent objects can be pulled back along refinements of surjective submersions this way, so that it is possible, without restriction of generality, to always work on contractible surjective submersions. Notice that for these the structure of

$$
\Omega^\bullet_{\mathrm{vert}}(Y) \twoheadleftarrow \Omega^\bullet(Y) \leftarrow \Omega^\bullet(X) \tag{4.59}
$$

is rather similar to that of

$$
\mathrm{CE}(\mathfrak{g}) \twoheadleftarrow \mathrm{W}(\mathfrak{g}) \leftarrow \mathrm{inv}(\mathfrak{g}) \ , \tag{4.60}
$$

since $\mathrm{W}(\mathfrak{g})$ is also contractible, according to proposition 5.7.

Vertical derivations on universal $\mathfrak{g}$-bundles. The other important example of vertical flows, those on DGCAs modeling universal $\mathfrak{g}$-bundle for $\mathfrak{g}$ an L_∞-algebra, is discussed at the beginning of section 5.3.

5. L_∞-algebras and their String-like extensions

L_∞-algebras are a generalization of Lie algebras, where the Jacobi identity is demanded to hold only up to higher coherent equivalence, as the category theorist would say, or "strongly homotopic", as the homotopy theorist would say.

5.1. L_∞-algebras

Definition 5.1. *Given a graded vector space V, the tensor space $T^\bullet(V) := \bigoplus_{n=0} V^{\otimes n}$ with V^0 being the ground field. We will denote by $T^a(V)$ the* tensor algebra *with the concatenation product on $T^\bullet(V)$:*

$$x_1 \otimes x_2 \otimes \cdots \otimes x_p \bigotimes x_{p+1} \otimes \cdots \otimes x_n \mapsto x_1 \otimes x_2 \otimes \cdots \otimes x_n \qquad (5.1)$$

and by $T^c(V)$ the tensor coalgebra *with the deconcatenation product on $T^\bullet(V)$:*

$$x_1 \otimes x_2 \otimes \cdots \otimes x_n \mapsto \sum_{p+q=n} x_1 \otimes x_2 \otimes \cdots \otimes x_p \bigotimes x_{p+1} \otimes \cdots \otimes x_n. \qquad (5.2)$$

The graded symmetric algebra $\wedge^\bullet(V)$ *is the quotient of the tensor algebra $T^a(V)$ by the graded action of the symmetric groups $\mathbf{S}_n$ on the components $V^{\otimes n}$. The* graded symmetric coalgebra $\vee^\bullet(V)$ *is the sub-coalgebra of the tensor coalgebra $T^c(V)$ fixed by the graded action of the symmetric groups $\mathbf{S}_n$ on the components $V^{\otimes n}$.*

Remark. $\vee^\bullet(V)$ is spanned by graded symmetric tensors

$$x_1 \vee x_2 \vee \cdots \vee x_p \qquad (5.3)$$

for $x_i \in V$ and $p \geq 0$, where we use $\vee$ rather than $\wedge$ to emphasize the coalgebra aspect, e.g.

$$x \vee y = x \otimes y \pm y \otimes x. \qquad (5.4)$$

     In characteristic zero, the graded symmetric algebra can be identified with a sub-algebra of $T^a(V)$ but that is unnatural and we will try to avoid doing so. The coproduct on $\vee^\bullet(V)$ is given by

$$\Delta(x_1 \vee x_2 \cdots \vee x_n) = \sum_{p+q=n} \sum_{\sigma \in Sh(p,q)} \epsilon(\sigma)(x_{\sigma(1)} \vee x_{\sigma(2)} \cdots x_{\sigma(p)}) \otimes (x_{\sigma(p+1)} \vee \cdots x_{\sigma(n)}).$$
$$(5.5)$$

The notation here means the following:

- $Sh(p,q)$ is the subset of all those bijections (the "unshuffles") of $\{1, 2, \cdots, p+q\}$ that have the property that $\sigma(i) < \sigma(i+1)$ whenever $i \neq p$;
- $\epsilon(\sigma)$, which is shorthand for $\epsilon(\sigma, x_1 \vee x_2, \cdots x_{p+q})$, the Koszul sign, defined by

$$x_1 \vee \cdots \vee x_n = \epsilon(\sigma)x_{\sigma(1)} \vee \cdots x_{\sigma(n)}. \qquad (5.6)$$

Definition 5.2 (L_∞-algebra). *An L_∞-algebra $\mathfrak{g} = (\mathfrak{g}, D)$ is a $\mathbb{N}_+$-graded vector space $\mathfrak{g}$ equipped with a degree -1 coderivation*

$$D : \vee^\bullet \mathfrak{g} \to \vee^\bullet \mathfrak{g} \qquad (5.7)$$

on the graded co-commutative coalgebra generated by $\mathfrak{g}$, such that $D^2 = 0$. This induces a differential

$$d_{\mathrm{CE}(\mathfrak{g})} : \mathrm{Sym}^\bullet(\mathfrak{g}) \to \mathrm{Sym}^{\bullet+1}(\mathfrak{g}) \tag{5.8}$$

on graded-symmetric multilinear functions on $\mathfrak{g}$. When $\mathfrak{g}$ is finite dimensional this yields a degree $+1$ differential

$$d_{\mathrm{CE}(\mathfrak{g})} : \wedge^\bullet \mathfrak{g}^* \to \wedge^\bullet \mathfrak{g}^* \tag{5.9}$$

on the graded-commutative algebra generated from $\mathfrak{g}^*$. This is the Chevalley-Eilenberg dg-algebra corresponding to the L_∞-algebra $\mathfrak{g}$. If $\mathfrak{g}^*$ is concentrated in degree $1, \ldots, n$, we also say that $\mathfrak{g}$ is a **Lie n-algebra**.

Remark. That the original definition of L_∞-algebras in terms of multibrackets yields a codifferential coalgebra as above was shown in [57]. That every such codifferential comes from a collection of multibrackets this way is due to [58].

Example. For $(\mathfrak{g}[-1], [\cdot, \cdot])$ an ordinary Lie algebra (meaning that we regard the vector space $\mathfrak{g}$ to be in degree 1), the corresponding Chevalley-Eilenberg qDGCA is

$$\mathrm{CE}(\mathfrak{g}) = (\wedge^\bullet \mathfrak{g}^*, d_{\mathrm{CE}(\mathfrak{g})}) \tag{5.10}$$

with

$$d_{\mathrm{CE}(\mathfrak{g})} : \quad \mathfrak{g}^* \xrightarrow{\;[\cdot,\cdot]^*\;} \mathfrak{g}^* \wedge \mathfrak{g}^* \ . \tag{5.11}$$

If we let $\{t_a\}$ be a basis of $\mathfrak{g}$ and $\{C^a{}_{bc}\}$ the corresponding structure constants of the Lie bracket $[\cdot, \cdot]$, and if we denote by $\{t^a\}$ the corresponding basis of $\mathfrak{g}^*$, then we get

$$d_{\mathrm{CE}(\mathfrak{g})} t^a = -\frac{1}{2} C^a{}_{bc} t^b \wedge t^c \ . \tag{5.12}$$

Notice that built in we have a shift of degree for convenience, which makes ordinary Lie 1-algebras be in degree 1 already. In much of the literature a Lie n-algebra would be based on a vector space concentrated in degrees 0 to $n-1$. An ordinary Lie algebra is a Lie 1-algebra. Here the coderivation differential $D = [\cdot, \cdot]$ is just the Lie bracket, extended as a coderivation to $\vee^\bullet \mathfrak{g}$, with $\mathfrak{g}$ regarded as being in degree 1.

In the rest of the paper we assume, just for simplicity and since it is sufficient for our applications, all $\mathfrak{g}$ to be finite-dimensional. Then, by the above, these L_∞-algebras are equivalently conceived of in terms of their dual Chevalley-Eilenberg algebras, $\mathrm{CE}(\mathfrak{g})$, as indeed every quasi-free differential graded commutative algebra ("qDGCA", meaning that it is free as a graded commutative algebra) corresponds to an L_∞-algebra. We will find it convenient to work entirely in terms of qDGCAs, which we will usually denote as $\mathrm{CE}(\mathfrak{g})$.

While not very interesting in themselves, truly free differential algebras are a useful tool for handling quasi-free differential algebras.

Definition 5.3. *We say a qDGCA is* free *(even as a differential algebra) if it is of the form*

$$F(V) := (\wedge^\bullet (V^* \oplus V^*[1]), d_{F(V)}) \tag{5.13}$$

with

$$d_{F(V)}|_{V^*} = \sigma : V^* \to V^*[1] \tag{5.14}$$

the canonical isomorphism and

$$d_{F(V)}|_{V^*[1]} = 0. \tag{5.15}$$

Remark. Such algebras are indeed free in that they satisfy the universal property: given any linear map $V \to W$, it uniquely extends to a morphism of qDGCAs $F(V) \to (\wedge^\bullet (W^*), d)$ for any choice of differential d.

Example. The free qDGCA on a 1-dimensional vector space in degree 0 is the graded commutative algebra freely generated by two generators, t of degree 0 and dt of degree 1, with the differential acting as $d : t \mapsto dt$ and $d : dt \mapsto 0$. In rational homotopy theory, this models the interval $I = [0, 1]$. The fact that the qDGCA is free corresponds to the fact that the interval is homotopy equivalent to the point.

We will be interested in qDGCAs that arise as mapping cones of morphisms of L_∞-algebras.

Definition 5.4 ("Mapping cone" of qDGCAs). *Let*

$$\mathrm{CE}(\mathfrak{h}) \xleftarrow{\;t^*\;} \mathrm{CE}(\mathfrak{g}) \tag{5.16}$$

be a morphism of qDGCAs. The mapping cone *of t^*, which we write $\mathrm{CE}(\mathfrak{h} \xrightarrow{t} g)$, is the qDGCA whose underlying graded algebra is*

$$\wedge^\bullet (\mathfrak{g}^* \oplus \mathfrak{h}^*[1]) \tag{5.17}$$

and whose differential d_{t^*} is such that it acts as

$$d_{t^*} = \begin{pmatrix} d_\mathfrak{g} & 0 \\ t^* & d_\mathfrak{h} \end{pmatrix}. \tag{5.18}$$

Here and in the following, we write morphisms of qDGCAs with a star to emphasize that they are dual to morphisms of the corresponding L_∞-algebras. We postpone a more detailed definition and discussion to section 7.1; see definition 7.3 and proposition 7.4. Strictly speaking, the more usual notion of mapping cones of chain complexes applies to $t : \mathfrak{h} \to \mathfrak{g}$, but then is extended as a derivation differential to the entire qDGCA.

Definition 5.5 (Weil algebra of an L_∞-algebra). *The mapping cone of the identity on $\mathrm{CE}(\mathfrak{g})$ is the* Weil algebra

$$W(\mathfrak{g}) := \mathrm{CE}(\mathfrak{g} \xrightarrow{\mathrm{Id}} \mathfrak{g}) \tag{5.19}$$

of $\mathfrak{g}$.

Proposition 5.6. *For $\mathfrak{g}$ an ordinary Lie algebra this does coincide with the ordinary Weil algebra of $\mathfrak{g}$.*

Proof. See the example in section 5.1.1. □

The Weil algebra has two important properties.

Proposition 5.7. *The Weil algebra $W(\mathfrak{g})$ of any L_∞-algebra $\mathfrak{g}$*
- *is isomorphic to a free differential algebra*

$$W(\mathfrak{g}) \simeq F(\mathfrak{g}), \qquad (5.20)$$

and hence is contractible;
- *has a canonical surjection*

$$CE(\mathfrak{g}) \xleftarrow{\;i^*\;} W(\mathfrak{g}) . \qquad (5.21)$$

Proof. Define a morphism

$$f : F(\mathfrak{g}) \to W(\mathfrak{g}) \qquad (5.22)$$

by setting

$$f \;:\; a \mapsto a \qquad (5.23)$$
$$f \;:\; (d_{F(V)}a = \sigma a) \mapsto (d_{W(\mathfrak{g})}a = d_{CE(\mathfrak{g})}a + \sigma a) \qquad (5.24)$$

for all $a \in \mathfrak{g}^*$ and extend as an algebra homomorphism. This clearly respects the differentials: for all $a \in V^*$ we find

$$
\begin{array}{ccc}
a \xrightarrow{\;d_{F(\mathfrak{g})}\;} \sigma a & & \sigma a \xrightarrow{\;d_{F(\mathfrak{g})}\;} 0 \\
\downarrow{\scriptstyle f} \qquad \downarrow{\scriptstyle f} & \text{and} & \downarrow{\scriptstyle f} \qquad \downarrow{\scriptstyle f} \\
a \xrightarrow[\;d_{W(\mathfrak{g})}\;]{} d_{CE(\mathfrak{g})}a + \sigma a & & d_{W(\mathfrak{g})}a \xrightarrow[\;d_{W(\mathfrak{g})}\;]{} 0
\end{array}
\qquad (5.25)
$$

One checks that the strict inverse exists and is given by

$$f^{-1}|_{\mathfrak{g}^*} \;:\; a \mapsto a \qquad (5.26)$$
$$f^{-1}|_{\mathfrak{g}^*[1]} \;:\; \sigma a \mapsto d_{F(\mathfrak{g})}a - d_{CE(\mathfrak{g})}a . \qquad (5.27)$$

Here $\sigma : \mathfrak{g}^* \to \mathfrak{g}^*[1]$ is the canonical isomorphism that shifts the degree. The surjection $CE(\mathfrak{g}) \xleftarrow{\;i^*\;} W(\mathfrak{g})$ simply projects out all elements in the shifted copy of $\mathfrak{g}$:

$$i^*|_{\wedge^\bullet \mathfrak{g}^*} \;=\; \text{id} \qquad (5.28)$$
$$i^*|_{\mathfrak{g}^*[1]} \;=\; 0 . \qquad (5.29)$$

This is an algebra homomorphism that respects the differential. □

As a corollary we obtain

Corollary 5.8. *For $\mathfrak{g}$ any L_∞-algebra, the cohomology of $W(\mathfrak{g})$ is trivial.*

Proposition 5.9. *The step from a Chevalley-Eilenberg algebra to the corresponding Weil algebra is functorial: for any morphism*

$$\mathrm{CE}(\mathfrak{h}) \xleftarrow{\;\;f^*\;\;} \mathrm{CE}(\mathfrak{g}) \tag{5.30}$$

we obtain a morphism

$$\mathrm{W}(\mathfrak{h}) \xleftarrow{\;\;\hat{f}^*\;\;} \mathrm{W}(\mathfrak{g}) \tag{5.31}$$

and this respects composition.

Proof. The morphism $\hat{f}^*$ acts as for all generators $a \in \mathfrak{g}^*$ as

$$\hat{f}^* : a \mapsto f^*(a) \tag{5.32}$$

and

$$\hat{f}^* : \sigma a \mapsto \sigma f^*(a) \,. \tag{5.33}$$

We check that this does respect the differentials

$$
\begin{array}{ccc}
a \xrightarrow{\;d_{\mathrm{W}(\mathfrak{g})}\;} d_{\mathrm{CE}(\mathfrak{g})}a + \sigma a & \qquad & \sigma a \xrightarrow{\;d_{\mathrm{W}(\mathfrak{g})}\;} -\sigma(d_{\mathrm{CE}(\mathfrak{g})}a) \\
\downarrow{\scriptstyle \hat{f}^*} \qquad\qquad \downarrow{\scriptstyle \hat{f}^*} & & \downarrow{\scriptstyle \hat{f}^*} \qquad\qquad \downarrow{\scriptstyle \hat{f}^*} \\
f^*(a) \xrightarrow{\;d_{\mathrm{W}(\mathfrak{h})}\;} d_{\mathrm{CE}(\mathfrak{h})}f^*(a) + \sigma f^*(a) & & \sigma f^*(a) \xrightarrow{\;d_{\mathrm{W}(\mathfrak{h})}\;} -\sigma(d_{\mathrm{CE}(\mathfrak{h})}a)
\end{array}
\tag{5.34}
$$

$\square$

Remark. As we will shortly see, $\mathrm{W}(\mathfrak{g})$ plays the role of the algebra of differential forms on the universal $\mathfrak{g}$-bundle. The surjection $CE(\mathfrak{g}) \xleftarrow{\;i^*\;} W(\mathfrak{g})$ plays the role of the restriction to the differential forms on the fiber of the universal $\mathfrak{g}$-bundle.

5.1.1. Examples. In section 5.4 we construct large families of examples of L_∞-algebras, based on the first two of the following examples:

1. Ordinary Weil algebras as Lie 2-algebras. What is ordinarily called the Weil algebra $\mathrm{W}(\mathfrak{g})$ of a Lie algebra $(\mathfrak{g}[-1], [\cdot, \cdot])$ can, since it is again a DGCA, also be interpreted as the Chevalley-Eilenberg algebra of a Lie 2-algebra. This Lie 2-algebra we call $\mathrm{inn}(\mathfrak{g})$. It corresponds to the Lie 2-group $\mathrm{INN}(G)$ discussed in [65]:

$$\mathrm{W}(\mathfrak{g}) = \mathrm{CE}(\mathrm{inn}(\mathfrak{g})) \,. \tag{5.35}$$

We have

$$\mathrm{W}(\mathfrak{g}) = (\wedge^\bullet(\mathfrak{g}^* \oplus \mathfrak{g}^*[1]), d_{\mathrm{W}(\mathfrak{g})}) \,. \tag{5.36}$$

Denoting by $\sigma : \mathfrak{g}^* \to \mathfrak{g}^*[1]$ the canonical isomorphism, extended as a derivation to all of $\mathrm{W}(\mathfrak{g})$, we have

$$d_{\mathrm{W}(\mathfrak{g})} : \mathfrak{g}^* \xrightarrow{\;\;[\cdot,\cdot]^*+\sigma\;\;} \mathfrak{g}^* \wedge \mathfrak{g}^* \oplus \mathfrak{g}^*[1] \tag{5.37}$$

and

$$dW_{(\mathfrak{g})} : \; \mathfrak{g}^*[1] \xrightarrow{\;-\sigma \circ d_{CE(\mathfrak{g})} \circ \sigma^{-1}\;} \mathfrak{g}^* \otimes \mathfrak{g}^*[1] \; . \tag{5.38}$$

With $\{t^a\}$ a basis for $\mathfrak{g}^*$ as above, and $\{\sigma t^a\}$ the corresponding basis of $\mathfrak{g}^*[1]$ we find

$$dW_{(\mathfrak{g})} : t^a \mapsto -\frac{1}{2}C^a{}_{bc}t^b \wedge t^c + \sigma t^a \tag{5.39}$$

and

$$dW_{(\mathfrak{g})} : \sigma t^a \mapsto -C^a{}_{bc}t^b \sigma t^c \; . \tag{5.40}$$

The Lie 2-algebra $\mathrm{inn}(\mathfrak{g})$ is, in turn, nothing but the strict Lie 2-algebra as in the third example below, which comes from the infinitesimal crossed module $(\mathfrak{g} \xrightarrow{\mathrm{Id}} \mathfrak{g} \xrightarrow{\mathrm{ad}} \mathrm{der}(\mathfrak{g}))$.

2. Shifted $\mathfrak{u}(1)$. By the above, the qDGCA corresponding to the Lie algebra $\mathfrak{u}(1)$ is simply

$$CE(\mathfrak{u}(1)) = (\wedge^\bullet \mathbb{R}[1], d_{CE(\mathfrak{u}(1))} = 0) \; . \tag{5.41}$$

We write

$$CE(b^{n-1}\mathfrak{u}(1)) = (\wedge^\bullet \mathbb{R}[n], d_{CE(b^{n-1}\mathfrak{u}(1))} = 0) \tag{5.42}$$

for the Chevalley-Eilenberg algebras corresponding to the Lie n-algebras $b^{n-1}\mathfrak{u}(1)$.

3. Infinitesimal crossed modules and strict Lie 2-algebras. An *infinitesimal crossed module* is a diagram

$$(\; \mathfrak{h} \xrightarrow{\;t\;} \mathfrak{g} \xrightarrow{\;\alpha\;} \mathrm{der}(\mathfrak{h}) \;) \tag{5.43}$$

of Lie algebras where t and α satisfy two compatibility conditions. These conditions are equivalent to the nilpotency of the differential on

$$CE(\mathfrak{h} \xrightarrow{\;t\;} \mathfrak{g}) := (\wedge^\bullet(\mathfrak{g}^* \oplus \mathfrak{h}^*[1]), d_t) \tag{5.44}$$

defined by

$$d_t|_{\mathfrak{g}^*} \;=\; [\cdot,\cdot]^*_{\mathfrak{g}} + t^* \tag{5.45}$$

$$d_t|_{\mathfrak{h}^*[1]} \;=\; \alpha^* \; , \tag{5.46}$$

where we consider the vector spaces underlying both $\mathfrak{g}$ and $\mathfrak{h}$ to be in degree 1. Here in the last line we regard α as a linear map $\alpha : \mathfrak{g} \otimes \mathfrak{h} \to \mathfrak{h}$. The Lie 2-algebras $(\mathfrak{h} \xrightarrow{\;t\;} \mathfrak{g})$ thus defined are called strict Lie 2-algebras: these are precisely those Lie 2-algebras whose Chevalley-Eilenberg differential contains at most co-binary components. As described later in section 7.1, in the case that t is *normal* this is a special case of the mapping cone construction.

4. Inner derivation L_∞-algebras. In straightforward generalization of the first example we find: for $\mathfrak{g}$ any L_∞-algebra, its Weil algebra $W(\mathfrak{g})$ is again a DGCA, hence the Chevalley-Eilenberg algebra of some other L_∞-algebra. This we address as the L_∞-algebra of inner derivations and write

$$CE(\mathrm{inn}(\mathfrak{g})) := W(\mathfrak{g}) \; . \tag{5.47}$$

This identification is actually useful for identifying the Lie ∞-groups that correspond to an integrated picture underlying our differential discussion. In [65] the Lie 3-group corresponding to $\mathrm{inn}(\mathfrak{g})$ for $\mathfrak{g}$ the strict Lie 2-algebra of any strict Lie 2-group is discussed. This 3-group is in particular the right codomain for incorporating the the non-fake flat nonabelian gerbes with connection considered in [17] into the integrated version of the picture discussed here. This is indicated in [74] and should be discussed elsewhere.

tangent category	inner automorphism $(n+1)$-group	inner derivation Lie $(n+1)$-algebra	Weil algebra	shifted tangent bundle

$$\mathrm{CE}(\mathrm{Lie}(T\mathbf{B}G)) = \mathrm{CE}(\mathrm{Lie}(\mathrm{INN}(G))) =\!\!= \mathrm{CE}(\mathrm{inn}(\mathfrak{g})) =\!\!= \mathrm{W}(\mathfrak{g}) = C^\infty(T[1]\mathfrak{g})$$

FIGURE 3. **A remarkable coincidence of concepts** relates the notion of tangency to the notion of universal bundles. The leftmost equality is discussed in [65]. The second one from the right is the identification 5.47. The rightmost equality is equation (5.59).

Proposition 5.10. *For $\mathfrak{g}$ any finite dimensional L_∞-algebra, the differential forms on the smooth space of morphisms from the Chevalley-Eilenberg algebra $\mathrm{CE}(\mathfrak{g})$ to the algebra of "functions on the superpoint", definition 4.7, i.e. the elements in* $\mathrm{maps}(\mathrm{CE}(\mathfrak{g}), C(\mathbf{pt}))$, *which come from currents as in definition 4.5, form the Weil algebra $\mathrm{W}(\mathfrak{g})$ of $\mathfrak{g}$:*

$$\mathrm{W}(\mathfrak{g}) \subset \mathrm{maps}(\mathrm{CE}(\mathfrak{g}), C(\mathbf{pt})). \tag{5.48}$$

Proof. For any test domain U, an element in $\mathrm{Hom}_{\mathrm{DGCAs}}(\mathrm{CE}(\mathfrak{g}), C(\mathbf{pt}) \otimes \Omega^\bullet(U))$ is specified by a degree 0 algebra homomorphism

$$\lambda : \mathrm{CE}(\mathfrak{g}) \to \Omega^\bullet(U) \tag{5.49}$$

and a degree $+1$ map

$$\omega : \mathrm{CE}(\mathfrak{g}) \to \Omega^\bullet(U) \tag{5.50}$$

by

$$a \mapsto \lambda(a) + c \wedge \omega(a) \tag{5.51}$$

for all $a \in \mathfrak{g}^*$ and for c denoting the canonical degree -1 generator of $C(\mathbf{pt})$; such that the equality in the bottom right corner of the diagram

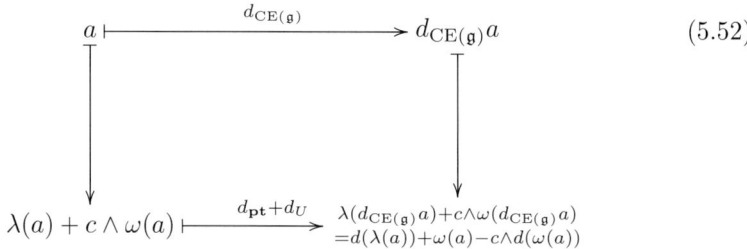

$$(5.52)$$

holds. Under the two canonical currents on $C(\mathbf{pt})$ of degree 0 and degree 1, respectively, this gives rise for each $a \in \mathfrak{g}^*$ of degree $|a|$ to an $|a|$-form and an $(|a|+1)$ form on $\mathrm{maps}(\mathrm{CE}(\mathfrak{g}), C(\mathbf{pt}))$ whose values on a given plot are $\lambda(a)$ and $\omega(a)$, respectively.

By the above diagram, the differential of these forms satisfies

$$d\lambda(a) = \lambda(d_{\mathrm{CE}(\mathfrak{g})}a) - \omega(a) \qquad (5.53)$$

and

$$d\omega(a) = -\omega(d_{\mathrm{CE}(\mathfrak{g})}a) . \qquad (5.54)$$

But this is precisely the structure of $\mathrm{W}(\mathfrak{g})$. $\square$

To see the last step, it may be helpful to consider this for a simple case in terms of a basis:

let $\mathfrak{g}$ be an ordinary Lie algebra, $\{t^a\}$ a basis of $\mathfrak{g}^*$ and $\{C^a{}_{bc}\}$ the corresponding structure constants. Then, using the fact that, since we are dealing with algebra homomorphisms, we have

$$\lambda(t^a \wedge t^b) = \lambda(t^a) \wedge \lambda(t^b) \qquad (5.55)$$

and

$$\omega(t^a \wedge t^b) = c \wedge (\omega(t^a) \wedge \lambda(t^b) - \lambda(t^a) \wedge \omega(t^b)) \qquad (5.56)$$

we find

$$d\lambda(t^a) = -\frac{1}{2}C^a{}_{bc}\lambda(t^b) \wedge \lambda(t^c) - \omega(t^a) \qquad (5.57)$$

and

$$d\omega(t^a) = -C^a{}_{bc}\lambda(t^b) \wedge \omega(t^c) . \qquad (5.58)$$

This is clearly just the structure of $\mathrm{W}(\mathfrak{g})$.

Remark. As usual, we may think of the superpoint as an "infinitesimal interval". The above says that the algebra of inner derivations of an L_∞-algebra consists of the maps from the infinitesimal interval to the supermanifold on which $\mathrm{CE}(\mathfrak{g})$ is the "algebra of functions". On the one hand this tells us that

$$\mathrm{W}(\mathfrak{g}) = C^\infty(T[1]\mathfrak{g}) \qquad (5.59)$$

in supermanifold language. On the other hand, this construction is clearly analogous to the corresponding discussion for Lie n-groups given in [65]: there the 3-group $\mathrm{INN}(G)$ of inner automorphisms of a strict 2-group G was obtained by mapping the "fat point" groupoid $\mathbf{pt} = \{\, \bullet \longrightarrow \circ \,\}$ into G. As indicated there, this is a special case of a construction of "tangent categories" which mimics the relation between $\mathrm{inn}(\mathfrak{g})$ and the shifted tangent bundle $T[1]\mathfrak{g}$ in the integrated world of Lie ∞-groups. This relation between these concepts is summarized in figure 3 (see page 336).

5.2. L_∞-algebra homotopy and concordance

Like cochain complexes, differential graded algebras can be thought of as being objects in a higher categorical structure, which manifests itself in the fact that there are not only morphisms between DGCAs, but also higher morphisms between these morphisms. It turns out that we need to consider a couple of slightly differing notions of morphisms and higher morphisms for these. While differing, these concepts are closely related among each other, as we shall discuss.

In section 4.2 we had already considered 2-morphisms of DGCAs obtained after forgetting their algebra structure and just remembering their differential structure. The 2-morphisms we present now crucially do know about the algebra structure.

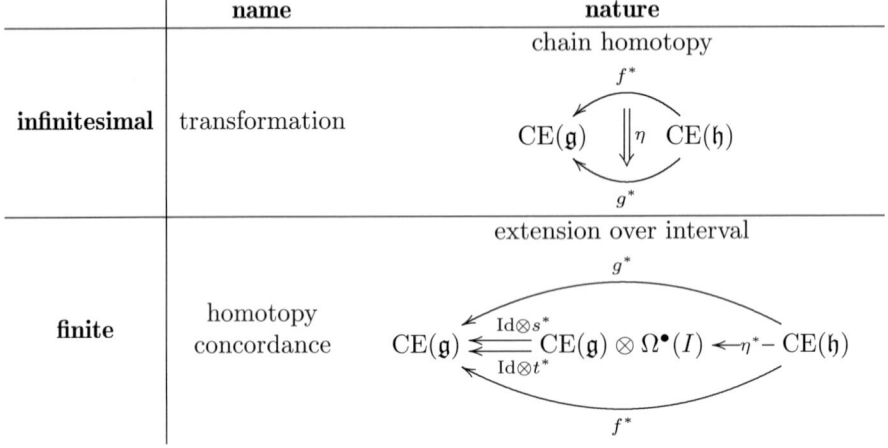

	name	nature
infinitesimal	transformation	chain homotopy
finite	homotopy concordance	extension over interval

TABLE 2. The two different notions of **higher morphisms** of qDGCAs.

Infinitesimal homotopies between dg-algebra homomorphisms. When we restrict attention to cochain maps between qDGCAs which respect not only the differentials but also the free graded commutative algebra structure, i.e. to qDGCA homomorphisms, it becomes of interest to express the cochain homotopies in terms of their action on generators of the algebra. We now define transformations (2-morphisms) between morphisms of qDGCAs by first defining them for the case when the domain is a Weil algebra, and then extending the definition to arbitrary qDGCAs.

Definition 5.11 (Transformation of morphisms of L_∞-algebras). *We define trans-formations between qDGCA morphisms in two steps*

- *A 2-morphism*

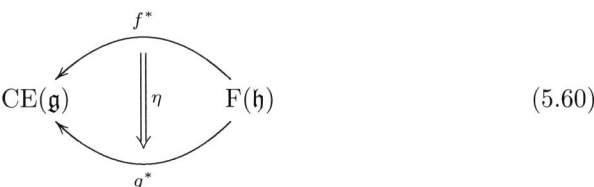

$$\hspace{8cm}(5.60)$$

  is defined by a degree -1 map $\eta : \mathfrak{h}^* \oplus \mathfrak{h}^*[1] \to CE(\mathfrak{g})$ which is extended to a linear degree -1 map $\eta : \wedge^\bullet(\mathfrak{h}^* \oplus \mathfrak{h}^*[1]) \to CE(\mathfrak{g})$ by defining it on all monomials of generators by the formula

$$\eta : x_1 \wedge \cdots \wedge x_n \;\longmapsto\; \frac{1}{n!}\sum_\sigma \epsilon(\sigma)\sum_{k=1}^n (-1)^{\sum_{i=1}^{k-1}|x_{\sigma(i)}|} g^*(x_{\sigma(1)}\wedge\cdots\wedge x_{\sigma(k-1)})\wedge$$
$$\wedge \eta(x_{\sigma(k)}) \wedge f^*(x_{\sigma(k+1)}\wedge\cdots\wedge x_{\sigma(n)}) \hspace{2cm}(5.61)$$

  for all $x_1, \cdots, x_n \in \mathfrak{h}^* \oplus \mathfrak{h}^*[1]$, such that this is a chain homotopy from f^* to g^*:

$$g^* = f^* + [d, \eta]\,. \hspace{4cm}(5.62)$$

- *A general 2-morphism*

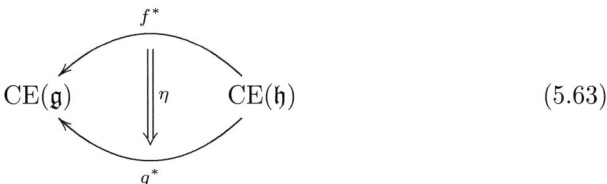

$$\hspace{8cm}(5.63)$$

  is a 2-morphism

$$\hspace{9cm}(5.64)$$

of the above kind that vanishes on the shifted generators, i.e. such that

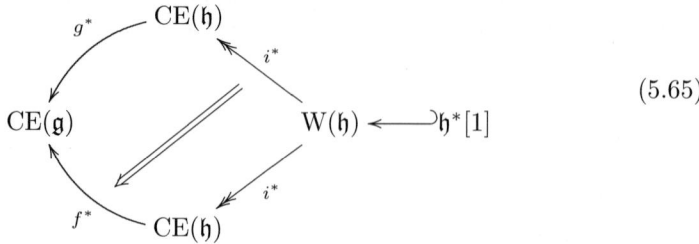

$$(5.65)$$

vanishes.

Proposition 5.12. *Formula 5.61 is consistent in that* $g^*|_{\mathfrak{h}^* \oplus \mathfrak{h}^*[1]} = (f^* + [d, \eta])|_{\mathfrak{h}^* \oplus \mathfrak{h}^*[1]}$ *implies that* $g^* = f^* + [d, \eta]$ *on all elements of* $F(\mathfrak{h})$.

Remark. Definition 5.11, which may look ad hoc at this point, has a practical and a deep conceptual motivation.

- **Practical motivation.** While it is clear that 2-morphisms of qDGCAs should be chain homotopies, it is not straightforward, in general, to characterize these by their action on generators. Except when the domain qDGCA is free, in which case our formula (5.11) makes sense. The prescription (5.64) then provides a systematic algorithm for extending this to arbitrary qDGCAs.

  In particular, using the isomorphism $W(\mathfrak{g}) \simeq F(\mathfrak{g})$ from proposition 5.7, the above yields the usual explicit description of the homotopy operator τ : $W(\mathfrak{g}) \to W(\mathfrak{g})$ with $\mathrm{Id}_{W(\mathfrak{g})} = [d_{W(\mathfrak{g})}, \tau]$. Among other things, this computes for us the transgression elements ("Chern-Simons elements") for L_∞-algebras in section 5.3.

- **Conceptual motivation.** As we will see in sections 5.3 and 5.5, the qDGCA $W(\mathfrak{g})$ plays an important twofold role: it is both the algebra of differential forms on the total space of the universal $\mathfrak{g}$-bundle – while $CE(\mathfrak{g})$ is that of forms on the fiber –, as well as the domain for $\mathfrak{g}$-valued differential forms, where the shifted component, that in $\mathfrak{g}^*[1]$, is the home of the corresponding curvature.

  In the light of this, the above restriction (5.65) can be understood as saying either that

  – *vertical* transformations induce transformations on the fibers;

  or

  – gauge transformations of $\mathfrak{g}$-valued forms are transformations under which the curvatures transform covariantly.

Finite transformations between qDGCA morphisms: concordances. We now consider the finite transformations of morphisms of DGCAs. What we called 2-morphisms or transformations for qDGCAs above would in other contexts possibly be called a homotopy. Also the following concept is a kind of homotopy, and appears as such in [75] which goes back to [14]. Here we wish to clearly distinguish these different kinds of homotopies and address the following concept as *concordance* – a finite notion of 2-morphism between dg-algebra morphisms.

Remark. In the following the algebra of forms $\Omega^\bullet(I)$ on the interval

$$I := [0, 1]$$

plays an important role. Essentially everything would also go through by instead using $F(\mathbb{R})$, the DGCA on a single degree 0 generator, which is the algebra of *polynomial* forms on the interval. This is the model used in [75].

Definition 5.13 (Concordance). *We say that two qDGCA morphisms*

$$\mathrm{CE}(\mathfrak{g}) \xleftarrow{\;\;g^*\;\;} \mathrm{CE}(\mathfrak{h}) \tag{5.66}$$

and

$$\mathrm{CE}(\mathfrak{g}) \xleftarrow{\;\;h^*\;\;} \mathrm{CE}(\mathfrak{h}) \tag{5.67}$$

are concordant, if there exists a dg-algebra homomorphism

$$\mathrm{CE}(\mathfrak{g}) \otimes \Omega^\bullet(I) \xleftarrow{\;\;\eta^*\;\;} \mathrm{CE}(\mathfrak{h}) \tag{5.68}$$

from the source $\mathrm{CE}(\mathfrak{h})$ *to the target* $\mathrm{CE}(\mathfrak{g})$ *tensored with forms on the interval, which restricts to the two given homomorphisms when pulled back along the two boundary inclusions*

$$\{\bullet\} \underset{t}{\overset{s}{\rightrightarrows}} I \ , \tag{5.69}$$

so that the diagram of dg-algebra morphisms

$$\mathrm{CE}(\mathfrak{g}) \underset{\mathrm{Id}\otimes t^*}{\overset{\mathrm{Id}\otimes s^*}{\rightrightarrows}} \mathrm{CE}(\mathfrak{g}) \otimes \Omega^\bullet(I) \xleftarrow{\;\;\eta^*\;\;} \mathrm{CE}(\mathfrak{h}) \tag{5.70}$$

commutes. See also table 2.

Notice that the above diagram is shorthand for two separate commuting diagrams, one involving g^* and s^*, the other involving f^* and t^*.

We can make precise the statement that definition 5.11 is the infinitesimal version of definition 5.13, as follows.

Proposition 5.14. *Concordances*

$$\mathrm{CE}(\mathfrak{g}) \otimes \Omega^\bullet(I) \xleftarrow{\;\;\eta^*\;\;} \mathrm{CE}(\mathfrak{h}) \tag{5.71}$$

are in bijection with 1-parameter families

$$\alpha : [0, 1] \to \mathrm{Hom}_{\mathrm{dg-Alg}}(\mathrm{CE}(\mathfrak{h}), \mathrm{CE}(\mathfrak{g})) \tag{5.72}$$

of morphisms whose derivatives with respect to the parameter is a chain homotopy, i.e. a 2-morphism

$$\forall t \in [0,1] \quad : \quad \mathrm{CE}(\mathfrak{g}) \quad \underset{\frac{d}{dt}\alpha(t)=[d,\rho]}{\overset{0}{\rightrightarrows}} \quad \mathrm{CE}(\mathfrak{h}) \tag{5.73}$$

in the 2-category of cochain complexes. For any such α, the morphisms f^* and g^* between which it defines a concordance are defined by the value of α on the boundary of the interval.

Proof. Writing $t : [0,1] \to \mathbb{R}$ for the canonical coordinate function on the interval $I = [0,1]$ we can decompose the dg-algebra homomorphism η^* as

$$\eta^* : \omega \mapsto (t \mapsto \alpha(\omega)(t) + dt \wedge \rho(\omega)(t)). \tag{5.74}$$

α is itself a degree 0 dg-algebra homomorphism, while ρ is degree -1 map. Then the fact that η^* respects the differentials implies that for all $\omega \in \mathrm{CE}(\mathfrak{h})$ we have

$$
\begin{array}{ccc}
\omega & \xrightarrow{\quad d_{\mathfrak{h}} \quad} & d_{\mathfrak{h}}\omega \\
\Big\downarrow{\eta^*} & & \Big\downarrow{\eta^*} \\
(t \mapsto (\alpha(\omega)(t) + dt \wedge \rho(\omega)(t))) & \xrightarrow{d_{\mathfrak{g}}+d_t} & \begin{array}{l} (t \mapsto (\alpha(d_{\mathfrak{h}}\omega)(t) + dt \wedge \rho(d_{\mathfrak{h}}\omega)(t)) \\ = (t \mapsto (d_{\mathfrak{g}}(\alpha(\omega))(t) \\ \quad + dt \wedge (\frac{d}{dt}\alpha(\omega) - d_{\mathfrak{g}}\rho(\omega))(t)) \end{array}
\end{array}
$$

$$\tag{5.75}$$

The equality in the bottom right corner says that

$$\alpha \circ d_{\mathfrak{h}} - d_{\mathfrak{g}} \circ \alpha = 0 \tag{5.76}$$

and

$$\forall \omega \in \mathrm{CE}(\mathfrak{g}) : \frac{d}{dt}\alpha(\omega) = \rho(d_{\mathfrak{h}}\omega) + d_{\mathfrak{g}}(\rho(\omega)). \tag{5.77}$$

But this means that α is a chain homomorphism whose derivative is given by a chain homotopy. $\square$

5.2.1. Examples.
Transformations between DGCA morphisms. We demonstrate two examples for the application of the notion of transformations of DGCA morphisms from definition 5.11 which are relevant for us.

Computation of transgression forms. As an example for the transformation in definition 5.11, we show how the usual Chern-Simons transgression form is computed using formula (5.61). The reader may wish to first skip to our discussion of Lie ∞-algebra cohomology in section 5.3 for more background. So let $\mathfrak{g}$ be an ordinary Lie algebra with invariant bilinear form P, which we regard as a $d_{\mathrm{W}(\mathfrak{g})}$-closed element $P \in \wedge^2\mathfrak{g}^*[1] \subset \mathrm{W}(\mathfrak{g})$. We would like to compute τP, where τ is the contracting homotopy of $\mathrm{W}(\mathfrak{g})$, such that

$$[d, \tau] = \mathrm{Id}_{\mathrm{W}(\mathfrak{g})}, \tag{5.78}$$

which according to proposition 5.7 is given on generators by

$$\tau \ : \quad a \mapsto 0 \tag{5.79}$$

$$\tau \ : \quad d_{\mathrm{W}(\mathfrak{g})}a \mapsto a \tag{5.80}$$

for all $a \in \mathfrak{g}^*$. Let $\{t^a\}$ be a chosen basis of $\mathfrak{g}^*$ and let $\{P_{ab}\}$ be the components of P in that basis, then

$$P = P_{ab}(\sigma t^a) \wedge (\sigma t^b). \tag{5.81}$$

In order to apply formula (5.61) we need to first rewrite this in terms of monomials in $\{t^a\}$ and $\{d_{\mathrm{W}(\mathfrak{g})}t^a\}$. Hence, using $\sigma t^a = d_{\mathrm{W}(\mathfrak{g})}t^a + \frac{1}{2}C^a{}_{bc}t^b \wedge t^c$, we get

$$\begin{aligned}
\tau P &= \tau\big(P_{ab}(d_{\mathrm{W}(\mathfrak{g})}t^a) \wedge (d_{\mathrm{W}(\mathfrak{g})}t^a) - P_{ab}(d_{\mathrm{W}(\mathfrak{g})}t^a) \wedge C^b{}_{cd}t^c \wedge t^d \\
&\quad + \frac{1}{4}P_{ab}C^a{}_{cd}C^b{}_{ef}t^c \wedge t^d \wedge t^e \wedge t^d\big).
\end{aligned} \tag{5.82}$$

Now equation (5.61) can be applied to each term. Noticing the combinatorial prefactor $\frac{1}{n!}$, which depends on the number of factors in the above terms, and noticing the sum over all permutations, we find

$$\begin{aligned}
\tau\left(P_{ab}(d_{\mathrm{W}(\mathfrak{g})}t^a) \wedge (d_{\mathrm{W}(\mathfrak{g})}t^a)\right) &= P_{ab}(d_{\mathrm{W}(\mathfrak{g})}t^a) \wedge t^b \\
\tau\left(-P_{ab}(d_{\mathrm{W}(\mathfrak{g})}t^a) \wedge C^b{}_{cd}t^c \wedge t^d\right) &= \frac{1}{3!} \cdot 2\, P_{ab}C^b_{cd}t^b \wedge t^c \wedge t^d \\
&= \frac{1}{3}C_{abc}t^a \wedge t^b \wedge t^c
\end{aligned} \tag{5.83}$$

where we write $C_{abc} := P_{ad}C^d{}_{bc}$ as usual. Finally $\tau\left(\frac{1}{4}P_{ab}C^a{}_{cd}C^b_{ef}t^c \wedge t^d \wedge t^e \wedge t^d\right)$
$= 0$. In total this yields

$$\tau P = P_{ab}(d_{\mathrm{W}(\mathfrak{g})}t^a) \wedge t^b + \frac{1}{3}C_{abc}t^a \wedge t^b \wedge t^c. \tag{5.84}$$

By again using $d_{\mathrm{W}(\mathfrak{g})}t^a = -\frac{1}{2}C^a{}_{bc}t^b \wedge t^c + \sigma t^a$ together with the invariance of P (hence the $d_{\mathrm{W}(\mathfrak{g})}$-closedness of P which implies that the constants C_{abc} are skew symmetric in all three indices), one checks that this does indeed satisfy

$$d_{\mathrm{W}(\mathfrak{g})}\tau P = P. \tag{5.85}$$

In section 5.5 we will see that after choosing a $\mathfrak{g}$-valued connection on the space Y the generators t^a here will get sent to components of a $\mathfrak{g}$-valued 1-form A, while the $d_{W(\mathfrak{g})}t^a$ will get sent to the components of dA. Under this map the element $\tau P \in W(\mathfrak{g})$ maps to the familiar Chern-Simons 3-form

$$\mathrm{CS}_P(A) := P(A \wedge dA) + \frac{1}{3}P(A \wedge [A \wedge A]) \qquad (5.86)$$

whose differential is the characteristic form of A with respect to P:

$$d\mathrm{CS}_P(A) = P(F_A \wedge F_A). \qquad (5.87)$$

Characteristic forms, for arbitrary Lie ∞-algebra valued forms, are discussed further in section 5.6.

2-Morphisms of Lie 2-algebras.

Proposition 5.15. *For the special case that $\mathfrak{g}$ is any Lie 2-algebra (any L_∞-algebra concentrated in the first two degrees) the 2-morphisms defined by definition 5.11 reproduce the 2-morphisms of Lie 2-algebras as stated in [5] and used in [7].*

     Proof. The proof is given in the appendix. □

     This implies in particular that with the 1- and 2-morphisms as defined above, Lie 2-algebras do form a 2-category. There is a rather straightforward generalization of definition 5.11 to higher morphisms, which one would expect yields correspondingly n-categories of Lie n-algebras. But this we shall not try to discuss here.

5.3. L_∞-algebra cohomology

The study of ordinary Lie algebra cohomology and of invariant polynomials on the Lie algebra has a simple formulation in terms of the qDGCAs $\mathrm{CE}(\mathfrak{g})$ and $W(\mathfrak{g})$. Furthermore, this has a straightforward generalization to arbitrary L_∞-algebras which we now state.

     For $\mathrm{CE}(\mathfrak{g}) \xleftarrow{\quad i^* \quad} W(\mathfrak{g})$ the canonical morphism from proposition 5.7, notice that

$$\mathrm{CE}(\mathfrak{g}) \simeq W(\mathfrak{g})/\ker(i^*) \qquad (5.88)$$

and that

$$\ker(i^*) = \langle \mathfrak{g}^*[1] \rangle_{W(\mathfrak{g})}, \qquad (5.89)$$

the ideal in $W(\mathfrak{g})$ generated by $\mathfrak{g}^*[1]$. Algebra derivations

$$\iota_X : W(\mathfrak{g}) \to W(\mathfrak{g}) \qquad (5.90)$$

for $X \in \mathfrak{g}$ are like (contractions with) vector fields on the space on which $W(\mathfrak{g})$ is like differential forms. In the case of an ordinary Lie algebra $\mathfrak{g}$, the corresponding inner derivations $[d_{W(\mathfrak{g})}, \iota_X]$ for $X \in \mathfrak{g}$ are of degree -1 and are known as the Lie derivative L_X. They generate flows $\exp([d_{W(\mathfrak{g})}, \iota_X]) : W(\mathfrak{g}) \to W(\mathfrak{g})$ along these vector fields.

Definition 5.16 (Vertical derivations). *We say an algebra derivation $\tau : \mathrm{W}(\mathfrak{g}) \to \mathrm{W}(\mathfrak{g})$ is* vertical *if it vanishes on the shifted copy $\mathfrak{g}^*[1]$ of $\mathfrak{g}^*$ inside $\mathrm{W}(\mathfrak{g})$,*

$$\tau|_{\mathfrak{g}^*[1]} = 0 \,. \tag{5.91}$$

Proposition 5.17. *The vertical derivations are precisely those that come from contractions*

$$\iota_X : \mathfrak{g}^* \mapsto \mathbb{R} \tag{5.92}$$

for all $X \in \mathfrak{g}$, extended to 0 on $\mathfrak{g}^*[1]$ and extended as algebra derivations to all of $\wedge^\bullet(\mathfrak{g}^* \oplus \mathfrak{g}^*[1])$.

The reader should compare this and the following definitions to the theory of vertical Lie derivatives and basic differential forms with respect to any surjective submersion $\pi : Y \to X$. This is discussed in section 4.3.1.

Definition 5.18 (Basic forms and invariant polynomials). *The algebra $\mathrm{W}(\mathfrak{g})_{\mathrm{basic}}$ of* **basic forms** *in $\mathrm{W}(\mathfrak{g})$ is the intersection of the kernels of all vertical derivations and Lie derivatives. i.e. all the contractions ι_X and Lie derivatives L_X for $X \in \mathfrak{g}$. Since $L_X = [d_{\mathrm{W}(\mathfrak{g})}, \iota_X]$, it follows that in the kernel of ι_X, the Lie derivative vanishes only if $\iota_X d_{\mathrm{W}(\mathfrak{g})}$ vanishes.*

As will be discussed in a moment, basic forms in $\mathrm{W}(\mathfrak{g})$ play the role of **invariant polynomials** on the L_∞-algebra $\mathfrak{g}$. Therefore we often write $\mathrm{inv}(\mathfrak{g})$ for $\mathrm{W}(\mathfrak{g})_{\mathrm{basic}}$:

$$\mathrm{inv}(\mathfrak{g}) := \mathrm{W}(\mathfrak{g})_{\mathrm{basic}} \,. \tag{5.93}$$

Using the obvious inclusion $\mathrm{W}(\mathfrak{g}) \xleftarrow{\ p^*\ } \mathrm{inv}(\mathfrak{g})$ we obtain the sequence

$$\mathrm{CE}(\mathfrak{g}) \xleftarrow{\ i^*\ } \mathrm{W}(\mathfrak{g}) \xleftarrow{\ p^*\ } \mathrm{inv}(\mathfrak{g}) \tag{5.94}$$

of dg-algebras that plays a major role in our analysis: it can be interpreted as coming from the universal bundle for the Lie ∞-algebra $\mathfrak{g}$. As shown in figure 4, we can regard vertical derivations on $\mathrm{W}(\mathfrak{g})$ as derivations along the *fibers* of the corresponding dual sequence.

Definition 5.19 (Cocycles, invariant polynomials and transgression elements). *Let $\mathfrak{g}$ be an L_∞-algebra. Then*

- *An L_∞-algebra* **cocycle** *on $\mathfrak{g}$ is a $d_{\mathrm{CE}(\mathfrak{g})}$-closed element of $\mathrm{CE}(\mathfrak{g})$.*

$$\mu \in \mathrm{CE}(\mathfrak{g}) \,, \qquad d_{\mathrm{CE}(\mathfrak{g})}\mu = 0 \,. \tag{5.95}$$

- *An L_∞-algebra* **invariant polynomial** *on $\mathfrak{g}$ is an element $P \in \mathrm{inv}(\mathfrak{g}) := \mathrm{W}(\mathfrak{g})_{\mathrm{basic}}$.*
- *An L_∞-algebra* **$\mathfrak{g}$-transgression element** *for a given cocycle μ and an invariant polynomial P is an element $\mathrm{cs} \in \mathrm{W}(\mathfrak{g})$ such that*

$$d_{\mathrm{W}(\mathfrak{g})}\mathrm{cs} = p^*P \tag{5.96}$$

$$i^*\mathrm{cs} = \mu \,. \tag{5.97}$$

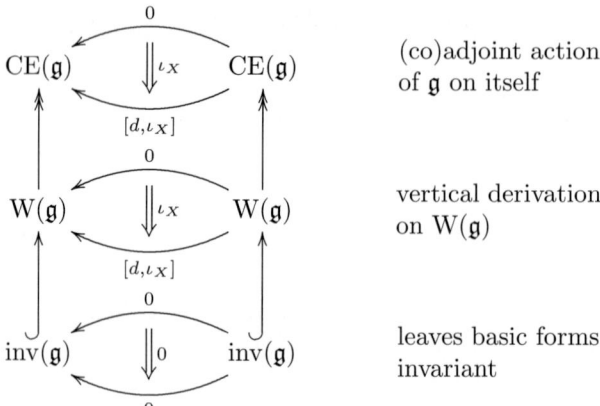

$$\begin{array}{ll}
\text{(co)adjoint action} \\
\text{of } \mathfrak{g} \text{ on itself}
\end{array}$$

$$\begin{array}{ll}
\text{vertical derivation} \\
\text{on } W(\mathfrak{g})
\end{array}$$

$$\begin{array}{ll}
\text{leaves basic forms} \\
\text{invariant}
\end{array}$$

FIGURE 4. **Interpretation of vertical derivations on** $W(\mathfrak{g})$**.** The algebra $CE(\mathfrak{g})$ plays the role of the algebra of differential forms on the Lie ∞-group that integrates the Lie ∞-algebra $\mathfrak{g}$. The coadjoint action of $\mathfrak{g}$ on these forms corresponds to Lie derivatives along the fibers of the universal bundle. These vertical derivatives leave the forms on the base of this universal bundle invariant. The diagram displayed is in the 2-category $\mathbf{Ch}^\bullet$ of cochain complexes, as described in the beginning of section 5.2.

If a transgression element for μ and P exists, we say that μ **transgresses** to P and that P **suspends** to μ. If $\mu = 0$ we say that P **suspends to** 0. The situation is illustrated diagrammatically in figure 5 and figure 6.

Definition 5.20 (Suspension to 0). *An element* $P \in \mathrm{inv}(\mathfrak{g})$ *is said to suspend to 0 if under the inclusion*

$$\mathrm{ker}(i^*) \xleftarrow{\ p^*\ } W(\mathfrak{g}) \tag{5.98}$$

it becomes a coboundary:

$$p^* P = d_{\mathrm{ker}(i^*)} \alpha \tag{5.99}$$

for some $\alpha \in \mathrm{ker}(i^*)$.

Remark. We will see that it is the intersection of $\mathrm{inv}(\mathfrak{g})$ with the cohomology of $\mathrm{ker}(i^*)$ that is a candidate, in general, for an algebraic model of the classifying space of the object that integrates the L_∞-algebra $\mathfrak{g}$. But at the moment we do not entirely clarify this relation to the integrated theory, unfortunately.

Proposition 5.21. *For the case that* $\mathfrak{g}$ *is an ordinary Lie algebra, the above definition reproduces the ordinary definitions of Lie algebra cocycles, invariant polynomials, and transgression elements. Moreover, all elements in* $\mathrm{inv}(\mathfrak{g})$ *are closed.*

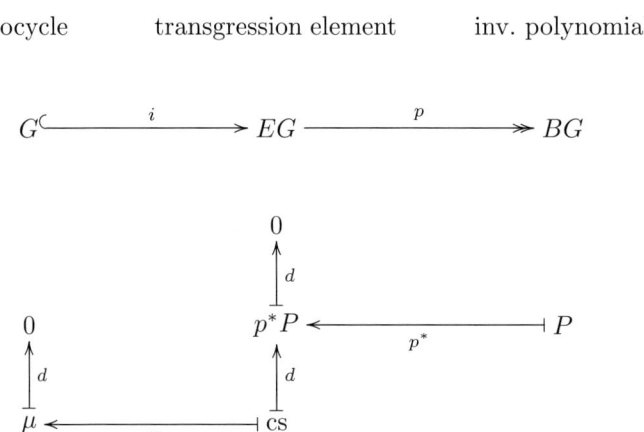

FIGURE 5. **Lie algebra cocycles, invariant polynomials and transgression forms** in terms of cohomology of the universal G-bundle. Let G be a simply connected compact Lie group with Lie algebra $\mathfrak{g}$. Then invariant polynomials P on $\mathfrak{g}$ correspond to elements in the cohomology $H^\bullet(BG)$ of the classifying space of G. When pulled back to the total space of the universal G-bundle $EG \to BG$, these classes become trivial, due to the contractability of EG: $p^*P = d(\mathrm{cs})$. Lie algebra cocycles, on the other hand, correspond to elements in the cohomology $H^\bullet(G)$ of G itself. A cocycle $\mu \in H^\bullet(G)$ is in transgression with an invariant polynomial $P \in H^\bullet(BG)$ if $\mu = i^*\mathrm{cs}$.

Proof. That the definitions of Lie algebra cocycles and transgression elements coincides is clear. It remains to be checked that $\mathrm{inv}(\mathfrak{g})$ really contains the invariant polynomials. In the ordinary definition a $\mathfrak{g}$-invariant polynomial is a $d_{\mathrm{W}(\mathfrak{g})}$-closed element in $\wedge^\bullet(\mathfrak{g}^*[1])$. Hence one only needs to check that all elements in $\wedge^\bullet(\mathfrak{g}^*[1])$ with the property that their image under $d_{\mathrm{W}(\mathfrak{g})}$ is again in $\wedge^\bullet(\mathfrak{g}^*[1])$ are in fact already closed. This can be seen for instance in components, using the description of $\mathrm{W}(\mathfrak{g})$ given in section 5.1.1. □

Remark. For ordinary Lie algebras $\mathfrak{g}$ corresponding to a simply connected compact Lie group G, the situation is often discussed in terms of the cohomology of the universal G-bundle. This is recalled in figure 5 and in section 5.3.1. The general definition above is a precise analog of that familiar situation: $\mathrm{W}(\mathfrak{g})$ plays the role of the algebra of (left invariant) differential forms on the universal $\mathfrak{g}$-bundle and $\mathrm{CE}(\mathfrak{g})$ plays the role of the algebra of (left invariant) differential forms on its fiber. Then $\mathrm{inv}(\mathfrak{g})$ plays the role of differential forms on the base, $BG = EG/G$. In fact,

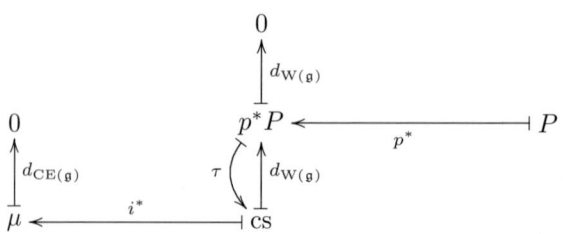

$$CE(\mathfrak{g}) \xleftarrow{\quad i^* \quad} W(\mathfrak{g}) \xleftarrow{\quad p^* \quad} inv(\mathfrak{g})$$

cocycle   transgression element   inv. polynomial

FIGURE 6. **The homotopy operator** τ is a contraction homotopy for $W(\mathfrak{g})$. Acting with it on a closed invariant polynomial $P \in inv(\mathfrak{g}) \subset \wedge^\bullet\mathfrak{g}[1] \subset W(\mathfrak{g})$ produces an element cs $\in W(\mathfrak{g})$ whose "restriction to the fiber" $\mu := i^*$cs is necessarily closed and hence a cocycle. We say that cs induces the *transgression* from μ to P, or that P *suspends* to μ.

for G a compact and simply connected Lie group and $\mathfrak{g}$ its Lie algebra, we have

$$H^\bullet(inv(\mathfrak{g})) \simeq H^\bullet(BG, \mathbb{R}). \tag{5.100}$$

In summary, the situation we thus obtain is that depicted in figure 1. Compare this to the following fact.

Proposition 5.22. *For* $p : P \to X$ *a principal* G-*bundle, let* vert$(P) \subset \Gamma(TP)$ *be the vertical vector fields on* P. *The horizontal differential forms on* P *which are invariant under* vert(P) *are precisely those that are pulled back along* p *from* X.

These are called the **basic differential forms** in [41].

Remark. We will see that, contrary to the situation for ordinary Lie algebras, in general invariant polynomials of L_∞ algebras are not $d_{W(\mathfrak{g})}$-closed (the $d_{W(\mathfrak{g})}$-differential of them is just horizontal). We will also see that those indecomposable invariant polynomials in $inv(\mathfrak{g})$, i.e. those that become exact in $\ker(i^*)$, are not characteristic for the corresponding $\mathfrak{g}$-bundles. This probably means that the real cohomology of the classifying space of the Lie ∞-group integrating $\mathfrak{g}$ is spanned by invariant polynomials modulo those suspending to 0. But here we do not attempt to discuss this further.

Proposition 5.23. *For every invariant polynomial* $P \in \wedge^\bullet\mathfrak{g}[1] \subset W(\mathfrak{g})$ *on an* L_∞-*algebra* $\mathfrak{g}$ *such that* $d_{W(\mathfrak{g})}p^*P = 0$, *there exists an* L_∞-*algebra cocycle* $\mu \in CS(\mathfrak{g})$ *that transgresses to* P.

Proof. This is a consequence of proposition 5.7 and proposition 5.8. Let $P \in W(\mathfrak{g})$ be an invariant polynomial. By proposition 5.7, p^*P is in the kernel of the restriction homomorphism $CE(\mathfrak{g}) \xleftarrow{\;i^*\;} W(\mathfrak{g}) : i^*P = 0$. By proposition 5.8, p^*P is the image under $d_{W(\mathfrak{g})}$ of an element $cs := \tau(p^*P)$ and by the algebra homomorphism property of i^* we know that its restriction, $\mu := i^*cs$, to the fiber is closed, because

$$d_{CE(\mathfrak{g})} i^* cs = i^* d_{W(\mathfrak{g})} cs = i^* p^* P = 0 . \tag{5.101}$$

Therefore μ is an L_∞-algebra cocycle for $\mathfrak{g}$ that transgresses to the invariant polynomial P. □

Remark. Notice that this statement is useful only for *indecomposable* invariant polynomials. All others trivially suspend to the 0 cocycle.

Proposition 5.24. *An invariant polynomial which suspends to a Lie ∞-algebra cocycle that is a coboundary also suspends to 0.*

Proof. Let P be an invariant polynomial, cs the corresponding transgression element and $\mu = i^*cs$ the corresponding cocycle, which is assumed to be a coboundary in that $\mu = d_{CE(\mathfrak{g})} b$ for some $b \in CE(\mathfrak{g})$. Then by the definition of $d_{W(\mathfrak{g})}$ it follows that $\mu = i^*(d_{W(\mathfrak{g})} b)$.

Now notice that

$$cs' := cs - d_{W(\mathfrak{g})} b \tag{5.102}$$

is another transgression element for P, since

$$d_{W(\mathfrak{g})} cs' = p^* P . \tag{5.103}$$

But now

$$i^*(cs') = i^*(cs - d_{W(\mathfrak{g})} b) = 0 . \tag{5.104}$$

Hence P suspends to 0. □

5.3.1. Examples.
The cohomologies of G and of BG in terms of qDGCAs. To put our general considerations for L_∞-algebras into perspective, it is useful to keep the following classical results for ordinary Lie algebras in mind.

A classical result of E. Cartan [22] [23] (see also [49]) says that for a connected finite dimensional Lie group G, the cohomology $H^\bullet(G)$ of the group is isomorphic to that of the Chevalley-Eilenberg algebra $CE(\mathfrak{g})$ of its Lie algebra $\mathfrak{g}$:

$$H^\bullet(G) \simeq H^\bullet(CE(\mathfrak{g})) , \tag{5.105}$$

namely to the algebra of Lie algebra cocycles on $\mathfrak{g}$. If we denote by Q_G the space of *indecomposable* such cocycles, and form the qDGCA $\wedge^\bullet Q_G = H^\bullet(\wedge^\bullet Q_G)$ with trivial differential, the above says that we have an isomorphism in cohomology

$$H^\bullet(G) \simeq H^\bullet(\wedge^\bullet Q_G) = \wedge^\bullet Q_G \tag{5.106}$$

which is realized by the canonical inclusion

$$i : \wedge^\bullet Q_G \hookrightarrow \mathrm{CE}(\mathfrak{g}) \tag{5.107}$$

of all cocycles into the Chevalley-Eilenberg algebra.

Subsequently, we have the classical result of Borel [13]: For a connected finite dimensional Lie group G, the cohomology of its classifying space BG is a finitely generated polynomial algebra on even dimensional generators:

$$H^\bullet(BG) \simeq \wedge^\bullet P_G . \tag{5.108}$$

Here P_G is the space of *indecomposable* invariant polynomials on $\mathfrak{g}$, hence

$$H^\bullet(BG) \simeq H^\bullet(\mathrm{inv}(\mathfrak{g})) . \tag{5.109}$$

In fact, P_G and Q_G are isomorphic after a shift:

$$P_G \simeq Q_G[1] \tag{5.110}$$

and this isomorphism is induced by *transgression* between indecomposable cocycles $\mu \in \mathrm{CE}(\mathfrak{g})$ and indecomposable invariant polynomials $P \in \mathrm{inv}(\mathfrak{g})$ via a transgression element $\mathrm{cs} = \tau P \in \mathrm{W}(\mathfrak{g})$.

Cohomology and invariant polynomials of $b^{n-1}\mathfrak{u}(1)$.

Proposition 5.25. *For every integer $n \geq 1$, the Lie n-algebra $b^{n-1}\mathfrak{u}(1)$ (the $(n-1)$-folded shifted version of ordinary $\mathfrak{u}(1)$) from section 5.1.1) we have the following:*

- *there is, up to a scalar multiple, a single indecomposable Lie ∞-algebra cocycle which is of degree n and* linear,

$$\mu_{b^{n-1}\mathfrak{u}(1)} \in \mathbb{R}[n] \subset \mathrm{CE}(b^{n-1}\mathfrak{u}(1)) , \tag{5.111}$$

- *there is, up to a scalar multiple, a single indecomposable Lie ∞-algebra invariant polynomial, which is of degree $(n+1)$*

$$P_{b^{n-1}\mathfrak{u}(1)} \in \mathbb{R}[n+1] \subset \mathrm{inv}(b^{n-1}\mathfrak{u}(1)) = \mathrm{CE}(b^n\mathfrak{u}(1)) . \tag{5.112}$$

- *The cocycle $\mu_{b^{n-1}\mathfrak{u}(1)}$ is in transgression with $P_{b^{n-1}\mathfrak{u}(1)}$.*

These statements are an obvious consequence of the definitions involved, but they are important. The fact that $b^{n-1}\mathfrak{u}(1)$ has a single invariant polynomial of degree $(n+1)$ will immediately imply, in section 6, that $b^{n-1}\mathfrak{u}(1)$-bundles have a single characteristic class of degree $(n+1)$: known (at least for $n = 2$, as the Dixmier-Douady class). Such a $b^{n-1}\mathfrak{u}(1)$-bundle classes appear in 7 as the obstruction classes for lifts of n-bundles through String-like extensions of their structure Lie n-algebra.

Cohomology and invariant polynomials of strict Lie 2-algebras. Let $\mathfrak{g}_{(2)} = (\mathfrak{h} \xrightarrow{t} \mathfrak{g} \xrightarrow{\alpha} \mathrm{der}(\mathfrak{h}))$ be a strict Lie 2-algebra as described in section 5.1. Notice that there is a canonical projection homomorphism

$$\mathrm{CE}(\mathfrak{g}) \xleftarrow{\quad j^* \quad} \mathrm{CE}(\mathfrak{h} \xrightarrow{t} \mathfrak{g}) \tag{5.113}$$

which, of course, extends to the Weil algebras

$$\mathrm{W}(\mathfrak{g}) \xleftarrow{\quad j^* \quad} \mathrm{W}(\mathfrak{h} \xrightarrow{t} \mathfrak{g}) \;. \tag{5.114}$$

Here j^* is simply the identity on $\mathfrak{g}^*$ and on $\mathfrak{g}^*[1]$ and vanishes on $\mathfrak{h}^*[1]$ and $\mathfrak{h}^*[2]$.

Proposition 5.26. *Every invariant polynomial $P \in \mathrm{inv}(\mathfrak{g})$ of the ordinary Lie algebra $\mathfrak{g}$ lifts to an invariant polynomial on the Lie 2-algebra $(\mathfrak{h} \xrightarrow{t} \mathfrak{g})$:*

$$
\begin{array}{ccc}
\mathrm{W}(\mathfrak{h} \xrightarrow{t} \mathfrak{g}) & & \\
\Big\downarrow{\scriptstyle i^*} & \diagdown & \\
\mathrm{W}(\mathfrak{g}) & \xleftarrow{\qquad} & \mathrm{inv}(\mathfrak{g})
\end{array}
\qquad . \tag{5.115}
$$

However, a closed invariant polynomial will not necessarily lift to a closed one.

Proof. Recall that $d_t := d_{\mathrm{CE}(\mathfrak{h} \xrightarrow{t} \mathfrak{g})}$ acts on $\mathfrak{g}^*$ as

$$d_t|_{\mathfrak{g}^*} = [\cdot, \cdot]_{\mathfrak{g}}^* + t^* \;. \tag{5.116}$$

By definition 5.4 and definition 5.5 it follows that $d_{\mathrm{W}(\mathfrak{h} \xrightarrow{t} \mathfrak{g})}$ acts on $\mathfrak{g}^*[1]$ as

$$d_{\mathrm{W}(\mathfrak{h} \xrightarrow{t} \mathfrak{g})}|_{\mathfrak{g}^*[1]} = -\sigma \circ [\cdot, \cdot]_{\mathfrak{g}}^* - \sigma \circ t^* \tag{5.117}$$

and on $\mathfrak{h}^*[1]$ as

$$d_{\mathrm{W}(\mathfrak{h} \xrightarrow{t} \mathfrak{g})}|_{\mathfrak{h}^*[1]} = -\sigma \circ \alpha^* \;. \tag{5.118}$$

Then notice that

$$(\sigma \circ t^*) : \mathfrak{g}^*[1] \to \mathfrak{h}^*[2] \;. \tag{5.119}$$

But this means that $d_{\mathrm{W}(\mathfrak{h} \xrightarrow{t} \mathfrak{g})}$ differs from $d_{\mathrm{W}(\mathfrak{g})}$ on $\wedge^\bullet(\mathfrak{g}^*[1])$ only by elements that are annihilated by vertical ι_X. This proves the claim. $\qquad\square$

It may be easier to appreciate this proof by looking at what it does in terms of a chosen basis.

Same discussion in terms of a basis. Let $\{t^a\}$ be a basis of $\mathfrak{g}^*$ and $\{b^i\}$ be a basis of $\mathfrak{h}^*[1]$. Let $\{C^a{}_{bc}\}$, $\{\alpha^i{}_{aj}\}$, and $\{t^a{}_i\}$, respectively, be the components of $[\cdot, \cdot]_\mathfrak{g}$, α and t in that basis. Then corresponding to $\mathrm{CE}(\mathfrak{g})$, $\mathrm{W}(\mathfrak{g})$, $\mathrm{CE}(\mathfrak{h} \xrightarrow{t} \mathfrak{g})$, and $\mathrm{W}(\mathfrak{h} \xrightarrow{t} \mathfrak{g})$, respectively, we have the differentials

$$d_{\mathrm{CE}(\mathfrak{g})} : t^a \mapsto -\frac{1}{2}C^a{}_{bc}t^b \wedge t^c, \tag{5.120}$$

$$d_{\mathrm{W}(\mathfrak{g})} : t^a \mapsto -\frac{1}{2}C^a{}_{bc}t^b \wedge t^c + \sigma t^a, \tag{5.121}$$

$$d_{\mathrm{CE}(\mathfrak{h} \xrightarrow{t} \mathfrak{g})} : t^a \mapsto -\frac{1}{2}C^a{}_{bc}t^b \wedge t^c + t^a{}_ib^i, \tag{5.122}$$

and

$$d_{\mathrm{W}(\mathfrak{h} \xrightarrow{t} \mathfrak{g})} : t^a \mapsto -\frac{1}{2}C^a{}_{bc}t^b \wedge t^c + t^a{}_ib^i + \sigma t^a. \tag{5.123}$$

Hence we get

$$d_{\mathrm{W}(\mathfrak{g})} : \sigma t^a \mapsto -\sigma(-\frac{1}{2}C^a{}_{bc}t^b \wedge t^c) = C^a{}_{bc}(\sigma t^b) \wedge t^c \tag{5.124}$$

as well as

$$d_{\mathrm{W}(\mathfrak{h} \xrightarrow{t} \mathfrak{g})} : \sigma t^a \mapsto -\sigma(-\frac{1}{2}C^a{}_{bc}t^b \wedge t^c + t^a{}_ib^i) = C^a{}_{bc}(\sigma t^b) \wedge t^c + t^a{}_i\sigma b^i. \tag{5.125}$$

Then if

$$P = P_{a_1 \cdots a_n}(\sigma t^{a_1}) \wedge \cdots \wedge (\sigma t^{a_n}) \tag{5.126}$$

is $d_{\mathrm{W}(\mathfrak{g})}$-closed, i.e. an invariant polynomial on $\mathfrak{g}$, it follows that

$$d_{\mathrm{W}(\mathfrak{h} \xrightarrow{t} \mathfrak{g})}P = nP_{a_1,a_2,\cdots a_n}(t^{a_1}{}_i\sigma b^i) \wedge (\sigma t^{a_2}) \wedge \cdots \wedge (\sigma t^{a_n}). \tag{5.127}$$

(all terms appearing are in the image of the shifting isomorphism σ), hence P is also an invariant polynomial on $(\mathfrak{h} \xrightarrow{t} \mathfrak{g})$.

Remark. Notice that the invariant polynomials P lifted from $\mathfrak{g}$ to $(\mathfrak{h} \xrightarrow{t} \mathfrak{g})$ this way are no longer *closed*, in general. This is a new phenomenon we encounter for higher L_∞-algebras. While, according to proposition 5.21, for $\mathfrak{g}$ an ordinary Lie algebra all elements in $\mathrm{inv}(\mathfrak{g})$ are closed, this is no longer the case here: the lifted elements P above vanish only after we hit with them with both $d_{\mathrm{W}(\mathfrak{h} \xrightarrow{t} \mathfrak{g})}$ *and a* vertical τ. □

We will see a physical application of this fact in 5.6.

Proposition 5.27. *Let P be any invariant polynomial on the ordinary Lie algebra $\mathfrak{g}$ in transgression with the cocycle μ on $\mathfrak{g}$. Regarded both as elements of $\mathrm{W}(\mathfrak{h} \xrightarrow{t} \mathfrak{g})$ and $\mathrm{CE}(\mathfrak{h} \xrightarrow{t} \mathfrak{g})$ respectively. Notice that $d_{\mathrm{CE}(\mathfrak{h} \xrightarrow{t} \mathfrak{g})}\mu$ is in general non-vanishing but is of course now an exact cocycle on $(\mathfrak{h} \xrightarrow{t} \mathfrak{g})$.*

We have : the $(\mathfrak{h} \xrightarrow{t} \mathfrak{g})$-cocycle $d_{\mathrm{CE}(\mathfrak{h} \xrightarrow{t} \mathfrak{g})}\mu$ transgresses to $d_{\mathrm{inv}(\mathfrak{h} \xrightarrow{t} \mathfrak{g})}P$.

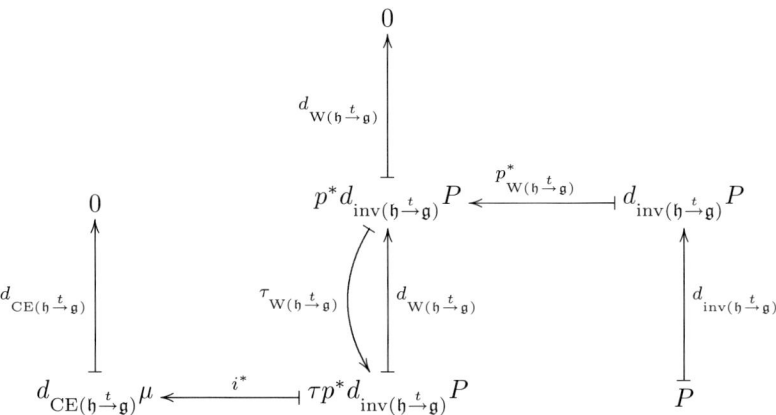

FIGURE 7. **Cocycles and invariant polynomials on strict Lie 2-algebras** $(\mathfrak{h} \xrightarrow{t} \mathfrak{g})$, induced from cocycles and invariant polynomials on $\mathfrak{g}$. An invariant polynomial P on $\mathfrak{g}$ in transgression with a cocycle μ on $\mathfrak{g}$ lifts to a generally non-closed invariant polynomial on $(\mathfrak{h} \xrightarrow{t} \mathfrak{g})$. The diagram says that its closure, $d_{\mathrm{inv}(\mathfrak{h}\xrightarrow{t}\mathfrak{g})}P$, suspends to the $d_{\mathrm{CE}(\mathfrak{h}\xrightarrow{t}\mathfrak{g})}$-closure of the cocycle μ. Since this $(\mathfrak{h} \xrightarrow{t} \mathfrak{g})$-cocycle $d_{(\mathfrak{h}\xrightarrow{t}\mathfrak{g})}\mu$ is hence a coboundary, it follows from proposition 5.24 that $d_{\mathrm{inv}(\mathfrak{h}\xrightarrow{t}\mathfrak{g})}P$ suspends also to 0. Nevertheless the situation is of interest, in that it governs the topological field theory known as BF theory. This is discussed in section 5.6.1.

The situation is illustrated by the diagram in figure 7.

Concrete Example: $\mathfrak{su}(5) \rightarrow \mathfrak{sp}(5)$. It is known that the cohomology of the Chevalley-Eilenberg algebras for $\mathfrak{su}(5)$ and $\mathfrak{sp}(5)$ are generated, respectively, by four and five indecomposable cocycles,

$$H^\bullet(\mathrm{CE}(\mathfrak{su}(5))) = \wedge^\bullet[a, b, c, d] \tag{5.128}$$

and

$$H^\bullet(\mathrm{CE}(\mathfrak{sp}(5))) = \wedge^\bullet[v, w, x, y, z], \tag{5.129}$$

which have degree as indicated in the following table:

$H^\bullet \mathrm{CE}(\mathfrak{su}(5))$	generator	degree
	a	3
	b	5
	c	7
	d	9
$H^\bullet(\mathrm{CE}(\mathfrak{sp}(5)))$	v	3
	w	7
	x	11
	y	15
	z	19

As discussed for instance in [41], the inclusion of groups

$$\mathrm{SU}(5) \hookrightarrow \mathrm{Sp}(5) \tag{5.130}$$

is reflected in the morphism of DGCAs

$$\mathrm{CE}(\mathfrak{su}(5)) \xleftarrow{\quad t^* \quad} \mathrm{CE}(\mathfrak{sp}(5)) \tag{5.131}$$

which acts, in cohomology, on v and w as

$$a \longleftarrow\!\mid v \tag{5.132}$$
$$c \longleftarrow\!\mid w$$

and which sends x, y and z to wedge products of generators.

We would like to apply the above reasoning to this situation. Now, $\mathfrak{su}(5)$ is not normal in $\mathfrak{sp}(5)$ hence $(\mathfrak{su}(5) \hookrightarrow \mathfrak{sp}(5))$ does not give a Lie 2-algebra. But we can regard the cohomology complexes $H^\bullet(\mathrm{CE}(\mathfrak{su}(5)))$ and $H^\bullet(\mathrm{CE}(\mathfrak{sp}(5)))$ as Chevalley-Eilenberg algebras of abelian L_∞-algebras in their own right. Their inclusion is normal, in the sense to be made precise below in definition 7.1. By useful abuse of notation, we write now $\mathrm{CE}(\mathfrak{su}(5) \hookrightarrow \mathfrak{sp}(5))$ for this inclusion at the level of cohomology.

Recalling from 5.45 that this means that in $\mathrm{CE}(\mathfrak{su}(5) \hookrightarrow \mathfrak{sp}(5))$ we have

$$d_{\mathrm{CE}(\mathfrak{su}(5) \hookrightarrow \mathfrak{sp}(5))} v := \sigma a \tag{5.133}$$

and

$$d_{\mathrm{CE}(\mathfrak{su}(5) \hookrightarrow \mathfrak{sp}(5))} w := \sigma c \tag{5.134}$$

we see that the generators σa and σb drop out of the cohomology of the Chevalley-Eilenberg algebra

$$\mathrm{CE}(\mathfrak{su}(5) \hookrightarrow \mathfrak{sp}(5)) = (\textstyle\bigwedge^\bullet (\mathfrak{sp}(5)^* \oplus \mathfrak{su}(5)^*[1]), d_t) \tag{5.135}$$

of the strict Lie 2-algebra coming from the infinitesimal crossed module $(t : \mathfrak{su}(5) \hookrightarrow \mathfrak{sp}(5))$.

A simple spectral sequence argument shows that products are not killed in $H^\bullet(\mathrm{CE}(\mathfrak{su}(5) \hookrightarrow \mathfrak{sp}(5)))$, but they may no longer be decomposable. Hence

$$H^\bullet(\mathrm{CE}(\mathfrak{su}(5) \hookrightarrow \mathfrak{sp}(5))) \tag{5.136}$$

is generated by classes in degrees 6 and 10 by σb and σd, and in degrees 21 and 25, which are represented by products in 5.135 involving σa and σc, with the only non zero product being

$$6 \wedge 25 = 10 \wedge 21 \,, \tag{5.137}$$

where 31 is the dimension of the manifold $\mathrm{Sp}(5)/\mathrm{SU}(5)$. Thus the strict Lie 2-algebra $(t : \mathfrak{su}(5) \hookrightarrow \mathfrak{sp}(5))$ plays the role of the quotient Lie 1-algebra $\mathfrak{sp}(5)/\mathfrak{su}(5)$. We will discuss the general mechanism behind this phenomenon in 7.1: the Lie 2-algebra $(\mathfrak{su}(5) \hookrightarrow \mathfrak{sp}(5))$ is the *weak cokernel*, i.e. the *homotopy cokernel* of the inclusion $\mathfrak{su}(5) \hookrightarrow \mathfrak{sp}(5)$.

The Weil algebra of $(\mathfrak{su}(5) \hookrightarrow \mathfrak{sp}(5))$ is

$$W(\mathfrak{su}(5) \hookrightarrow \mathfrak{sp}(5)) = (\wedge^\bullet(\mathfrak{sp}(5)^* \oplus \mathfrak{su}(5)^*[1] \oplus \mathfrak{sp}(5)^*[1] \oplus \mathfrak{su}(5)^*[2]), d_{W(\mathfrak{su}(5) \hookrightarrow \mathfrak{sp}(5))}) \,. \tag{5.138}$$

Recall the formula 5.40 for the action of $d_{W(\mathfrak{su}(5) \hookrightarrow \mathfrak{sp}(5))}$ on generators in $\mathfrak{sp}(5)^*[1] \oplus \mathfrak{su}(5)^*[2]$. By that formula, σv and σw are invariant polynomials on $\mathfrak{sp}(5)$ which lift to non-closed invariant polynomials on $\mathfrak{su}(5) \hookrightarrow \mathfrak{sp}(5)$:

$$d_{W(\mathfrak{su}(5) \hookrightarrow \mathfrak{sp}(5))}) : \sigma v \mapsto -\sigma(d_{\mathrm{CE}(\mathfrak{su}(5) \hookrightarrow \mathfrak{sp}(5))} v) = -\sigma\sigma a \tag{5.139}$$

by equation (5.133); and

$$d_{W(\mathfrak{su}(5) \hookrightarrow \mathfrak{sp}(5))}) : \sigma w \mapsto -\sigma(d_{\mathrm{CE}(\mathfrak{su}(5) \hookrightarrow \mathfrak{sp}(5))} w) = -\sigma\sigma c \tag{5.140}$$

by equation (5.134). Hence σv and σw are not closed in $\mathrm{CE}(\mathfrak{su}(5) \hookrightarrow \mathfrak{sp}(5))$, but they are still invariant polynomials according to definition 5.18, since their differential sits entirely in the shifted copy $(\mathfrak{sp}(5)^* \oplus \mathfrak{su}(5)^*[1])[1]$.

On the other hand, notice that we do also have closed invariant polynomials on $(\mathfrak{su}(5) \hookrightarrow \mathfrak{sp}(5))$, for instance $\sigma\sigma b$ and $\sigma\sigma d$.

5.4. L_∞-algebras from cocycles: String-like extensions

We now consider the main object of interest here: families of L_∞-algebras that are induced from L_∞-cocycles and invariant polynomials. First we need the following

Definition 5.28 (String-like extensions of L_∞-algebras). *Let $\mathfrak{g}$ be an L_∞-algebra.*

- *For each degree $(n + 1)$-cocycle μ on $\mathfrak{g}$, let $\mathfrak{g}_\mu$ be the L_∞-algebra defined by*

$$\mathrm{CE}(\mathfrak{g}_\mu) = (\wedge^\bullet(\mathfrak{g}^* \oplus \mathbb{R}[n]), d_{\mathrm{CE}(\mathfrak{g}_\mu)}) \tag{5.141}$$

  with differential given by

$$d_{\mathrm{CE}(\mathfrak{g}_\mu)}|_{\mathfrak{g}^*} := d_{\mathrm{CE}(\mathfrak{g})}, \tag{5.142}$$

  and

$$d_{\mathrm{CE}(\mathfrak{g}_\mu)})|_{\mathbb{R}[n]} : b \mapsto -\mu \,, \tag{5.143}$$

  *where $\{b\}$ denotes the canonical basis of $\mathbb{R}[n]$. This we call the **String-like extension** of $\mathfrak{g}$ with respect to μ, because, as described below in 5.4.1, it generalizes the construction of the String Lie 2-algebra.*

- For each degree n invariant polynomial P on $\mathfrak{g}$, let $\mathrm{ch}_P(\mathfrak{g})$ be the L_∞-algebra defined by

$$\mathrm{CE}(\mathrm{ch}_P(\mathfrak{g})) = (\wedge^\bullet(\mathfrak{g}^* \oplus \mathfrak{g}^*[1]) \oplus \mathbb{R}[2n-1]), d_{\mathrm{CE}(\mathrm{ch}_P(\mathfrak{g}))}) \qquad (5.144)$$

  with the differential given by

$$d_{\mathrm{CE}(\mathrm{ch}_P(\mathfrak{g}))}|_{\mathfrak{g}^* \oplus \mathfrak{g}^*[1]} := d_{\mathrm{W}(\mathfrak{g})} \qquad (5.145)$$

  and

$$d_{\mathrm{CE}(\mathrm{ch}_P(\mathfrak{g}))})|_{\mathbb{R}[2n-1]} : c \mapsto P \,, \qquad (5.146)$$

  where $\{c\}$ denotes the canonical basis of $\mathbb{R}[2n-1]$. This we call the **Chern** L_∞-**algebra** corresponding to the invariant polynomial P, because, as described below in 5.5.1, connections with values in it pick out the Chern-form corresponding to P.
- For each degree $2n-1$ transgression element cs, let $\mathrm{cs}_P(\mathfrak{g})$ be the L_∞-algebra defined by

$$\mathrm{CE}(\mathrm{cs}_P(\mathfrak{g})) = (\wedge^\bullet(\mathfrak{g}^* \oplus \mathfrak{g}^*[1]) \oplus \mathbb{R}[2n-2] \oplus \mathbb{R}[2n-1]), d_{\mathrm{CE}(\mathrm{ch}_P(\mathfrak{g}))}) \qquad (5.147)$$

  with

$$d_{\mathrm{CE}(\mathrm{cs}_P(\mathfrak{g}))}|_{\wedge^\bullet(\mathfrak{g}^* \oplus \mathfrak{g}^*[1])} = d_{\mathrm{W}(\mathfrak{g})} \qquad (5.148)$$

$$d_{\mathrm{CE}(\mathrm{cs}_P(\mathfrak{g}))}|_{\mathbb{R}[2n-2]} : b \mapsto -\mathrm{cs} + c \qquad (5.149)$$

$$d_{\mathrm{CE}(\mathrm{ch}_p(\mathfrak{g}))}|_{\mathbb{R}[2n-1]} : c \mapsto P \,, \qquad (5.150)$$

  where $\{b\}$ and $\{c\}$ denote the canonical bases of $\mathbb{R}[2n-2]$ and $\mathbb{R}[2n-1]$, respectively. This we call the **Chern-Simons** L_∞-**algebra** with respect to the transgression element cs, because, as described below in 5.5.1, connections with values in these come from (generalized) Chern-Simons forms.

The nilpotency of these differentials follows directly from the very definition of L_∞-algebra cocycles and invariant polynomials.

Proposition 5.29 (The String-like extensions). *For each L_∞-cocycle $\mu \in \wedge^n(\mathfrak{g}^*)$ of degree n, the corresponding String-like extension sits in an exact sequence*

$$0 \longleftarrow \mathrm{CE}(b^{n-1}\mathfrak{u}(1)) \longleftarrow \mathrm{CE}(\mathfrak{g}_\mu) \longleftarrow \mathrm{CE}(\mathfrak{g}) \longleftarrow 0 \qquad (5.151)$$

Proof. The morphisms are the canonical inclusion and projection.   □

Proposition 5.30. *For cs $\in \mathrm{W}(\mathfrak{g})$ any transgression element interpolating between the cocycle $\mu \in \mathrm{CE}(\mathfrak{g})$ and the invariant polynomial $P \in \wedge^\bullet(\mathfrak{g}[1]) \subset \mathrm{W}(\mathfrak{g})$, we obtain a homotopy-exact sequence*

$$\mathrm{CE}(\mathfrak{g}_\mu) \longleftarrow \mathrm{CE}(\mathrm{cs}_P(\mathfrak{g})) \longleftarrow \mathrm{CE}(\mathrm{ch}_P(\mathfrak{g})) \,. \qquad (5.152)$$

$$\Big\downarrow \simeq$$

$$\mathrm{W}(\mathfrak{g}_\mu)$$

Here the isomorphism

$$f : W(\mathfrak{g}_\mu) \xrightarrow{\;\simeq\;} CE(cs_P(\mathfrak{g})) \qquad (5.153)$$

is the identity on $\mathfrak{g}^* \oplus \mathfrak{g}^*[1] \oplus \mathbb{R}[n]$

$$f|_{\mathfrak{g}^* \oplus \mathfrak{g}^*[1] \oplus \mathbb{R}[n]} = \mathrm{Id} \qquad (5.154)$$

and acts as

$$f|_{\mathbb{R}[n+1]} : b \mapsto c + \mu - cs \qquad (5.155)$$

for b the canonical basis of $\mathbb{R}[n]$ and c that of $\mathbb{R}[n+1]$. We check that this does respect the differentials

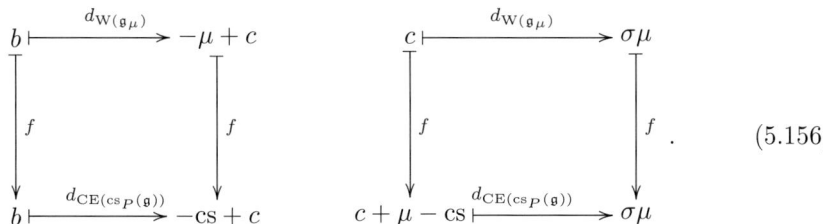

$$(5.156)$$

Recall from definition 7.3 that σ is the canonical isomorphism $\sigma : \mathfrak{g}^* \to \mathfrak{g}^*[1]$ extended by 0 to $\mathfrak{g}^*[1]$ and then as a derivation to all of $\wedge^\bullet(\mathfrak{g}^* \oplus \mathfrak{g}^*[1])$. In the above the morphism between the Weil algebra of $\mathfrak{g}_\mu$ and the Chevalley-Eilenberg algebra of $cs_P(\mathfrak{g})$ is indeed an *isomorphism* (not just an equivalence). This isomorphism exhibits one of the main points to be made here: it makes manifest that the invariant polynomial P that is related by transgression to the cocycle μ which induces $\mathfrak{g}_\mu$ becomes exact with respect to $\mathfrak{g}_\mu$. This is the statement of proposition 5.32 below.

L_∞-algebra cohomology and invariant polynomials of String-like extensions. The L_∞-algebra $\mathfrak{g}_\mu$ obtained from an L_∞-algebra $\mathfrak{g}$ with an L_∞-algebra cocycle $\mu \in H^\bullet(CE(\mathfrak{g}))$ can be thought of as being obtained from $\mathfrak{g}$ by "killing" a cocycle μ. This is familiar from Sullivan models in rational homotopy theory.

Proposition 5.31. *Let $\mathfrak{g}$ be an ordinary semisimple Lie algebra and μ a cocycle on it. Then*

$$H^\bullet(CE(\mathfrak{g}_\mu)) = H^\bullet(CE(\mathfrak{g}))/\langle\mu\rangle . \qquad (5.157)$$

Accordingly, one finds that in cohomology the invariant polynomials on $\mathfrak{g}_\mu$ are those of $\mathfrak{g}$ except that the polynomial in transgression with μ now suspends to 0.

Proposition 5.32. *Let $\mathfrak{g}$ be an L_∞-algebra and $\mu \in CE(\mathfrak{g})$ in transgression with the invariant polynomial $P \in \mathrm{inv}(\mathfrak{g})$. Then with respect to the String-like extension $\mathfrak{g}_\mu$ the polynomial P suspends to 0.*

Proof. Since μ is a coboundary in $\mathrm{CE}(\mathfrak{g}_\mu)$, this is a corollary of proposition 5.24. □

Remark. We will see in section 6 that those invariant polynomials which suspend to 0 do actually not contribute to the characteristic classes. As we will also see there, this can be understood in terms of the invariant polynomials not with respect to the projection $\mathrm{CE}(\mathfrak{g}) \longleftarrow \mathrm{W}(\mathfrak{g})$ but with respect to the projection

$$\mathrm{CE}(\mathfrak{g}) \longleftarrow \mathrm{CE}(\mathrm{cs}_P(\mathfrak{g}_\mu)),$$ recalling from 5.30 that $\mathrm{W}(\mathfrak{g})$ is isomorphic to $\mathrm{cs}_P(\mathfrak{g})$.

Proposition 5.33. *For $\mathfrak{g}$ any L_∞-algebra with cocycle μ of degree $2n+1$ in transgression with the invariant polynomial P, denote by $\mathrm{cs}_P(\mathfrak{g})_{\mathrm{basic}}$ the DGCA of basic forms with respect to the canonical projection*

$$\mathrm{CE}(\mathfrak{g}) \longleftarrow \mathrm{CE}(\mathrm{cs}_P(\mathfrak{g}_\mu)), \qquad (5.158)$$

according to the general definition 4.10.

Then the cohomology of $\mathrm{cs}_P(\mathfrak{g})_{\mathrm{basic}}$ is that of $\mathrm{inv}(\mathfrak{g})$ modulo P:

$$H^\bullet(\mathrm{cs}_P(\mathfrak{g})_{\mathrm{basic}}) \simeq H^\bullet(\mathrm{inv}(\mathfrak{g}))/\langle P \rangle . \qquad (5.159)$$

Proof. One finds that the vertical derivations on $\mathrm{CE}(\mathrm{cs}_P(\mathfrak{g})) = \wedge^\bullet(\mathfrak{g}^* \oplus \mathfrak{g}^*[1] \oplus \mathbb{R}[n] \oplus \mathbb{R}[n+1])$ are those that vanish on everything except the unshifted copy of $\mathfrak{g}^*$. Therefore the basic forms are those in $\wedge^\bullet(\mathfrak{g}^*[1] \oplus \mathbb{R}[n] \oplus \mathbb{R}[n+1])$ such that also their $d_{\mathrm{cs}_P(\mathfrak{g})}$-differential is in that space. Hence all invariant $\mathfrak{g}$-polynomials are among them. But one of them now becomes exact, namely P. □

Remark. The first example below, definition 5.34, introduces the String Lie 2-algebra of an ordinary semisimple Lie algebra $\mathfrak{g}$, which gave all our String-like extensions its name. It is known, corollary 2 in [8], that the real cohomology of the classifying space of the 2-group integrating it is that of $G = \exp(\mathfrak{g})$, modulo the ideal generated by the class corresponding to P. Hence $\mathrm{CE}(\mathrm{cs}_P(\mathfrak{g}))$ is an algebraic model for this space.

5.4.1. Examples.
Ordinary central extensions. Ordinary central extensions coming from a 2-cocycle $\mu \in H^2(\mathrm{CE}(\mathfrak{g}))$ of an ordinary Lie algebra $\mathfrak{g}$ are a special case of the "String-like" extensions we are considering:

By definition 5.28 the L_∞-algebra $\mathfrak{g}_\mu$ is the Lie 1-algebra whose Chevalley-Eilenberg algebra is

$$\mathrm{CE}(\mathfrak{g}_\mu) = (\wedge^\bullet(\mathfrak{g}^* \oplus \mathbb{R}[1]), d_{\mathrm{CE}(\mathfrak{g}_\mu)}) \qquad (5.160)$$

where

$$d_{\mathrm{CE}(\mathfrak{g}_\mu)}|_{\mathfrak{g}^*} = d_{\mathrm{CE}(\mathfrak{g})} \qquad (5.161)$$

and

$$d_{\mathrm{CE}(\mathfrak{g}_\mu)}|_{\mathbb{R}[1]} : b \mapsto \mu \qquad (5.162)$$

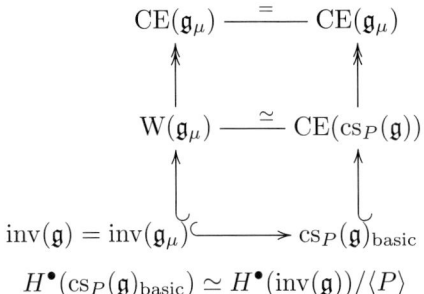

$$H^\bullet(\mathrm{cs}_P(\mathfrak{g})_\mathrm{basic}) \simeq H^\bullet(\mathrm{inv}(\mathfrak{g}))/\langle P \rangle$$

FIGURE 8. The DGCA sequence playing the role of differential forms on the **universal (higher) String n-bundle** for a String-like extension $\mathfrak{g}_\mu$, definition 5.28, of an L_∞-algebra $\mathfrak{g}$ by a cocycle μ of odd degree in transgression with an invariant polynomial P. Compare with figure 1. In $H^\bullet(\mathrm{inv}(\mathfrak{g}_\mu)) = H^\bullet(\mathrm{W}(\mathfrak{g}_\mu)_\mathrm{basic})$ the class of P is still contained, but suspends to 0, according to proposition 5.32. In $H^\bullet(\mathrm{cs}_P(\mathfrak{g})_\mathrm{basic})$ the class of P vanishes, according to proposition 5.33. The isomorphism $\mathrm{W}(\mathfrak{g}_\mu) \simeq \mathrm{CE}(\mathrm{cs}_P(\mathfrak{g}))$ is from proposition 5.30. For $\mathfrak{g}$ an ordinary semisimple Lie algebra and $\mathfrak{g}_\mu$ the ordinary String extension coming from the canonical 3-cocycle, this corresponds to the fact that the classifying space of the String 2-group [7, 43] has the cohomology of the classifying space of the underlying group, modulo the first Pontrajagin class [8].

for b the canonical basis of $\mathbb{R}[1]$. (Recall that in our conventions $\mathfrak{g}$ is in degree 1).

This is indeed the Chevalley-Eilenberg algebra corresponding to the Lie bracket

$$[(x,c),(x',c')] = ([x,x'],\mu(x,x')) \tag{5.163}$$

(for all $x,x' \in \mathfrak{g}$, $c,c' \in \mathbb{R}$) on the centrally extended Lie algebra.

The String Lie 2-algebra.

Definition 5.34. *Let $\mathfrak{g}$ be a semisiple Lie algebra and $\mu = \langle \cdot, [\cdot,\cdot] \rangle$ the canonical 3-cocycle on it. Then*

$$\mathrm{string}(\mathfrak{g}) \tag{5.164}$$

is defined to be the strict Lie 2-algebra coming from the crossed module

$$(\hat{\Omega}\mathfrak{g} \to P\mathfrak{g}), \tag{5.165}$$

where $P\mathfrak{g}$ is the Lie algebra of based paths in $\mathfrak{g}$ and $\hat{\Omega}\mathfrak{g}$ the Lie algebra of based loops in $\mathfrak{g}$, with central extension induced by μ. Details are in [7].

Proposition 5.35 ([7]). *The Lie 2-algebra $\mathfrak{g}_\mu$ obtained from $\mathfrak{g}$ and μ as in definition 5.28 is equivalent to the strict string Lie 2-algebra*

$$\mathfrak{g}_\mu \simeq \text{string}(\mathfrak{g})\,. \tag{5.166}$$

This means there are morphisms $\mathfrak{g}_\mu \to \text{string}(\mathfrak{g})$ and $\text{string}(\mathfrak{g}) \to \mathfrak{g}_\mu$ whose composite is the identity only up to homotopy

$$\mathfrak{g}_\mu \longrightarrow \text{string}(\mathfrak{g}) \longrightarrow \mathfrak{g}_\mu \qquad \text{string}(\mathfrak{g}) \longrightarrow \mathfrak{g}_\mu \longrightarrow \text{string}(\mathfrak{g}) \tag{5.167}$$

We call $\mathfrak{g}_\mu$ the *skeletal* and $\text{string}(\mathfrak{g})$ the *strict* version of the String Lie 2-algebra.

The Fivebrane Lie 6-algebra.

Definition 5.36. *Let $\mathfrak{g} = \mathfrak{so}(n)$ and μ the canonical 7-cocycle on it. Then*

$$\text{fivebrane}(\mathfrak{g}) \tag{5.168}$$

is defined to be the strict Lie 7-algebra which is equivalent to $\mathfrak{g}_\mu$

$$\mathfrak{g}_\mu \simeq \text{fivebrane}(\mathfrak{g})\,. \tag{5.169}$$

A Lie n-algebra is *strict* if it corresponds to a differential graded Lie algebra on a vector space in degree 1 to n. (Recall our grading conventions from 5.1.) The strict Lie n-algebras corresponding to certain weak Lie n-algebras can be found by first integrating to strict Lie n-groups and then differentiating the result again. This shall not concern us here, but discussion of this point can be found in [71] and [70].

The BF-theory Lie 3-algebra.

Definition 5.37. *For $\mathfrak{g}$ any ordinary Lie algebra with bilinear invariant symmetric form $\langle \cdot, \cdot \rangle \in \text{inv}(\mathfrak{g})$ in transgression with the 3-cocycle μ, and for $\mathfrak{h} \xrightarrow{t} \mathfrak{g}$ a strict Lie 2-algebra based on $\mathfrak{g}$, denote by*

$$\hat{\mu} := d_{\text{CE}(\mathfrak{h} \xrightarrow{t} \mathfrak{g})} \mu \tag{5.170}$$

the corresponding exact 4-cocycle on $(\mathfrak{h} \xrightarrow{t} \mathfrak{g})$ discussed in 5.3.1. Then we call the String-like extended Lie 3-algebra

$$\mathfrak{bf}(\mathfrak{h} \xrightarrow{t} \mathfrak{g}) := (\mathfrak{h} \xrightarrow{t} \mathfrak{g})_{\hat{\mu}} \tag{5.171}$$

the corresponding BF-theory Lie 3-algebra.

The terminology here will become clear once we describe in 7.3.1 and 8.1.1 how the BF-theory action functional discussed in 5.6.1 arises as the parallel 4-transport given by the $b^3\mathfrak{u}(1)$-4-bundle which arises as the obstruction to lifting $(\mathfrak{h} \xrightarrow{t} \mathfrak{g})$-2-descent objects to $\mathfrak{bf}(\mathfrak{h} \xrightarrow{t} \mathfrak{g})$-3-descent objects.

5.5. L_∞-algebra valued forms

Consider an ordinary Lie algebra $\mathfrak{g}$ valued connection form A regarded as a linear map $\mathfrak{g}^* \to \Omega^1(Y)$. Since $CE(\mathfrak{g})$ is free as a graded commutative algebra, this linear map extends uniquely to a morphism of graded commutative algebras, though not in general of differential graded commutative algebra. In fact, the deviation is measured by the *curvature* F_A of the connection. However, the differential in $W(\mathfrak{g})$ is precisely such that the connection does extend to a morphism of differential graded-commutative algebras

$$W(\mathfrak{g}) \to \Omega^\bullet(Y). \tag{5.172}$$

This implies that a good notion of a $\mathfrak{g}$-valued differential form on a smooth space Y, for $\mathfrak{g}$ any L_∞-algebra, is a morphism of differential graded-commutative algebras from the Weil algebra of $\mathfrak{g}$ to the algebra of differential forms on Y.

Definition 5.38 ($\mathfrak{g}$-valued forms). *For Y a smooth space and $\mathfrak{g}$ an L_∞-algebra, we call*

$$\Omega^\bullet(Y, \mathfrak{g}) := \mathrm{Hom}_{\mathrm{dgc-Alg}}(W(\mathfrak{g}), \Omega^\bullet(Y)) \tag{5.173}$$

the space of $\mathfrak{g}$-valued differential forms on X.

Definition 5.39 (Curvature). *We write $\mathfrak{g}$-valued differential forms as*

$$\left(\Omega^\bullet(Y) \xleftarrow{\;(A, F_A)\;} W(\mathfrak{g}) \right) \in \Omega^\bullet(Y, \mathfrak{g}), \tag{5.174}$$

where F_A denotes the restriction to the shifted copy $\mathfrak{g}^*[1]$ given by

$$\mathrm{curv} : \left(\Omega^\bullet(Y) \xleftarrow{\;(A, F_A)\;} W(\mathfrak{g}) \right) \mapsto \left(\Omega^\bullet(Y) \xleftarrow[\;(A, F_A)\;]{\overset{F_A}{\frown}} W(\mathfrak{g}) \longleftarrow \mathfrak{g}^*[1] \right). \tag{5.175}$$

F_A we call the curvature of A.

Proposition 5.40. *The $\mathfrak{g}$-valued differential form $\Omega^\bullet(Y) \xleftarrow{\;(A, F_A)\;} W(\mathfrak{g})$ factors through $CE(\mathfrak{g})$ precisely when its curvature F_A vanishes.*

$$
\begin{array}{ccc}
CE(\mathfrak{g}) & \xleftarrow{\quad\quad} & W(\mathfrak{g}) \\[1mm]
\Big\downarrow{\scriptstyle (A, F_A=0)} & & \Big\downarrow{\scriptstyle (A, F_A)} \\[1mm]
\Omega^\bullet(Y) & \xequal{\quad\quad} & \Omega^\bullet(Y)
\end{array}
\tag{5.176}
$$

In this case we say that A is **flat**. Hence the space of flat $\mathfrak{g}$-valued forms is

$$\Omega^\bullet_{\mathrm{flat}}(Y, \mathfrak{g}) \simeq \mathrm{Hom}_{\mathrm{dgc-Alg}}(CE(\mathfrak{g}), \Omega^\bullet(Y)). \tag{5.177}$$

Bianchi identity. Recall from 5.1 that the Weil algebra $W(\mathfrak{g})$ of an L_∞-algebra $\mathfrak{g}$ is the same as the Chevalley-Eilenberg algebra $CE(\mathrm{inn}(\mathfrak{g}))$ of the L_∞-algebra of inner derivation of $\mathfrak{g}$. It follows that $\mathfrak{g}$-valued differential forms on Y are the same as *flat* $\mathrm{inn}(\mathfrak{g})$-valued differential forms on Y:

$$\Omega^\bullet(Y,\mathfrak{g}) = \Omega^\bullet_{\mathrm{flat}}(\mathrm{inn}(\mathfrak{g})).\tag{5.178}$$

By the above definition of curvature, this says that the curvature F_A of a $\mathfrak{g}$-valued connection (A, F_A) is itself a flat $\mathrm{inn}(\mathfrak{g})$-valued connection. This is the generalization of the ordinary *Bianchi identity* to L_∞-algebra valued forms.

Definition 5.41. *Two $\mathfrak{g}$-valued forms $A, A' \in \Omega^\bullet(Y,\mathfrak{g})$ are called* **(gauge) equivalent** *precisely if they are related by a* vertical *concordance, i.e. by a concordance, such that the corresponding derivation ρ from proposition 5.14 is vertical, in the sense of definition 4.9.*

5.5.1. Examples.
1. Ordinary Lie-algebra valued 1-forms. We have already mentioned ordinary Lie algebra valued 1-forms in this general context in 2.1.3.

2. Forms with values in shifted $b^{n-1}\mathfrak{u}(1)$. A $b^{n-1}\mathfrak{u}(1)$-valued form is nothing but an ordinary n-form $A \in \Omega^n(Y)$:

$$\Omega^\bullet(b^{n-1}\mathfrak{u}(1), Y) \simeq \Omega^n(Y).\tag{5.179}$$

A flat $b^{n-1}\mathfrak{u}(1)$-valued form is precisely a closed n-form.

3. Crossed module valued forms. Let $\mathfrak{g}_{(2)} = (\mathfrak{h} \xrightarrow{t} \mathfrak{g})$ be a strict Lie 2-algebra coming from a crossed module. Then a $\mathfrak{g}_{(2)}$-valued form is an ordinary $\mathfrak{g}$-valued 1-form A and an ordinary $\mathfrak{h}$-valued 2-form B. The corresponding curvature is an ordinary $\mathfrak{g}$-valued 2-form $\beta = F_A + t(B)$ and an ordinary $\mathfrak{h}$-valued 3-form $H = d_A B$. This is denoted by the right vertical arrow in the following diagram.

(5.180)

Precisely if the curvature components β and H vanish, does this morphism on the right factor through $\mathrm{CE}(\mathfrak{h} \xrightarrow{t} \mathfrak{g})$, which is indicated by the left vertical arrow of the above diagram.

4. String Lie n-algebra valued forms. For $\mathfrak{g}$ an ordinary Lie algebra and μ a degree $(2n + 1)$-cocycle on $\mathfrak{g}$ the situation is captured by the following diagram

$$\tag{5.181}$$

String-like	Chern-Simons	Chern
1	$2n$	$2n + 1$	$2n + 1$

$$\mathrm{CE}(\mathfrak{g}) \hookrightarrow \mathrm{CE}(\mathfrak{g}_\mu) \longleftarrow \mathrm{CE}(\mathrm{cs}_P(\mathfrak{g})) \longleftarrow \mathrm{CE}(\mathrm{ch}_P(\mathfrak{g}))$$

$$
\begin{array}{cccc}
(A) & (A,B) & (A,B,C) & (A,C) \\
F_A=0 & \begin{array}{c} F_A=0 \\ dB+\mathrm{CS}_k(A)=0 \end{array} & C=dB+\mathrm{CS}_P(A) & dC=k((F_A)^{n+1}) \\
\end{array}
$$

$$\Omega^\bullet(Y) \overset{=}{\longrightarrow} \Omega^\bullet(Y) \overset{=}{\longrightarrow} \Omega^\bullet(Y) \overset{=}{\longrightarrow} \Omega^\bullet(Y)$$

Here $\mathrm{CS}_P(A)$ denotes the Chern-Simons form such that $d\mathrm{CS}_P(A) = P(F_A)$, given by the specific contracting homotopy.

The standard example is that corresponding to the ordinary String-extension.

$$\mathrm{CE}(\mathfrak{g}) \hookrightarrow \mathrm{CE}(\mathrm{string}(\mathfrak{g})) \longleftarrow W(\mathrm{string}_k(\mathfrak{g})) \tag{5.182}$$

$$
\begin{array}{ccc}
\| & \|{\simeq} & \|{\simeq} \\
\end{array}
$$

$$\mathrm{CE}(\mathfrak{g}) \hookrightarrow \mathrm{CE}(\mathfrak{g}_\mu) \longleftarrow \mathrm{CE}(\mathrm{cs}_k(\mathfrak{g})) \longleftarrow \mathrm{CE}(\mathrm{ch}_P(\mathfrak{g}))$$

$$
\begin{array}{cccc}
(A) & (A,B) & (A,B,C) & (A,C) \\
F_A=0 & \begin{array}{c} F_A=0 \\ dB+\mathrm{CS}_P(A)=0 \end{array} & C=dB+\mathrm{CS}_P(A) & dC=\langle F_A \wedge F_A \rangle \\
\end{array}
$$

$$\Omega^\bullet(Y) \overset{=}{\longrightarrow} \Omega^\bullet(Y) \overset{=}{\longrightarrow} \Omega^\bullet(Y) \overset{=}{\longrightarrow} \Omega^\bullet(Y)$$

In the above, $\mathfrak{g}$ is semisimple with invariant bilinear form $P = \langle \cdot, \cdot \rangle$ related by transgression to the 3-cocycle $\mu = \langle \cdot, [\cdot, \cdot] \rangle$. Then the Chern-Simons 3-form for any $\mathfrak{g}$-valued 1-form A is

$$\mathrm{CS}_{\langle \cdot, \cdot \rangle}(A) = \langle A \wedge dA \rangle + \frac{1}{3}\langle A \wedge [A \wedge A] \rangle. \tag{5.183}$$

5. Fields of 11-dimensional supergravity. While we shall not discuss it in detail here, it is clear that the entire discussion we gave has a straightforward generalization to *super L_∞-algebras*, obtained simply by working entirely within the category of super vector spaces (the category of $\mathbb{Z}_2$-graded vector spaces equipped with the unique non-trivial symmetric braiding on it, which introduces a sign whenever two odd-graded vector spaces are interchanged).

A glance at the definitions shows that, up to mere differences in terminology, the theory of "FDA"s ("free differential algebras") considered in [4, 24] is nothing but that of what we call qDGCAs here: quasi-free differential graded commutative algebras.

Using that and the interpretation of qDGCAs in terms of L_∞-algebras, one can translate everything said in [4, 24] into our language here to obtain the following statement:

The field content of 11-dimensional supergravity is nothing but a $\mathfrak{g}$-valued form, for

$$\mathfrak{g} = \mathfrak{sugra}(10,1) \tag{5.184}$$

the Lie 3-algebra which is the String-like extension

$$0 \to b^2\mathfrak{u}(1) \to \mathfrak{sugra}(10,1) \to \mathfrak{siso}(10,1) \to 0 \tag{5.185}$$

of the *super-Poincaré* Lie algebra in 10+1 dimensions, coming from a certain 4-cocycle on that.

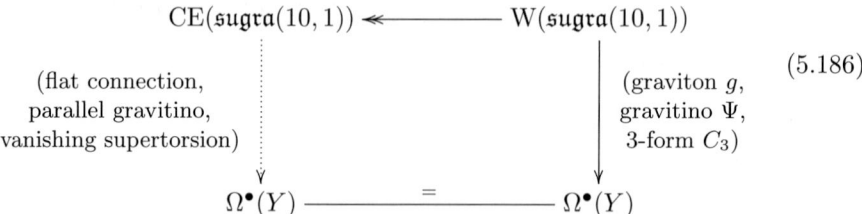

$$\tag{5.186}$$

While we shall not further pursue this here, this implies the following two interesting issues.

- It is known in string theory [28] that the supergravity 3-form in fact consists of three parts: two Chern-Simons parts for an $\mathfrak{e}_8$ and for a $\mathfrak{so}(10,1)$-connection, as well as a further fermionic part, coming precisely from the 4-cocycle that governs $\mathfrak{sugra}(10,1)$. As we discuss in 7 and 8, the two Chern-Simons components can be understood in terms of certain Lie 3-algebra connections coming from the Chern-Simons Lie 3-algebra $\mathrm{cs}_P(\mathfrak{g})$ from definition 5.28. It hence seems that there should be a Lie n-algebra which nicely unifies $\mathrm{cs}_P(\mathfrak{e}_8)$, $\mathrm{cs}_P(\mathfrak{so}(10,1)_8)$ and $\mathfrak{sugra}(10,1)$. This remains to be discussed.
- The discussion in section 6 shows how to obtain from $\mathfrak{g}$-valued forms globally defined connections on possibly nontrivial $\mathfrak{g}$-n-bundles. Applied to $\mathfrak{sugra}(10,1)$ this should yield a *global* description of the supergravity field content, which extends the local field content considered in [4, 24] in the way a connection in a possibly nontrivial Yang-Mills bundle generalizes a Lie algebra valued 1-form. This should for instance allow to discuss supergravity *instanton* solutions.

5.6. L_∞-algebra characteristic forms

Definition 5.42. *For $\mathfrak{g}$ any L_∞ algebra and*

$$\Omega^\bullet(Y) \xleftarrow{(A,F_A)} W(\mathfrak{g}) \tag{5.187}$$

any $\mathfrak{g}$-valued differential form, we call the composite

$$\{P(F_A)\}$$
$$\Omega^\bullet(Y) \xleftarrow{(A,F_A)} W(\mathfrak{g}) \xleftarrow{\quad} \text{inv}(\mathfrak{g}) \tag{5.188}$$

the collection of **invariant forms** *of the $\mathfrak{g}$-valued form A. We call the de Rham classes $[P(F_A)]$ of the characteristic forms arising as the image of closed invariant polynomials*

$$\{P_i(F_A)\}$$
$$\Omega^\bullet(Y) \xleftarrow{(A,F_A)} W(\mathfrak{g}) \xleftarrow{\quad} \text{inv}(\mathfrak{g}) \tag{5.189}$$

$$H^\bullet_{\mathrm{dR}}(Y) \xleftarrow{\{[P(F_A)]\}} H^\bullet(\text{inv}(\mathfrak{g}))$$

the collection of **characteristic classes** *of the $\mathfrak{g}$-valued form A.*

Recall from section 5.3 that for ordinary Lie algebras all invariant polynomials are closed, while for general L_∞-algebras it is only true that their $d_{W(\mathfrak{g})}$-differential is horizontal. Notice that Y will play the role of a cover of some space X soon, and that characteristic forms really live down on X. We will see shortly a constraint imposed which makes the characteristic forms descend down from the Y here to such an X.

Proposition 5.43. *Under gauge transformations as in definition 5.41, characteristic classes are invariant.*

Proof. This follows from proposition 5.14: By that proposition, the derivative of the concordance form $\hat{A}$ along the interval $I = [0,1]$ is a chain homotopy

$$\frac{d}{dt}\hat{A}(P) = [d, \iota_X]P = d\tau(P) + \iota_X(d_{W(\mathfrak{g})}P). \tag{5.190}$$

By definition of gauge-transformations, ι_X is vertical. By definition of basic forms, P is both in the kernel of ι_X as well as in the kernel of $\iota_X \circ d$. Hence the right hand vanishes. $\qquad\square$

5.6.1. Examples.
Characteristic forms of $b^{n-1}\mathfrak{u}(1)$-valued forms.

Proposition 5.44. *A $b^{n-1}\mathfrak{u}(1)$-valued form $\Omega^\bullet(Y) \xleftarrow{\quad A \quad} W(b^{n-1}\mathfrak{u}(1))$ is precisely an n-form on Y:*

$$\Omega^\bullet(Y, b^{n-1}\mathfrak{u}(1)) \simeq \Omega^n(Y). \tag{5.191}$$

If two such $b^{n-1}\mathfrak{u}(1)$-valued forms are gauge equivalent according to definition 5.41, then their curvatures coincide

$$(\Omega^\bullet(Y) \xleftarrow{\quad A \quad} W(b^{n-1}\mathfrak{u}(1))) \sim (\Omega^\bullet(Y) \xleftarrow{\quad A' \quad} W(b^{n-1}\mathfrak{u}(1))) \quad \Rightarrow \quad dA = dA'. \tag{5.192}$$

BF-theory. We demonstrate that the expression known in the literature as the *action functional for BF-theory with cosmological term* is the integral of an invariant polynomial for $\mathfrak{g}$-valued differential forms where $\mathfrak{g}$ is a Lie 2-algebra. Namely, let $\mathfrak{g}_{(2)} = (\mathfrak{h} \xrightarrow{\ t\ } \mathfrak{g})$ be any strict Lie 2-algebra as in section 5.1. Let

$$P = \langle \cdot, \cdot \rangle \tag{5.193}$$

be an invariant bilinear form on $\mathfrak{g}$, hence a degree 2 invariant polynomial on $\mathfrak{g}$. According to proposition 5.26, P therefore also is an invariant polynomial on $\mathfrak{g}_{(2)}$.

Now for (A, B) a $\mathfrak{g}_{(2)}$-valued differential form on X, as in the example in section 5.5,

$$\Omega^\bullet(Y) \xleftarrow{\ ((A,B),(\beta,H))\ } W(\mathfrak{g}_{(2)}) \ , \tag{5.194}$$

one finds

$$\Omega^\bullet(Y) \xleftarrow{\ ((A,B),(\beta,H))\ } W(\mathfrak{g}_{(2)}) \xleftarrow{\qquad} \mathrm{inv}(\mathfrak{g}_{(2)}) \tag{5.195}$$
$$P \mapsto \langle \beta, \beta \rangle$$

so that the corresponding characteristic form is the 4-form

$$P(\beta, H) = \langle \beta \wedge \beta \rangle = \langle (F_A + t(B)) \wedge (F_A + t(B)) \rangle. \tag{5.196}$$

Collecting terms as

$$P(\beta, H) = \underbrace{\langle F_A \wedge F_A \rangle}_{\text{Pontryagin term}} + 2\underbrace{\langle t(B) \wedge F_A \rangle}_{\text{BF-term}} + \underbrace{\langle t(B) \wedge t(B) \rangle}_{\text{``cosmological constant''}} \tag{5.197}$$

we recognize the Lagrangian for topological Yang-Mills theory and BF theory with cosmological term.

For X a compact 4-manifold, the corresponding action functional

$$S : \Omega^\bullet(X, \mathfrak{g}_{(2)}) \to \mathbb{R} \tag{5.198}$$

sends $\mathfrak{g}_{(2)}$-valued 2-forms to the integral of this 4-form

$$(A, B) \mapsto \int_X (\langle F_A \wedge F_A \rangle + 2\langle t(B) \wedge F_A \rangle + \langle t(B) \wedge t(B) \rangle) . \tag{5.199}$$

The first term here is usually not considered an intrinsic part of BF-theory, but its presence does not affect the critical points of S.

The critical points of S, i.e. the $\mathfrak{g}_{(2)}$-valued differential forms on X that satisfy the equations of motion defined by the action S, are given by the equation

$$\beta := F_A + t(B) = 0 \,. \tag{5.200}$$

Notice that this implies

$$d_A t(B) = 0 \tag{5.201}$$

but does not constrain the full 3-curvature

$$H = d_A B \tag{5.202}$$

to vanish. In other words, the critical points of S are precisely the *fake flat* $\mathfrak{g}_{(2)}$-valued forms which precisely integrate to strict parallel transport 2-functors [38, 73, 9].

While the 4-form $\langle \beta \wedge \beta \rangle$ looks similar to the Pontrjagin 4-form $\langle F_A \wedge F_A \rangle$ for an ordinary connection 1-form A, one striking difference is that $\langle \beta \wedge \beta \rangle$ is, in general, not closed. Instead, according to equation (5.127), we have

$$d\langle \beta \wedge \beta \rangle = 2\langle \beta \wedge t(H) \rangle \,. \tag{5.203}$$

Remark. Under the equivalence [7] of the skeletal String Lie 2-algebra to its strict version, recalled in proposition 5.35, the characteristic forms for strict Lie 2-algebras apply also to one of our central objects of interest here, the String 2-connections. But a little care needs to be exercised here, because the strict version of the String Lie 2-algebra is no longer finite dimensional.

Remark. Our interpretation above of BF-theory as a gauge theory for Lie 2-algebras is not unrelated to, but different from the one considered in [38, 39]. There only the Lie 2-algebra coming from the infinitesimal crossed module $(|\mathfrak{g}| \xrightarrow{0} \mathfrak{g} \xrightarrow{\mathrm{ad}} \mathrm{der}(\mathfrak{g}))$ (for $\mathfrak{g}$ any ordinary Lie algebra and $|\mathfrak{g}|$ its underlying vector space, regarded as an abelian Lie algebra) is considered, and the action is restricted to the term $\int \langle F_A \wedge B \rangle$. We can regard the above discussion as a generalization of this approach to arbitrary Lie 2-algebras. Standard BF-theory (with "cosmological" term) is reproduced with the above Lagrangian by using the Lie 2-algebra $\mathrm{inn}(\mathfrak{g})$ corresponding to the infinitesimal crossed module $(\mathfrak{g} \xrightarrow{\mathrm{Id}} \mathfrak{g} \xrightarrow{\mathrm{ad}} \mathrm{der}(\mathfrak{g}))$ discussed in section 5.1.1.

6. L_∞-algebra Cartan-Ehresmann connections

We will now combine all of the above ingredients to produce a definition of $\mathfrak{g}$-valued connections. As we shall explain, the construction we give may be thought of as a generalization of the notion of a Cartan-Ehresmann connection, which is given by a Lie algebra-valued 1-form on the total space of a bundle over base space satisfying two conditions:

- first Cartan-Ehresmann condition: on the fibers the connection form restricts to a *flat* canonical form
- second Cartan-Ehresmann condition: under vertical flows the connections transforms nicely, in such a way that its characteristic forms descend down to base space.

We will essentially interpret these two conditions as a pullback of the universal $\mathfrak{g}$-bundle, in its DGC-algebraic incarnation as given in equation (5.94).

The definition we give can also be seen as the Lie algebraic image of a similar construction involving locally trivializable transport n-functors [9, 74], but this shall not be further discussed here.

6.1. $\mathfrak{g}$-bundle descent data

Definition 6.1 ($\mathfrak{g}$-bundle descent data). *Given a Lie n-algebra $\mathfrak{g}$, a $\mathfrak{g}$-bundle descent object on X is a pair (Y, A_{vert}) consisting of a choice of surjective submersion $\pi : Y \to X$ with connected fibers (this condition will be dropped when we extend to $\mathfrak{g}$-connection descent objects in (6.2) together with a morphism of dg-algebras*

$$\Omega^\bullet_{\mathrm{vert}}(Y) \xleftarrow{\quad A_{\mathrm{vert}} \quad} CE(\mathfrak{g}) \ . \tag{6.1}$$

Two such descent objects are taken to be equivalent

$$(\ \Omega^\bullet_{\mathrm{vert}}(Y) \xleftarrow{A_{\mathrm{vert}}} CE(\mathfrak{g}) \ ) \sim (\ \Omega^\bullet_{\mathrm{vert}}(Y') \xleftarrow{A'_{\mathrm{vert}}} CE(\mathfrak{g}) \ ) \tag{6.2}$$

precisely if their pullbacks $\pi_1^* A_{\mathrm{vert}}$ and $\pi_2^* A'_{\mathrm{vert}}$ to the common refinement

$$
\begin{array}{ccc}
Y \times_X Y' & \xrightarrow{\ \pi_1\ } & Y \\
{\scriptstyle \pi_2}\downarrow & & \downarrow{\scriptstyle \pi} \\
Y' & \xrightarrow{\ \pi'\ } & X
\end{array}
\tag{6.3}
$$

are concordant in the sense of definition 5.13.

Thus two such descent objects A_{vert}, A'_{vert} on the same Y are equivalent if there is η^*_{vert} such that

$$\Omega^\bullet_{\mathrm{vert}}(Y) \underset{\underset{t^*}{\xleftarrow{\ s^*\ }}}{} \Omega^\bullet_{\mathrm{vert}}(Y \times I) \xleftarrow{\ \eta^*_{\mathrm{vert}}\ } CE(\mathfrak{g}) \ . \tag{6.4}$$

Recall from the discussion in section 2.2.1 that the surjective submersions here play the role of open covers of X.

6.1.1. Examples.

Ordinary G-bundles. The following example is meant to illustrate how the notion of descent data with respect to a Lie algebra $\mathfrak{g}$ as defined here can be related to the ordinary notion of descent data with respect to a Lie group G. Consider the case where $\mathfrak{g}$ is an ordinary Lie (1-)algebra. A $\mathfrak{g}$-cocycle then is a surjective submersion $\pi : Y \to X$ together with a $\mathfrak{g}$-valued flat vertical 1-form A_{vert} on Y. Assume the fiber of $\pi : Y \to X$ to be simply connected. Then for any two points $(y, y') \in Y \times_X Y$ in the same fiber we obtain an element $g(y, y') \in G$, where G is the simply connected Lie group integrating $\mathfrak{g}$, by choosing any path $y \xrightarrow{\gamma} y'$ in the fiber connecting y with y' and forming the parallel transport determined by A_{vert} along this path

$$g(y, y') := P \exp\left(\int_\gamma A_{\mathrm{vert}} \right). \tag{6.5}$$

By the flatness of A_{vert} and the assumption that the fibers of Y are simply connected

- $g : Y \times_X Y \to G$ is well defined (does not depend on the choice of paths), and
- satisfies the cocycle condition for G-bundles

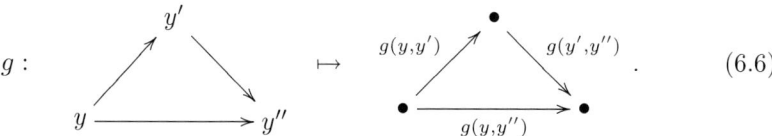

$$\tag{6.6}$$

Any such cocycle g defines a G-principal bundle. Conversely, every G-principal bundle $P \to X$ gives rise to a structure like this by choosing $Y := P$ and letting A_{vert} be the canonical invariant $\mathfrak{g}$-valued vertical 1-form on $Y = P$. Then suppose (Y, A_{vert}) and (Y, A'_{vert}) are two such cocycles defined on the same Y, and let $(\hat{Y} := Y \times I, \hat{A}_{\mathrm{vert}})$ be a concordance between them. Then, for every path

$$y \times \{0\} \xrightarrow{\gamma} y \times \{1\} \tag{6.7}$$

connecting the two copies of a point $y \in Y$ over the endpoints of the interval, we again obtain a group element

$$h(y) := P \exp\left(\int_\gamma \hat{A}_{\mathrm{vert}} \right). \tag{6.8}$$

By the flatness of $\hat{A}$, this is

- well defined in that it is independent of the choice of path;

- has the property that for all $(y, y') \in Y \times_X Y$ we have

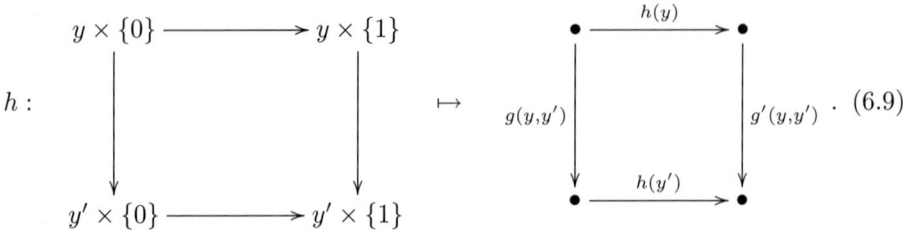

$$(6.9)$$

Therefore h is a gauge transformation between g and g', as it should be.

Note that there is no holonomy since the fibers are assumed to be simply connected in this example.

Abelian gerbes, Deligne cohomology and $(b^{n-1}\mathfrak{u}(1))$-descent objects. For the case that the L_∞-algebra in question is shifted $\mathfrak{u}(1)$, i.e. $\mathfrak{g} = b^{n-1}\mathfrak{u}(1)$, classes of $\mathfrak{g}$-descent objects on X should coincide with classes of "line n-bundles", i.e. with classes of abelian $(n-1)$-gerbes on X, hence with elements in $H^n(X, \mathbb{Z})$. In order to understand this, we relate classes of $b^{n-1}\mathfrak{u}(1))$-descent objects to Deligne cohomology. We recall Deligne cohomology for a fixed surjective submersion $\pi : Y \to X$. For comparison with some parts of the literature, the reader should choose Y to be the disjoint union of sets of a good cover of X. More discussion of this point is in section 4.3.1.

The following definition should be thought of this way: a collection of p-forms on fiberwise intersections of a surjective submersion $Y \to X$ are given. The 0-form part defines an n-bundle (an $(n-1)$-gerbe) itself, while the higher forms encode a connection on that n-bundle.

Definition 6.2 (Deligne cohomology). *Deligne cohomology can be understood as the cohomology on differential forms on the simplicial space $Y^\bullet$ given by a surjective submersion $\pi : Y \to X$, where the complex of forms is taken to start as*

$$0 \longrightarrow C^\infty(Y^{[n]}, \mathbb{R}/\mathbb{Z}) \xrightarrow{d} \Omega^1(Y^{[n]}, \mathbb{R}) \xrightarrow{d} \Omega^2(Y^{[n]}, \mathbb{R}) \xrightarrow{d} \cdots, \quad (6.10)$$

where the first differential, often denoted dlog in the literature, is evaluated by acting with the ordinary differential on any $\mathbb{R}$-valued representative of a $U(1) \simeq \mathbb{R}/\mathbb{Z}$-valued function.

More in detail, given a surjective submersion $\pi : Y \to X$, we obtain the augmented simplicial space

$$Y^\bullet = \left(\cdots Y^{[3]} \underset{\substack{\longrightarrow \\ \pi_3}}{\overset{\substack{\pi_1 \\ \longrightarrow}}{\underset{-\pi_2 \longrightarrow}{}}} Y^{[2]} \underset{\substack{\longrightarrow \\ \pi_2}}{\overset{\pi_1}{\longrightarrow}} Y \xrightarrow{\pi} Y^{[0]} \right) \qquad (6.11)$$

of fiberwise cartesian powers of Y, $Y^{[n]} := \underbrace{Y \times_X Y \times_X \cdots \times_X Y}_{n \text{ factors}}$, with $Y^{[0]} := X$.

The double complex of differential forms

$$\Omega^\bullet(Y^\bullet) = \bigoplus_{n \in \mathbb{N}} \Omega^n(Y^\bullet) = \bigoplus_{n \in \mathbb{N}} \bigoplus_{\substack{r,s \in \mathbb{N} \\ r+s=n}} \Omega^r(Y^{[s]}) \tag{6.12}$$

on $Y^\bullet$ has the differential $d \pm \delta$ coming from the de Rham differential d and the alternating pullback operation

$$\delta : \Omega^r(Y^{[s]}) \quad \to \quad \Omega^r(Y^{[s+1]})$$
$$\delta : \omega \quad \mapsto \quad \pi_1^* \omega - \pi_2^* \omega + \pi_3^* \omega + \cdots - (-1)^{s+1} \ . \tag{6.13}$$

Here we take 0-forms to be valued in $\mathbb{R}/\mathbb{Z}$. The map $\Omega^0(Y) \xrightarrow{\ d\ } \Omega^1(Y)$ takes any $\mathbb{R}$-valued representative f of an $\mathbb{R}/\mathbb{Z}$-valued form and sends that to the ordinary df. This operation is often denoted $\Omega^0(Y) \xrightarrow{\ dlog\ } \Omega^1(Y)$. Writing $\Omega^\bullet_k(Y^\bullet)$ for the space of forms that vanish on $Y^{[l]}$ for $l < k$ we define (everything with respect to Y):

- *A **Deligne n-cocycle** is a closed element in $\Omega^n(Y^\bullet)$;*
- *a **flat Deligne n-cocycle** is a closed element in $\Omega^n_1(Y^\bullet)$;*
- *a **Deligne coboundary** is an element in $(d \pm \delta)\Omega^\bullet_1(Y^\bullet)$ (i.e. no component in $Y^{[0]} = X$);*
- *a **shift of connection** is an element in $(d \pm \delta)\Omega^\bullet(Y^\bullet)$ (i.e. with possibly a contribution in $Y^{[0]} = X$).*

The 0-form part of a Deligne cocycle is like the transition function of a $U(1)$-bundle. Restricting to this part yields a group homomorphism

$$[\cdot] : \ H^n(\Omega^\bullet(Y^\bullet)) \longrightarrow H^n(X, \mathbb{Z}) \tag{6.14}$$

to the integral cohomology on X. (Notice that the degree on the right is indeed as given, using the total degree on the double complex $\Omega^\bullet(Y^\bullet)$ as given.)

Addition of a Deligne coboundary is a gauge transformation. Using the fact [62] that the "fundamental complex"

$$\Omega^r(X) \xrightarrow{\ \delta\ } \Omega^r(Y) \xrightarrow{\ \delta\ } \Omega^r(Y^{[2]}) \cdots \tag{6.15}$$

is exact for all $r \geq 1$, one sees that Deligne cocycles with the same class in $H^n(X, \mathbb{Z})$ differ by elements in $(d \pm \delta)\Omega^\bullet(Y^\bullet)$. Notice that they do not, in general, differ by an element in $\Omega^\bullet_1(Y^\bullet)$: two Deligne cochains which differ by an element in $(d \pm \delta)\Omega^\bullet_1(Y^\bullet)$ describe equivalent line n-bundles *with equivalent connections*, while those that differ by something in $(d \pm \delta)\Omega^\bullet_0(Y^\bullet)$ describe equivalent line n-bundles with possibly inequivalent connections on them.

Let

$$v : \Omega^\bullet(Y^\bullet) \to \Omega^\bullet_{\text{vert}}(Y) \tag{6.16}$$

be the map which sends each Deligne n-cochain a with respect to Y to the vertical part of its $(n-1)$-form on $Y^{[1]}$

$$\nu : a \mapsto a|_{\Omega^{n-1}_{\text{vert}}(Y^{[1]})} . \tag{6.17}$$

(Recall the definition 4.11 of $\Omega^{\bullet}_{\text{vert}}(Y)$.) Then we have

Proposition 6.3. *If two Deligne n-cocycles a and b over Y have the same class in $H^n(X, \mathbb{Z})$, then the classes of $\nu(a)$ and $\nu(b)$ coincide.*

Proof. As mentioned above, a and b have the same class in $H^n(X, \mathbb{Z})$ if and only if they differ by an element in $(d \pm \delta)(\Omega^{\bullet}(Y^{\bullet}))$. This means that on $Y^{[1]}$ they differ by an element of the form

$$d\alpha + \delta\beta = d\alpha + \pi^*\beta . \tag{6.18}$$

Since $\pi^*\beta$ is horizontal, this is exact in $\Omega^{\bullet}_{\text{vert}}(Y^{[1]})$. □

Proposition 6.4. *If the $(n-1)$-form parts $B, B' \in \Omega^{n-1}(Y)$ of two Deligne n-cocycles differ by a $d \pm \delta$-exact part, then the two Deligne cocycles have the same class in $H^n(X, \mathbb{Z})$.*

Proof. If the surjective submersion is not yet contractible, we pull everything back to a contractible refinement, as described in section 4.3.1. So assume without restriction of generality that all $Y^{[n]}$ are contractible. This implies that $H^{\bullet}_{\text{deRham}}(Y^{[n]}) = H^0(Y^{[n]})$, which is a vector space spanned by the connected components of $Y^{[n]}$. Now assume

$$B - B' = d\beta + \delta\alpha \tag{6.19}$$

on Y. We can immediately see that this implies that the real classes in $H^n(X, \mathbb{R})$ coincide: the Deligne cocycle property says

$$d(B - B') = \delta(H - H') \tag{6.20}$$

hence, by the exactness of the de Rham complex we have now,

$$\delta(H - H') = \delta(d\alpha) \tag{6.21}$$

and by the exactness of δ we get $[H] = [H']$.

To see that also the integral classes coincide we use induction over k in $Y^{[k]}$. For instance on $Y^{[2]}$ we have

$$\delta(B - B') = d(A - A') \tag{6.22}$$

and hence

$$\delta d\beta = d(A - A') . \tag{6.23}$$

Now using again the exactness of the de Rham differential d this implies

$$A - A' = \delta\beta + d\gamma . \tag{6.24}$$

This way we work our way up to $Y^{[n]}$, where it then follows that the 0-form cocycles are coboundant, hence that they have the same class in $H^n(X, \mathbb{Z})$. □

Proposition 6.5. $b^{n-1}\mathfrak{u}(1)$-*descent objects with respect to a given surjective submersion Y are in bijection with closed vertical n-forms on Y:*

$$\left\{ \Omega^\bullet_{\mathrm{vert}}(Y) \xleftarrow{\ A_{\mathrm{vert}}\ } \mathrm{CE}(b^{n-1}\mathfrak{u}(1)) \right\} \ \leftrightarrow\ \{A_{\mathrm{vert}} \in \Omega^n_{\mathrm{vert}}(Y),\ dA_{\mathrm{vert}} = 0\}\,.$$
(6.25)

Two such $b^{n-1}\mathfrak{u}(1)$ descent objects on Y are equivalent precisely if these forms represent the same cohomology class

$$(A_{\mathrm{vert}} \sim A'_{\mathrm{vert}}) \ \Leftrightarrow\ [A_{\mathrm{vert}}] = [A'_{\mathrm{vert}}] \in H^n(\Omega^\bullet_{\mathrm{vert}}(Y))\,.$$
(6.26)

Proof. The first statement is a direct consequence of the definition of $b^{n-1}\mathfrak{u}(1)$ in section 5.1. The second statement follows from proposition 5.14 using the reasoning as in proposition 5.43. $\qquad\square$

Hence two Deligne cocycles with the same class in $H^n(X, \mathbb{Z})$ indeed specify the same class of $b^{n-1}\mathfrak{u}(1)$-descent data.

6.2. Connections on $\mathfrak{g}$-bundles: the extension problem

It turns out that a useful way to conceive of the curvature on a non-flat $\mathfrak{g}$ n-bundle is, essentially, as the $(n+1)$-*bundle with connection obstructing* the existence of a flat connection on the original $\mathfrak{g}$-bundle. This superficially trivial statement is crucial for our way of coming to grips with non-flat higher bundles with connection.

Definition 6.6 (Descent object for $\mathfrak{g}$-connection). *Given $\mathfrak{g}$-bundle descent object*

$$\Omega^\bullet_{\mathrm{vert}}(Y) \xleftarrow{\ A_{\mathrm{vert}}\ } \mathrm{CE}(\mathfrak{g})$$
(6.27)

as above, a $\mathfrak{g}$-connection on it is a completion of this morphism to a diagram

$$
\begin{array}{ccc}
\Omega^\bullet_{\mathrm{vert}}(Y) & \xleftarrow{\ A_{\mathrm{vert}}\ } & \mathrm{CE}(\mathfrak{g}) \\
\big\uparrow{\scriptstyle i^*} & & \big\uparrow \\
\Omega^\bullet(Y) & \xleftarrow{\ (A, F_A)\ } & W(\mathfrak{g}) \\
\big\uparrow{\scriptstyle \pi^*} & & \big\uparrow \\
\Omega^\bullet(X) & \xleftarrow{\ \{K_i\}\ } & \mathrm{inv}(\mathfrak{g})
\end{array}
$$
(6.28)

As before, two $\mathfrak{g}$-connection descent objects are taken to be equivalent, if their pullbacks to a common refinement are concordant.

The top square can always be completed: any representative $A \in \Omega^\bullet(Y)$ of $A_{\mathrm{vert}} \in \Omega^\bullet_{\mathrm{vert}}(Y)$ will do. The curvature F_A is then uniquely fixed by the dg-algebra homomorphism property. The existence of the top square then says that we have a 1-form on a total space which restricts to a canonical flat 1-form on the fibers. The commutativity of the lower square means that for all invariant polynomials P of $\mathfrak{g}$, the form $P(F_A)$ on Y is a form pulled back from X and is the differential of a form cs that vanishes on vertical vector fields

$$P(F_A) = \pi^* K . \tag{6.29}$$

The completion of the bottom square is hence an extra condition: it demands that A has been chosen such that its curvature F_A has the property that the form $P(F_A) \in \Omega^\bullet(Y)$ for all invariant polynomials P are lifted from base space, up to that exact part.

- The commutativity of the top square generalizes the **first Cartan-Ehresmann condition**: the connection form on the total space restricts to a nice form on the fibers.
- The commutativity of the lower square generalizes the **second Cartan-Ehresmann condition**: the connection form on the total space has to behave in such a way that the invariant polynomials applied to its curvature descend down to the base space.

The pullback

$$f^*(Y, (A, F_A)) = (Y', (f^*A, f^*F_A)) \tag{6.30}$$

of a $\mathfrak{g}$-connection descent object $(Y, (A, F_A))$ on a surjective submersion Y along a morphism

$$\tag{6.31}$$

is the $\mathfrak{g}$-connection descent object depicted in figure 9.

Notice that the characteristic forms remain unaffected by such a pullback. This way, any two $\mathfrak{g}$-connection descent objects may be pulled back to a common surjective submersion. A concordance between two $\mathfrak{g}$-connection descent objects on the same surjective submersion is depicted in figure 10.

Suppose (A, F_A) and $(A', F_{A'})$ are descent data for $\mathfrak{g}$-bundles with connection over the same Y (possibly after having pulled them back to a common refinement). Then a concordance between them is a diagram as in figure 10.

Definition 6.7 (Equivalence of $\mathfrak{g}$-connections). *We say that two $\mathfrak{g}$-connection descent objects are* equivalent as $\mathfrak{g}$-connection descent objects *if they are connected by a* vertical *concordance namely one for which the derivation part of η^* (according to proposition 5.14) vanishes on the shifted copy $\mathfrak{g}^*[1] \hookrightarrow W(\mathfrak{g})$.*

We have a closer look at concordance and equivalence of $\mathfrak{g}$-connection descent objects in section 6.3.

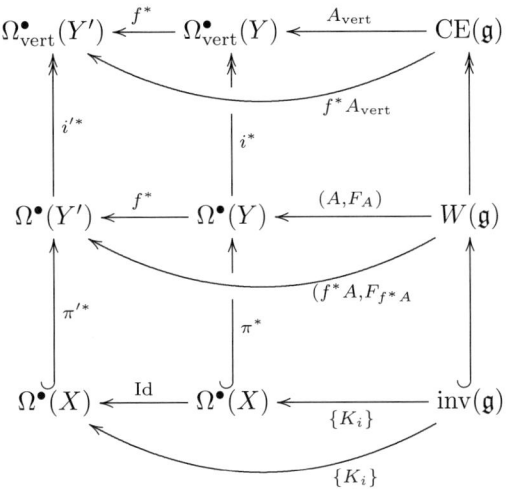

FIGURE 9. **Pullback of a $\mathfrak{g}$-connection descent object** $(Y, (A, F_A))$ along a morphism $f : Y' \to Y$ of surjective submersions, to $f^*(Y, (A, F_A)) = (Y', (f^*A, F_{f^*A}))$.

6.2.1. Examples.

Ordinary Cartan-Ehresmann connection. Let $P \to X$ be a principal G-bundle and consider the descent object obtained by setting $Y = P$ and letting A_{vert} be the canonical invariant vertical flat 1-form on fibers P. Then finding the morphism

$$\Omega^\bullet(Y) \xleftarrow{\quad (A,F_A) \quad} W(\mathfrak{g}) \tag{6.32}$$

such that the top square commutes amounts to finding a 1-form on the total space of the bundle which restricts to the canonical 1-form on the fibers. This is the first of the two conditions on a Cartan-Ehresmann connection. Then requiring the lower square to commute implies requiring that the $2n$-forms $P_i(F_A)$, formed from the curvature 2-form F_A and the degree n-invariant polynomials P_i of $\mathfrak{g}$, have to descend to $2n$-forms K_i on the base X. But that is precisely the case when $P_i(F_A)$ is invariant under flows along vertical vector fields. Hence it is true when A satisfies the second condition of a Cartan-Ehresmann connection, the one that says that the connection form transforms nicely under vertical flows.

Further examples appear in section 7.3.1.

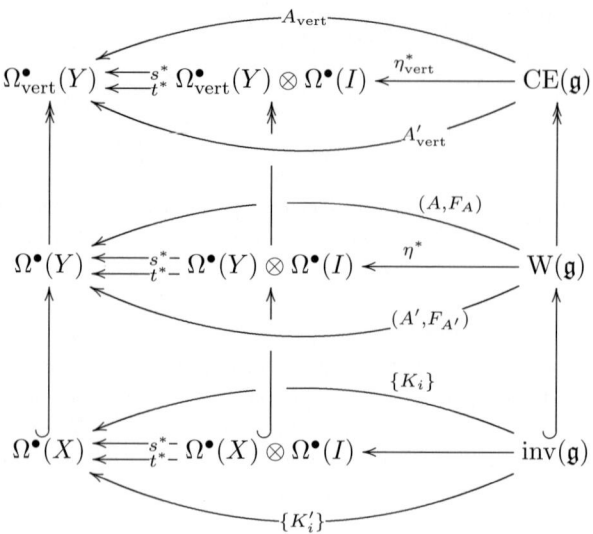

FIGURE 10. **Concordance between $\mathfrak{g}$-connection descent objects**
$(Y, (A, F_A))$ and $(Y, (A', F_{A'}))$ defined on the same surjective submersion $\pi : Y \to X$. Concordance between descent objects not on
the same surjective submersion is reduced to this case by pulling
both back to a common refinement, as in figure 9.

6.3. Characteristic forms and characteristic classes

Definition 6.8. *For any $\mathfrak{g}$-connection descent object $(Y, (A, F_A))$ we say that in*

$$
\begin{array}{ccc}
\Omega^\bullet_{\mathrm{vert}}(Y) & \xleftarrow{\quad A_{\mathrm{vert}} \quad} & \mathrm{CE}(\mathfrak{g}) \\
\uparrow{\scriptstyle i^*} & & \uparrow \\
\Omega^\bullet(Y) & \xleftarrow{\quad (A, F_A) \quad} & W(\mathfrak{g}) \\
\uparrow{\scriptstyle \pi^*} & & \uparrow \\
\Omega^\bullet(X) & \xleftarrow{\quad \{K_i\} \quad} & \mathrm{inv}(\mathfrak{g}) \\
\\
H^\bullet_{\mathrm{dR}}(X) & \xleftarrow{\quad \{[K_i]\} \quad} & H^\bullet(\mathrm{inv}(\mathfrak{g}))
\end{array}
\tag{6.33}
$$

the $\{K_i\}$ are the characteristic forms, *while their de Rham classes $[K_i] \in H^\bullet_{\mathrm{deRham}}(X)$ are the* characteristic classes *of $(Y, (A, F_A))$.*

Proposition 6.9. *If two $\mathfrak{g}$-connection descent objects $(Y, (A, F_A))$ and $(Y', (A', F_{A'}))$ are related by a concordance as in figure 9 and figure 10 then they have the same* *characteristic classes:*

$$(Y, (A, F_A)) \sim (Y', (A', F_{A'})) \;\Rightarrow\; \{[K_i]\} = \{[K_i']\}. \tag{6.34}$$

Proof. We have seen that pullback does not change the characteristic forms. It follows from proposition 5.43 that the characteristic classes are invariant under concordance. □

Corollary 6.10. *If two $\mathfrak{g}$-connection descent objects are equivalent according to definition 6.7, then they even have the same characteristic forms.*

Proof. For concordances between equivalent $\mathfrak{g}$-connection descent objects the derivation part of η^* is vertical and therefore vanishes on $\mathrm{inv}(\mathfrak{g}) = \mathrm{W}(\mathfrak{g})_{\mathrm{basic}}$. □

Remark (Shifts of $\mathfrak{g}$-connections). We observe that, by the very definition of $\mathrm{W}(\mathfrak{g})$, any shift in the connection A,

$$A \mapsto A' = A + D \in \Omega^\bullet(Y, \mathfrak{g})$$

can be understood as a transformation

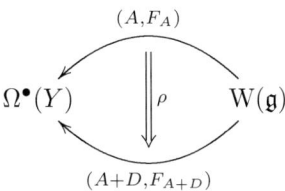

with the property that ρ vanishes on the *non*-shifted copy $\mathfrak{g}^* \hookrightarrow \mathrm{W}(\mathfrak{g})$ and is nontrivial only on the shifted copy $\mathfrak{g}^*[1] \hookrightarrow \mathrm{W}(\mathfrak{g})$: in that case for $a \in \mathfrak{g}^*$ any element in the unshifted copy, we have

$$
\begin{aligned}
(A + D)(a) &= A(a) + [d, \rho](a) = A(a) + \rho(d_{\mathrm{W}(\mathfrak{g})} a) \\
&= A(a) + \rho(d_{\mathrm{CE}(\mathfrak{g})} a + \sigma a) = A(a) + \rho(\sigma a).
\end{aligned}
$$

and hence $D(a) = \rho(\sigma)$, which uniquely fixes ρ in terms of D and vice versa.

Therefore concordances which are not purely vertical describe homotopies between $\mathfrak{g}$-connection descent objects in which the connection is allowed to vary.

Remark (Gauge transformations versus shifts of the connections). We therefore obtain the following picture.

- Vertical concordances relate gauge equivalent $\mathfrak{g}$-connections (compare definition 5.41 of gauge transformations of $\mathfrak{g}$-valued forms)

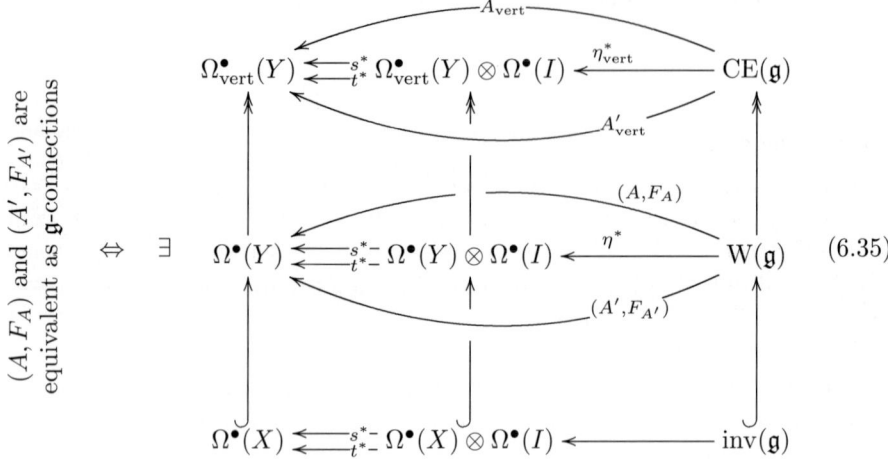

(A, F_A) and $(A', F_{A'})$ are equivalent as $\mathfrak{g}$-connections $\qquad \Leftrightarrow \quad \exists \qquad$ (6.35)

- Non-vertical concordances relate $\mathfrak{g}$-connection descent objects whose underlying $\mathfrak{g}$-descent object – the underlying $\mathfrak{g}$-n-bundles – are equivalent, but which possibly differ in the choice of connection on these $\mathfrak{g}$-bundles:

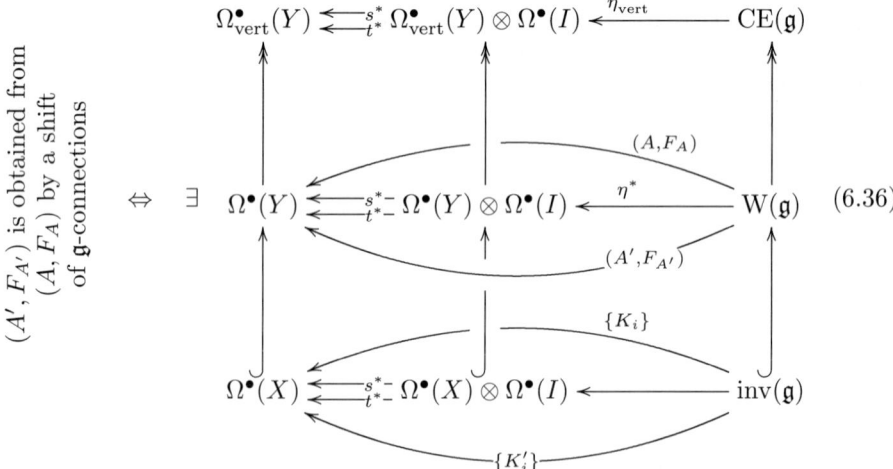

$(A', F_{A'})$ is obtained from (A, F_A) by a shift of $\mathfrak{g}$-connections $\qquad \Leftrightarrow \quad \exists \qquad$ (6.36)

Remark. This in particular shows that for a given $\mathfrak{g}$-descent object

$$\Omega^\bullet_{\mathrm{vert}}(Y) \xleftarrow{\;A_{\mathrm{vert}}\;} \mathrm{CE}(\mathfrak{g})$$

the corresponding characteristic classes obtained by choosing a connection (A, F_A) does not depend on continuous variations of that choice of connection.

In the case of ordinary Lie (1-)algebras $\mathfrak{g}$ it is well known that any two connections on the same bundle may be continuously connected by a path of connections: the space of 1-connections is an affine space modeled on $\Omega^1(X, \mathfrak{g})$. If we had an analogous statement for $\mathfrak{g}$-connections for higher L_∞-algebras, we could strengthen the above statement.

6.3.1. Examples.

Ordinary characteristic classes of $\mathfrak{g}$-bundles. Let $\mathfrak{g}$ be an ordinary Lie algebra and $(Y, (A, F_A))$ be a $\mathfrak{g}$-descent object corresponding to an ordinary Cartan-Ehresmann connection as in section 6.2.1. Using proposition 5.21 we know that $\mathrm{inv}(\mathfrak{g})$ contains all the ordinary invariant polynomials P on $\mathfrak{g}$. Hence the characteristic classes $[P(F_A)]$ are precisely the standard characteristic classes (in de Rham cohomology) of the G-bundle with connection.

Characteristic classes of $b^{n-1}\mathfrak{u}(1)$-bundles. For $\mathfrak{g} = b^{n-1}\mathfrak{u}(1)$ we have, according to proposition 5.25, $\mathrm{inv}(b^{n-1}\mathfrak{u}(1)) = \mathrm{CE}(b^n\mathfrak{u}(1))$ and hence a single degree $n+1$ characteristic class: the curvature itself.

This case we had already discussed in the context of Deligne cohomology in section 6.1.1. In particular, notice that in definition 6.2 we had already encountered the distinction between homotopies of L_∞-algebra that are or are not pure gauge transformations, in that they do or do not shift the connection: what is called a *Deligne coboundary* in definition 6.2 corresponds to an equivalence of $b^{n-1}\mathfrak{u}$-connection descent objects as in (6.35), while what is called a *shift of connection* there corresponds to a concordance that involves a shift as in (6.36).

6.4. Universal and generalized $\mathfrak{g}$-connections

We can generalize the discussion of $\mathfrak{g}$-bundles with connection on spaces X, by

- allowing all occurrences of the algebra of differential forms to be replaced with more general differential graded algebras; this amounts to admitting generalized smooth spaces as in section 4.1;
- by allowing all Chevalley-Eilenberg and Weil algebras of L_∞-algebras to be replaced by DGCAs which may be nontrivial in degree 0. This amounts to allowing not just structure ∞-groups but also structure ∞-groupoids.

Definition 6.11 (Generalized $\mathfrak{g}$-connection descent objects). *Given any L_∞-algebra $\mathfrak{g}$, and given any DGCA A, we say a $\mathfrak{g}$-connection descent object for A is*

- *a surjection $F \xleftarrow{\ i^*\ } P$ such that $A \simeq P_{\mathrm{basic}}$;*

- *a choice of horizontal morphisms in the diagram*

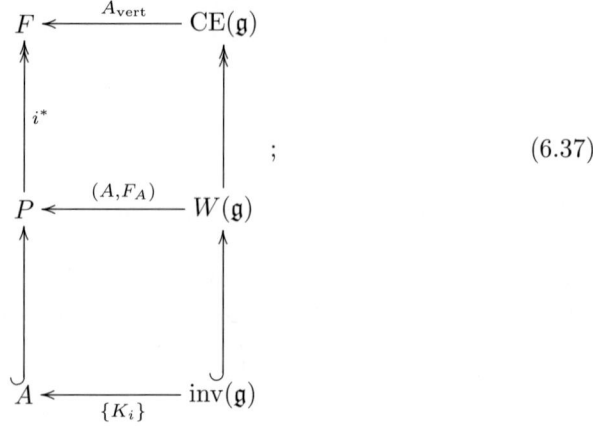

$$\tag{6.37}$$

The notion of equivalence of these descent objects is as before.

Horizontal forms. Given any algebra surjection

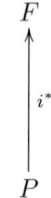

we know from definition 4.9 what the "vertical directions" on P are. After we have chosen a $\mathfrak{g}$-connection on P, we obtain also notion of *horizontal* elements in P:

Definition 6.12 (Horizontal elements). *Given a $\mathfrak{g}$-connection (A, F_A) on P, the algebra of horizontal elements*

$$\mathrm{hor}_A(P) \subset P$$

of P with respect to this connection are those elements not *in the ideal generated by the image of A.*

Notice that $\mathrm{hor}_A(P)$ is in general just a graded-commutative algebra, not a differential algebra. Accordingly the inclusion $\mathrm{hor}_A(P) \subset P$ is meant just as an inclusion of algebras.

6.4.1. Examples.

The universal $\mathfrak{g}$-connection. The tautological example is actually of interest: for any L_∞-algebra $\mathfrak{g}$, there is a *canonical* $\mathfrak{g}$-connection descent object on $\mathrm{inv}(\mathfrak{g})$. This comes from choosing

$$(\ F \xleftarrow{\ i^*\ } P\ ) := (\ \mathrm{CE}(\mathfrak{g}) \xleftarrow{\ i^*\ } \mathrm{W}(\mathfrak{g})\ ) \tag{6.38}$$

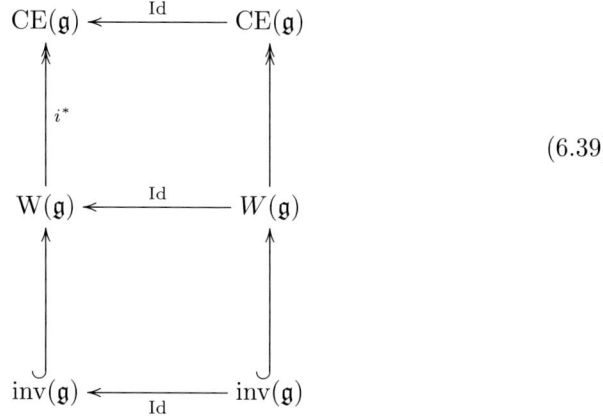

(6.39)

FIGURE 11. **The universal $\mathfrak{g}$-connection** descent object.

and then taking the horizontal morphisms to be all identities, as shown in figure 11.

We can then finally give an *intrinsic* interpretation of the decomposition of the generators of the Weil algebra $W(\mathfrak{g})$ of any L_∞-algebra into elements in $\mathfrak{g}^*$ and elements in the shifted copy $\mathfrak{g}^*[1]$, which is crucial for various of our constructions (for instance for the vanishing condition in (5.65)).

Proposition 6.13. *The horizontal elements of* $W(\mathfrak{g})$ *with respect to the universal $\mathfrak{g}$-connection (A, F_A) on $W(\mathfrak{g})$ are precisely those generated entirely from the shifted copy $\mathfrak{g}^*[1]$:*

$$\mathrm{hor}_A(W(\mathfrak{g})) = \wedge^\bullet(\mathfrak{g}^*[1]) \subset W(\mathfrak{g}).$$

Line n-bundles on classifying spaces.

Proposition 6.14. *Let $\mathfrak{g}$ be any L_∞-algebra and $P \in \mathrm{inv}(\mathfrak{g})$ any closed invariant polynomial on $\mathfrak{g}$ of degree $n+1$. Let $\mathrm{cs} := \tau P$ be the transgression element and $\mu := i^*\mathrm{cs}$ the cocycle that P transgresses to according to proposition 5.23. Then we* *canonically obtain a $b^{n-1}\mathfrak{u}(1)$-connection descent object in $\mathrm{inv}(\mathfrak{g})$:*

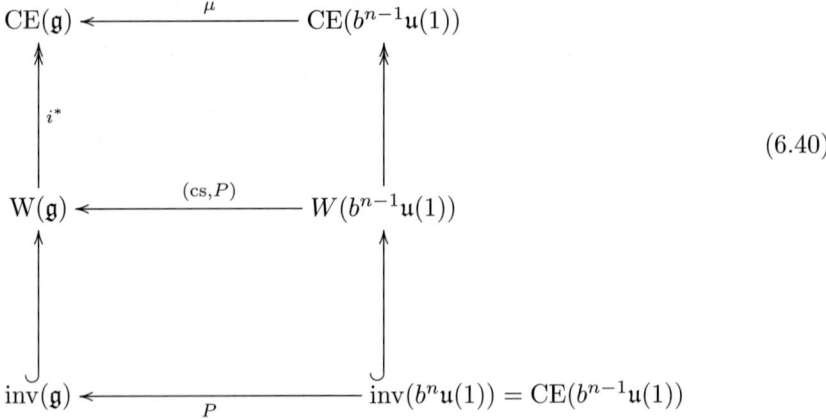

$$(6.40)$$

Remark. For instance for $\mathfrak{g}$ an ordinary semisimple Lie algebra and μ its canonical 3-cocylce, we obtain a descent object for a Lie 3-bundle which plays the role of what is known as the canonical 2-gerbe on the classifying space BG of the simply connected group G integrating $\mathfrak{g}$ [21]. From the above and using section 5.5.1 we read off that its connection 3-form is the canonical Chern-Simons 3-form. We will see this again in 8.3.1, where we show that the 3-particle (the 2-brane) coupled to the above $\mathfrak{g}$-connection descent object indeed reproduces Chern-Simons theory.

7. Higher string- and Chern-Simons n-bundles: the lifting problem

We discuss the general concept of weak cokernels of morphisms of L_∞-algebras. Then we apply this to the special problem of lifts of differential $\mathfrak{g}$-cocycles through String-like extensions.

7.1. Weak cokernels of L_∞-morphisms

After introducing the notion of a mapping cone of qDGCAs, the main point here is proposition 7.8, which establishes the existence of the weak inverse f^{-1} that was mentioned in section 2.3. It will turn out to be that very weak inverse which picks up the information about the existence or non-existence of the lifts discussed in section 7.3. We can define the weak cokernel for *normal L_∞-subalgebras*:

Definition 7.1 (Normal L_∞-subalgebra). *We say a Lie ∞-algebra $\mathfrak{h}$ is a normal sub L_∞-algebra of the L_∞-algebra $\mathfrak{g}$ if there is a morphism*

$$\mathrm{CE}(\mathfrak{h}) \xleftarrow{\;\;t^*\;\;} \mathrm{CE}(\mathfrak{g}) \tag{7.1}$$

which the property that

- *on $\mathfrak{g}^*$ it restricts to a surjective linear map $\mathfrak{h}^* \xleftarrow{\;t_1^*\;} \mathfrak{g}^*$;*
- *if $a \in \ker(t^*)$ then $d_{\mathrm{CE}(\mathfrak{g})}a \in \wedge^\bullet(\ker(t_1^*))$.*

Proposition 7.2. *For $\mathfrak{h}$ and $\mathfrak{g}$ ordinary Lie algebras, the above notion of normal sub L_∞-algebra coincides with the standard notion of normal sub Lie algebras.*

Proof. If $a \in \ker(t^*)$ then for any $x, y \in \mathfrak{g}$ the condition says that $(d_{\mathrm{CE}(\mathfrak{g})}a)(x \vee y) = -a(D[x \vee y]) = -a([x,y])$ vanishes when x or y are in the image of t. But $a([x,y])$ vanishes when $[x,y]$ is in the image of t. Hence the condition says that if at least one of x and y is in the image of t, then their bracket is. $\square$

Definition 7.3 (Mapping cone of qDGCAs; Crossed module of normal sub L_∞-algebras). *Let $t : \mathfrak{h} \hookrightarrow \mathfrak{g}$ be an inclusion of a normal sub L_∞-algebra $\mathfrak{h}$ into $\mathfrak{g}$. The mapping cone of t^* is the qDGCA whose underlying graded algebra is*

$$\wedge^\bullet(\mathfrak{g}^* \oplus \mathfrak{h}^*[1]) \tag{7.2}$$

and whose differential d_t is such that it acts on generators schematically as

$$d_t = \begin{pmatrix} d_{\mathfrak{g}} & 0 \\ t^* & d_{\mathfrak{h}} \end{pmatrix}. \tag{7.3}$$

In more detail, d_{t^*} is defined as follows. We write σt^* for the degree $+1$ derivation on $\wedge^\bullet(\mathfrak{g}^* \oplus \mathfrak{h}^*[1])$ which acts on $\mathfrak{g}^*$ as t^* followed by a shift in degree and which acts on $\mathfrak{h}^*[1]$ as 0. Then, for any $a \in \mathfrak{g}^*$, we have

$$d_t a := d_{\mathrm{CE}(\mathfrak{g})}a + \sigma t^*(a). \tag{7.4}$$

and

$$d_t \sigma t^*(a) := -\sigma t^*(d_{\mathrm{CE}(\mathfrak{g})}a) = -d_t d_{\mathrm{CE}(\mathfrak{g})}a. \tag{7.5}$$

Notice that the last equation

- defines d_t on all of $\mathfrak{h}^*[1]$ since t^* is surjective;
- is well defined in that it agrees for a and a' if $t^*(a) = t^*(a')$, since t is normal.

Proposition 7.4. *The differential d_t defined this way indeed satisfies $(d_t)^2 = 0$.*

Proof. For $a \in \mathfrak{g}^*$ we have

$$d_t d_t a = d_t(d_{\mathrm{CE}(\mathfrak{g})}a + \sigma t^*(a)) = \sigma t^*(d_{\mathrm{CE}(\mathfrak{g})}a) - \sigma t^*(d_{\mathrm{CE}(\mathfrak{g})}a) = 0. \tag{7.6}$$

Hence $(d_t)^2$ vanishes on $\wedge^\bullet(\mathfrak{g}^*)$. Since

$$d_t d_t \sigma t^*(a) = -d_t d_t d_{\mathrm{CE}(\mathfrak{g})}a \tag{7.7}$$

and since $d_{\mathrm{CE}(\mathfrak{g})}a \in \wedge^\bullet(\mathfrak{g}^*)$ this implies $(d_t)^2 = 0$. $\square$

We write $\mathrm{CE}(\mathfrak{h} \overset{t}{\hookrightarrow} \mathfrak{g}) := (\wedge^\bullet(\mathfrak{g}^* \oplus \mathfrak{h}^*[1]), d_t)$ for the resulting qDGCA and $(\mathfrak{h} \overset{t}{\hookrightarrow} \mathfrak{g})$ for the corresponding L_∞-algebra.

The next proposition asserts that $\mathrm{CE}(\mathfrak{h} \overset{t}{\hookrightarrow} \mathfrak{g})$ is indeed a (weak) kernel of t^*.

Proposition 7.5. *There is a canonical morphism* $\mathrm{CE}(\mathfrak{g}) \longleftarrow \mathrm{CE}(\mathfrak{h} \overset{t}{\hookrightarrow} \mathfrak{g})$ *with the property that*

$$\mathrm{CE}(\mathfrak{h}) \overset{t^*}{\longleftarrow} \mathrm{CE}(\mathfrak{g}) \longleftarrow \mathrm{CE}(\mathfrak{h} \overset{t}{\hookrightarrow} \mathfrak{g}) \,. \tag{7.8}$$

$$\Uparrow_\tau$$

$$0$$

Proof. On components, this morphism is the identity on $\mathfrak{g}^*$ and 0 on $\mathfrak{h}^*[1]$. One checks that this respects the differentials. The homotopy to the 0-morphism sends

$$\tau : \sigma t^*(a) \mapsto t^*(a) \,. \tag{7.9}$$

Using definition 5.11 one checks that then indeed

$$[d, \tau] : a \mapsto \tau(d_{\mathrm{CE}(\mathfrak{g})} a + \sigma t^* a) = a \tag{7.10}$$

and

$$[d, \tau] : \sigma t^* a \mapsto d_{\mathrm{CE}(\mathfrak{g})} a + \tau(-\sigma t^*(d_{\mathrm{CE}(\mathfrak{g})} a)) = 0 \,. \tag{7.11}$$

Here the last step makes crucial use of the condition (5.65) which demands that

$$\tau(d_{\mathrm{W}(\mathfrak{h} \overset{t}{\hookrightarrow} \mathfrak{g})} \sigma t^* a - d_{\mathrm{CE}(\mathfrak{h} \overset{t}{\hookrightarrow} \mathfrak{g})} \sigma t^* a) = 0 \tag{7.12}$$

and the formula (5.61) which induces precisely the right combinatorial factors. $\square$

Notice that not only is $\mathrm{CE}(\mathfrak{h} \overset{t}{\hookrightarrow} \mathfrak{g})$ in the kernel of t^*, it is indeed the universal object with this property, hence is *the* kernel of t^* (of course up to equivalence).

Proposition 7.6. *Let* $\mathrm{CE}(\mathfrak{h}) \overset{t^*}{\longleftarrow} \mathrm{CE}(\mathfrak{g}) \overset{u^*}{\longleftarrow} \mathrm{CE}(\mathfrak{f})$ *be a sequence of qDGCAs with t^* normal, as above, and with the property that u^* restricts, on the underlying vector spaces of generators, to the kernel of the linear map underlying t^*. Then there is a unique morphism $f : \mathrm{CE}(\mathfrak{f}) \to \mathrm{CE}(\mathfrak{h} \overset{t}{\hookrightarrow} \mathfrak{g})$ such that*

$$\mathrm{CE}(\mathfrak{h}) \overset{t^*}{\longleftarrow} \mathrm{CE}(\mathfrak{g}) \longleftarrow \mathrm{CE}(\mathfrak{h} \overset{t}{\hookrightarrow} \mathfrak{g}) \,. \tag{7.13}$$

$$\uparrow_{u^*} \qquad \nearrow_f$$

$$\mathrm{CE}(\mathfrak{f})$$

Proof. The morphism f has to be in components the same as $\mathrm{CE}(\mathfrak{g}) \leftarrow \mathrm{CE}(\mathfrak{f})$. By the assumption that this is in the kernel of t^*, the differentials are respected. $\square$

Remark. There should be a generalization of the entire discussion where u^* is not restricted to be the kernel of t^* on generators. However, for our application here, this simple situation is all we need.

Proposition 7.7. *For a String-like extension $\mathfrak{g}_\mu$ from definition 5.28, the morphism*

$$\mathrm{CE}(b^{n-1}\mathfrak{u}(1)) \xleftarrow{\quad t^* \quad} \mathrm{CE}(\mathfrak{g}_\mu) \tag{7.14}$$

is normal in the sense of definition 7.1.

Proposition 7.8. *In the case that the sequence*

$$\mathrm{CE}(\mathfrak{h}) \xleftarrow{\quad t^* \quad} \mathrm{CE}(\mathfrak{g}) \xleftarrow{\quad u^* \quad} \mathrm{CE}(\mathfrak{f}) \tag{7.15}$$

above is a String-like extension

$$\mathrm{CE}(b^{n-1}\mathfrak{u}(1)) \xleftarrow{\quad t^* \quad} \mathrm{CE}(\mathfrak{g}_\mu) \xleftarrow{\quad u^* \quad} \mathrm{CE}(\mathfrak{g}) \tag{7.16}$$

from proposition 5.29 or the corresponding Weil-algebra version

$$
\begin{array}{ccccc}
\mathrm{W}(b^{n-1}\mathfrak{u}(1)) & \xleftarrow{\quad t^* \quad} & \mathrm{W}(\mathfrak{g}_\mu) & \xleftarrow{\quad u^* \quad} & \mathrm{W}(\mathfrak{g}) \\
\Big\downarrow {\scriptstyle =} & & \Big\downarrow {\scriptstyle =} & & \Big\downarrow {\scriptstyle =} \\
\mathrm{CE}(\mathrm{inn}(b^{n-1}\mathfrak{u}(1))) & \xleftarrow{\quad t^* \quad} & \mathrm{CE}(\mathrm{inn}(\mathfrak{g}_\mu)) & \xleftarrow{\quad u^* \quad} & \mathrm{CE}(\mathrm{inn}(\mathfrak{g}))
\end{array} \tag{7.17}
$$

the morphisms $f : \mathrm{CE}(\mathfrak{f}) \to \mathrm{CE}(\mathfrak{h} \xhookrightarrow{t} \mathfrak{g})$ and $\hat{f} : \mathrm{W}(\mathfrak{f}) \to \mathrm{W}(\mathfrak{h} \xhookrightarrow{t} \mathfrak{g})$ have weak inverses $f^{-1} : \mathrm{CE}(\mathfrak{h} \xhookrightarrow{t} \mathfrak{g}) \to \mathrm{CE}(\mathfrak{f})$ and $\hat{f}^{-1} : \mathrm{W}(\mathfrak{h} \xhookrightarrow{t} \mathfrak{g}) \to \mathrm{W}(\mathfrak{f})$, respectively.

Proof. We first construct a morphism f^{-1} and then show that it is weakly inverse to f. The statement for $\hat{f}$ then follows from the functoriality of forming the Weil algebra, proposition 5.9. Start by choosing a splitting of the vector space V underlying $\mathfrak{g}^*$ as

$$V = \ker(t^*) \oplus V_1 . \tag{7.18}$$

This is the non-canonical choice we need to make. Then take the component map of f^{-1} to be the identity on $\ker(t^*)$ and 0 on V_1. Moreover, for $a \in V_1$ set

$$f^{-1} : \sigma t^*(a) \mapsto -(d_{\mathrm{CE}(\mathfrak{g})}a)|_{\wedge^\bullet \ker(t^*)} , \tag{7.19}$$

where the restriction is again with respect to the chosen splitting of V. We check that this assignment, extended as an algebra homomorphism, does respect the differentials.

For $a \in \ker(t^*)$ we have

$$
\begin{array}{ccc}
a & \xmapsto{\quad d_t \quad} & d_{\mathrm{CE}(\mathfrak{g})}a \\
{\scriptstyle f^{-1}}\Big\downarrow & & \Big\uparrow {\scriptstyle f^{-1}} \\
a & \xmapsto[\quad d_{\mathrm{CE}(\mathfrak{f})} \quad]{} & d_{\mathrm{CE}(\mathfrak{g})}a
\end{array} \tag{7.20}
$$

using the fact that t^* is normal. For $a \in V_1$ we have

$$
\begin{array}{ccc}
a & \xrightarrow{\;d_t\;} & d_{\mathrm{CE}(\mathfrak{g})}a + \sigma t^*(a) \\
{\scriptstyle f^{-1}}\downarrow & & \uparrow{\scriptstyle f^{-1}} \\
0 & \xrightarrow{\;d_{\mathrm{CE}(\mathfrak{f})}\;} & (d_{\mathrm{CE}(\mathfrak{g})}a)|_{\wedge^\bullet \ker(t^*)} - (d_{\mathrm{CE}(\mathfrak{g})}a)|_{\wedge^\bullet \ker(t^*)}
\end{array}
\tag{7.21}
$$

and

$$
\begin{array}{ccc}
\sigma t^*(a) & \xrightarrow{\;d_t\;} & -\sigma t^*(d_{\mathrm{CE}(\mathfrak{g})}a) \\
{\scriptstyle f^{-1}}\downarrow & & \uparrow{\scriptstyle f^{-1}} \\
-(d_{\mathrm{CE}(\mathfrak{g})}a)|_{\wedge^\bullet \ker(t^*)} & \xrightarrow{\;d_{\mathrm{CE}(\mathfrak{f})}\;} & -d_{\mathrm{CE}(\mathfrak{f})}((d_{\mathrm{CE}(\mathfrak{g})}a)|_{\ker(t^*)})
\end{array}
\tag{7.22}
$$

This last condition happens to be satisfied for the examples stated in the proposition. The details for that are discussed in 7.1.1 below. By the above, f^{-1} is indeed a morphism of qDGCAs.

Next we check that f^{-1} is a weak inverse of f. Clearly

$$
\mathrm{CE}(\mathfrak{f}) \longleftarrow \mathrm{CE}(\mathfrak{h} \xrightarrow{t} \mathfrak{g}) \longleftarrow \mathrm{CE}(\mathfrak{f})
\tag{7.23}
$$

is the identity on $\mathrm{CE}(\mathfrak{f})$. What remains is to construct a homotopy

$$
\mathrm{CE}(\mathfrak{h} \xrightarrow{t} \mathfrak{g}) \longleftarrow \mathrm{CE}(\mathfrak{f}) \longleftarrow \mathrm{CE}(\mathfrak{h} \xrightarrow{t} \mathfrak{g}) \;.
\tag{7.24}
$$

One checks that this is accomplished by taking τ to act on σV_1 as $\tau : \sigma V_1 \xrightarrow{\cong} V_1$ and extended suitably. □

7.1.1. Examples.
Weak cokernel for the String-like extension. Let our sequence

$$
\mathrm{CE}(\mathfrak{h}) \xleftarrow{\;t^*\;} \mathrm{CE}(\mathfrak{g}) \xleftarrow{\;u^*\;} \mathrm{CE}(\mathfrak{f})
\tag{7.25}
$$

be a String-like extension

$$
\mathrm{CE}(b^{n-1}\mathfrak{u}(1)) \xleftarrow{\;t^*\;} \mathrm{CE}(\mathfrak{g}_\mu) \xleftarrow{\;u^*\;} \mathrm{CE}(\mathfrak{g})
\tag{7.26}
$$

from proposition 5.29. Then the mapping cone Chevalley-Eilenberg algebra

$$
\mathrm{CE}(b^{n-1}\mathfrak{u}(1) \hookrightarrow \mathfrak{g}_\mu)
\tag{7.27}
$$

is

$$
\wedge^\bullet(\mathfrak{g}^* \oplus \mathbb{R}[n] \oplus \mathbb{R}[n+1])
\tag{7.28}
$$

with differential given by

$$d_t|_{\mathfrak{g}^*} = d_{\mathrm{CE}(\mathfrak{g})} \tag{7.29}$$

$$d_t|_{\mathbb{R}[n]} = -\mu + \sigma \tag{7.30}$$

$$d_t|_{\mathbb{R}[n+1]} = 0. \tag{7.31}$$

(As always, σ is the canonical degree shifting isomorphism on generators extended as a derivation.) The morphism

$$\mathrm{CE}(\mathfrak{g}) \xleftarrow[\simeq]{f^{-1}} \mathrm{CE}(b^{n-1}\mathfrak{u}(1) \hookrightarrow \mathfrak{g}_\mu) \tag{7.32}$$

acts as

$$f^{-1}|_{\mathfrak{g}^*} = \mathrm{Id} \tag{7.33}$$

$$f^{-1}|_{\mathbb{R}[n]} = 0 \tag{7.34}$$

$$f^{-1}|_{\mathbb{R}[n+1]} = \mu. \tag{7.35}$$

To check the condition in equation (7.22) explicitly in this case, let $b \in \mathbb{R}[n]$ and write $b := t^* b$ for simplicity (since t^* is the identity on $\mathbb{R}[n]$). Then

$$
\begin{array}{ccc}
\sigma b & \xrightarrow{\;d_t\;} & 0 \\
\downarrow{\scriptstyle f^{-1}} & & \downarrow{\scriptstyle f^{-1}} \\
\mu & \xrightarrow{\;d_{\mathrm{CE}(\mathfrak{g})}\;} & 0
\end{array}
\tag{7.36}
$$

does commute.

Weak cokernel for the String-like extension in terms of the Weil algebra. We will also need the analogous discussion not for the Chevalley-Eilenberg algebras, but for the corresponding Weil algebras. To that end consider now the sequence

$$W(b^{n-1}\mathfrak{u}(1)) \xleftarrow{\;t^*\;} W(\mathfrak{g}_\mu) \xleftarrow{\;u^*\;} W(\mathfrak{g}). \tag{7.37}$$

This is handled most conveniently by inserting the isomorphism

$$W(\mathfrak{g}_\mu) \simeq \mathrm{CE}(\mathrm{cs}_P(\mathfrak{g})) \tag{7.38}$$

from proposition 5.30 as well as the identification

$$W(\mathfrak{g}) = \mathrm{CE}(\mathrm{inn}(\mathfrak{g})) \tag{7.39}$$

such that we get

$$\mathrm{CE}(\mathrm{inn}(b^{n-1}\mathfrak{u}(1))) \xleftarrow{\;t^*\;} \mathrm{CE}(\mathrm{cs}_P(\mathfrak{g})) \xleftarrow{\;u^*\;} \mathrm{CE}(\mathrm{inn}(\mathfrak{g})). \tag{7.40}$$

Then we find that the mapping cone algebra $\mathrm{CE}(b^{n-1}\mathfrak{u}(1) \hookrightarrow \mathrm{cs}_P(\mathfrak{g}))$ is

$$\wedge^\bullet(\mathfrak{g}^* \oplus \mathfrak{g}^*[1] \oplus (\mathbb{R}[n] \oplus \mathbb{R}[n+1]) \oplus (\mathbb{R}[n+1] \oplus \mathbb{R}[n+2])). \tag{7.41}$$

Write b and c for the canonical basis elements of $\mathbb{R}[n] \oplus \mathbb{R}[n+1]$, then the differential is characterized by

$$d_t|_{\mathfrak{g}^* \oplus \mathfrak{g}^*} \quad = \quad d_{W(\mathfrak{g})} \tag{7.42}$$
$$d_t \quad : \quad b \mapsto c - \mathrm{cs} + \sigma b \tag{7.43}$$
$$d_t \quad : \quad c \mapsto P + \sigma c \tag{7.44}$$
$$d_t \quad : \quad \sigma b \mapsto -\sigma c \tag{7.45}$$
$$d_t \quad : \quad \sigma c \mapsto 0 \,. \tag{7.46}$$

Notice above the relative sign between σb and σc. This implies that the canonical injection

$$\mathrm{CE}(b^{n-1}\mathfrak{u}(1) \hookrightarrow \mathrm{cs}_P(\mathfrak{g})) \xleftarrow{\ i\ } W(b^n\mathfrak{u}(1)) \tag{7.47}$$

also carries a sign: if we denote the degree $n+1$ and $n+2$ generators of $W(b^n\mathfrak{u}(1))$ by h and dh, then

$$i \quad : \quad h \mapsto \sigma b \tag{7.48}$$
$$i \quad : \quad dh \mapsto -\sigma c \,. \tag{7.49}$$

This sign has no profound structural role, but we need to carefully keep track of it, for instance in order for our examples in 7.3.1 to come out right. The morphism

$$\mathrm{CE}(b^{n-1}\mathfrak{u}(1) \hookrightarrow \mathrm{cs}_P(\mathfrak{g})) \xleftarrow[\simeq]{\ f^{-1}\ } W(\mathfrak{g}) \tag{7.50}$$

acts as

$$f^{-1}|_{\mathfrak{g}^* \oplus \mathfrak{g}^*[1]} \quad = \quad \mathrm{Id} \tag{7.51}$$
$$f^{-1} : \sigma b \quad \mapsto \quad \mathrm{cs} \tag{7.52}$$
$$f^{-1} : \sigma c \quad \mapsto \quad -P \,. \tag{7.53}$$

Again, notice the signs, as they follow from the general prescription in proposition 7.8. We again check explicitly equation (7.22):

$$\begin{array}{ccc} \sigma b & \xmapsto{\ d_t\ } & -\sigma c \\ {\scriptstyle f^{-1}}\downarrow & & \downarrow{\scriptstyle f^{-1}} \\ \mathrm{cs} & \xmapsto{\ d_{W(\mathfrak{g})}\ } & P \end{array} \,. \tag{7.54}$$

7.2. Lifts of $\mathfrak{g}$-descent objects through String-like extensions

We need the above general theory for the special case where we have the mapping cone $\mathrm{CE}(b^{n-1}\mathfrak{u}(1) \hookrightarrow \mathfrak{g}_\mu)$ as the weak kernel of the left morphism in a String-like extension

$$\mathrm{CE}(b^{n-1}\mathfrak{u}(1)) \longleftarrow \mathrm{CE}(\mathfrak{g}_\mu) \longleftarrow \mathrm{CE}(\mathfrak{g}) \tag{7.55}$$

coming from an $(n+1)$ cocycle μ on an ordinary Lie algebra $\mathfrak{g}$. In this case $\mathrm{CE}(b^{n-1}\mathfrak{u}(1) \hookrightarrow \mathfrak{g}_\mu)$ looks like

$$\mathrm{CE}(b^{n-1}\mathfrak{u}(1) \hookrightarrow \mathfrak{g}_\mu) = (\wedge^\bullet(\mathfrak{g}^* \oplus \mathbb{R}[n] \oplus \mathbb{R}[n+1]), d_t) \,. \tag{7.56}$$

By chasing this through the above definitions, we find

Proposition 7.9. *The morphism*

$$f^{-1} : \mathrm{CE}(b^{n-1}\mathfrak{u}(1) \hookrightarrow \mathfrak{g}_\mu) \to \mathrm{CE}(\mathfrak{g}) \tag{7.57}$$

acts as the identity on $\mathfrak{g}^*$

$$f^{-1}|_{\mathfrak{g}^*} = \mathrm{Id}\,, \tag{7.58}$$

vanishes on $\mathbb{R}[n]$

$$f^{-1}|_{\mathbb{R}[n]} : b \mapsto 0, \tag{7.59}$$

and satisfies

$$f^{-1}|_{\mathbb{R}[n+1]} : \sigma t^* b \mapsto \mu\,. \tag{7.60}$$

Therefore we find the $(n+1)$-cocycle

$$\Omega^\bullet_{\mathrm{vert}}(Y) \xleftarrow{\hat{A}_{\mathrm{vert}}} \mathrm{CE}(b^n\mathfrak{u}(1)) \tag{7.61}$$

obstructing the lift of a $\mathfrak{g}$-cocycle

$$\Omega^\bullet_{\mathrm{vert}}(Y) \xleftarrow{A_{\mathrm{vert}}} \mathrm{CE}(\mathfrak{g})\,, \tag{7.62}$$

according to 2.3 given by

$$\mathrm{CE}(b^{n-1}\mathfrak{u}(1) \hookrightarrow \mathfrak{g}_\mu) \xleftarrow{\;j\;} \mathrm{CE}(b^n\mathfrak{u}(1))\,, \tag{7.63}$$

to be the $(n+1)$-form

$$\mu(A_{\mathrm{vert}}) \in \Omega^{n+1}_{\mathrm{vert}}(Y)\,. \tag{7.64}$$

Proposition 7.10. *Let* $A_{\mathrm{vert}} \in \Omega^1_{\mathrm{vert}}(Y, \mathfrak{g})$ *be the cocycle of a G-bundle $P \to X$ for $\mathfrak{g}$ semisimple and let $\mu = \langle\cdot, [\cdot,\cdot]\rangle$ be the canonical 3-cocycle. Then $\mathfrak{g}_\mu$ is the standard String Lie 3-algebra and the obstruction to lifting P to a String 2-bundle, i.e. lifting to a $\mathfrak{g}_\mu$-cocycle, is the Chern-Simons 3-bundle with cocycle given by the vertical 3-form*

$$\langle A_{\mathrm{vert}} \wedge [A_{\mathrm{vert}} \wedge A_{\mathrm{vert}}]\rangle \in \Omega^3_{\mathrm{vert}}(Y)\,. \tag{7.65}$$

In the following we will express these obstruction in a more familiar way in terms of their characteristic classes. In order to do that, we first need to generalize the discussion to differential $\mathfrak{g}$-cocycle. But that is now straightforward.

7.2.1. Examples. The continuation of the discussion of 5.3.1 to coset spaces gives a classical illustration of the lifting construction considered here.

Cohomology of coset spaces. The above relation between the cohomology of groups and that of their Chevalley-Eilenberg qDGCAs generalizes to coset spaces. This also illustrates the constructions which are discussed later in 7. Consider the case of an ordinary extension of (compact connected) Lie groups:

$$1 \to H \to G \to G/H \to 1 \tag{7.66}$$

or even the same sequence in which G/H is only a homogeneous space and not itself a group. For a closed connected subgroup $t : H \hookrightarrow G$, there is the induced map $Bt : BH \to BG$ and a commutative diagram

$$\begin{CD} W(\mathfrak{g}) @>{dt^*}>> W(\mathfrak{h}) \\ @AAA @AAA \\ \wedge^\bullet P_G @>{dt^*}>> \wedge^\bullet P_H \end{CD} \tag{7.67}$$

By analyzing the fibration sequence

$$G/H \to EG/H \simeq BH \to BG, \tag{7.68}$$

Halperin and Thomas [42] show there is a morphism

$$\wedge^\bullet (P_G \oplus Q_H) \to \Omega^\bullet (G/K) \tag{7.69}$$

inducing an isomorphism in cohomology. It is not hard to see that their morphism factors through

$$\wedge^\bullet (\mathfrak{g}^* \oplus \mathfrak{h}^*[1]). \tag{7.70}$$

In general, the homogeneous space G/H itself is not a group, but in case of an extension $H \to G \to K$, we also have BK and the sequences $K \to BH \to BG$ and $BH \to BG \to BK$. Up to homotopy equivalence, the fiber of the bundle $BH \to BG$ is K and that of $BG \to BK$ is BH. In particular, consider an extension of $\mathfrak{g}$ by a String-like Lie ∞-algebra

$$\mathrm{CE}(b^{n-1}\mathfrak{u}(1)) \xleftarrow{\ i\ } \mathrm{CE}(\mathfrak{g}_\mu) \xleftarrow{\quad\quad} \mathrm{CE}(\mathfrak{g}) \ . \tag{7.71}$$

Regard $\mathfrak{g}$ now as the quotient $\mathfrak{g}_\mu/b^{n-1}\mathfrak{u}(1)$ and recognize that corresponding to BH we have $b^n\mathfrak{u}(1)$. Thus we have a quasi-isomorphism

$$\mathrm{CE}(b^{n-1}\mathfrak{u}(1) \hookrightarrow \mathfrak{g}_\mu) \simeq \mathrm{CE}(\mathfrak{g}) \tag{7.72}$$

and hence a morphism

$$\mathrm{CE}(b^n\mathfrak{u}(1)) \to \mathrm{CE}(\mathfrak{g}). \tag{7.73}$$

Given a $\mathfrak{g}$-bundle cocycle

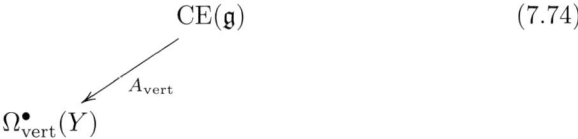

$$\text{CE}(\mathfrak{g}) \qquad (7.74)$$

$$A_{\text{vert}}$$

$$\Omega^\bullet_{\text{vert}}(Y)$$

and given an extension of $\mathfrak{g}$ by a String-like Lie ∞-algebra

$$\text{CE}(b^{n-1}\mathfrak{u}(1)) \xleftarrow{\quad i \quad} \text{CE}(\mathfrak{g}_\mu) \xleftarrow{\qquad\qquad} \text{CE}(\mathfrak{g}) \qquad (7.75)$$

we ask if it is possible to *lift the cocycle* through this extension, i.e. to find a dotted arrow in

$$\text{CE}(b^{n-1}\mathfrak{u}(1)) \xleftarrow{\quad} \text{CE}(\mathfrak{g}_\mu) \xleftarrow{\qquad\qquad} \text{CE}(\mathfrak{g}) \;. \qquad (7.76)$$

$$A_{\text{vert}}$$

$$\Omega^\bullet_{\text{vert}}(Y)$$

In general this is not possible. Indeed, consider the map A'_{vert} given by $\text{CE}(b^n\mathfrak{u}(1)) \to CE(\mathfrak{g})$ composed with A_{vert}. The nontriviality of the $b^n\mathfrak{u}(1)$-cocycle A'_{vert} is the obstruction to constructing the desired lift.

7.3. Lifts of $\mathfrak{g}$-connections through String-like extensions

In order to find the obstructing characteristic classes, we would like to extend the above lift 7.76 of $\mathfrak{g}$-descent objects to a lift of $\mathfrak{g}$-connection descent objects extending them, according to 6.2. Hence we would like first to *extend* A_{vert} to (A, F_A)

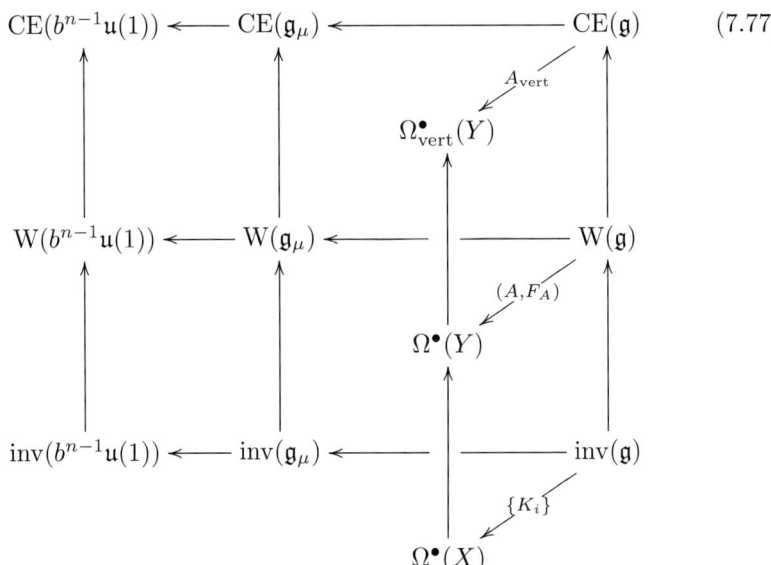

$$(7.77)$$

and then *lift* the resulting $\mathfrak{g}$-connection descent object (A, F_A) to a $\mathfrak{g}_\mu$-connection object $(\hat{A}, F_{\hat{A}})$

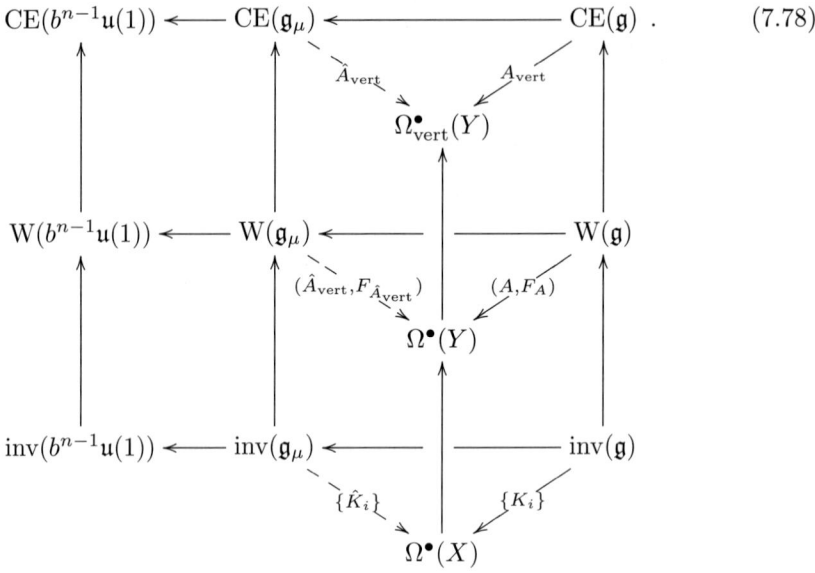

$$(7.78)$$

The situation is essentially an obstruction problem as before, only that instead of single morphisms, we are now lifting an entire sequence of morphisms. As before, we measure the obstruction to the existence of the lift by precomposing everything with the a map from a weak cokernel:

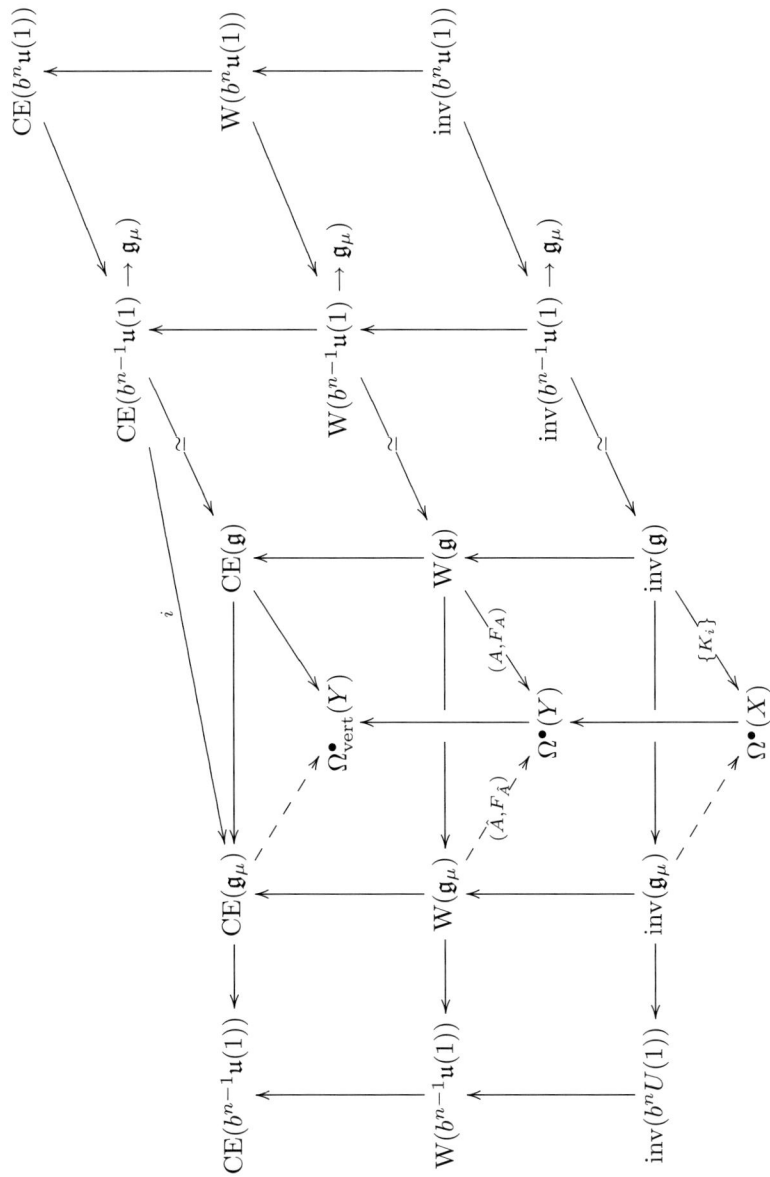

The result is a $b^n\mathfrak{u}(1)$-connection object. We will call (the class of) this the **generalized Chern-Simons** $(n+1)$-**bundle** obstructing the lift.

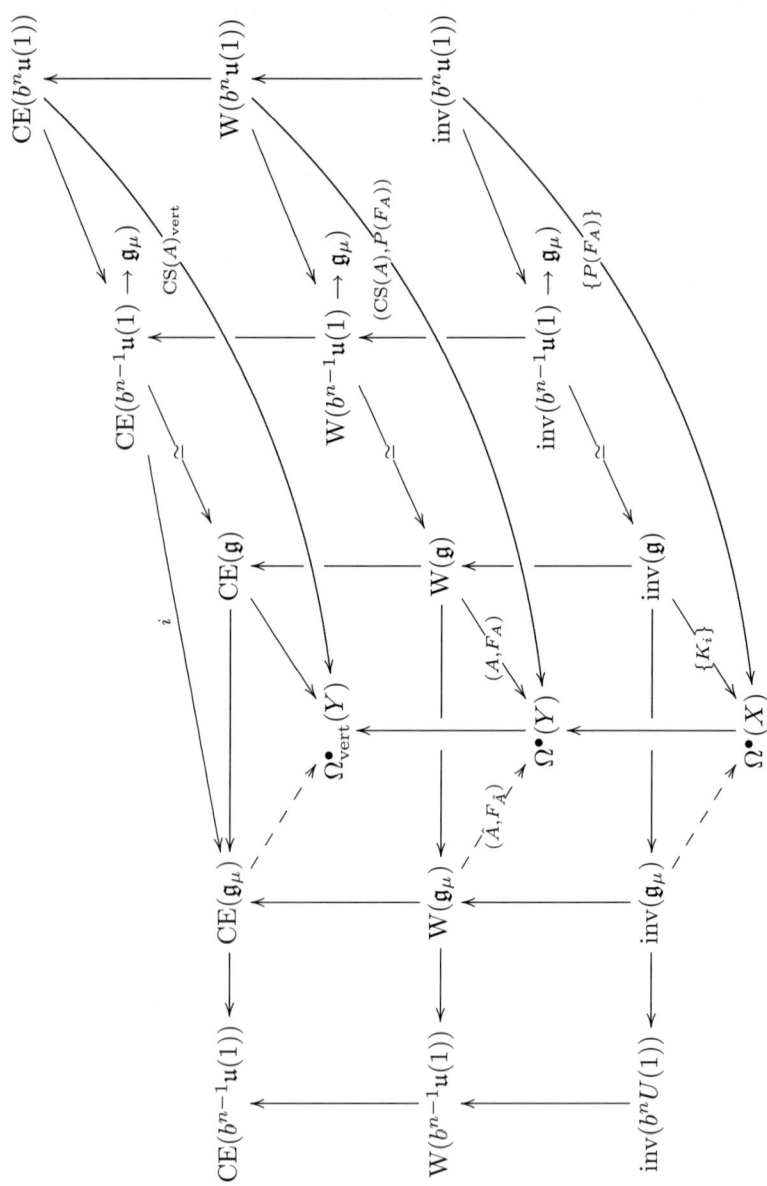

FIGURE 12. The **generalized Chern-Simons** $b^n\mathfrak{u}(1)$-**bundle that obstructs the lift of a given** $\mathfrak{g}$-**bundle to a** $\mathfrak{g}_\mu$-**bundle,** or rather the descent object representing it.

In order to construct the lift it is convenient, for similar reasons as in the proof of proposition 5.32, to work with $\mathrm{CE}(\mathrm{cs}_P(\mathfrak{g}))$ instead of the isomorphic $\mathrm{W}(\mathfrak{g}_\mu)$, using the isomorphism from proposition 5.30. Furthermore, using the identity

$$\mathrm{W}(\mathfrak{g}) = \mathrm{CE}(\mathrm{inn}(\mathfrak{g})) \tag{7.79}$$

mentioned in 5.1, we can hence consider instead of

$$\mathrm{W}(b^{n-1}) \longleftarrow\!\!\!\longleftarrow \mathrm{W}(\mathfrak{g}_\mu) \longleftarrow\!\!\!\supset \mathrm{W}(\mathfrak{g}) \tag{7.80}$$

the sequence

$$\mathrm{CE}(\mathrm{inn}(b^{n-1})) \longleftarrow\!\!\!\longleftarrow \mathrm{CE}(\mathrm{cs}_P(\mathfrak{g})) \longleftarrow\!\!\!\supset \mathrm{CE}(\mathrm{inn}(\mathfrak{g})) \ . \tag{7.81}$$

Fortunately, this still satisfies the assumptions of proposition 7.6. So in complete analogy, we find the extension of proposition 7.9 from $\mathfrak{g}$-bundle cocycles to differential $\mathfrak{g}$-cocycles:

Proposition 7.11. *The morphism*

$$f^{-1} : \mathrm{CE}(\mathrm{inn}(b^{n-1}\mathfrak{u}(1)) \hookrightarrow \mathrm{CE}(\mathrm{cs}_P(\mathfrak{g})) \to \mathrm{CE}(\mathrm{inn}(\mathfrak{g})) \tag{7.82}$$

constructed as in proposition 7.9 acts as the identity on $\mathfrak{g}^* \oplus \mathfrak{g}^*[1]$

$$f^{-1}|_{\mathfrak{g}^* \oplus \mathfrak{g}^*[1]} = \mathrm{Id} \tag{7.83}$$

and satisfies

$$f^{-1}|_{\mathbb{R}[n+2]} : c \mapsto P \ . \tag{7.84}$$

This means that, as an extension of proposition 7.10, we find the differential $b^n\mathfrak{u}(1)$ $(n+1)$-cocycle

$$\Omega^\bullet(Y) \xleftarrow{\ \hat{A}\ } \mathrm{W}(b^n\mathfrak{u}(1)) \tag{7.85}$$

obstructing the lift of a differential $\mathfrak{g}$-cocycle

$$\Omega^\bullet(Y) \xleftarrow{\ (A,F_A)\ } \mathrm{W}(\mathfrak{g}) \ , \tag{7.86}$$

according to the above discussion

$$
\begin{array}{c}
\mathrm{CE}(\mathrm{inn}(b^{n-1}\mathfrak{u}(1)) \hookrightarrow \mathrm{inn}(\mathfrak{g}_\mu)) \xleftarrow{\ j\ } \mathrm{W}(b^n\mathfrak{u}(1)) \ , \\[4pt]
\Big\downarrow {\scriptstyle f^{-1}} \\[4pt]
\mathrm{W}(b^{n-1}\mathfrak{u}(1)) \xleftarrow{\ i^*\ } \mathrm{W}(\mathfrak{g}_\mu) \xleftarrow{\ \ \ } \mathrm{W}(\mathfrak{g}) \qquad {\scriptstyle (\hat{A},F_{\hat{A}})} \\[4pt]
\Big\downarrow {\scriptstyle (A,F_A)} \\[4pt]
\Omega^\bullet(Y)
\end{array}
\tag{7.87}
$$

to be the connection $(n+1)$-form

$$\hat{A} = \mathrm{CS}(A) \in \Omega^{n+1}(Y) \tag{7.88}$$

with the corresponding curvature $(n+2)$-form

$$F_{\hat{A}} = P(F_A) \in \Omega^{n+2}(Y).$$ (7.89)

Then we finally find, in particular,

Proposition 7.12. *For μ a cocycle on the ordinary Lie algebra $\mathfrak{g}$ in transgression with the invariant polynomial P, the obstruction to lifting a $\mathfrak{g}$-bundle cocycle through the String-like extension determined by μ is the characteristic class given by P.*

Remark. Notice that, so far, all our statements about characteristic classes are in de Rham cohomology. Possibly our construction actually holds for integral cohomology classes, but if so, we have not extracted that yet. A more detailed consideration of this will be the subject of [68].

7.3.1. Examples.
Chern-Simons 3-bundles obstructing lifts of G-bundles to $\mathrm{String}(G)$-bundles. Consider, on a base space X for some semisimple Lie group G, with Lie algebra $\mathfrak{g}$ a principal G-bundle $\pi : P \to X$. Identify our surjective submersion with the total space of this bundle

$$Y := P.$$ (7.90)

Let P be equipped with a connection, (P, ∇), realized in terms of an Ehresmann connection 1-form

$$A \in \Omega^1(Y, \mathfrak{g})$$ (7.91)

with curvature

$$F_A \in \Omega^2(Y, \mathfrak{g})$$ (7.92)

i.e. a dg-algebra morphism

$$\Omega^\bullet(Y) \xleftarrow{\quad (A, F_A) \quad} W(\mathfrak{g})$$ (7.93)

satisfying the two Ehresmann conditions. By the discussion in 6.2.1 this yields a $\mathfrak{g}$-connection descent object $(Y, (A, F_A))$ in our sense.

We would like to compute the obstruction to lifting this G-bundle to a String 2-bundle, i.e. to lift the $\mathfrak{g}$-connection descent object to a $\mathfrak{g}_\mu$-connection descent object, for

$$0 \to b u(1) \to \mathfrak{g}_\mu \to \mathfrak{g} \to 0$$ (7.94)

the ordinary String extension from definition 5.34. By the above discussion in 7.3, the obstruction is the (class of the) $b^2 u(1)$-connection descent object

$(Y, (H_{(3)}, G_{(4)}))$ whose connection and curvature are given by the composite

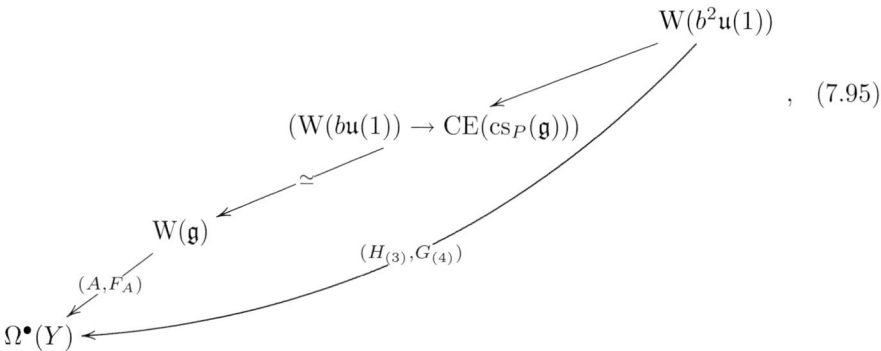

,   (7.95)

where, as discussed above, we are making use of the isomorphism $W(\mathfrak{g}_\mu) \simeq CE(cs_P(\mathfrak{g}))$ from proposition 5.30. The crucial aspect of this composite is the isomorphism

$$W(\mathfrak{g}) \xleftarrow[\simeq]{f^{-1}} (W(b\mathfrak{u}(1)) \to CE_P(\mathfrak{g})) \qquad (7.96)$$

from proposition 7.8. This is where the obstruction data is picked up. The important formula governing this is equation (7.19), which describes how the shifted elements coming from $W(b\mathfrak{u}(1))$ in the mapping cone $(W(b\mathfrak{u}(1)) \to CE_P(\mathfrak{g}))$ are mapped to $W(\mathfrak{g})$.

Recall that $W(b^2\mathfrak{u}(1)) = F(\mathbb{R}[3])$ is generated from elements (h, dh) of degree 3 and 4, respectively, that $W(b\mathfrak{u}(1)) = F(\mathbb{R}[2])$ is generated from elements (c, dc) of degree 2 and 3, respectively, and that $CE(cs_P(\mathfrak{g}))$ is generated from $\mathfrak{g}^* \oplus \mathfrak{g}^*[1]$ together with elements b and c of degree 2 and 3, respectively, with

$$d_{CE(cs_P(\mathfrak{g}))}b = c - cs \qquad (7.97)$$

and

$$d_{CE(cs_P(\mathfrak{g}))}c = P, \qquad (7.98)$$

where $cs \in \wedge^3(\mathfrak{g}^* \oplus \mathfrak{g}^*[1])$ is the transgression element interpolating between the cocycle $\mu = \langle \cdot, [\cdot, \cdot] \rangle \in \wedge^3(\mathfrak{g}^*)$ and the invariant polynomial $P = \langle \cdot, \cdot \rangle \in \wedge^2(\mathfrak{g}^*[1])$. Hence the map f^{-1} acts as

$$f^{-1} : \sigma b \mapsto -(d_{CE(cs_P(\mathfrak{g}))}b)|_{\wedge^\bullet(\mathfrak{g}^* \oplus \mathfrak{g}^*[1])} = +cs \qquad (7.99)$$

and

$$f^{-1} : \sigma c \mapsto -(d_{CE(cs_P(\mathfrak{g}))}c)|_{\wedge^\bullet(\mathfrak{g}^* \oplus \mathfrak{g}^*[1])} = -P. \qquad (7.100)$$

Therefore the above composite $(H_{(3)}, G_{(4)})$ maps the generators (h, dh) of $W(b^2\mathfrak{u}(1))$ as

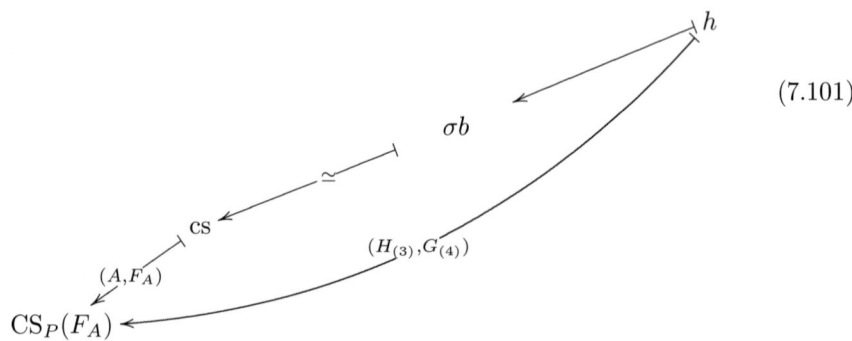

$$(7.101)$$

and

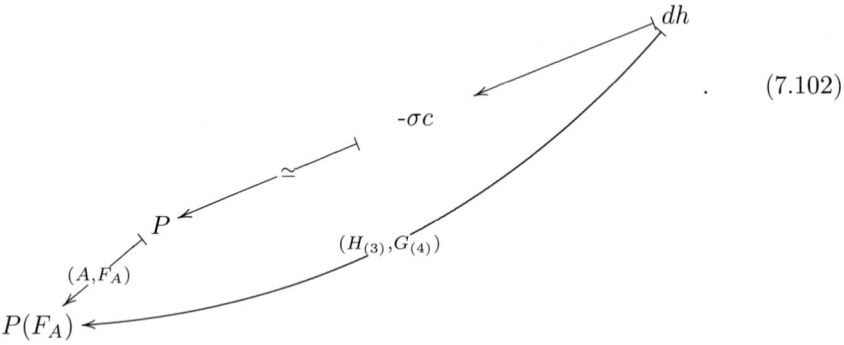

$$(7.102)$$

Notice the signs here, as discussed around equation (7.47). We then have that the connection 3-form of the Chern-Simons 3-bundle given by our obstructing $b^2\mathfrak{u}(1)$-connection descent object is the Chern-Simons form

$$H_{(3)} = -\mathrm{CS}(A, F_A) = -\langle A \wedge dA \rangle - \frac{1}{3}\langle A \wedge [A \wedge A] \rangle \in \Omega^3(Y) \qquad (7.103)$$

of the original Ehresmann connection 1-form A, and its 4-form curvature is therefore the corresponding 4-form

$$G_{(4)} = -P(F_A) = \langle F_A \wedge F_A \rangle \in \Omega^4(Y). \qquad (7.104)$$

This descends down to X, where it constitutes the characteristic form which classifies the obstruction. Indeed, noticing that $\mathrm{inv}(b^2\mathfrak{u}(1)) = \wedge^\bullet(\mathbb{R}[4])$, we see that (this works the same for all line n-bundles, i.e., for all $b^{n-1}\mathfrak{u}(1)$-connection descent objects) the characteristic forms of the obstructing Chern-Simons 3-bundle

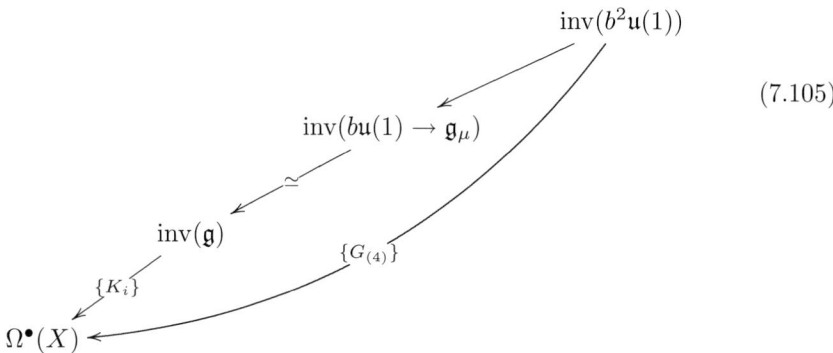

$$(7.105)$$

consist only and precisely of this curvature 4-form: the second Chern-form of the original G-bundle P.

8. L_∞-algebra parallel transport

One of the main points about a connection is that it allows to do parallel transport. Connections on ordinary bundles give rise to a notion of parallel transport along curves, known as holonomy if these curves are closed.

Higher connections on n-bundles should yield a way to obtain a notion of parallel transport over n-dimensional spaces. In physics, this assignment plays the role of the gauge coupling term in the non-kinetic part of the action functional: the action functional of the charged particle is essentially its parallel transport with respect to an ordinary (1-)connection, while the action functional of the string contains the parallel transport of a 2-connection (the Kalb-Ramond field). Similarly the action functional of the membrane contains the parallel transport of a 3-connection (the supergravity "C-field").

There should therefore be a way to assign to any one of our $\mathfrak{g}$-connection descent objects for $\mathfrak{g}$ any Lie n-algebra

- a prescription for parallel transport over n-dimensional spaces;
- a configuration space for the n-particle coupled to that transport;
- a way to transgress the transport to an action functional on that configuration space;
- a way to obtain the corresponding quantum theory.

Each point deserves a separate discussion, but in the remainder we shall quickly give an impression for how each of these points is addressed in our context.

8.1. L_∞-parallel transport

In this section we indicate briefly how our notion of $\mathfrak{g}$-connections give rise to a notion of parallel transport over n-dimensional spaces. The abelian case (meaning here that $\mathfrak{g}$ is an L_∞ algebra such that CE($\mathfrak{g}$) has trivial differential) is comparatively easy to discuss. It is in fact the only case considered in most of the literature. Nonabelian parallel n-transport in the integrated picture for n up to 2 is discussed in [9, 72, 73, 74]. There is a close relation between all differential concepts we develop here and the corresponding integrated concepts, but here we will not attempt to give a comprehensive discussion of the translation.

Given an $(n-1)$-brane ("n-particle") whose n-dimensional worldvolume is modeled on the smooth parameter space Σ (for instance $\Sigma = T^2$ for the closed string) and which propagates on a target space X in that its configurations are given by maps

$$\phi : \Sigma \to X \qquad (8.1)$$

hence by dg-algebra morphisms

$$\Omega^\bullet(\Sigma) \xleftarrow{\quad\phi^*\quad} \Omega^\bullet(X) \qquad (8.2)$$

we can couple it to a $\mathfrak{g}$-descent connection object $(Y, (A, F_A))$ over X pulled back to Σ if Y is such that for every map

$$\phi : \Sigma \to X \qquad (8.3)$$

the pulled back surjective submersion has a global section

$$(8.4)$$

Definition 8.1 (Parallel transport). *Given a $\mathfrak{g}$-descent object $(Y, (A, F_A))$ on a target space X and a parameter space Σ such that for all maps $\phi : \Sigma \to X$ the pullback $\phi^* Y$ has a global section, we obtain a map*

$$\mathrm{tra}_{(A)} : \mathrm{Hom}_{\mathrm{DGCA}}(\Omega^\bullet(X), \Omega^\bullet(\Sigma)) \to \mathrm{Hom}_{\mathrm{DGCA}}(\mathrm{W}(\mathfrak{g}), \Omega^\bullet(\Sigma)) \qquad (8.5)$$

by precomposition with

$$\Omega^\bullet(Y) \xleftarrow{\quad(A, F_A)\quad} \mathrm{W}(\mathfrak{g}) \ . \qquad (8.6)$$

This is essentially the parallel transport of the $\mathfrak{g}$-connection object $(Y, (A, F_A))$. A full discussion is beyond the scope of this article, but for the special case that our L_∞-algebra is $(n-1)$-fold shifted $\mathfrak{u}(1)$, $\mathfrak{g} = b^{n-1}\mathfrak{u}(1)$, the elements in

$$\mathrm{Hom}_{\mathrm{DGCA}}(\mathrm{W}(\mathfrak{g}), \Omega^\bullet(\Sigma)) = \Omega^\bullet(\Sigma, b^{n-1}\mathfrak{u}(1)) \simeq \Omega^n(\Sigma) \qquad (8.7)$$

are in bijection with n-forms on Σ. Therefore they can be integrated over Σ. Then the functional

$$\int_\Sigma \mathrm{tra}_A : \mathrm{Hom}_{\mathrm{DGCA}}(\Omega^\bullet(Y), \Omega^\bullet(\Sigma)) \to \mathbb{R} \qquad (8.8)$$

is the full parallel transport of A.

Proposition 8.2. *The map* $\mathrm{tra}_{(A)}$ *is indeed well defined, in that it depends at most on the homotopy class of the choice of global section* $\hat{\phi}$ *of* ϕ.

Proof. Let $\hat{\phi}_1$ and $\hat{\phi}_2$ be two global sections of ϕ^*Y. Let $\hat{\phi} : \Sigma \times I \to \phi^*Y$ be a homotopy between them, i.e. such that $\hat{\phi}|_0 = \hat{\phi}_1$ and $\hat{\phi}|_1 = \hat{\phi}_2$. Then the difference in the parallel transport using $\hat{\phi}_1$ and $\hat{\phi}_2$ is the integral of the pullback of the curvature form of the $\mathfrak{g}$-descent object over $\Sigma \times I$. But that vanishes, due to the commutativity of

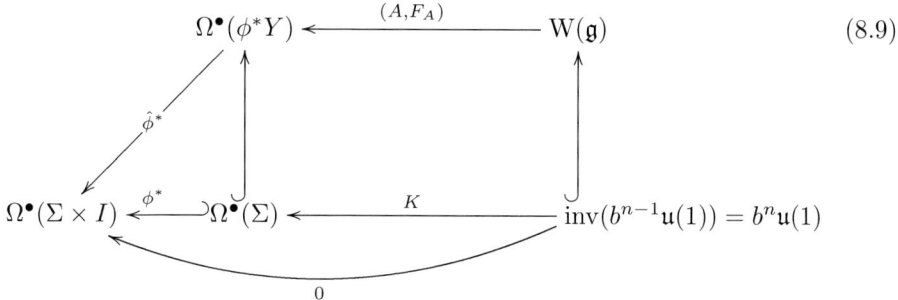

$$(8.9)$$

The composite of the morphisms on the top boundary of this diagram send the single degree $(n+1)$-generator of $\mathrm{inv}(b^{n-1}\mathfrak{u}(1)) = \mathrm{CE}(b^n\mathfrak{u}(1))$ to the curvature form of the $\mathfrak{g}$-connection descent object pulled back to Σ. It is equal to the composite of the horizontal morphisms along the bottom boundary by the definition of $\mathfrak{g}$-descent objects. These vanish, as there is no nontrivial $(n+1)$-form on the n-dimensional Σ. □

8.1.1. Examples.
Chern-Simons and higher Chern-Simons action functionals.

Proposition 8.3. *For G simply connected, the parallel transport coming from the Chern-Simons 3-bundle discussed in 7.3.1 for $\mathfrak{g} = \mathrm{Lie}(G)$ reproduces the familiar Chern-Simons action functional* [31]

$$\int_\Sigma \left(\langle A \wedge dA \rangle + \frac{1}{3} \langle A \wedge [A \wedge A] \rangle \right) \qquad (8.10)$$

over 3-dimensional Σ.

Proof. Recall from section 7.3.1 that we can build the connection descent object for the Chern-Simons connection on the surjective submersion Y coming from the total space P of the underlying G-bundle $P \to X$. Then $\phi^*Y = \phi^*P$ is simply the pullback of that G-bundle to Σ. For G simply connected, BG is 3-connected and hence any G-bundle on Σ is trivializable. Therefore the required lift $\hat{\phi}$ exists and we can construct the above diagram. By equation (7.103) one sees that the integral which gives the parallel transport is indeed precisely the Chern-Simons action functional.   □

Higher Chern-Simons n-bundles, coming from obstructions to fivebrane lifts or still higher lifts, similarly induce higher dimensional generalizations of the Chern-Simons action functional.

BF-theoretic functionals. From proposition 5.27 it follows that we can similarly obtain the action functional of BF theory, discussed in 5.6, as the parallel transport of the 4-connection descent object which arises as the obstruction to lifting a 2-connection descent object for a strict Lie 2-algebra $(\mathfrak{h} \xrightarrow{t} \mathfrak{g})$ through the String-like extension

$$b^2 \mathfrak{u}(1) \to (\mathfrak{h} \xrightarrow{t} \mathfrak{g})_{d_{CE(\mathfrak{h} \xrightarrow{t} \mathfrak{g})}\mu} \to (\mathfrak{h} \xrightarrow{t} \mathfrak{g}) \qquad (8.11)$$

for μ the 3-cocycle on μ which transgresses to the invariant polynomial P on $\mathfrak{g}$ which appears in the BF-action functional.

8.2. Transgression of L_∞-transport

An important operation on parallel transport is its *transgression* to mapping spaces. This is familiar from simple examples, where for instance n-forms on some space transgress to $(n-1)$-forms on the corresponding loop space. We should think of the n-form here as a $b^{n-1}\mathfrak{u}(1)$-connection which transgresses to an $b^{n-2}\mathfrak{u}(1)$ connection on loop space.

This modification of the structure L_∞-algebra under transgression is crucial. In [73] it is shown that for parallel transport n-functors ($n = 2$ there), the operation of transgression is a very natural one, corresponding to acting on the transport functor with an inner hom operation. As shown there, this operation automatically induces the familiar pull-back followed by a fiber integration on the corresponding differential form data, and also automatically takes care of the modification of the structure Lie n-group.

The analogous construction in the differential world of L_∞ algebras we state now, without here going into details about its close relation to [73].

Definition 8.4 (Transgression of $\mathfrak{g}$-connections). *For any $\mathfrak{g}$-connection descent object*

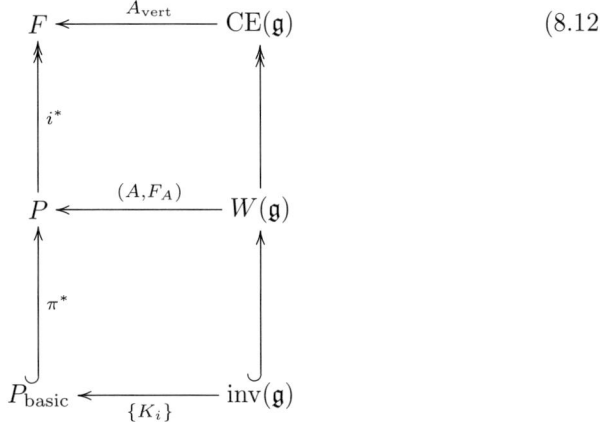

$$\begin{array}{ccc} F & \xleftarrow{\quad A_{\mathrm{vert}} \quad} & \mathrm{CE}(\mathfrak{g}) \\ \Big\uparrow{\scriptstyle i^*} & & \Big\uparrow \\ P & \xleftarrow{\quad (A,F_A) \quad} & W(\mathfrak{g}) \\ \Big\uparrow{\scriptstyle \pi^*} & & \Big\uparrow \\ P_{\mathrm{basic}} & \xleftarrow{\quad \{K_i\} \quad} & \mathrm{inv}(\mathfrak{g}) \end{array} \qquad (8.12)$$

and any smooth space par, *we can form the image of the above diagram under the functor*

$$\mathrm{maps}(-, \Omega^\bullet(\mathrm{par})) : \mathrm{DGCAs} \to \mathrm{DGCAs} \qquad (8.13)$$

from definition 4.4 to obtain the generalized $\mathfrak{g}$-connection descent object (according to definition 6.11)

$$\begin{array}{ccc} \mathrm{maps}(F,\Omega^\bullet(\mathrm{par})) & \xleftarrow{\;\mathrm{tg}_{\mathrm{par}}(A_{\mathrm{vert}})\;} & \mathrm{maps}(\mathrm{CE}(\mathfrak{g}),\Omega^\bullet(\mathrm{par})) \\ \Big\uparrow{\scriptstyle \mathrm{tg}_{\mathrm{par}}\,i^*} & & \Big\uparrow \\ \mathrm{maps}(P,\Omega^\bullet(\mathrm{par})) & \xleftarrow{\;\mathrm{tg}_{\mathrm{par}}(A,F_A)\;} & \mathrm{maps}(W(\mathfrak{g}),\Omega^\bullet(\mathrm{par})) \\ \Big\uparrow{\scriptstyle \mathrm{tg}_{\mathrm{par}}(\pi^*)} & & \Big\uparrow \\ \mathrm{maps}(P_{\mathrm{basic}},\Omega^\bullet(\mathrm{par})) & \xleftarrow{\;\mathrm{tg}_{\mathrm{par}}(\{K_i\})\;} & \mathrm{maps}(\mathrm{inv}(\mathfrak{g}),\Omega^\bullet(\mathrm{par})) \end{array} \qquad (8.14)$$

This new $\mathrm{maps}(\mathrm{CE}(\mathfrak{g}),\Omega^\bullet(\mathrm{par}))$-*connection descent object we call the transgression of the original one to* par.

The operation of transgression is closely related to that of integration.

8.2.1. Examples.

Transgression of $b^{n-1}\mathfrak{u}(1)$-connections. Let $\mathfrak{g}$ be an L_∞-algebra of the form shifted $\mathfrak{u}(1)$, $\mathfrak{g} = b^{n-1}\mathfrak{u}(1)$. By proposition 5.25 the Weil algebra $W(b^{n-1}\mathfrak{u}(1))$ is the free DGCA on a single degree n-generator b with differential $c := db$. Recall from 5.5.1 that a DGCA morphism $W(b^{n-1}\mathfrak{u}(1)) \to \Omega^\bullet(Y)$ is just an n-form on Y. For every point $y \in$ par and for every multivector $v \in \wedge^n T_y$par we get a 0-form on the smooth space

$$\mathrm{maps}(W(b^{n-1}\mathfrak{u}(1)), \Omega^\bullet(\mathrm{par})) \tag{8.15}$$

of all n-forms on par, which we denote

$$A(v) \in \Omega(\mathrm{maps}(W(b^{n-1}\mathfrak{u}(1)), \Omega^\bullet(\mathrm{par}))). \tag{8.16}$$

This is the 0-form on this space of maps obtained from the element $b \in W(b^{n-1}\mathfrak{u}(1))$ and the current δ_y (the ordinary delta-distribution on 0-forms) according to proposition 4.6. Its value on any n-form ω is the value of that form evaluated on v.

Since this, and its generalizations which we discuss in 8.3.1, is crucial for making contact with standard constructions in physics, it may be worthwhile to repeat that statement more explicitly in terms of components: Assume that par $= \mathbb{R}^k$ and for any point y let v be the unit in $\wedge^n T^n \mathbb{R}^n \simeq \mathbb{R}$. Then $A(v)$ is the 0-form on the space of forms which sends any form $\omega = \omega_{\mu_1\mu_2\ldots\mu_n} dx^{\mu_1} \wedge \cdots \wedge dx^{\mu_n}$ to its component

$$A(v) : \omega \mapsto \omega(y)_{12\cdots n}. \tag{8.17}$$

This implies that when a $b^{n-1}\mathfrak{u}(1)$-connection is transgressed to the space of maps from an n-dimensional parameter space par, it becomes a map that pulls back functions on the space of n-forms on par to the space of functions on maps from parameter space into target space. But such pullbacks correspond to *functions* (0-forms) on the space of maps par $\to$ tar with values in the space of n-forms on tra.

8.3. Configuration spaces of L_∞-transport

With the notion of $\mathfrak{g}$-connections and their parallel transport and transgression in hand, we can say what it means to *couple an n-particle/$(n-1)$-brane to a $\mathfrak{g}$-connection*.

Definition 8.5 (The charged n-particle/$(n-1)$-brane). *We say a charged n-particle/$(n-1)$-brane is a tuple* $(\mathrm{par}, (A, F_A))$ *consisting of*

- **parameter space** par: *a smooth space*
- **a background field** (A, F_A): *a $\mathfrak{g}$-connection descent object involving*
  - **target space** tar: *the smooth space*
  - *space!target that the $\mathfrak{g}$-connection (A, F_A) lives over;*
  - **space of phases** phas: *the smooth space such that $\Omega^\bullet(\mathrm{phas}) \simeq \mathrm{CE}(\mathfrak{g})$*

From such a tuple we form

- **configuration space**
- *configuration space* conf $= \hom_{S^\infty}(\mathrm{par}, \mathrm{tar})$;

- *the **action functional** $\exp(S) := \mathrm{tg}_{\mathrm{par}}$: the transgression of the background field to configuration space.*

The configuration space thus defined automatically comes equipped with a notion of vertical derivations as described in 4.3.

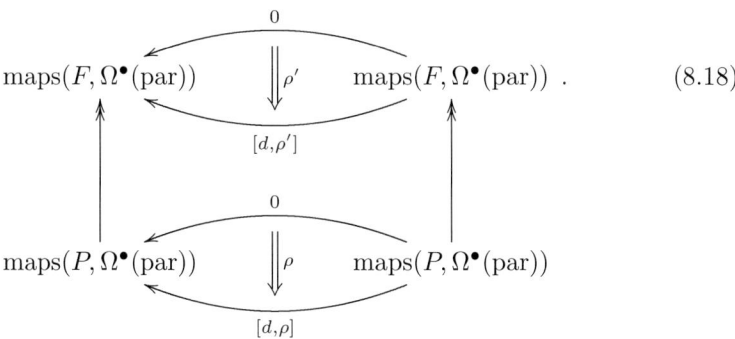

$$\text{(8.18)}$$

These form

- **the gauge symmetries** $\mathfrak{g}_{\mathrm{gauge}}$: an L_∞-algebra.

These act on the horizontal elements of configuration space, which form

- the **anti-fields and anti-ghosts**

in the language of BRST-BV-quantization [78].

We will not go into further details of this here, except for spelling out, as the archetypical example, some details of the computation of the configuration space of ordinary gauge theory.

8.3.1. Examples.

Configuration space of ordinary gauge theory. We compute here the configuration space of ordinary gauge theory on a manifold par with respect to an ordinary Lie algebra $\mathfrak{g}$. A configuration of such a theory is a $\mathfrak{g}$-valued differential form on par, hence, according to 5.5, an element in $\mathrm{Hom}_{\mathrm{DGCAs}}(\mathrm{W}(\mathfrak{g}), \Omega^\bullet(\mathrm{par}))$. So we are interested in understanding the smooth space

$$\mathrm{maps}(\mathrm{W}(\mathfrak{g}), \Omega^\bullet(\mathrm{par})) =: \Omega^\bullet(\mathrm{par}, \mathfrak{g}) \tag{8.19}$$

according to definition 4.4, and the differential graded-commutative algebra

$$\mathrm{maps}(\mathrm{W}(\mathfrak{g}), \Omega^\bullet(\mathrm{par})) =: \Omega^\bullet(\Omega^\bullet(\mathrm{par}, \mathfrak{g})) \tag{8.20}$$

of differential forms on it.

To make contact with the physics literature, we describe everything in components. So let $\mathrm{par} = \mathbb{R}^n$ and let $\{x^\mu\}$ be the canonical set of coordinate functions on par. Choose a basis $\{t_a\}$ of $\mathfrak{g}$ and let $\{t^a\}$ be the corresponding dual basis of $\mathfrak{g}^*$. Denote by

$$\delta_y \iota_{\frac{\partial}{\partial x^\mu}} \tag{8.21}$$

the delta-current on $\Omega^\bullet(\mathrm{par})$, according to definition 4.5, which sends a 1-form ω to

$$\omega_\mu(y) := \omega(\frac{\partial}{\partial x^\mu})(y)\,. \tag{8.22}$$

Summary of the structure of forms on configuration space of ordinary gauge theory. Recall that the Weil algebra $\mathrm{W}(\mathfrak{g})$ is generated from the $\{t^a\}$ in degree 1 and the σt^a in degree 2, with the differential defined by

$$dt^a \;=\; -\frac{1}{2}C^a{}_{bc}t^b \wedge t^c + \sigma t^a \tag{8.23}$$

$$d(\sigma t^a) \;=\; -C^a{}_{bc}t^b \wedge (\sigma t^c)\,. \tag{8.24}$$

We will find that $\mathrm{maps}(\mathrm{W}(\mathfrak{g}), \Omega^\bullet(\mathrm{par}))$ does look pretty much entirely like this, only that all generators are now forms on par. See table 3.

fields	$\{A^a_\mu(y), (F_A)_{\mu\nu}(y) \in \Omega^0(\Omega(\mathrm{par}, \mathfrak{g}))\ \|$ $\quad y \in \mathrm{par}, \mu, \nu \in \{1, \cdots, \dim(\mathrm{par}), a \in \{1, \ldots, \dim(\mathfrak{g})\}\}\}$
ghosts	$\{c^a(y) \in \Omega^1(\Omega(\mathrm{par}, \mathfrak{g}))\ \|$ $\quad y \in \mathrm{par}, a \in \{1, \ldots, \dim(\mathfrak{g})\}\}\}$
antifields	$\{\iota_{(\delta A^a_\mu(y))} \in \mathrm{Hom}(\Omega^1(\Omega(\mathrm{par}, \mathfrak{g})), \mathbb{R})\ \|$ $\quad y \in \mathrm{par}, \mu \in \{1, \cdots, \dim(\mathrm{par}), a \in \{1, \ldots, \dim(\mathfrak{g})\}\}\}$
anti-ghosts	$\{\iota_{(\beta^a(y))} \in \mathrm{Hom}(\Omega^2(\Omega(\mathrm{par}, \mathfrak{g})), \mathbb{R})\ \|$ $\quad y \in \mathrm{par}, \dim(\mathrm{par}), a \in \{1, \ldots, \dim(\mathfrak{g})\}\}$

TABLE 3. **The BRST-BV field content of gauge theory** obtained from our almost internal hom of dg-algebras, definition 4.4. The dgc-algebra $\mathrm{maps}(\mathrm{W}(\mathfrak{g}), \Omega^\bullet(\mathrm{par}))$ is the algebra of differential forms on a smooth space of maps from par to the smooth space underlying $\mathrm{W}(\mathfrak{g})$. In the above table β is a certain 2-form that one finds in this algebra of forms on the space of $\mathfrak{g}$-valued forms.

Remark. Before looking at the details of the computation, recall from 4.1 that an n-form ω in $\mathrm{maps}(\mathrm{W}(\mathfrak{g}), \Omega^\bullet(\mathrm{par}))$ is an assignment

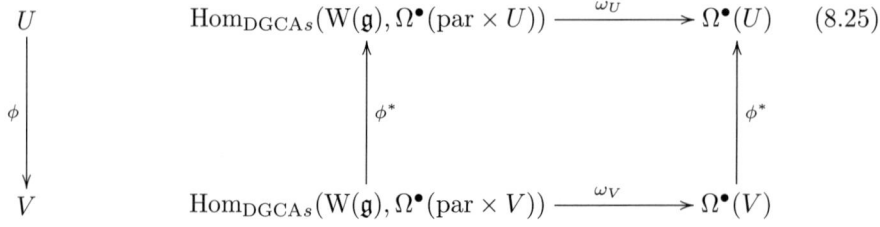

$$ \tag{8.25}$$

of forms on U to $\mathfrak{g}$-valued forms on par $\times\, U$ for all plot domains U (subsets of $\mathbb{R} \cup \mathbb{R}^2 \cup \cdots$ for us), natural in U. We concentrate on those n-forms ω which arise in the way of proposition 4.6.

0-Forms. The 0-forms on the space of $\mathfrak{g}$-value forms are constructed as in proposition 4.6 from an element $t^a \in \mathfrak{g}^*$ and a current $\delta_y \iota_{\frac{\partial}{\partial x^\mu}}$ using

$$t^a \delta_y \iota_{\frac{\partial}{\partial x^\mu}} \tag{8.26}$$

and from an element $\sigma t^a \in \mathfrak{g}^*[1]$ and a current

$$\delta_y \iota_{\frac{\partial}{\partial x^\mu}} \iota_{\frac{\partial}{\partial x^\nu}} . \tag{8.27}$$

This way we obtain the families of functions (0-forms) on the space of $\mathfrak{g}$-valued forms:

$$A_\mu^a(y) : (\Omega^\bullet(\text{par} \times U) \leftarrow W(\mathfrak{g}) : A) \mapsto (u \mapsto \iota_{\frac{\partial}{\partial x^\mu}} A(t^a)(y, u)) \tag{8.28}$$

and

$$F_{\mu\nu}^a(y) : (\Omega^\bullet(\text{par} \times U) \leftarrow W(\mathfrak{g}) : F_A) \mapsto (u \mapsto \iota_{\frac{\partial}{\partial x^\mu}} \iota_{\frac{\partial}{\partial x^\nu}} F_A(\sigma t^a)(y, u)) \tag{8.29}$$

which pick out the corresponding components of the $\mathfrak{g}$-valued 1-form and of its curvature 2-form, respectively. These are the *fields* of ordinary gauge theory.

1-Forms. A 1-form on the space of $\mathfrak{g}$-valued forms is obtained from either starting with a degree 1 element and contracting with a degree 0 delta-current

$$t^a \delta_y \tag{8.30}$$

or starting with a degree 2 element and contracting with a degree 1 delta current:

$$(\sigma t^a)\delta_y \frac{\partial}{\partial x^\mu} . \tag{8.31}$$

To get started, consider first the case where $U = I$ is the interval. Then a DGCA morphism

$$(A, F_A) : W(\mathfrak{g}) \to \Omega^\bullet(\text{par}) \otimes \Omega^\bullet(I) \tag{8.32}$$

can be split into its components proportional to $dt \in \Omega^\bullet(I)$ and those not containing dt. We can hence write the general $\mathfrak{g}$-valued 1-form on par $\times\, I$ as

$$(A, F_A) : t^a \mapsto A^a(y, t) + g^a(y, t) \wedge dt \tag{8.33}$$

and the corresponding curvature 2-form as

$$
\begin{aligned}
(A, F_A) : \sigma t^a \;\mapsto\; & (d_{\text{par}} + d_t)(A^a(y, t) + g^a(y, t) \wedge dt) \\
& + \frac{1}{2} C^a{}_{bc}(A^a(y, t) + g^a(y, t) \wedge dt) \wedge (A^b(y, t) + g^b(y, t) \wedge dt) \\
=\; & F_A^a(y, t) + (\partial_t A^a(y, t) + d_{\text{par}} g^a(y, t) + [g, A]^a) \wedge dt .
\end{aligned} \tag{8.34}
$$

By contracting this again with the current $\delta_y \frac{\partial}{\partial x^\mu}$ we obtain the 1-forms

$$t \mapsto g^a(y, t)dt \tag{8.35}$$

and

$$t \mapsto (\partial_t A_\mu^a(y, t) + \partial_\mu g^a(y, t) + [g, A_\mu]^a)dt \tag{8.36}$$

on the interval. We will identify the first one with the component of the 1-forms on the space of $\mathfrak{g}$-valued forms on par called the *ghosts* and the second one with the 1-forms which are killed by the objects called the *anti-fields*.

To see more of this structure, consider now $U = I^2$, the unit square. Then a DGCA morphism

$$(A, F_A) : W(\mathfrak{g}) \to \Omega^\bullet(\mathrm{par}) \otimes \Omega^\bullet(I^2) \tag{8.37}$$

can be split into its components proportional to $dt^1, dt^2 \in \Omega^\bullet(I^2)$. We hence can write the general $\mathfrak{g}$-valued 1-form on $Y \times I$ as

$$(A, F_A) : t^a \mapsto A^a(y, t) + g_i^a(y, t) \wedge dt^i , \tag{8.38}$$

and the corresponding curvature 2-form as

$$(A, F_A) : \sigma t^a \mapsto (d_Y + d_{I^2})(A^a(y, t) + g_i^a(y, t) \wedge dt^i)$$

$$+ \frac{1}{2} C^a{}_{bc}(A^a(y, t) + g_i^a(y, t) \wedge dt^i) \wedge (A^b(y, t) + g_i^b(y, t) \wedge dt^i)$$

$$= F_A^a(y, t) + (\partial_{t^i} A^a(y, t) + d_Y g_i^a(y, t) + [g_i, A]^a) \wedge dt^i$$

$$+ (\partial_i g_j^a + [g_i, g_j]^a) dt^i \wedge dt^j . \tag{8.39}$$

By contracting this again with the current $\delta_y \frac{\partial}{\partial x^\mu}$ we obtain the 1-forms

$$t \mapsto g_i^a(y, t) dt^i \tag{8.40}$$

and

$$t \mapsto (\partial_t A_\mu^a(y, t) + \partial_\mu g_i^a(y, t) + [g_i, A_\mu]^a) dt^i \tag{8.41}$$

on the unit square. These are again the local values of our

$$c^a(y) \in \Omega^1(\Omega^\bullet(\mathrm{par}, \mathfrak{g})) \tag{8.42}$$

and

$$\delta A_\mu^a(Y) \in \Omega^1(\Omega^\bullet(\mathrm{par}, \mathfrak{g})) . \tag{8.43}$$

The second 1-form vanishes in directions in which the variation of the $\mathfrak{g}$-valued 1-form A is a pure gauge transformation induced by the function g^a which is measured by the first 1-form. Notice that it is the sum of the exterior derivative of the 0-form $A_\mu^a(y)$ with another term.

$$\delta A_\mu^a(y) = d(A_\mu^a(y)) + \delta_g A_\mu^a(y) . \tag{8.44}$$

The first term on the right measures the change of the connection, the second subtracts the contribution to this change due to gauge transformations. So the 1-form $\delta A_\mu^a(y)$ on the space of $\mathfrak{g}$-valued forms vanishes along all directions along which the form A is modified purely by a gauge transformation. The $\delta A_\mu^a(y)$ are the 1-forms the pairings dual to which will be the *antifields*.

2-Forms. We have already seen the 2-form appear on the standard square. We call this 2-form

$$\beta^a \in \Omega^2(\Omega^\bullet(\mathrm{par}, \mathfrak{g})), \tag{8.45}$$

corresponding on the unit square to the assignment

$$\beta^a : (\Omega^\bullet(\mathrm{par} \times I^2) \leftarrow W(\mathfrak{g}) : A) \mapsto (\partial_i g_j^a + [g_i, g_j]^a)dt^i \wedge dt^j. \tag{8.46}$$

There is also a 2-form coming from $(\sigma t^a)\delta_y$. Then one immediately sees that our forms on the space of $\mathfrak{g}$-valued forms satisfy the relations

$$dc^a(y) = -\frac{1}{2}C^a{}_{bc}c^b(y) \wedge c^c(y) + \beta^a(y) \tag{8.47}$$

$$d\beta^a(y) = -C^a{}_{bc}c^b(y) \wedge \beta^c(y). \tag{8.48}$$

The 2-form β on the space of $\mathfrak{g}$-valued forms is what is being contracted by the horizontal pairings called the *antighosts*. We see, in total, that $\Omega^\bullet(\Omega^\bullet(\mathrm{par}, \mathfrak{g}))$ is the Weil algebra of a DGCA, which is obtained from the above formulas by setting $\beta = 0$ and $\delta A = 0$. This DGCA is the algebra of the gauge groupoid, that where the only morphisms present are gauge transformations.

The computation we have just performed are over $U = I^2$. However, it should be clear how this extends to the general case.

Chern-Simons theory. One can distinguish two ways to set up Chern-Simons theory. In one approach one regards principal G-bundles on abstract 3-manifolds, in the other approach one fixes a given principal G-bundle $P \to X$ on some base space X, and pulls it back to 3-manifolds equipped with a map into X. Physically, the former case is thought of as Chern-Simons theory proper, while the latter case arises as the gauge coupling part of the membrane propagating on X. One tends to want to regard the first case as a special case of the second, obtained by letting $X = BG$ be the classifying space for G-bundles and P the universal G-bundle on that.

In our context this is realized by proposition 6.14, which gives the canonical Chern-Simons 3-bundle on BG in terms of a $b^2\mathfrak{u}(1)$-connection descent object on $W(\mathfrak{g})$. Picking some 3-dimensional parameter space manifold par, we can transgress this $b^2\mathfrak{u}(1)$-connection to the configuration space maps$(W(\mathfrak{g}), \Omega^\bullet(\mathrm{par}))$, which we learned is the configuration space of ordinary gauge theory.

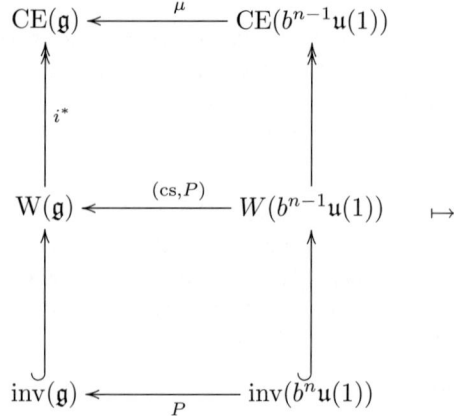

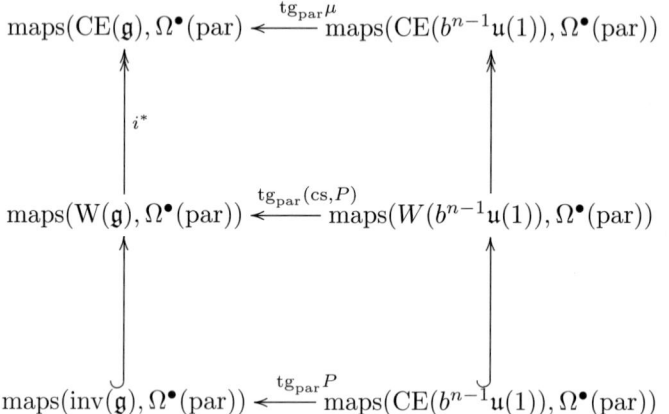

Proposition 8.3 says that the transgressed connection is the Chern-Simons action functional.

Further details of this should be discussed elsewhere.

Transgression of p-brane structures to loop space. It is well known that obstructions to String structures on a space X – for us: Chern-Simons 3-bundles as in 7 – can be conceived

- either in terms of a 3-bundle on X classified by a four class on X obstructing the lift of a 1-bundle on X to a 2-bundle;
- or in terms of a 2-bundle on LX classified by a 3-class on LX obstructing the lift of a 1-bundle on LX to another 1-bundle, principal for a Kac-Moody central extension of the loop group.

In the second case, one is dealing with the transgression of the first case to loop space.

The relation between the two points of views is carefully described in [56]. Essentially, the result is that *rationally* both obstructions are equivalent.

Remark. Unfortunately, there is no universal agreement on the convention of the direction of the operation called transgression. Both possible conventions are used in the literature relevant for our purpose here. For instance [18] say transgression for what [2] calls the inverse of transgression (which, in turn, should be called suspension).

We will demonstrate in the context of L_∞-algebra connections how Lie algebra $(n+1)$-cocycles related to p-brane structures on X transgress to loop Lie algebra n-cocycles on loop space. One can understand this also as an alternative proof of the strictification theorem of the String Lie 2-algebra (proposition 5.35), but this will not be further discussed here.

So let $\mathfrak{g}$ be an ordinary Lie algebra, μ an $(n+1)$-cocycle on it in transgression with an invariant polynomial P, where the transgression is mediated by the transgression element cs as described in section 5.3.

According to proposition 6.14 the corresponding universal obstruction structure is the $b^n\mathfrak{u}(1)$-connection

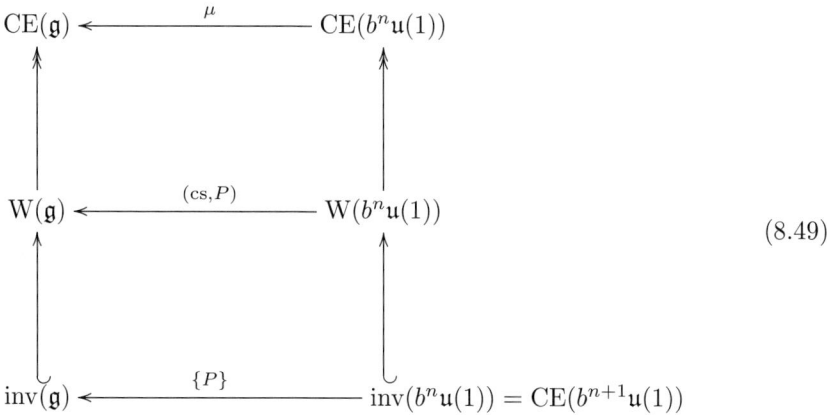

$$(8.49)$$

to be thought of as the universal higher Chern-Simons $(n+1)$-bundle with connection on the classifying space of the simply connected Lie group integrating $\mathfrak{g}$.

We transgress this to loops by applying the functor maps$(-, \Omega^\bullet(S^1))$ from definition 4.4 to it, which can be thought of as computing for all DGC algebras the DGC algebra of differential forms on the space of maps from the circle into the space that the original DGCA was the algebra of differential forms of:

$$
\begin{array}{ccc}
\text{maps}(\text{CE}(\mathfrak{g}), \Omega^\bullet(S^1)) & \xleftarrow{\quad \text{tg}_{S^1}\mu \quad} & \text{maps}(\text{CE}(b^n\mathfrak{u}(1)), \Omega^\bullet(S^1)) \\
\big\uparrow & & \big\uparrow \\
\text{maps}(\text{W}(\mathfrak{g}), \Omega^\bullet(S^1)) & \xleftarrow{\quad \text{tg}_{S^1}(\text{cs},P) \quad} & \text{maps}(\text{W}(b^n\mathfrak{u}(1)), \Omega^\bullet(S^1)) \\
\big\uparrow & & \big\uparrow \\
\text{maps}(\text{inv}(\mathfrak{g}), \Omega^\bullet(S^1)) & \xleftarrow{\quad \{\text{tg}_{S^1}P\} \quad} & \text{maps}(\text{CE}(b^{n+1}\mathfrak{u}(1)), \Omega^\bullet(S^1))
\end{array}
\qquad . \qquad (8.50)
$$

We want to think of the result as a $b^{n-1}\mathfrak{u}(1)$-bundle. This we can achieve by pulling back along the inclusion

$$
\text{CE}(b^{n-1}\mathfrak{u}(1)) \hookrightarrow \text{maps}(\text{CE}(b^n\mathfrak{u}(1)), \Omega^\bullet(S^1)) \qquad (8.51)
$$

which comes from the integration current $\int_{S^1}$ on $\Omega^\bullet(S^1)$ according to proposition 4.6.

(This restriction to the integration current can be understood from looking at the basic forms of the loop bundle descent object, which induces *integration without integration* essentially in the sense of [51]. But this we shall not further go into here.)

We now show that the transgressed cocycles $\text{tg}_{S^1}\mu$ are the familiar cocycles on loop algebras, as appearing for instance in Lemma 1 of [2]. For simplicity of exposition, we shall consider explicitly just the case where $\mu = \langle \cdot, [\cdot, \cdot] \rangle$ is the canonical 3-cocycle on a Lie algebra with bilinear invariant form $\langle \cdot, \cdot \rangle$.

Proposition 8.6. *The transgressed cocycle in this case is the 2-cocycle of the Kac-Moody central extension of the loop Lie algebra* $\Omega\mathfrak{g}$

$$
\text{tg}_{S^1}\mu : (f, g) \mapsto \int_{S^1} \langle f(\sigma), g'(\sigma) \rangle d\sigma + (\text{a coboundary}) \qquad (8.52)
$$

for all $f, g \in \Omega\mathfrak{g}$.

Proof. We compute $\text{maps}(\text{CE}(\mathfrak{g}), \Omega^\bullet(S^1))$ as before from proposition 4.6 along the same lines as in the above examples: for $\{t_a\}$ a basis of $\mathfrak{g}$ and U any test domain, a DGCA homomorphism

$$
\phi : \text{CE}(\mathfrak{g}) \to \Omega^\bullet(S^1) \otimes \Omega^\bullet(U) \qquad (8.53)
$$

sends

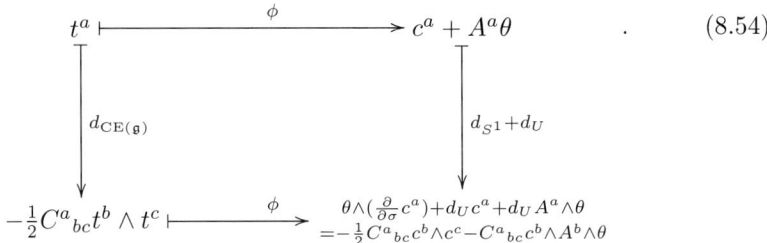

$$t^a \xmapsto{\phi} c^a + A^a \theta \qquad\qquad . \qquad (8.54)$$

Here $\theta \in \Omega^1(S^1)$ is the canonical 1-form on S^1 and $\frac{\partial}{\partial \sigma}$ the canonical vector field; moreover $c^a \in \Omega^0(S^1) \otimes \Omega^1(U)$ and $A^a \theta \in \Omega^1(S^1) \otimes \Omega^0(U)$.

By contracting with δ-currents on S^1 we get 1-forms $c^a(\sigma)$, $\frac{\partial}{\partial \sigma} c^a(\sigma)$ and 0-forms $A^a(\sigma)$ for all $\sigma \in S^1$ on $\mathrm{maps}(\mathrm{CE}(\mathfrak{g}), \Omega^\bullet(S^1))$ satisfying

$$d_{\mathrm{maps}(\cdots)} c^a(\sigma) + \frac{1}{2} C^a{}_{bc} c^b(\sigma) \wedge c^c(\sigma) = 0 \qquad (8.55)$$

and

$$d_{\mathrm{maps}(\cdots)} A^a(\sigma) - C^a{}_{bc} A^b(\sigma) \wedge c^c(\sigma) = \frac{\partial}{\partial \sigma} c^a(\sigma). \qquad (8.56)$$

Notice the last term appearing here, which is the crucial one responsible for the appearance of derivatives in the loop cocycles, as we will see now.

So $A^a(\sigma)$ (a "field") is the function on (necessarily flat) $\mathfrak{g}$-valued 1-forms on S^1 which sends each such 1-form for its t^a-component along θ at σ, while $c^a(\sigma)$ (a "ghost") is the 1-form which sends each tangent vector field to the space of flat $\mathfrak{g}$-valued forms to the gauge transformation in t^a direction which it induces on the given 1-form at $\sigma \in S^1$.

Notice that the transgression of our 3-cocycle

$$\mu = \mu_{abc} t^a \wedge t^b \wedge t^c = C_{abc} t^a \wedge t^b \wedge t^c \in H^3(\mathrm{CE}(\mathfrak{g})) \qquad (8.57)$$

is

$$\mathrm{tg}_{S^1} \mu = \int_{S^1} C_{abc} A^a(\sigma) c^b(\sigma) \wedge c^c(\sigma)\, d\sigma \ \in \ \Omega^2(\Omega^1_{\mathrm{flat}}(S^1, \mathfrak{g}). \qquad (8.58)$$

We can rewrite this using the identity

$$d_{\mathrm{maps}(\cdots)} \left(\int_{S^1} P_{ab} A^a(\sigma) c^b(\sigma) d\sigma \right) = \int_{S^1} P_{ab} \left(\partial_\sigma c^a(\sigma) \right) \wedge c^b(\sigma)$$
$$+ \frac{1}{2} \int_{S^1} C_{abc} A^a(\sigma) c^b(\sigma) \wedge c^c(\sigma), (8.59)$$

which follows from 8.55 and 8.56, as

$$\mathrm{tg}_{S^1} \mu = \int_{S^1} P_{ab} \left(\partial_\sigma c^a(\sigma) \right) \wedge c^b(\sigma) + d_{\mathrm{maps}(\cdots)}(\cdots). \qquad (8.60)$$

Then notice that

- equation (8.55) is the Chevalley-Eilenberg algebra of the loop algebra $\Omega\mathfrak{g}$;

- the term $\int_{S^1} P_{ab}(\partial_\sigma c^a(\sigma)) \wedge c^b(\sigma)$ is the familiar 2-cocycle on the loop algebra obtained from transgression of the 3-cocycle $\mu = \mu_{abc} t^a \wedge t^b \wedge t^c = C_{abc} t^a \wedge t^b \wedge t^c$.

□

9. Physical applications: string-, fivebrane- and p-brane structures

We can now discuss physical applications of the formalism that we have developed. What we describe is a useful way to handle obstructing n-bundles of various kinds that appear in string theory. In particular, we can describe generalizations of string structure in string theory. In the context of p-branes, such generalizations have been suggested based on p-loop spaces [30] [10] [64] and, more generally, on the space of maps $\mathrm{Map}(M, X)$ from the brane worldvolume M to spacetime X [60]. The statements in this section will be established in detail in [68].

From the point of view of supergravity, all branes, called p-branes in that setting, are a priori treated in a unified way. In tracing back to string theory, however, there is a distinction in the form-fields between the Ramond-Ramond (RR) and the Neveu-Schwarz (NS) forms. The former live in generalized cohomology and the latter play two roles: they act as twist fields for the RR fields and they are also connected to the geometry and topology of spacetime. The H-field H_3 plays the role of a twist in K-theory for the RR fields [50] [15] [59]. The twist for the degree seven dual field H_7 is observed in [67] at the rational level.

The ability to define fields and their corresponding partition functions puts constraints on the topology of the underlying spacetime. The most commonly understood example is that of fermions where the ability to define them requires spacetime to be spin, and the ability to describe theories with chiral fermions requires certain restrictions coming from the index theorem. In the context of heterotic string theory, the Green-Schwarz anomaly cancelation leads to the condition that the difference between the Pontrjagin classes of the tangent bundle and that of the gauge bundle be zero. This is called the string structure, which can be thought of as a spin structure on the loop space of spacetime [52] [27]. In M-theory, the ability to define the partition function leads to an anomaly given by the integral seventh-integral Stiefel-Whitney class of spacetime [29] whose cancelation requires spacetime to be orientable with respect to generalized cohomology theories beyond K-theory [55] .

In all cases, the corresponding structure is related to the homotopy groups of the orthogonal group: the spin structure amounts to killing the first homotopy group, the string structure and – to some extent– the W_7 condition to killing the third homotopy group. Note that when we say that the n-th homotopy group is killed, we really mean that all homotopy groups up to and including the n-th one are killed. For instance, a String structure requires killing everything up to and including the third, hence everything through the sixth, since there are no homotopy groups in degrees four, five or six.

The Green-Schwarz anomaly cancelation condition for the heterotic string can be translated to the language of n-bundles as follows. We have two bundles, the spin bundle with structure group $G = \mathrm{Spin}(10)$, and the gauge bundle with structure group G' being either $\mathrm{SO}(32)/\mathbb{Z}_2$ or $E_8 \times E_8$. Considering the latter, we have one copy of E_8 on each ten-dimensional boundary component, which can be viewed as an end-of-the-world nine-brane, or M9-brane [47]. The structure of the four-form on the boundary which we write as

$$G_4|_\partial = dH_3 \qquad\qquad (9.1)$$

implies that the 3-bundle (2-gerbe) becomes the trivializable lifting 2-gerbe of a String($\mathrm{Spin}(10) \times E_8$) bundle over the M9-brane. As the four-form contains the difference of the Pontrjagin classes of the bundles with structure groups G and G', the corresponding three-form will be a difference of Chern-Simons forms. The bundle aspect of this has been studied in [12] and will be revisited in the current context in [68].

The NS fields play a special role in relation to the homotopy groups of the orthogonal group. The degree three class $[H_3]$ plays the role of a twist for a spin structure. Likewise, the degree seven class plays a role of a twist for a higher structure related to $BO\langle 9 \rangle$, the 8-connected cover of BO, which we might call a *Fivebrane*-structure on spacetime. We can talk about such a structure once the spacetime already has a string structure. The obstructions are given in table 4, where A is the connection on the G' bundle and ω is a connection on the G bundle.

n	$\begin{array}{c}2\\ = 4 \cdot 0 + 2\end{array}$	$\begin{array}{c}6\\ = 4 \cdot 1 + 2\end{array}$
fundamental object $(n-1)$-brane n-particle	string	5-brane
target space structure	string structure $\mathrm{ch}_2(A) - p_1(\omega) = 0$	fivebrane structure $\mathrm{ch}_4(A) - \frac{1}{48}p_2(\omega) = 0$

TABLE 4. **Higher dimensional extended objects** and the corresponding topological structures.

In the above we alluded to how the brane structures are related to obstructions to having spacetimes with connected covers of the orthogonal groups as structures. The obstructing classes here may be regarded as classifying the corresponding obstructing n-bundles, after we apply the general formalism that we outlined earlier. The main example of this general mechanism that will be of interest to us here is the case where $\mathfrak{g}$ is an ordinary semisimple Lie algebra. In particular, we consider $\mathfrak{g} = \mathfrak{spin}(n)$. For $\mathfrak{g} = \mathfrak{spin}(d)$ and μ a $(2n+1)$-cocycle on $\mathfrak{spin}(d)$, we call $\mathfrak{spin}(d)_\mu$ the (skeletal version of the) $(2n-1)$-**brane Lie** $(2n) - algebra$. Thus, the case of String structure and Fivebrane structure occurring in the fundamental string and NS fivebrane correspond to the cases $n = 1$ and $n = 3$ respectively. Now applying our formalism for $\mathfrak{g} = \mathfrak{spin}(d)$, and μ_3, μ_7 the canonical 3- and 7-cocycle, respectively, we have:

- the obstruction to lifting a $\mathfrak{g}$-bundle descent object to a String 2-bundle (a $\mathfrak{g}_{\mu_3}$-bundle descent object) is the first Pontryagin class of the original $\mathfrak{g}$-bundle cocycle;
- the obstruction to lifting a String 2-bundle descent object to a Fivebrane 6-bundle cocycle (a $\mathfrak{g}_{\mu_7}$-bundle descent object) is the second Pontryagin class of the original $\mathfrak{g}$-bundle cocycle.

The cocyles and invariant polynomials corresponding to the two structures are given in the following table

p-brane	cocycle	invariant polynomial	
$p = 1 = 4 \cdot 0 + 1$	$\mu_3 = \langle \cdot, [\cdot, \cdot] \rangle$	$P_1 = \langle \cdot, \cdot \rangle$	first Pontrjagin
$p = 5 = 4 \cdot 1 + 1$	$\mu_7 = \langle \cdot, [\cdot, \cdot], [\cdot, \cdot], [\cdot, \cdot] \rangle$	$P_2 = \langle \cdot, \cdot, \cdot, \cdot \rangle$	second Pontrjagin

TABLE 5. **Lie algebra cohomology governing NS p-branes.**

In case of the fundamental string, the obstruction to lifting the $PU(H)$ bundles to $U(H)$ bundles is measured by a gerbe or a line 2-bundle. In the language of E_8 bundles this corresponds to lifting the loop group LE_8 bundles to the central extension $\hat{L}E_8$ bundles [59]. The obstruction for the case of the String structure is a 2-gerbe and that of a Fivebrane structure is a 6-gerbe. The structures are summarized in the following table

obstruction	$\rightarrow$	G-bundle	$\Rightarrow$	$\hat{G}$-bundle
1-gerbes / line 2-bundles		$PU(H)$-bundles		$U(H)$-bundles
2-gerbes / line 3-bundles	$\rightarrow$	$Spin(n)$-bundles	$\Rightarrow$	$String(n)$-2-bundles
6-gerbes / line 7-bundles		$String(n)$-2-bundles		$FiveBrane(n)$-6-bundles

TABLE 6. **Obstructing line n-bundles** appearing in string theory, where
$\rightarrow$ equals the phrase "obstruct the lift of", and
$\Rightarrow$ equals the phrase "to".

A description can also be given in terms of (higher) loop spaces, generalizing the known case where a String structure on a space X can be viewed as a Spin structure on the loop space LX. A fuller discussion of the ideas of this section is given in [68] and [69].

Appendix A. Explicit formulas for 2-morphisms of L_∞-algebras

To the best of our knowledge, the only place in the literature where 2-morphisms between 1-morphisms of L_∞-algebras have been spelled out in detail is [5], which gives a definition of 2-morphisms for Lie 2-algebras, i.e. for L_∞-algebras concentrated in the lowest two degrees. Our definition 5.11 provides an algorithm for computing 2-morphisms between morphisms of arbitrary (finite dimensional) L_∞-algebras. We had already demonstrated in 5.2 one application of that algorithm, showing explicitly how it allows to compute transgression elements (Chern-Simons forms).

   For completeness, we demonstrate that the formulas given in [5] for the special case of Lie 2-algebras also follow as a special case from our general definition 5.11. This is of relevance to our discussion of the String Lie 2-algebra, since the equivalence of its strict version with its weak skeletal version, mentioned in our proposition 5.35, has been established in [7] using these very formulas. First we quickly recall the relevant definitions from [5, 7]: A "2-term" L_∞-algebra is an L_∞-algebra concentrated in the lowest two degrees. A morphism

$$\varphi : \mathfrak{g} \to \mathfrak{h} \tag{A.1}$$

of 2-term L_∞-algebras $\mathfrak{g}$ and $\mathfrak{h}$ is a pair of maps

$$\phi_0 \quad : \quad \mathfrak{g}_1 \to \mathfrak{h}_1 \tag{A.2}$$
$$\phi_1 \quad : \quad \mathfrak{g}_2 \to \mathfrak{h}_2 \tag{A.3}$$

together with a skew-symmetric map

$$\phi_2 : \mathfrak{g}_1 \otimes \mathfrak{g}_1 \to \mathfrak{h}_2 \tag{A.4}$$

satisfying

$$\phi_0(d(h)) = d(\phi_1(h)) \tag{A.5}$$

as well as

$$d(\phi_2(x,y)) \quad = \quad \phi_0(l_2(x,y)) - l_2(\phi_0(x),\phi_0(y)) \tag{A.6}$$
$$\phi_2(x,dh) \quad = \quad \phi_1(l_2(x,h)) - l_2(\phi_0(x),\phi_1(h)) \tag{A.7}$$

and finally

$$l_3(\phi_0(x),\phi_0(y),\phi_0(z)) - \phi_1(l_3(x,y,z)) = \phi_2(x,l_2(y,z)) + \phi_2(y,l_2(z,x))$$
$$+\phi_2(z,l_2(x,y)) + l_2(\phi_0(x),\phi_2(y,z)) + l_2(\phi_0(y),\phi_2(z,x)) + l_2(\phi_0(z),\phi_2(x,y)) \,.$$

for all $x,y,z \in \mathfrak{g}_1$ and $h \in \mathfrak{g}_2$. This follows directly from the requirement that morphisms of L_∞-algebras be homomorphisms of the corresponding codifferential coalgebras, according to definition 5.2. The not quite so obvious aspect are the analogous formulas for 2-morphisms:

Definition A.1 (Baez-Crans). *A 2-morphism*

$$(A.8)$$

of 1-morphisms of 2-term L_∞-algebras is a linear map

$$\tau : \mathfrak{g}_1 \to \mathfrak{h}_2 \tag{A.9}$$

such that

$$\psi_0 - \phi_0 = t_W \circ \tau \tag{A.10}$$
$$\psi_1 - \phi_1 = \tau \circ t_v \tag{A.11}$$

and

$$\phi_2(x,y) - \psi_2(x,y) = l_2(\phi_0(x), \tau(y)) + l_2(\tau(x), \psi_0(y)) - \tau(l_2(x,y)) \tag{A.12}$$

Notice that $[d,\tau] := d_\mathfrak{h} \circ \tau + \tau \circ d_\mathfrak{g}$ and that it restricts to $d_\mathfrak{h} \circ \tau$ on $\mathfrak{g}_1$ and to $\tau \circ d_\mathfrak{g}$ on $\mathfrak{g}_2$.

Proposition A.2. *For finite dimensional L_∞-algebras, definition A.1 is equivalent to the restriction of our definition 5.11 to 2-term L_∞-algebras.*

Proof. Let $\mathfrak{g} = \mathfrak{g}_1 \oplus \mathfrak{g}_2$ and $\mathfrak{h} = \mathfrak{h}_1 \oplus \mathfrak{h}_2$ be any two 2-term L_∞-algebras. Then take

$$\psi, \phi : \mathfrak{g} \to \mathfrak{h} \tag{A.13}$$

to be any two L_∞ morphisms with

$$\mathrm{CE}(\mathfrak{g}) \xleftarrow{\;\psi^*, \phi^*\;} \mathrm{CE}(\mathfrak{h}) \tag{A.14}$$

the corresponding DGCA morphisms. We would like to describe the collection of all 2-morphisms

$$\mathrm{CE}(\mathfrak{g}) \quad \substack{\phi^* \\ \Big\Downarrow \tau \\ \psi^*} \quad \mathrm{CE}(\mathfrak{h}) \tag{A.15}$$

according to definition 5.11. We do this in terms of a basis. With $\{t^a\}$ a basis for $\mathfrak{h}_1$ and $\{b^i\}$ a basis for $\mathfrak{h}_2$, and accordingly $\{t'^a\}$ and $\{b'^i\}$ a basis of $\mathfrak{g}_1$ and $\mathfrak{g}_2$, respectively, this comes from a map

$$\tau^* : \mathfrak{h}_1^* \oplus \mathfrak{h}_2^* \oplus \mathfrak{h}_1^*[1] \oplus \mathfrak{h}_2^*[1] \to \wedge^\bullet \mathfrak{g}^* \tag{A.16}$$

of degree -1 which acts on these basis elements as

$$\tau^* : b^i \mapsto \tau^i{}_a t'^a \tag{A.17}$$

and
$$\tau^* : a^a \mapsto 0 \qquad (\text{A.18})$$
for some coefficients $\{\tau^i{}_a\}$. Now the crucial requirement (5.65) of definition 5.11 is that (5.64) *vanishes* when restricted

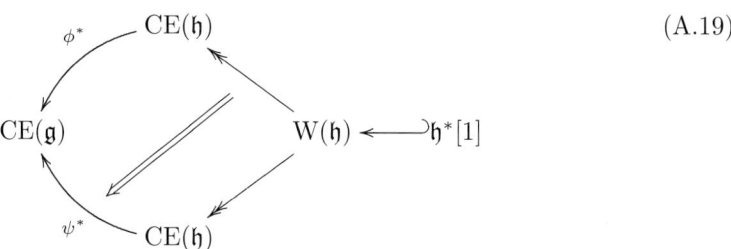

$$(\text{A.19})$$

to generators in the shifted copy of the Weil algebra. This implies the following. For τ^* to vanish on all σt^a we find that its value on $d_{W(\mathfrak{h})} t^a = -\frac{1}{2} C^a{}_{bc} t^a \wedge t^b - t^a{}_i b^i + \sigma t^a$ is fixed to be
$$\tau^* : d_{W(\mathfrak{g})} t^a \mapsto -t^a{}_i \tau^i{}_b t'^b \qquad (\text{A.20})$$
and on $d_{W(\mathfrak{h})} b^i = -\alpha^i{}_{aj} t^a \wedge b^j + c^i$ to be
$$\tau^*(db^i) = \tau^*(-\alpha^i{}_{aj} t^a \wedge b^j). \qquad (\text{A.21})$$
The last expression needs to be carefully evaluated using formula (5.61). Doing so we get
$$[d, \tau^*] : t^a \mapsto -t^a{}_i \tau^i{}_b t'^b \qquad (\text{A.22})$$
and
$$[d, \tau^*] : b^i \mapsto -\frac{1}{2} \tau^i{}_a C'^a{}_{bc} t'^b t'^c - \tau^i{}_a t'^a{}_j b'^j + \alpha^i{}_{aj} \frac{1}{2}(\phi + \psi)^a{}_b \tau^j{}_c t'^b t'^c. \qquad (\text{A.23})$$
Then the expression
$$\phi^* - \psi^* = [d, \tau^*] \qquad (\text{A.24})$$
is equivalent to the following ones
$$(\psi^a{}_b - \phi^a{}_b) t'^b \;=\; t^a{}_i \tau^i{}_b t'^b \qquad (\text{A.25})$$
$$(\psi^i{}_j - \phi^i{}_j) b'^j \;=\; \tau^i{}_a t'^a{}_j b'^j \qquad (\text{A.26})$$
$$\frac{1}{2}(\phi^i{}_{ab} - \psi^i{}_{ab}) t'^a t'^b \;=\; -\frac{1}{2} \tau^i{}_a C'^a{}_{bc} t'^b t'^c + \alpha^i{}_{aj} \frac{1}{2}(\phi + \psi)^a{}_b \tau^j{}_c t'^b t'^c. \qquad (\text{A.27})$$
The first two equations express the fact that τ is a chain homotopy with respect to t and t'. The last equation is equivalent to
$$\phi_2(x, y) - \psi_2(x, y) \;=\; -\tau([x, y]) + [q(x) + \frac{1}{2} t(\tau(x)), \tau(y)]$$
$$-[q'(y) - \frac{1}{2} t(\tau(y)), \tau(x)]$$
$$=\; -\tau([x, y]) + [q(x), \tau(y)] + [\tau(x), q'(y)] \qquad (\text{A.28})$$
This is indeed the Baez-Crans condition on a 2-morphism. $\qquad \square$

Acknowledgements

We thank Danny Stevenson for many helpful comments. Among other things, the discussion of the cohomology of the string Lie 2-algebra has been greatly influenced by conversation with him. We acknowledge stimulating remarks by John Baez on characteristic classes of string bundles. H.S. thanks Matthew Ando and U.S. thanks Stephan Stolz and Peter Teichner for discussions at an early stage of this work.

U.S. thankfully acknowledges invitations to the University of Oxford by Nigel Hitchin; to the conference "Lie algebroids and Lie groupoids in differential geometry" in Sheffield by Kirill Mackenzie and Ieke Moerdijk; to the Erwin Schrödinger Institute in the context of the program "Poisson σ-models, Lie algebroids, deformations and higher analogues"; to "Categories in Geometry and Physics" by Zoran Škoda and Igor Baković; and to Yale University which led to collaboration with H.S. and to useful discussions with Mikhail Kapranov. U.S. thanks Todd Trimble for discussion of DGCAs, and their relation to smooth spaces, thanks Mathieu Dupont and Larry Breen for discussion about weak cokernels and thanks Simon Willerton and Bruce Bartlett for helpful discussion of the notion of transgression. H.S. thanks Akira Asada and Katsuhiko Kuribayashi for useful correspondence on their work on (higher) string structures.

We thank the contributors to the weblog *The n-Category Café* for much interesting, often very helpful and sometimes outright invaluable discussion concerning the ideas presented here and plenty of related issues.

We have gratefully received useful comments on earlier versions of this document from Gregory Ginot and David Roberts. Evan Jenkins and Todd Trimble have provided useful help with references to literature on smooth function algebras.

While this article was being written, the preprints [80], [54] and [36] (based on [37]) appeared, which are closely related to our discussion here in that they address the issues of characteristic classes and extensions of n-bundles in one way or another. We are thankful to Amnon Yekutieli and to Gregory Ginot for pointing out their work to us and for further discussion.

Finally, we thank Bertfried Fauser and the organizers of the conference "Recent developments in QFT" in Leipzig.

U.S. thankfully acknowledges financial support from the DAAD German-Croatian bilateral project "Nonabelian cohomology and applications".

References

[1] M. Alexandrov, M. Kontsevich, A. Schwarz and O. Zaboronsky, *The Geometry of the Master Equation and Topological Quantum Field Theory*, [http://arxiv.org/refs/hep-th/0608150]

[2] A. Asada, *Characteristic classes of loop group bundles and generalized string classes*, Differential geometry and its applications (Eger, 1989), 33-66, Colloq. Math. Soc. János Bolyai, 56, North-Holland, Amsterdam, (1992)

[3] P. Aschieri and B. Jurco, *Gerbes, M5-brane anomalies and E_8 gauge theory* J. High Energy Phys. **0410** (2004) 068, [arXiv:hep-th/0409200v1].

[4] R. D'Auria and P. Fré, *Geometric supergravity in $D = 11$ and its hidden supergroup*, Nuclear Physics B, 201 (1982)

[5] J. Baez and A. Crans, *Higher-dimensional algebra VI: Lie 2-algebras*, Theory and Applications of Categories **12** (2004) 492–528, [arXiv:math/0307263v5].

[6] T. Bartels, *2-Bundles*, [arXiv:math/0410328v3]

[7] J. Baez, A. Crans, U. Schreiber and D. Stevenson, *From Loop Groups to 2-Groups*, Homology, Homotopy and Applications, Vol. 9 (2007), No. 2, pp.101–135, [arXiv:math/0504123v2].

[8] J. Baez and D. Stevenson, *The Classifying Space of a Topological 2-Group*, [arXiv:0801.3843v1]

[9] J. Baez and U. Schreiber, *Higher gauge theory*, in Contemporary Mathematics, 431, *Categories in Algebra, Geometry and Mathematical Physics*, [arXiv:math/0511710].

[10] E. Bergshoeff, R. Percacci, E. Sezgin, K.S. Stelle, and P.K. Townsend, *U(1)-extended gauge algebras in p-loop space*, Nucl. Phys. **B 398** (1993) 343, [arXiv:hep-th/9212037v1].

[11] E. Bergshoeff, E. Sezgin, and P.K. Townsend, *Properties of the eleven-dimensional super membrane theory*, Annals Phys.**185** (1988) 330.

[12] L. Bonora, P. Cotta-Ramusino, M. Rinaldi, and J. Stasheff, *The evaluation map in field theory, sigma-models and strings I*, Commun. Math. Phys. **112** (1987) 237.

[13] A. Borel, *Sur la cohomologie des espaces fibrés principaux et des espaces homogńes de groupes de Lie compacts*, Ann. Math. **(2) 57** (1953), 115–207.

[14] A.K. Bousfield and V.K.A.M. Gugenheim, *On PL de Rham theory and rational homotopy type*, Mem. Amer. Math. Soc.**8** (1976), no. 179.

[15] P. Bouwknegt and V. Mathai, *D-branes, B-fields and twisted K-theory*, J. High Energy Phys. **03** (2000) 007, [arXiv:hep-th/0002023].

[16] P. Bouwknegt, A. Carey, V. Mathai, M. Murray and D. Stevenson, *Twisted K-theory and K-theory of bundle gerbes*, Commun. Math. Phys. **228** (2002) 17, [arXiv:hep-th/0106194].

[17] L. Breen and W. Messing, *Differential geometry of gerbes*, Adv. Math. **198** (2005), no. 2, 732–846, [arXiv:math/0106083v3].

[18] J.-L. Brylinski and D.A. McLaughlin, *Čech cocycles for characteristic classes*, Comm. Math. Phys. 178, 225–236 (1996)

[19] J.-L. Brylinski and D.A. McLaughlin, *A geometric construction of the first Pontryagin class*, Quantum Topology, 209-220, World Scientific (1993)

[20] A. Carey, S. Johnson, and M. Murray, *Holonomy on D-branes*, J. Geom. Phys. **52** (2004), no. 2, 186–216, [arXiv:hep-th/0204199v3].

[21] A. Carey, S. Johnson, M. Murray, D. Stevenson, and B.-L. Wang, *Bundle gerbes for Chern-Simons and Wess-Zumino-Witten theories* , Commun. Math. Phys. 259 (2005) 577, [arXiv:math/0410013].

[22] H. Cartan, *Notions d'algébre différentielle; application aux groupes de Lie et aux variétés ou opère un groupe de Lie*, Colloque de topologie (espaces fibrs), Bruxelles, 1950, pp. 15–27.

[23] H. Cartan, *Cohomologie reélle d'un espace fibré principal diffrentielle I, II*, Séminaire Henri Cartan, 1949/50, pp. 19-01 – 19-10 and 20-01 – 20-11, CBRM, 1950.

[24] L. Castellani, R. D'Auria and P. Fré, *Supergravity and superstrings: a geometric perspective*, World Scientific, Singapore (1991)

[25] J. Cheeger and J. Simons, *Differential characters and geometric invariants*, in Geometry and Topology, 50–80, Lecture Notes in Math., 1167, Springer-Verlag, 1985.

[26] A. Clingher, *Heterotic string data and theta functions*, Adv. Theor. Math. Phys. **9** (2005) 173, [arXiv:math/0110320v2].

[27] R. Coquereaux and K. Pilch, *String structures on loop bundles*, Commun. Math. Phys. **120** (1989) 353.

[28] E. Diaconescu, D. Freed, and G. Moore, *The M theory three form and E_8 gauge theory*, [arXiv:hep-th/0312069].

[29] E. Diaconescu, G. Moore and E. Witten, *E_8 gauge theory, and a derivation of K-theory from M-theory* Adv. Theor. Math. Phys. **6** (2003) 1031, [arXiv:hep-th/0005090].

[30] J.A. Dixon, M.J. Duff, and E. Sezgin, *The coupling of Yang-Mills to extended objects*, Phys. Lett. **B 279** (1992) 265, [arXiv:hep-th/9201019v1].

[31] D.S. Freed, *Classical Chern-Simons Theory, Part I*, [hep-th/9206021v1]

[32] D.S. Freed, *Higher algebraic structures and quantization*, [arXiv:hep-th/9212115v2]

[33] D.S. Freed, *Quantum groups from path integrals*, [arXiv:q-alg/9501025v1]

[34] D.S. Freed and E. Witten, *Anomalies in string theory with D-Branes*, Asian J. Math. **3** (1999) 819, [arXiv:hep-th/9907189].

[35] E. Getzler, *Lie theory for nilpotent L_∞-algebras*, Ann. Math. [arXiv:math/0404003v4]

[36] G. Ginot and M. Stiénon, *Groupoid extensions, principal 2-group bundles and characteristic classes*, [arXiv:0801.1238v1].

[37] G. Ginot and P. Xu *Cohomology of Lie 2-groups*, [arXiv:0712.2069v1].

[38] F. Girelli and H. Pfeiffer, *Higher gauge theory – differential versus integral formulation*, J. Math. Phys. **45** (2004) 3949-3971, [arXiv:hep-th/0309173v2].

[39] F. Girelli, H. Pfeiffer, and E. Popescu, *Topological higher gauge theory - from BF to BFCG theory*, [arXiv:0708.3051v1].

[40] M.B. Green and J.H. Schwarz, *Anomaly cancellation in supersymmetric $D = 10$ gauge theory and superstring theory*, Phys. Lett. **B149** (1984) 117.

[41] W. Greub, S. Halperin, and R. Vanstone, *Connections, curvature, and cohomology. Vol. II: Lie groups, principal bundles, and characteristic classes*, Academic Press, New York-London, 1973.

[42] S. Halperin and J.-C. Thomas, *Rational equivalence of fibrations with fibre G/K*, Canad. J. Math. **34** (1982), no. 1, 31–43.

[43] A. Henriques, *Integrating L_∞ algebras*, [arXiv:math/0603563v2].

[44] K. Hess, *Rational Homotopy Theory: A Brief Introduction*, [http://www.math.uic.edu/~bshipley/hess_ratlhtpy.pdf]

[45] M.J. Hopkins, *Topological aspects of topological field theories*, Andrejewski lecture, Göttingen (2006)

[46] M.J. Hopkins, I.M. Singer, *Quadratic functions in geometry, topology, and M-theory*, J. Diff. Geom. **70** (2005) 329-452, [arXiv:math.AT/0211216].

[47] P. Horava and E. Witten, *Eleven-dimensional supergravity on a manifold with boundary*, Nucl.Phys. **B475** (1996) 94, [arXiv:hep-th/9603142].

[48] P.T. Johnstone, *Stone Spaces*, Cambridge studies in advanced mathematica (1986)

[49] J. Kalkman, *BRST model for equivariant cohomology and representatives for the equivariant Thom class*, Comm. Math. Phys. **153** (1993), no. 3, 447–463.

[50] A. Kapustin, *D-branes in a topologically nontrivial B-field*, Adv. Theor. Math. Phys. **4** (2000) 127, [arXiv:hep-th/9909089].

[51] L.H. Kauffman, *Vassiliev invariants and functional integration without integration*, in *Stochastic analysis and mathematical physics*, p 91-114

[52] T.P. Killingback, *World-sheet anomalies and loop geometry*, Nucl. Phys. B 288 (1987) 578.

[53] I. Kolár̆ r, J. Slová k and P.W. Michor, *Natural operations in differential geometry*, Springer-Verlag (1993)

[54] A. Kotov and Th. Strobl, *Characteristic classes associated to Q-bundles*, [arXiv:0711.4106v1]

[55] I. Kriz and H. Sati, *M Theory, type IIA superstrings, and elliptic cohomology*, Adv. Theor. Math. Phys. **8** (2004) 345, [arXiv:hep-th/0404013].

[56] K. Kuribayashi, *On the vanishing problem of String classes*, J. Austr. Math. Soc. (Series A) 61, 258-266 (1996)

[57] T. Lada and J. Stasheff, *Introduction to sh Lie algebras for physicists*, Int. J. Theor. Phys. **32** (1993) 1087, [arXiv:hep-th/9209099].

[58] T. Lada, M. Markl, *Strongly homotopy Lie algebras*, Comm. Algebra **23** (1995), no. 6, 2147–2161, [arXiv:hep-th/9406095v1].

[59] V. Mathai and H. Sati, *Some relations between twisted K-theory and E_8 gauge theory*, J. High Energy Phys. **0403** (2004) 016, [arXiv:hep-th/0312033v4].

[60] J. Mickelsson and R. Percacci, *Global aspects of p-branes*, J. Geom. and Phys. **15** (1995) 369-380, [arXiv:hep-th/9304054v1].

[61] I. Moerdijk and G. Reyes, *Models for smooth infinitesimal analysis*, Springer-Verlag (1991)

[62] M. Murray, *Bundle gerbes*, J. London Math. Soc. **(2) 54** (1996), no. 2, 403–416.

[63] M. Murray and D. Stevenson, *Higgs fields, bundle gerbes and string structures*, Commun. Math. Phys. **243** (2003) 541–555, [arXiv:math/0106179v1].

[64] R. Percacci and E. Sezgin, *Symmetries of p-branes*, Int. J. Theor. Phys. **A 8** (1993) 5367, [arXiv:hep-th/9210061v1].

[65] D. Roberts and U. Schreiber, *The inner automorphism 3-group of a strict 2-group*, to appear in Journal of Homotopy and Related Structures, [arXiv:0708.1741].

[66] D. Roytenberg, *AKSZ-BV Formalism and Courant Algebroid-induced Topological Field Theories*, [http://arxiv.org/abs/hep-th/0608150]

[67] H. Sati, *On Higher twists in string theory*, [arXiv:hep-th/0701232v2].

[68] H. Sati, U. Schreiber, J. Stasheff *Fivebrane structures: topology*, [arXiv:0805.0564]

[69] H. Sati, U. Schreiber, J. Stasheff *Fivebrane structures: geometry*, (in preparation)

[70] U. Schreiber, *On nonabelian differential cohomology*, lecture notes, [http://www.math.uni-hamburg.de/home/schreiber/ndclecture.pdf]

[71] U. Schreiber, *On ∞-Lie theory*, [http://www.math.uni-hamburg.de/home/schreiber/action.pdf]

[72] U. Schreiber and K. Waldorf, *Parallel transport and functors*, [arXiv:0705.0452v1].

[73] U. Schreiber and K. Waldorf, *2-Functors vs. differential forms*, [arXiv:0802.0663v1]

[74] U. Schreiber and K. Waldorf, *Connections on non-abelian Gerbes and their Holonomy*, to appear.

[75] J. Stasheff and M. Schlessinger, *Deformation Theory and Rational Homotopy Type*, unpublished.

[76] St. Stolz and P. Teichner, *What is an elliptic object?* in Topology, geometry and quantum field theory, London Math. Soc. LNS 308, Cambridge University Press 2004, 247-343.

[77] Sullivan, *Infinitesimal computations in topology* , Publications mathématique de l I.H.É.S., tome 47 (1977), p. 269–331

[78] C. Teitelboim and M. Henneaux, *Quantization of gauge systems* Princeton University Press, 1992.

[79] E. Witten, *On Flux quantization in M-Theory and the effective action*, J. Geom. Phys. **22** (1997) 1–13, [arXiv:hep-th/9609122].

[80] A. Yekutieli, *Central Extensions of Gerbes*, [arXiv:0801.0083v1].

[81] T. Trimble, private communication

Hisham Sati
Department of Mathematics
Yale University
10 Hillhouse Avenue
New Haven, CT 06511
e-mail: hisham.sati@yale.edu

Jim Stasheff
Department of Mathematics
University of Pennsylvania
209 South 33rd Street
Philadelphia, PA 19104-6395
e-mail: jds@math.upenn.edu

Urs Schreiber
Fachbereich Mathematik
Universität Hamburg
Bundesstraße 55
DE–20146 Hamburg
e-mail: schreiber@math.uni-hamburg.de

Index